Massachusetts Breeding Bird Atlas

With gratitude to
Robert C. Hooper
Ardent birder and conservationist

We also wish to acknowledge the generous support of
William C. Ackerly

Massachusetts BREEDING BIRD ATLAS

edited by
Wayne R. Petersen and
W. Roger Meservey

illustrated by
John Sill and Barry W. Van Dusen

NATURAL HISTORY of NEW ENGLAND SERIES
Christopher W. Leahy, General Editor

Distributed by the University of Massachusetts Press/Amherst & Boston

Copyright © 2003 edited by Wayne R. Petersen and W. Roger Meservey. Published by the Massachusetts Audubon Society. Distributed by the University of Massachusetts Press, P.O. Box 429, Amherst, Massachusetts 01004-0429.

All rights reserved. No part of this book may be reproduced or transmitted in any form by any means, electronic or mechanical, including photocopying and recording, or by any information or retrieval system, without permission in writing from the Massachusetts Audubon Society, publisher.

Library of Congress Cataloging-in-Publication Data
Massachusetts breeding bird atlas / edited by Wayne R. Petersen
and W. Roger Meservey; illustrated by John Sill and Barry W. Van Dusen
   p. cm. — (Natural history of New England series).
Includes bibliographical references (p. 434).
   ISBN 1-55849-420-0 (hard cover)
1. Birds—Massachusetts. 2. Birds—Massachusetts—Geographical distribution. 3. Birds—Massachusetts—Geographical distribution—Maps I. Petersen, Wayne R., 1944— II. Meservey, W. Roger. III. Series.
   QL684.M4M37 2003
   598'.09744—dc21
                              2003014390

Petersen, Wayne R., and W. Roger Meservey
   Massachusetts Breeding Bird Atlas
Illustrated by John Sill and Barry W. Van Dusen
   p. cm. — (Natural History of New England Series)
Includes bibliographical references (p. 434)
   ISBN 1-55849-420-0 (hard cover)

Printed in Canada

To Richard A. Forster,
ornithologist, naturalist, and mentor,
who will never see the product
of his significant efforts on behalf of the Atlas project.
I hope you would have been pleased, my friend.

W.R.P.

Contents

Foreword by Chandler S. Robbins viii
Acknowledgements x
Atlas Volunteers and Contributors xii
History of the Massachusetts Breeding Bird Atlas Project 2
Massachusetts Breeding Bird Distribution 5
Massachusetts Ecoregions 12
Atlas Methods and Criteria 19
Maps and Species Accounts 23
Appendix 1: Additional Breeding Bird Atlas Species Accounts 422
Appendix 2: Plant and Animal Species Mentioned in the Text 430
Artists 441
Production 441

Foreword

Chandler S. Robbins
Research Wildlife Biologist
Biological Resources Division
USGS Patuxent Wildlife Research Center

Ovenbird

An atlas is, fundamentally, a method of depicting and understanding some component of the world; and, usually, in addition to providing information purely for intellectual interest, it serves as a means to some end. The atlases of the early navigators were crucial aids to the development of global commerce, the effects of which—both good and bad—can hardly be exaggerated. Like navigational charts, grid-based biological atlases tell us where organisms are—namely that this particular species lives (or does *not* live) in this particular place.

But the greatest value of atlases depends on the addition of another key element: time. For naturalists, it is interesting to know that some rare creature may be rearing young in the neighboring town. And, if you volunteered some hard-earned leisure time on hot summer afternoons to detect birds carrying fecal sacs, there is inordinate pleasure in recognizing "your" dots on an authoritative distribution map. But the true and worthy rationale for undertaking a project as complex and labor-intensive as a breeding bird atlas is nothing less than depicting the status of life on one's "own" portion of the planet—and the measure of this status is change over time.

The need for this kind of measure—the ability to read the vital signs of life on earth—has become critical. In most instances, we exercise these godlike powers with the best of declared intentions: developing new sources of energy, eradicating disease, improving living standards. We live at a moment in the earth's biological evolution when a single species—our own—has acquired the ability to significantly alter fundamental components of the biosphere such as climate, sea level, vegetation, and animal populations. And to give proper credit to the unique capabilities of the human beings, we have been immensely successful in realizing many of our wildest dreams.

If there turns out to be a tragic flaw in the headlong course of human progress, it may be that our power to invent has dangerously outdistanced our ability to predict the consequences of our most "earth-shattering" creations. We did not *mean* to perforate the ozone layer or to set Lake Erie on fire or to deny Bald Eagles and Peregrine Falcons the ability to make eggshells. But there is no dodging our responsibility for these lamentable and portentous results.

The hope of course is that our unique ability to consider how we should direct our activities to affect the future will be our saving grace—and that of our fellow creatures. This will entail putting a much greater emphasis on accurately predicting and then making necessary allowances for the consequences of our planet-altering capabilities. An optimist might say that there is already a preservative trend, and that this book is evidence of it since a biological atlas can also be described as a chronicle of consequences.

The data for the Massachusetts Breeding Bird Atlas was collected 25 years ago; and, though many Massachusetts atlas volunteers have waited a long time to see the tangible fruit of their efforts, these records—far from being in any sense outdated—have set a baseline for measuring the consequences of a quarter-century of human progress. The value of this historic baseline increases with the passage of time.

For someone like me, accustomed to assessing the state of global health through the lens of bird populations, pressing questions about the environmental consequences of 25 years of rapid change in the United States and Neotropics leap from at least a third of the maps in this Atlas.

What is happening to the exquisite American Kestrel, still apparently on the increase when the Atlas data was collected but now declining alarmingly as a breeding bird? Is this decrease simply a consequence of reforestation of open habitats or the result of something less visible and more insidious?

Is the Whip-poor-will still widespread and locally common as the Atlas showed it to be in the 1970s, or is this voice of the summer night continuing to fade with expanding subdivisions and the attendant pressures such as habitat loss and house-pet persecution of ground-nesting birds?

Is the Bicknell's Thrush, having just become a species in its own right, about to go extinct as a result of destruction of its winter habitat in Hispaniola? It was one of my favorite nesting songsters on Mount Greylock in the 1930s, but it had already disappeared from Massachusetts as a breeding bird by the time of the Atlas.

Will the Blue-winged Warbler eventually swallow up its kissing cousin, the lovely Golden-winged Warbler, possibly as a consequence of global warming?

These and countless other questions beg to be answered with future surveys, not only because they are interesting to a handful of ornithologists but also because they are *of consequence to everyone.*

Not long ago I might have felt obliged in this foreword to defend the notion that birds matter to people or that they can serve more dramatically than we ever would have predicted as "canaries in the mine," foretelling danger on a global scale. But not today.

In 1999, the Department of the Environment for the United Kingdom published a white paper entitled *A Better Quality of Life.* One of the "headline" indicators of the sustainability of lifestyles in the UK is based on population trends of breeding birds. That birds should appear alongside more traditional economic and social indicators of the quality of human life is appropriate for a number of reasons: (1) their broad distribution and varied ecology make them excellent indicators of environmental change; (2) they have deep and increasingly wide appeal among the general public; (3) they are of direct economic importance because of the burgeoning expenditures associated with birdwatching; and (4) we have good data on them, including atlases like this one, against which to measure change.

Should you doubt the need for following such trends, let me cite another British example. During the past 25 years, the Skylark—once an abundant and ubiquitous avian emblem of the British countryside—has declined by more than half largely as a result of "improved" agricultural practices, and similar trends are documented for over twenty other "common" breeding birds. If you doubt the relevance of this trend to Massachusetts and other parts of North America, I direct your attention to the maps and species accounts in this handsome book for Pied-billed Grebe, American Bittern, Red-shouldered Hawk, Prairie Warbler, and Vesper Sparrow.

Some of the canaries are not faring well; it's time for the miners to pay attention.

Acknowledgements

A project such as the Massachusetts Breeding Bird Atlas was clearly a cooperative effort that could not have been attempted without the help, hard work, and support of literally hundreds of people. In a separate section of this volume, over 650 field volunteers, contributors, and organizations are specifically acknowledged for the thousands of hours that they invested in compiling information pertaining to the status and breeding distribution of Massachusetts birds during the years 1974 to 1979. While it should be abundantly obvious that without the labors of all these people this publication would not have been possible, it is equally important to recognize that a number of other individuals and institutions deserve specific recognition. To appreciate a few of the complexities involved in producing the Atlas, a brief chronology follows.

A complete history of breeding bird atlases in North America has been traced elsewhere in this volume; however, the genesis of the Massachusetts Atlas was primarily a result of the inspiration and vision of James Baird, director of Natural History Services at the Massachusetts Audubon Society in the 1970s. During that period, with the support of Massachusetts Audubon's president at the time, Allen H. Morgan, the Atlas project was launched. With cooperation from the Massachusetts Division of Fisheries and Wildlife, largely in the person of then Massachusetts State Ornithologist Bradford G. Blodget, responsibility for the initial Atlas effort was placed under the direction and coordination of Deborah Howard, a staff member working under the supervision of James Baird at the time.

Within a year of the seminal organizational efforts of Deborah Howard, primary responsibility for coordinating the Atlas project was turned over to Richard A. Forster, then Massachusetts Audubon field ornithologist, whose efforts in seeing the fieldwork for the Atlas through to completion cannot be overemphasized. Throughout the data-gathering phase of the project, he received significant administrative assistance from Evelyn Smith, whose attention to detail and cheerful personality did much to keep the project on track during the 1970s.

Following the final year of fieldwork in 1979, a year in which Richard S. Heil and Richard R. Veit specifically assisted Richard A. Forster in gathering breeding bird information for previously uncovered Atlas blocks, the next priority was to get the previous six years' breeding-bird data computerized and mapped. This responsibility ultimately fell to David Stemple at the University of Massachusetts, without whose expertise and efforts in developing a program to map the Atlas data, along with his willingness to serve as a technical advisor on other matters throughout the project, the present publication might never have become a reality.

After a suitable mapping program for the Atlas was created, the task of assigning authors to write species accounts, followed by the initial editing of these accounts, again fell to Richard A. Forster and James Baird. As noted in the section on the History of the Massachusetts Breeding Bird Atlas Project, the contributions of 90 participating authors instilled a heightened sense of ownership to the final product yet added considerably to the task of producing the uniformity of text desirable in a reference such as the Atlas. Through the course of a series of administrative delays, the Atlas enjoyed the scrutiny of other discerning eyes, most notably those of W. Roger Meservey. If it were not for his artful integration of data from the Cornell Nest Record Program and his finesse at smoothing out the overall manuscript, the final species accounts would never have attained their present uniformity of style and overall congruence. W. Roger Meservey wishes to particularly acknowledge assistance and support provided by Bradford G. Blodget and Francis McMenemy.

In addition to W. Roger Meservey's outstanding editorial endeavors, Massachusetts Audubon's Ann Prince Hecker has to be singled out as the final word on all matters pertaining to grammar, readability, and style. Without her unfailing and meticulous copy editor's eye, the book would not be the finely crafted document that it is. As always, Ann remained cheerful, patient, and optimistic throughout the many years it took to finalize publication of the Atlas.

In the final stages of production, when the book was nearly complete and laid out in page form, Stephanie Jones contributed her highly competent skills to the process. Her efficiency and accuracy in proofing the entire book from beginning to end are much appreciated.

Betty A. Graham also helped with editorial details when the project was near completion. Her willing, careful, skillful assistance was invaluable.

Along with the individuals already mentioned belongs the name of Mary Hopkins at Massachusetts Audubon for her help with any number of details related to the final production of the Atlas publication. Without Mary's constant encouragement and assistance in working with editors, typesetters, printers, artists, and authors, along with her irrepressible sense of humor, the senior editor could easily have lost his mind by the project's end!

As final production plans for the Atlas gelled, a decision was made to utilize a selection of watercolor bird paintings by talented bird artist John Sill to accompany the species accounts. Without the cooperation of Orion Barber at Orion Book Services, Amy T. Montague at Massachusetts Audubon's Visual Arts Center, and the Massachusetts Audubon Society, the *Massachusetts Breeding Bird Atlas* would not have the visual appeal that John Sill's stunning renditions bring to the book.

For assistance with the final production of the Atlas species maps, as well as for making maps available to be used as transparent overlays, special thanks go to Stephen D. McRae, director, and Dorothy Graskamp of the Massachusetts Department of Fisheries, Wildlife, and Environmental Law Enforcement Geographic Information Systems Program. The willingness of these two individuals to assist with a number of map-related questions, along with providing technical assistance in numerous other ways, is gratefully acknowledged.

The quality of the design, layout, supplemental illustrations, and cover art for the Atlas can be attributed to the skill and craftsmanship of artist and illustrator Barry W. Van Dusen. His enthusiasm for the project over the course of many years, his tireless energy, and his ability to coordinate and integrate all aspects of the completed work in a cheerful manner are aspects that would have been virtually impossible to find in any other individual. To Barry W. Van Dusen goes the lion's share of the credit for the final look and feel of the *Massachusetts Breeding Bird Atlas*. Without Barry W. Van Dusen, this volume might never have been published. The editors offer their sincerest thanks and admiration to him for his friendship, advice, and support throughout the years.

To Chandler S. Robbins, "Father of North American Breeding Bird Atlas Projects," go warm thanks for writing such an insightful foreword and for providing a global perspective on the value of bird atlas efforts. Chandler S. Robbins is an inspiration to all who follow the lives of North American birds.

Special kudos go to Simon A. Perkins, the senior editor's colleague, whose unfailing enthusiasm for the Atlas project provided continual motivation for seeing the publication through to completion.

Specific thanks for making the publication of this book possible would not be complete without full acknowledgement to Christopher W. Leahy for his incalculable assistance and support during every phase of this project from its inception to its conclusion. His editorial and literary acumen were called upon at virtually every stage in the final production of the Atlas, and his pen crafted or enhanced any number of passages throughout the text. Without his optimism, enthusiasm, and commitment, along with that of Massachusetts Audubon President Emeritus Gerard A. Bertrand and current President Laura A. Johnson, this publication might still be a manuscript in waiting.

Finally and with deepest sincerity, thanks go to Betty Petersen, spouse of the senior editor, for her steadfast support and companionship through the many years that this project has been in production.

Atlas Volunteers and Contributors

Over 650 volunteers and contributors participated in the Massachusetts Breeding Bird Atlas project. Some of these individuals gave up many days and weeks of their time during the course of the five-year Atlas period, while others participated only for a few hours during a single year. Regardless of the amount of energy invested, the project could not have been successfully completed without the combined effort of these many dedicated volunteers and contributors.

Because of unforeseen delays in the publication of the Massachusetts Breeding Bird Atlas, reconstructing the volunteer list of all those who participated in the project was not easy. As a consequence, if there are individuals who participated in the Atlas effort whose names do not appear in the list that follows, they can be assured that the omission of their names was not deliberate. The contributions of any such missing individuals were valued no less than those of individuals whose names appear in the following list.

*indicates Atlas Quad Coordinator
bold typeface indicates Species Accounts Author

Northern Waterthrush

Leslie Abbey, Robert Abbey, *Trescott Abele, Rock Agostino, Andrew Agush, Virginia Albee, *Don Alexander, David Allard, Craig Allen, Harvey Allen, Becky Anderson, **Kathleen S. Anderson**, Jonathan Andrew, *****Edith F. Andrews**, John Andrews, **Ralph Andrews**, Joe Antos, **Steven M. Arena**, **Margaret W. Argue**, *Herbert Armitt, L. Colette Armitt, Widge Arms, Madeleine Arnaudet, **Dorothy Rodwell Arvidson**, *Eleanor Athearn, Elizabeth Atherton, Hope Atkinson, Theodore Atkinson, Frederick Atwood, **Thomas Aversa**, Janet Aylward

Leona Babbitt, **Priscilla Bailey**, *****Wallace Bailey**, Doris Baines, Fay Baird, *****James Baird**, *****Mary Baird**, Timothy Baird, Stan Baker, Roan Barber, John Barclay, Ellsworth Barnard, Mary Barnard, Glad Bartlet, Lucy Batchelder, **Helen C. Bates**, *Moreton Bates, Allan Beale, Peg Becker, Robert Bednarek, Ellen Beekman, Susan Bellows, Winifred Bellows, Robert Bellville, **Augustus Ben David II**, Major Benton, *****James Berry**, Robert Bieda, Richard Bierregaard, Jr., Lewis Black, Chester Blasczak, **Bradford G. Blodget**, Ada Bohlke, *Margarette Bon, Mrs. Robert Booth, Diane Bouchard, Richard Bowen, Frederick Bowes III, Elizabeth Boyd, *Denise Braunhardt, Clarence Briggs, *Dorothy Briggs, Leo Brodeur, George Brooks, Jr., **Arnold K. Brown**, David Brown, Joseph Broyles, *David Brule, Janet Bryan, Jeff Bryant, M.P. Buck, P.A. Buckley, John Buffinton, Andrew Bunker, James Bunton, James Burke, John Burnap, Richard Burrell, William Burt

Carl CaPobianco, M. Campbell, **James E. Cardoza**, R.W. Carpenter, **Thomas L. Carrolan**, Josette Carter, David Casoni, **Brian E. Cassie**, James Cavanaugh, June Chamberlain-Auger, George Champoux, Helen Champoux, Robert Chapell, Clifford Chausse, **Nancy A. Claflin**, *****David E. Clapp**, Carolyn Clark, *Anne Clarke, Ken Clarkson, Nancy Clayton, F. Thomas Collins, Florence Colman, Richard Comeau, Alicia Conklin, Marc Connelly, Mr. and Mrs. Edward Connolly, Harold Connor, **Hamilton Coolidge**, Lee Cooper, Charlotte Corwin, Richard Cowles, *****Robert B. Coyle**, Victoria Crain, E. Mildred Crane, Davis Crompton, Donald Crooks, Pearl Crooks, Richard Cunningham, Stephen Curtis, Charles Merrill Cushman III, Eric Cutler

Mario Danieli, Sarah Daniels, Mrs. Philip Dater, Russ Davenport, *Fred David, Mark David, **William E. Davis, Jr.**, **William J. Davis**, Ward De Haro, Mary Ellen Delger, Gordon Dennis, John Dennis, Edward Densmore, Miriam Dickey, Sue Dixey, Mrs. David Dixon, *Thomas Dodd, William Dornbusch, **Christina Dowd**, Lee Drickamer, Ruth Drinker, William Drury, Edna Dunbar, Barbara Durling, Larry Dwight, Jr.

Joan Eastman, Carl Eckel, Ruth Edwards, Barbara Eliades, David Ellis, Benjamin Ellison, **Steven Ells**, **David L. Emerson**, **Ruth P. Emery**, Richard Enser, William Ervin, Essex County Ornithological Club (ECOC)

James Felkel, B. Ferguson, **Gilbert Fernandez**, **Josephine Fernandez**, ***Richard L. Ferren**, Carolee Ferris, Russell Fessenden, Edna Fielding, **Lillian Lund Files**, Albert Fischer III, *David Fischer, Mark Fisette, Elaine Fisher, **Erma J. Fisk**, Jeffrey Fiske, John Fitzgerald, John Fitzpatrick, George Flagg, Gerald Flaherty, Marilyn Flor, Karen Fogerty, Jean Foley, ***Richard A. Forster**, William Forward, **Patricia Noyes Fox**, ***Robert P. Fox**, Richard Franz, Robert Franzen, **Thomas W. French**, Joan Friend, Granger Frost, John Fuller, J. Fulton, George Furst, Jean Furst

Thomas H. Gagnon, Richard Galat, *David Galvin, Aileen Gardner, Frank Gardner, Eleanor Garfield, Pat Garrey, James Garvey, Mrs. Lester Garvin, **George W. Gavutis, Jr.**, Anthony Gawienowski, Miriam Gholson, Malcolm Gibb, Grace Gibson, Thomas Gilgut, Joan Giovanella, Ida Giriunas, Randolph Gleason, Barbara Godard, **Anthony A. Gola**, Carl Goodrich, Tim Goodwin, Tom Goodwin, *Michael Gooley, George Gove, Robert Grant, Frances Gray, Daniel Gray III, Roberta Greene, Marc Gross

Joseph A. Hagar, Wendy Haggerty, Jennifer Hall, Robert Halstead, Frank Hammond, Jr., Dave Hansroth, Richard Harlow, Brian Harrington, H. Warren Harrington, **Winthrop W. Harrington**, Nancy Harris, William Harris, Frank Hart, Karsten Hartel, ***Jeremy Hatch**, Adam Heard, Scott Hecker, **Richard S. Heil**, Edith Henderson, **Bartlett Hendricks**, Edwin Hendrickson, Anne Henneke, David Herrick, John Herrmann, ***H.W. Heusmann**, *Sibley Higginbotham, **Norman P. Hill**, James Hinds, *Chester Hobbs, Michael Hoffman, Elizabeth Holden, Suzanne Holloran, Karen Holmes, **Denver W. Holt**, Anthony Horn, Barbara Horn, Natalie Houghton, *Deborah Howard, Edwin Howes, Robert Humphrey, William Hunt, Dorothy Hunter, Rita Hureau

Lucy Ingalls, *Sarah Ingalls

Cary James, Lee Jameson, Max Jarvis, *Rodney Jarvis, Warren Jewell, *David Johnson, Edith Johnson, *Gordon Johnson, Sophie Johnson, Tad Johnston, Carol Jones, Helen Jones, Melissa Jones, Ann Jorn, John Joy, Alberta Juergens

Walter Kaminski, Barton Kamp, Carl Kamp, Eva Kampits, Natalie Karl, Mark Kasprzyk, Betty Keil, Dana Keil, Barker Keith, *****Seth Kellogg**, **Joseph F. Kenneally, Jr.**, James Kenney, Jr., Christopher Kent, *Charles Kieweg, Bruce Kindseth, Dot Kindseth, Mrs. Victor Kloppenburg, Beverly Klunk, William Klunk, **Oliver Komar**, **John C. Kricher**, Donald Kroodsma, Vertene Kuehn

Marie Ladd, Rich Lafountaine, Susan Lagace, Mr. and Mrs. Alexander Laird II, David Lamont, Michael Lamontagne, **Elissa M. Landre**, Ruth Langley, Linda LaPlume, L. Donald Laramee, **E. Vernon Laux, Jr.**, Linda Lawson, Louise Lawson, *****Christopher W. Leahy**, Richard Lent, Willard Leshure, Robert Lessard, Scott Lesure, Viola Lesure, Nancy Leue, Tom Leue, Frances Libby, Steve Lima, Gordon Lindaminneli, **Thomas Lipsky**, Bev Litchfield, Myron Litchfield, Donald Livermore, **Trevor L. Lloyd-Evans**, William Locke, *Dotsie Long, Raymond Longley, John Lortie, *Jay Loughlin, John Ludovina, *Bruce Lund, Joseph Lund, **Mark C. Lynch**, John Lynes, **Paul J. Lyons**

Joseph MacDonald, Sherrard MacDonald, J.S. MacDougall, Richard Mack, Sally Maloney, Joan Manley, Manomet Center for Conservation Sciences (MCCS), Sally Mansfield, Clifton Marks, *John Marshall, Jr., Massachusetts Division of Fisheries and Wildlife (MDFW), Denise Masson, Mrs. Sargis Matson, Reginald Maxim, Mrs. Wilho Mayranen, Robert McAllester, Richard McAllister, **Martha H. McClellan**, Keith McClelland, Mary Mcdonough, Linda McGraw, Charles McLaughlin, Jr., Bonnie McLean, *Francis McMenemy, Doug McNair, Sally McNair, Paul Meleski, Lorene Melvin, **Scott M. Melvin**, Harold Merriman, **W. Roger Meservey**, L. Michaels, Julia Michaelyan, Paula Miller, **Elwood O. Mills, Jr.**, Lydia Mills, *George Mock, Mr. and Mrs. Hugh Montgomery, Johnes Moore, Ruth Moore, Margaret Moran, **Allen H. Morgan**, Edward Morrier, George Morrow, Paul Motts, Russell Mroczek, Paul Mugford, Christina Murley, John Murphy, Joshua Murphy, Marilyn Murphy, David Murray, Irene Murray

Peter Nagorniuk, William Natti, *William Newbury, *Jean Nichols, Albert Nickerson, **Paul R. Nickerson**, *****Blair Nikula**, **Ian C.T. Nisbet**, John Nove, William Nutting, Mrs. C.H. Nyberg

Nancy Ober, Thomas O'Connell, Robert O'Hara, **Michael Olmstead**, Larry Olson, Lynn O'Neil, Blanche Orrell, Bob Orrell, *Neil Osborne, Sharon Osborne, Richard Osterberg

Atlas Volunteers and Contributors xiii

B. Parker, Donald Parsons, Katherine Parsons, Stuart Parsons, Ethel Pearson, ***Robert Pease**, Stephen Pelikan, Fred Pelletier, Simon Perkins, **Wayne R. Petersen**, *Bruce Peterson, David Peterson, Mason Phelps, Mr. and Mrs. Wayne Philbrook, *Leslie Pierce, Alfred Pillsbury, Joanne Pleskowicz, Edward Plotkin, **Mark Pokras**, **E. Michael Pollack**, Howard Polonsky, Lois Polonsky, Alan Poole, Mary Poore, *Lawrence Porter, John Portnoy, R.S. Portnoy, *Raymond Pothier, Nancy Powell, L.B. Pratt, **Robert Prescott**, L. Charles Priest, Jr. Rachel Priestman, Ruth Proctor, Joyce Provost

Mrs. Raymond Quick, Charles Quinlan, Fred Quinn

Ernest Randall, Eliose Randolph, **Edward H. Raymond**, Cloyce Reed, Ramoth Rees, Robert Reeve, Phyllis Regan, Charlotte Reid, William Reid, Louis Resmini, Bessie Rice, Joan Rice, Howard Rich, Carroll Richard, *Allan Richards, Robert Richards, Althea Richardson, Scott Richardson, *George Rickards, Margaret Rinsma, Virginia Riorden, Henry Ritzer, George Robert, Julie Roberts, Nancy Roberts, **Paul M. Roberts**, Ross Roberts, John Robertson, ***Leif J. Robinson**, Mary Roelke, Ebba Rogers, John Root, Daniel Roser, **Charles E. Roth**, *Miriam Rowell, A. Runeman

Ellie Sabin, *Stewart Sanders, Michael Santner, Virginia Saunders, G. Schoeler, Mrs. Howard Schuck, *Albert Schwab, Shirley Scott, Margaret Scribner, James Seamans, Harriet Sears, Joan Sease, Larry Seidl, June Seitz, *Elizabeth Sewall, *Gary Shampang, *Wayne Shampang, Mrs. Gerald Shepherd, Audrey Sherman, *Albert Short, Benjamin Shreve, Mrs. J. Edward Sibel, Evian Simcovitz, Albin Slesiwski, David Small, Melinda Smallwood, **Charlotte E. Smith**, Evelyn Smith, Florrie Smith, Julius Smith, Katherine Smith, Norman Smith, **P. William Smith**, Susan Smith, Richard Smulski, *Betty Smyth, David Snoekenbos, Howard Snow, Nancy Solari, **Bruce A. Sorrie**, South Shore Bird Club (SSBC), Donald Southall, Mary Souza, Barbara Spector, David Spector, Richard Spedding, Tony Spinelli, Jayne Stairs, *Robert Stanhope, Steve Stanne, Tilley Steenmetz, *Esther Steinmetz, Daniel Steven, **Donald W. Stokes**, **Lillian Q. Stokes**, ***Rudolph H. Stone**, Cherrie Stout, Peter Stowe, Jane Stroup, ***Robert H. Stymeist**, Phyllis Swift

Robert Tamarin, Vivian Tamp, Lois Tarlow, Randy Tate, Deb Taylor, Don Taylor, **Eliot Taylor**, Lee Taylor, Jean Tewksbury, Gladys Thatcher, Douglas Thoen, Edna Thoen, *Cynthia Thomas, Dave Thomas, Mrs. Ray Thomas, Steve Thompson, Mrs. W.D. Ticknor, Mary Tilley, Stephen Tilley, Dorothy Townsley, Barbara Treat, **Peter Trull**, Mrs. Lee Turley, Dennis Turner, **Richard E. Turner**, *Frank Tuttle, David Tyler, **Thomas F. Tyning**

Barry Van Dusen, Paul Vassalotti, Richard R. Veit, *Don Verger, Robert Vernon

William Wadas, Emily Wade, Eleanor Waldron, **Richard K. Walton**, Florence Warner, Thomas Warren, Pamela Weatherbee, Mertys Webb, Sarah Webb, Phyllis Webber, Donald Weber, Laurence Webster, Rosemary Webster, Susan Weeks, Kathleen Weinheimer, Lawrence Weinheimer, Harry Weiss, Marcia Weld, Francis Wells, Peg Wert, Don Wheeler, Louise Wheeler, *Strickland Wheelock, Susan Whiting, Gail Wiedmann, **Henry Wiggin**, John Williams, Lyle Williams, Robert Williams, Ruth Williams, J.F. Willis, Mark Wilson, Kenneth Winston, Corky Witt, Elizabeth Wogan, LaVern Wolcott, Sara Wollaston, Robert Wood, Andrew Woodruff, Robert Woodruff, Courtenay Worthington, Calvin Wright, Richard Wright, Charlotte Wyman

E. Yaeger, Jean Yarusites, Mrs. Kenneth Yates, John Young, *Cindy Youngstrom

Soheil Zendeh, Arlene Zuchowski

History of the Massachusetts Breeding Bird Atlas Project

Edward Howe Forbush in his monumental *Birds of Massachusetts and Other New England States* (1925, 1927, 1929) was the first ornithologist to use maps with overlying symbols to depict the precise breeding localities of selected bird species in Massachusetts. Since then, interest in creating accurate depictions of bird distribution within defined areas and an ability to do so using relatively sophisticated techniques have grown dramatically.

Davis (1997) has summarized the recent evolution of North American breeding bird atlases—currently the ultimate in bird mapping schemes. He noted that the seminal effort in grid-based biological mapping occurred in Great Britain during the 1950s, which initially resulted in the publication of *Atlas of the British Flora* (Perring & Walters 1962). Fourteen years later, this botanical atlas was followed by publication of *The Atlas of Breeding Birds in Britain and Ireland* (Sharrock 1976).

During the same period, efforts to produce the first breeding bird atlases in North America were in their infancy. Although several western states (e.g., Colorado and Montana) attempted to map breeding bird distribution on grids of one degree of latitude by one degree of longitude (latilongs), such an enormous scale made it difficult to detect changes and patterns in bird distribution over time. It was not until the early 1970s, when Maryland undertook a small-scale grid-based atlas for a couple of Maryland counties (Klimkiewicz & Solem 1978), that bird atlasing in North America began in earnest.

In 1974 the Massachusetts Audubon Society, with assistance from the Massachusetts Division of Fisheries and Wildlife (MDFW), launched a statewide, five-year atlas project—the first such comprehensive effort in North America—to map the breeding distribution of birds in the Commonwealth of Massachusetts. A sixth year was eventually added to allow for completion of poorly covered and unsampled blocks.

Patterned after the previous efforts in Great Britain and Maryland, the Massachusetts Breeding Bird Atlas Project utilized volunteers to survey areas defined by the US Geological Survey (USGS) topographical quadrangle map grid for Massachusetts. Each of the 189 "topo"-map quadrangles in Massachusetts represents an area of slightly less than 60 square miles. To create a convenient and broadly comparable grid, each quadrangle was divided into six equal blocks, every one representing an area of approximately 10 square miles. The overall statewide grid was made up of 989 blocks (since certain quadrangles contained blocks not in Massachusetts).

Following the final year of fieldwork in 1979, the breeding bird data gathered was computerized and converted into a map format that would depict the breeding status of every species encountered within the 989 blocks surveyed in Massachusetts during the five-year Atlas period. With comprehensive and readily available USGS topo maps being the basis for the Atlas grid, it was easy to provide volunteers with precise coverage boundaries, as well as basic topographic and logistical information about each coverage area. The grid system also made it easier to convert distribution data from handwritten records to a computerized mapping program.

Before launching a full-scale atlas effort, Massachusetts Audubon and MDFW initiated a statewide pilot project in 1973 to test the Atlas protocols. Six trial quadrangles (Ayer, Gardner, Manomet, Newton, Springfield South, and Westport) were surveyed during the trial year. Through the use of carefully defined breeding bird criteria (see page 19), the status of every species represented on the completed Atlas maps was indicated as a "possible," "probable," or "confirmed" breeding species in every block where the species was recorded. During the trial year, 139 species were identified in 24 blocks with the following results: 91 bird species were recorded as "confirmed" breed-

House Finch

ers, 25 species as "probable" breeders, and 23 species as "possible" breeders.

This trial effort clearly revealed the merit of undertaking an atlas project to systematically gather and document useful information about breeding bird distribution in Massachusetts. Although no species was confirmed breeding in every block covered during the pilot year, American Robin was confirmed in 22 out of 24 blocks and European Starling in 21 out of 24. Other species noted in most blocks, even if they were not actually confirmed, included Downy Woodpecker, Northern Flicker, Eastern Phoebe, Eastern Kingbird, Blue Jay, Red-eyed Vireo, Tree Swallow, Black-capped Chickadee, Song Sparrow, and House Sparrow. In the Westport quadrangle, Osprey was confirmed in 5 out of 6 blocks and Carolina Wren in 4 out of 6.

Bird species expanding their ranges to include southern New England were well represented in the pilot year. House Finches were found in single blocks in both Springfield South and Newton, and Tufted Titmice were confirmed in 9 out of 19 blocks; Northern Mockingbirds in 6 out of 16 blocks; and Northern Cardinals in 15 out of 18 blocks. The sole Eastern Screech-Owl breeding confirmation occurred in Newton, as did the only Eastern Bluebird. The most notable confirmations in 1973 were a first state breeding record of a Fish Crow in the Newton quad and a "Lawrence's" Warbler in the Springfield South quad.

The success of the pilot year provided the impetus to begin a comprehensive, statewide atlas effort in 1974. In the five years between 1974 and the completion of the project in 1979, volunteer birders established "probable" or "confirmed" breeding of 198 bird species in Massachusetts.

One of the chief values of such an extensive undertaking is that the resulting data creates the necessary baseline for assessing changes in the status of the Commonwealth's breeding birds. In a world increasingly impacted by human activity, especially in the third most populous state in the nation, it is not surprising to find significant changes in bird distribution related to phenomena such as fragmentation of habitats through development, drainage and degradation of wetlands, and decline of agriculture, as well as natural succession of field and forest habitats and long-term climatic trends. Furthermore, the Atlas maps make it possible to look for correlations between breeding-bird distribution and features associated with the state's physiographic regions, as well as with variations in forest cover, elevation, and even temperature and rainfall.

Another important aspect of establishing an extensive distributional record like that provided by the Atlas is to use it as a basis for comparison with future surveys. The potential for accurate repeatability was ensured in the 1974 to 1979 effort by using a standard map grid and a standardized atlas methodology for determining avian breeding status. Use of standardized breeding bird atlas protocols in future efforts will make it possible to determine the changes in the distribution of Massachusetts breeding birds during the interval since the initial atlas effort was completed. The Massachusetts Breeding Bird Atlas project provides a clear snapshot of the distribution of the Commonwealth's breeding birds during the 1970s—a snapshot with which all future Massachusetts atlas efforts can be compared. In addition to representing a valuable resource for all who have an interest in Massachusetts birds, the Atlas should also be helpful in guiding land-use policy by directing future development away from some of our rarest and most valuable natural features and bird habitats.

Today, use of data obtained from resources as sophisticated and diverse as satellite imagery, Geographic Information Systems strata, and gap analysis makes it possible to generate computerized maps of faunal and floral distributions on a scale never before imaginable. Already, successful attempts at mapping and depicting North American bird distribution on a continental scale have been undertaken for both winter and summer bird populations using Christmas Bird Count data (Root 1988) and Breeding Bird Survey data (Price et al 1995), respectively. With the development of computer links to continental data-gathering schemes, such as the US Fish and Wildlife Service Breeding Bird Survey and the National Audubon

Eastern Screech-Owl

Society's BirdSource, it will be possible to literally watch as volunteer birders from all over North America enter their data on-line to sites on the Worldwide Web.

While techniques for mapping and portraying the distribution of organisms have reached a level of sophistication that Forbush and the other early twentieth-century ornithologists could have hardly imagined, projects such as the Massachusetts Breeding Bird Atlas make it abundantly clear that there will never be a substitute for dedicated field observers gathering information about living organisms in the wild. Without the contribution of thousands of hours of volunteer time, the data needed to complete the Atlas could never have been assembled. Only through diligent exploration and careful observation by people knowledgeable about birds and their behavior was it possible to amass the volume of information recorded in the Atlas.

Over six hundred observers contributed breeding bird information to the Massachusetts Breeding Bird Atlas project. In addition, 90 volunteers wrote species accounts to accompany the maps that appear in the present volume. These accounts were prepared by individuals who were especially knowledgeable about, or who had specific interest in, particular bird species. Each account represents many hours of research, the sifting of much superfluous bird lore, and the discipline of confining one's personal writing style to an abbreviated and standardized text format. Although the use of multiple authors added significantly to the editorial task of preparing the Atlas manuscript for publication, the final text is far richer for capturing the voices, as well as the labor and expertise, of all those who made this ambitious undertaking possible.

Cedar Waxwing

Massachusetts Breeding Bird Distribution

To understand the distribution of Massachusetts breeding birds requires an appreciation of a number of factors that apply almost universally to the distribution of living organisms worldwide, past and present.

Distribution Over Geological Time.
Since a fundamental purpose of any biological atlas is to document both distribution and distributional *change*, it is important to remember that fossil and geomorphological evidence represents a scope of change in the global avifauna that dwarfs the entire span of human existence on the planet.

Vertebrate fossils are scarce in New England—in fact, no bird fossils at all are known from the region. Consequently, it is impossible to say with certainty whether Massachusetts can lay any claim to a resident fauna of the Theropod dinosaurs that evolved into the earliest birds (as well as into *Velociraptor*) during the Jurassic period about 200 million years ago, or to the Cretaceous forms, such as the gull-like *Icthyornis*, originally known from Kansas. However, a mere 50 million years ago, seven-foot-tall avian predators roamed their territories at least as nearby as present-day New Jersey. About 12 million years ago, modern genera and species of birds began to emerge. Indeed, the Pliocene is thought to have represented the zenith of avian diversity—alas, predating the emergence of our own species by 4 to 6 million years. While some of these Pliocene species remain largely unchanged today, many more have passed into oblivion; in fact, more bird species have become extinct through the ages than share the planet with us today.

Needless to say, not all of the diversity of bird life referred to previously passed through the Commonwealth. As recently as 12,000 years ago, there were no birds living in Massachusetts at all because what is now New England lay under a sheet of glacial ice as much as two miles thick. In fact, one theory that attempts to explain the evolution of present-day bird migration patterns holds that bird species that were once resident in northern regions were pushed south by the Pleistocene glaciers while insectivorous species were forced into the equatorial zone. As the ice retreated, at least some of these species reoccupied their former ranges, yet they continued to return to their tropical refuges each year with the onset of winter (Dorst 1962). While such events occurred thousands of years before the beginning of recorded human history, the birds involved were all modern forms—including many of the same species that today continue to return from the Neotropics to breed in Massachusetts each spring.

Geographical Distribution.
Each of the world's approximately 9,700 bird species (Sibley & Monroe 1990) lives within a unique and well-defined geographical range on the land masses and oceans of the earth. The boundaries of these distributional ranges can be viewed as both relatively stable and ever changing. Though this seems contradictory, anyone who has studied birds in a given locality for several decades recognizes that, while modest range extensions and retreats are common in many species, especially those that are migratory, "macrodistribution" of most species tends to be relatively stable, at least within a human-scale time frame.

While there is doubtless an evolutionary imperative in some species to occupy as much of the landscape as the species' physical attributes will permit, it is unusual for a species to fundamentally alter the size and shape of its range in a short time span unless it is within the context of direct human influence such as extirpation, habitat destruction, or long-distance transportation.

There are, however, some notable exceptions. The Cattle Egret, once limited in its distribution to the Old World, bridged the gap between West Africa and Brazil sometime during the late 1800s—apparently without human assistance (Telfair 1994). The species then spread rapidly in Latin America, and the first North American specimen was collected in Wayland, Massachusetts, in 1952 (though Cattle Egrets had

Carolina Wren

House Sparrow

certainly reached Florida some years earlier) and have been breeding in the Commonwealth since 1974.

During the last century, there has been a pronounced northward expansion of the breeding ranges of a number of more southern species into the northeastern United States. These fall into two main categories: wading birds whose breeding ranges are centered in the Neotropics and a diverse assortment of land birds whose only common trait would seem to be their ability to thrive in suburbia. The first group (with dates of first recorded nesting in Massachusetts) includes the Great Egret (1954), Snowy Egret (1955), Little Blue Heron (1940), Tricolored Heron (1976), and Glossy Ibis (1974). Examples of the second group include the Turkey Vulture (1954), Red-bellied Woodpecker (1977), Tufted Titmouse (1958), Carolina Wren (formerly nesting almost exclusively in southeastern Massachusetts and expanding north and west c1970s), Blue-gray Gnatcatcher (1961), Northern Mockingbird (gradually increasing during the 1900s with a marked increase in the 1960s and 1970s), and Northern Cardinal (1961).

Explanations offered for these comparatively sudden range expansions include rebound from hunting pressure (e.g., waterbirds), climate change (i.e., global warming), prolonged southern drought (e.g., wading birds), creation of expansion corridors with the merging of the Boston-New York-Washington DC suburbs (e.g., several passerine species), and an increase in winter bird feeding (Veit & Petersen 1993). While none of these may be convincing explanations by themselves, all are plausible contributing factors.

Other notable breeding-range changes in Massachusetts within the last half-century include the following.

- Willet. This shorebird formerly bred in coastal salt marshes from Nova Scotia to Florida and Texas. It was extirpated in the Northeast by hunters by the turn of the century; then, when it was protected, it reestablished itself in southeastern Nova Scotia and southern New Jersey by the 1950s, but nowhere in between. It reestablished itself as a breeder in Massachusetts by 1976 and is once again common in favorable habitat from Cape Cod to Essex County.
- Wilson's Phalarope. Typically breeding in prairie marshes, this species has probably nested continuously in Essex County salt marshes since 1979.
- Forster's Tern. This species, which nests in western freshwater marshes and southeastern salt marshes, has possibly nested, at least irregularly, in the Plum Island salt marshes since 1990 or 1991.
- Cerulean Warbler. A globally declining forest breeder, this Midwestern species began nesting in very small numbers in Massachusetts in 1989 (Veit & Petersen 1993). Cerulean Warblers require mature deciduous forests with sparse understories and closed or semiopen canopies for nesting (Robbins et al 1992). It is presently most common as a breeder in the Mississippi and Ohio river valleys (Dunn & Garrett 1997). Given the global rate of decline of this species, whether it continues to increase or be present as a breeder in Massachusetts, and elsewhere in New England, is a matter of pure conjecture.

It is difficult to guess whether these last examples, involving at most only a few pairs of individuals, represent the early stages of true range expansions or just "vagrant breeders" that may eventually vanish rather than steadily increase.

Another category comprises those species (e.g., Canada Goose, European Starling, House Finch, and House Sparrow) that have undergone successful trans- or intercontinental range expansions through deliberate introduction by humans.

Range changes that may have occurred as a result of the rapid and wholesale deforestation in the Northeast and the opening of corridors between New England and the Great Plains during the early years of European colonization are more difficult to document. While there can be little doubt that grassland birds such as Bobolinks, Eastern Meadowlarks, and several sparrow species became more widespread and abundant in New England as settlers turned the eastern forests into fields and pastures, they do not appear to be recent arrivals.

There is historical evidence that in pre-Columbian times these species occupied marginal grasslands in river valleys and the edges of salt marshes, as well as the meadows that evolved from abandoned Beaver ponds and areas cleared by wildfires and those set by Native Americans. Therefore, while their status may have changed due to ecological changes caused by people, there is no evidence that they actually expanded their range.

The status of the Upland Sandpiper in New England before deforestation is even less clear. Because this species requires extensive areas of habitat in which to breed successfully, it now survives here mainly in large-scale tracts of managed grasslands, especially airports. It continues to decline in New England even though it is presently increasing in the Midwest. This suggests that Upland Sandpipers were unable to colonize New England until European settlers created pastures of prairielike proportions and did, in fact, extend their breeding range early in the colonial period.

Even though the range of a given bird species tends to be relatively stable, variation in the size, form, and movement within the range is often great among different species. For example, of the world's various bird species having "cosmopolitan" breeding ranges, meaning that they nest in all four hemispheres or in all six of the major zoogeographic regions,[1] seven of them—the Great Egret, Black-crowned Night-Heron, Osprey, Peregrine Falcon, Roseate Tern, Barn Owl, and Barn Swallow—nest in Massachusetts.

It is possible to depict the breeding ranges of the remaining Massachusetts nesting birds as a series of geographic polygons with different shapes and areas, with the following examples.

- Twenty-six (13 percent) of our native breeders have Holarctic breeding ranges, nesting in Europe (and in some cases northern Asia), as well as in North America. A few other species (e.g., Mute Swan, Ring-necked Pheasant, Rock Dove, European Starling, and House Sparrow) have recently acquired a Holarctic breeding range as a result of introductions from one continent to another.
- Seventy-nine species (40 percent) that nest in the Commonwealth are confined to the New World but breed in the Neotropics as well as in North America.
- One hundred twenty-five (61 percent) of Massachusetts breeding birds nest continentwide.
- Approximately 55 species (28 percent) can be described as birds of primarily northern distribution, that is, species that nest mainly to the north (and west) within the zone of coniferous and mixed forest and approach the southeastern limit of their range in Massachusetts (though many follow their preferred habitat south into the Appalachian highlands).
- Approximately 36 species (18 percent) can be described as species of primarily southern distribution, that is, species near the northern limits of their breeding range in Massachusetts, including those noted above that have begun to nest here only within the last half-century.

While a number of eastern bird species meet ecological limits to their breeding range in the Great Plains or Rocky Mountains, there is no such thing as a Northeastern (much less a New England or Massachusetts) endemic breeding bird. This is in notable contrast to central and western North America where a significant number of species nest only in well-circumscribed mountain or lowland habitats (e.g., Gunnison Sage-Grouse, Yellow-billed Magpie).

Seasonal Distribution.
For a large majority of migratory species occurring in Massachusetts, the breeding range is but a small portion of their overall distribution, a distribution that also includes a winter range and stopover areas used during migration. With some species, the area covered by their spring passage is completely different from that used in the fall (e.g., American Golden-Plover, Connecticut Warbler), which greatly increases the overall distributional area used by the species.

There is frequently great variation in the amount of time any given species spends within its immediate breeding range. So-called permanent residents, such as the Black-capped Chickadee, may both nest and winter in a restricted area, though it has been demonstrated that other populations of chickadees that breed in Massachusetts move south for the winter

---

[1] Nearctic (New World temperate zone), Neotropical (New World tropics), Palearctic (Old World temperate zone), Ethiopian (Sub-Saharan Africa), Oriental (India through southern Asia), and Australasian. The Holarctic region combines the Nearctic and Palearctic regions.

Black-capped Chickadee

while still others arrive from the North to spend the winter here. Sixty-four Massachusetts breeding bird species (31 percent) are regarded as permanent residents in the state; that is, species that maintain year-round populations in the Commonwealth.

The *duration* of the breeding season also varies among species. Resident Louisiana Waterthrushes typically arrive in Massachusetts by mid-April and depart for their Neotropical wintering grounds by the end of July. Gray Catbirds, by comparison, arrive by early May, rear two broods, and typically remain in Massachusetts through September, with a few birds overwintering along the coast every year. Catbirds spend a far greater proportion of their life cycle on their breeding grounds than waterthrushes. Another pattern followed by many migratory colonial nesting birds is to disperse from the immediate nesting area in all directions once the young are fledged, only to gather together again at traditional staging areas before eventually moving south on a more or less synchronized timetable (e.g., Great Egret, Snowy Egret).

If we believe that migration evolved in response to glaciation, then it should not be surprising to realize that most Massachusetts breeding birds migrate on a more or less north-south axis. One hundred-thirty (63 percent) of these migratory species winter in the Neotropics, principally in Mexico, Central America, and the Caribbean Basin, with only a few species reaching South America. The remainder winter in North America, mainly in the southeastern United States.

Habitat Preference.
It is well established that birds, like other organisms, are closely associated with specific habitats, however broad and variable these habitats may be during different portions of their annual cycle. For highly mobile birds, which are endowed with the power of flight, habitat requirements often change with the seasons. However, during the breeding season at least, most bird species are predictably associated with specific environments. Even highly adaptable species such as the American Crow, American Robin, and Common Grackle—species found in practically every terrestrial and wetland habitat in Massachusetts at one time of year or another—exhibit more or less distinct habitat preferences during the nesting season. Other species, such as the Piping Plover, Alder Flycatcher, and Louisiana Waterthrush, are so specific in their habitat requirements that it is difficult to imagine them in habitats other than those in which they prefer to nest. These nesting preferences are of course yet another variable determining the distribution of breeding birds in Massachusetts.

This is not the forum for speculating on how and why any given bird species acquired certain habitat preferences. Suffice to say that each species has evolved separately, and each has acquired its breeding habits in response to a variety of opportunities and limiting factors including food availability, spatial requirements, regional topography, vegetation type, structural characteristics, competition from other species, and climate. The collective effect of these factors on a given species in a particular place, such as Massachusetts, may be to severely limit its distribution, or else to ensure that it will be widespread.

For example, in the Northeast, Red-tailed Hawks require an area of enclosed forest with tall trees (ideally, White Pines) in which to nest, coupled with extensive open areas where they can readily find an abundance of their preferred small rodent prey. As a result, this large raptor is able to range throughout Massachusetts, particularly where highways and river corridors create the ideal combination of forest stands for nesting and for scanning for prey in adjacent open areas.

However, as the Atlas map confirms, Red-tailed Hawks are likely to be absent as breeding species from areas where there is extensive forest cover with few large openings, such as the north-central parts of the state. The habitat requirements of the Red-tailed Hawk in Massachusetts demonstrate how the natural forest cover of the region potentially limits the species' distribution, and also how human land-use patterns that create a patchwork of forest and open spaces have significantly increased breeding opportunities for this species. Indeed, Red-tailed Hawks have steadily moved into even the largest cities where parks and "urban canyons" supply their need for perches and nearby open spaces inhabited by an abundance of suitable prey.

Gray Catbird

In striking contrast to the Red-tailed Hawk, the Seaside Sparrow has evolved to fit into a very narrow ecological niche in coastal salt marshes. Within these intertidal grasslands, the sparrows nest very locally and semicolonially in marginal areas where taller grasses or shrubs grow above all but the highest tides. They have also shifted their feeding habits from ground foraging for seeds, which most sparrows favor, to capturing more-readily-available salt marsh invertebrates. As a result of these strict habitat requirements (and the fact that the species reaches the northern extent of its range in the Commonwealth), Seaside Sparrows in Massachusetts are rare and local in the state, confined to only a few salt marsh locations within a narrow coastal margin.

As noted, many factors contribute to the occurrence of different types of habitat across the Massachusetts landscape, and each Massachusetts breeding bird species has a unique set of breeding requirements that govern the parts of the state that it is able to occupy. The Blackpoll Warbler is essentially a bird of Canadian spruce-fir forests, a type of vegetation that has evolved to thrive in the harsh boreal climate that prevails mainly north of Massachusetts. However, at the summit of Mount Greylock, relatively high altitude (i.e., 3,491 feet) replicates the climatic conditions of more northern latitudes, which in turn gives rise to enough spruce-fir habitat to support a breeding population of Blackpolls.

Unlike Blackpoll Warblers, Bank Swallows nest throughout the Northern Hemisphere in the temperate zone (and throughout Massachusetts), regardless of the type of vegetative cover, but only where there are riverbanks, dune faces, or gravel pits in which to excavate their nest holes. Their distribution is therefore localized not only by topography, soil chemistry, climate, and the resulting plant community but also by the occurrence of a physical force—erosion—acting on a soil of a suitable texture. Similar profiles of breeding habitat requirements and preferences could be articulated for every bird species in Massachusetts—and every bird species in the world.

In conclusion, it is perhaps worth emphasizing that (1) every profile consists of a combination of factors so that bird habitat is never defined solely by climate, topography, vegetation, or any other single element, and (2) no two species have identical habitat profiles.

Describing Massachusetts Breeding Bird Habitats.
The maps in this volume provide a clear picture of the distribution of Massachusetts breeding birds and, with the accompanying acetate overlays, show some of the factors that influence these patterns. What the maps do not do is indicate relative abundance within each survey block. The species accounts, however, add many insights into the complex inquiry of why different species live where they do.

The interested reader should be aware that there are also other key publications and mapping systems, both current and historical, that have addressed the biogeography of Massachusetts, especially its avifauna. A number of previous authors have provided extensive and detailed descriptions of the topography, faunal areas, and some of the more important ornithological regions of the Bay State (e.g., Howe & Allen 1901, Forbush 1927, Veit & Petersen 1993).

Other publications describe regional distribution for specific areas in the Commonwealth (e.g., Faxon & Hoffmann 1900, Townsend 1905, Brewster 1906, Bagg & Eliot 1937, Wetherbee 1945, Griscom & Folger 1948, Griscom 1949, Griscom & Emerson 1959, and Hill 1965). Most recently, Leahy et al (1996) have produced an elegant literary and visual overview of the primary ecosystems and habitat types in Massachusetts, thereby providing detailed descriptions of virtually every avian habitat in the state.

Additions to our scientific knowledge, and especially our ability to analyze and graphically depict multiple layers of data using Geographical Information Systems, have resulted in new schemes for defining ecological communities. These modern techniques define the landscape in global terms rather than by politically drawn boundaries. One of the most important of these ecological characterization schemes describes North America in terms of ecoregions, defined as "regions of relative homogeneity in ecological systems and relationships between organisms and their environments" (Griffith et al 1994).

Originally established to assist scientists in structuring their research, assessment, monitoring, and

Red-tailed Hawk

ultimately the management of environmental resources, the ecoregional approach permits analysis of patterns of homogeneity in ecosystem components such as soils, vegetation, climate, geology, and patterns of human uses. This approach provides another lens through which to examine bird distribution.

Defining ecoregions (e.g., Omernik 1987) has rapidly been refined through a number of collaborative projects among state and federal environmental agencies and private conservation organizations. In the early nineties, the Massachusetts Department of Environmental Protection, Division of Water Pollution Control, and the United States Environmental Protection Agency jointly produced *The Massachusetts Ecological Regions Project* (Griffith et al 1994). This document places Massachusetts in the context of two major ecoregions—the Northeastern Coastal Zone and the Northeastern Highlands, which in turn are divided into 13 subregions (see page 12).

The subregional delineations are derived from maps of soil types, physiography, vegetation, and land use. Other strata incorporated into the classification criteria are bedrock and surficial geology, hydrology, and climate information. It is probable that, ultimately, climate has the greatest influence on breeding bird distribution. In order to set scientifically defensible conservation priorities, The Nature Conservancy has further refined ecological subregions into polygons that define individual biological communities, as well as their "quality" (i.e., their ability to provide for the long-term conservation of biodiversity in the region).

While not all natural community definitions are directly relevant to describing patterns of bird distribution, there are many instances when the application of gap analysis (i.e., assessing the relationship between the distribution of bird life and plant communities) can reveal hitherto unrecognized patterns that may prove useful in guiding management and avian habitat protection efforts.

Finally, as this volume goes to press, two other major projects are underway in Massachusetts that both build upon, and add to, the definitions of the Commonwealth's breeding bird fauna contained in this Atlas. They are the Massachusetts Important Bird Areas (IBA) Program, coordinated by the Massachusetts Audubon Society, and the Massachusetts BioMap Project, organized through the Commonwealth's Executive Office of Environmental Affairs and the Massachusetts Natural Heritage and Endangered Species Program.

By soliciting nominations and developing consensus among the state's most knowledgeable bird authorities, the IBA program will designate specific localities deemed to be critical to the conservation of bird life in Massachusetts. The program identifies important wintering and migratory sites as well as breeding areas. The even more ambitious BioMap Project is attempting to document and map *all* of the state's most important biotic elements, including, but by no means restricted to, rare or otherwise significant breeding bird habitat. The comprehensive scope of this project is likely to yield the kinds of insights referred to previously in which the incidence of rare species occurrences will help highlight conservation priorities.

Laughing Gull

Figure 1
Principal Waterbird Colonies on the Massachusetts Coast

Massachusetts Breeding Bird Distribution

Massachusetts Ecoregions:
A Landscape Approach to Delineating Bird Distribution

Figure 2 illustrates the location of the two Massachusetts ecoregions, the Northeastern Highlands and the Northeastern Coastal Zone, as well as the Commonwealth's 13 subregions. When the overlay map showing Major Forest Types of Massachusetts is superimposed on the Northeastern Highlands Ecoregion, it is obvious that the major forest cover in this sector is northern hardwood (maple-beech-birch) and northern hardwood-spruce. Northeastern spruce-fir forests occur in the areas of highest elevation, and the soils throughout this ecoregion tend to be frigid and cryic spodosols (Omernik 1987, Griffith et al 1994). Major Massachusetts river systems in the Northeastern Highlands include the Hoosic, Housatonic, Deerfield, Westfield, Farmington, Millers, and small parts of the Connecticut, upper Nashua, and upper Chicopee.

The most extensive ecoregion in Massachusetts—the Northeastern Coastal Zone—is dominated by glacial plains covered with low to high hills that support land uses of woodland and forest, along with some cropland and pasture, and some sizable areas of urbanization (Anderson 1970). The predominant natural forest cover is Appalachian oak forest and northeastern oak-pine forest (Kuchler 1970), with mostly inceptisol soils (Griffith et al 1994). Primary drainage basins in the Northeastern Coastal Zone are the Connecticut, Chicopee, Quinebaug, French, Blackstone, Nashua, Concord, Merrimack, Shawsheen, Parker, Ipswich, North Coastal, Boston Harbor, Charles, Ten Mile, Taunton, South Coastal, Narragansett Bay, Mount Hope Bay, Buzzards Bay, Cape Cod Bay, and the Islands.

Yellow-rumped Warbler

Northeastern Highlands Subregions

Taconic Mountains—This subregion, dominated by Mount Greylock (3,491 feet), the highest point in the Commonwealth, is the primary breeding station for the Blackpoll Warbler, Mourning Warbler, and Bicknell's Thrush (historically) in the state. The forest is mainly northern hardwood (maple-beech-birch) with some spruce and fir remaining at higher elevations.

Western New England Marble Valleys (i.e., Berkshire Valley and Stockbridge Valley in Massachusetts)—This scenic region is characterized by a number of suburban and semiurban communities, cropland and pastures, and both transition (oak-hickory) and northern hardwood forests, depending upon the latitude and elevation. The Turkey Vulture and Yellow-throated Vireo are two species typical of this subregion in Massachusetts.

Green Mountains/Berkshire Highlands—This subregion is relatively high (for Massachusetts), with elevations ranging from 1,000 to more than 2,500 feet, and is cloaked with spruce-fir and northern hardwood forest types. This area—often referred to as the Berkshire Plateau—is the only subregion hosting three rare and local Massachusetts breeding species that have been recorded nowhere else in the Commonwealth: Ruby-crowned Kinglet, Lincoln's Sparrow, and Rusty Blackbird.

Lower Berkshire Hills—This region is similar to the Berkshire Highlands to the north, but the average elevation is lower, (1,000 to 1,700 feet) with a forest cover comprised primarily of northern hardwoods and lacking the high-elevation spruce-fir; it also contains an element of the more southern transition hardwood (oak-hickory) forest. No breeding species are unique to this subregion.

Berkshire Transition—This region is not unlike the Lower Berkshire Hills, except its average elevation is lower (400 to 1,400 feet) and it possesses primarily transition hardwood forest, with patches of northern hardwoods at higher elevations. No breeding species are unique to this subregion.

Vermont Piedmont—This subregion is characterized by hills, some having steep slopes and elevations ranging from 400 to 1,400 feet in Massachusetts. The forest cover resembles that of the Berkshire Transition subregion, the primary differences being soil types. No breeding species are unique to this subregion, although, not unlike the Berkshire Highlands to the

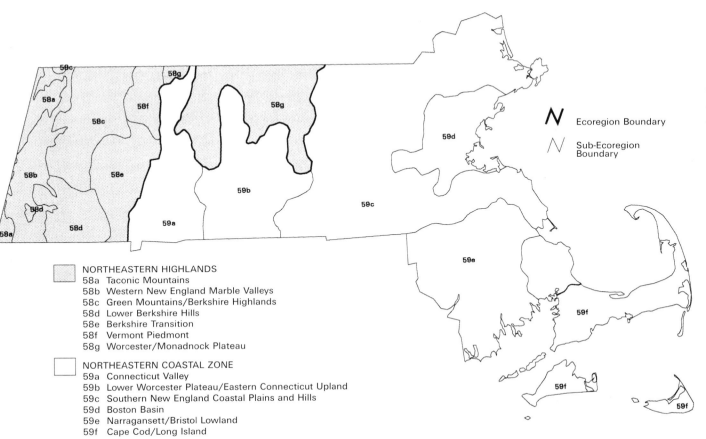

*Figure 2
Ecoregions of
Massachusetts*

west, a number of breeding species with northern affinities (e.g., the Yellow-bellied Sapsucker, Black-throated Blue Warbler, Yellow-rumped Warbler) are notably more numerous as breeders in this region than in other regions of the state.

Worcester/Monadnock Plateau—This is a large and important subregion that includes most of the mountainous and hilly areas of the central uplands of Massachusetts. Elevations range from 500 to 1,400 feet, with the exception of Mount Watatic (1,832 feet) and Mount Wachusett (2,006 feet), both of which provide easterly outposts for northern breeding species more typical of the higher elevations of western Massachusetts (e.g., Yellow-bellied Sapsucker,

Olive-sided Flycatcher [rare], Golden-crowned Kinglet, Dark-eyed Junco). Subsequent to the Atlas period, Common Loons and Common Ravens had also established themselves as breeding residents in this subregion. Forest types are largely transition and northern hardwoods, with pockets of spruce on the higher hilltops.

Northeastern Coastal Zone Subregions

Connecticut Valley—A clearly discernible subregion, the Connecticut Valley has a milder climate than the adjacent hill country; relatively level terrain, ranging from 100 to 500 feet on the highest ridges; and land comprised mostly of urban and built-up areas, croplands, and pasture. Deciduous forests of central hardwoods (oak-hickory) and transition hardwoods (maple-beech-birch, oak-hickory) cover most of the ridges. Breeding birds that showed a correlation with this subregion during the Atlas period included the Tufted Titmouse, Northern Mockingbird, Vesper Sparrow, and House Finch, particularly when compared with the hill subregions to the east and west. The combined effects of milder temperatures, lower elevation, and an overall more southerly forest type in this subregion were quite likely among the factors influencing the distribution of the Tufted Titmouse, Northern Mockingbird, and House Finch as they initiated their colonization of New England during the 1970s.

Lower Worcester Plateau/Eastern Connecticut Upland—This region and the Southern New England Coastal Plains and Hills subregion are in many respects quite similar. In this sector, elevation ranges from 300 to 500 feet, and the landscape is described as open hills where 50 to 75 percent of the gentle slope is on the upland (Hammond 1970). The forest cover is mainly a mixture of transition and central hardwoods (oak-hickory), as well as some elm-ash-Red Maple and White Pine-Red Pine. There are no breeding bird species specifically associated with this subregion.

Southern New England Coastal Plains and Hills—This is the largest subregion in Massachusetts and is perhaps the most diverse in its characteristics and

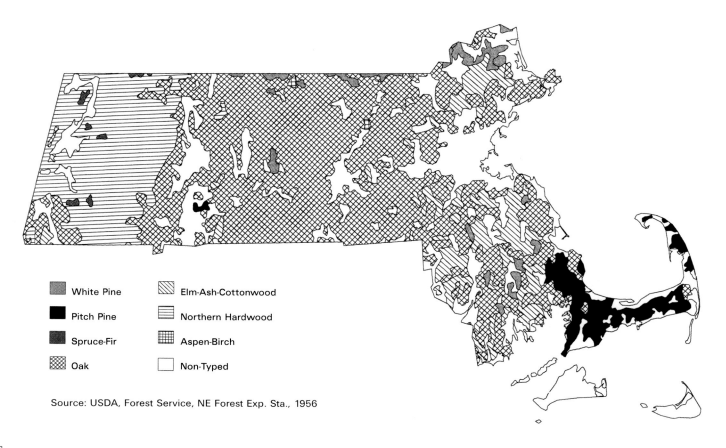

Figure 3
Major Forest Types of Massachusetts

Source: USDA, Forest Service, NE Forest Exp. Sta., 1956

habitats. Elevation ranges from sea level to 800 feet, with Great Blue Hill representing the highest point in eastern Massachusetts. The principal forest types are central hardwoods, along with transition hardwoods on the richer soils, White Pine-Red Pine, and elm-ash-Red Maple, particularly in southeastern Massachusetts. Although no breeding species are closely tied to this extensive ecoregion, it nonetheless contains a wide representation of the bird species regularly found nesting in Massachusetts. Curiously, the Ring-necked Pheasant showed an affinity for this subregion more than any other species tallied during the Atlas, undoubtedly an artifact produced by the large numbers of pheasants released in eastern Massachusetts for hunting purposes.

Boston Basin—A heavily developed subregion, the Boston Basin contains a number of ponds, lakes, and reservoirs and is drained primarily by the Neponset, Charles, Mystic, and Saugus rivers. The topography is low and rolling, studded with drumlins and glacial-till-covered bedrock hills. Forests are mainly transition hardwoods (oak-hickory), often heavily mixed with White Pine. Few bird species are specifically associated with this subregion; nearly every block surveyed in this sector during the Atlas period confirmed the nesting of those widespread species routinely associated with suburbia (e.g., American Robin, Northern Cardinal, House Finch). One species showing an interesting affinity for this subregion was the Common Nighthawk, a fact readily explained by its nesting on roofs in urban surroundings.

Narragansett/Bristol Lowland—This subregion is characterized by flat to gently rolling irregular plains with elevations generally lower than 200 feet. Land cover is mostly mixed forest with numerous wetlands, including lakes and ponds, and small areas of cropland and pastures. Cranberry bogs are notably abundant. The forest type is mainly central hardwood, with a liberal presence of elm-ash-Red Maple and White Pine-Red Pine (Kingsley 1974). Breeding birds showing an affinity for this subregion include Northern Bobwhite, Yellow-billed Cuckoo, and Prairie Warbler.

Cape Cod/Long Island—One of the most distinctive subregions, this area is defined by its low elevation (generally below 200 feet) and prominent glacial features such as terminal moraines, outwash plains, kettle ponds, and coastal sand dunes. The most abundant forest cover is oak-Pitch Pine, and there are a number of distinctive aquatic habitats including salt- and freshwater marshes, swamps, and bogs. Characteristic breeding birds of this subregion are many, some of the more typical being Piping Plover, American Oystercatcher, Roseate Tern, and Pine Warbler.

Additional Factors: Maps and Transparent Overlays

The ecoregion scheme integrates selected physical and vegetational strata in order to describe certain ecological characteristics of Massachusetts, as well as those of the greater New England region. Despite the importance of the ecoregion concept in defining the ecological makeup of the state, readers are also encouraged to carefully examine both the printed maps and the overlay strata maps that accompany the Atlas. In some cases, a correlation between a particular environmental factor such as forest cover or elevation and a species' distribution in the state may be more obvious by using these transparencies than by using the Ecoregions of Massachusetts map exclusively.

As noted in the section on Massachusetts Breeding Bird Distribution, a number of environmental factors contribute to defining breeding bird distribution; and, by examining the strata maps separately, it is sometimes possible to establish which of these factors may have the highest correlative value. A fine example of how correlative habitat information can be useful in guiding future conservation planning at the habitat level rather than the ecoregion level may be found in the *Classification of the Natural Communities of Massachusetts* (Swain & Kearsley 2000).

Towns and Counties of Massachusetts—The towns and counties map outlines the 14 counties in Massachusetts as well as delineating the boundaries of all the towns and cities in the Commonwealth. The map

Black Skimmer

is especially useful for determining the precise location of a county or township mentioned in the text or for checking to see if a particular bird species was confirmed at a specific locality.

Massachusetts Quads Index—The quads index map gives the location of each of the 189 map quadrangles representing Massachusetts. Along with the Towns and Counties of Massachusetts map, the Quads Index is useful in determining the locations of places described in the text or for determining on which topographical quadrangle sheet a particular town is located.

Ecoregions of Massachusetts—The ecoregions map delineates the extent and location of the Northeastern Highlands and Northeastern Coastal Zone, as well as the 13 subregions that make up these ecoregions. This map is especially helpful when it is used in conjunc-

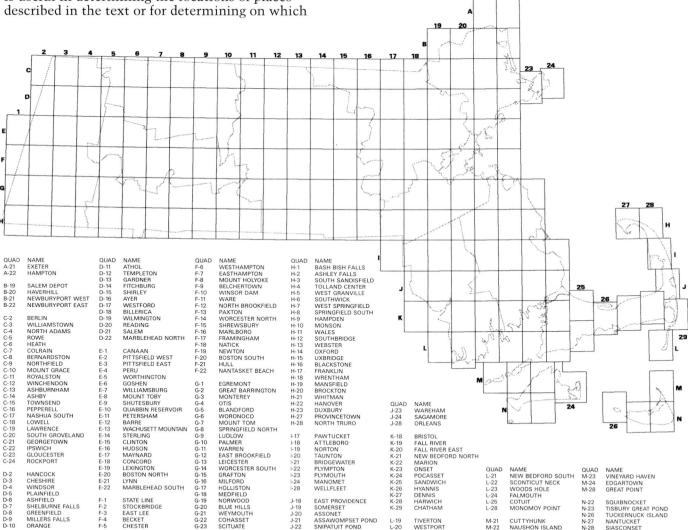

Figure 4
Quads Index

QUAD	NAME	QUAD	NAME	QUAD	NAME	QUAD	NAME
A-21	EXETER	D-11	ATHOL	F-6	WESTHAMPTON	H-1	BASH BISH FALLS
A-22	HAMPTON	D-12	TEMPLETON	F-7	EASTHAMPTON	H-2	ASHLEY FALLS
		D-13	GARDNER	F-8	MOUNT HOLYOKE	H-3	SOUTH SANDISFIELD
B-19	SALEM DEPOT	D-14	FITCHBURG	F-9	BELCHERTOWN	H-4	TOLLAND CENTER
B-20	HAVERHILL	D-15	SHIRLEY	F-10	WINSOR DAM	H-5	WEST GRANVILLE
B-21	NEWBURYPORT WEST	D-16	AYER	F-11	WARE	H-6	SOUTHWICK
B-22	NEWBURYPORT EAST	D-17	WESTFORD	F-12	NORTH BROOKFIELD	H-7	WEST SPRINGFIELD
		D-18	BILLERICA	F-13	PAXTON	H-8	SPRINGFIELD SOUTH
C-2	BERLIN	D-19	WILMINGTON	F-14	WORCESTER NORTH	H-9	HAMPDEN
C-3	WILLIAMSTOWN	D-20	READING	F-15	SHREWSBURY	H-10	MONSON
C-4	NORTH ADAMS	D-21	SALEM	F-16	MARLBORO	H-11	WALES
C-5	ROWE	D-22	MARBLEHEAD NORTH	F-17	FRAMINGHAM	H-12	SOUTHBRIDGE
C-6	HEATH			F-18	NATICK	H-13	WEBSTER
C-7	COLRAIN	E-1	CANAAN	F-19	NEWTON	H-14	OXFORD
C-8	BERNARDSTON	E-2	PITTSFIELD WEST	F-20	BOSTON SOUTH	H-15	UXBRIDGE
C-9	NORTHFIELD	E-3	PITTSFIELD EAST	F-21	HULL	H-16	BLACKSTONE
C-10	MOUNT GRACE	E-4	PERU	F-22	NANTASKET BEACH	H-17	FRANKLIN
C-11	ROYALSTON	E-5	WORTHINGTON			H-18	WRENTHAM
C-12	WINCHENDON	E-6	GOSHEN	G-1	EGREMONT	H-19	MANSFIELD
C-13	ASHBURNHAM	E-7	WILLIAMSBURG	G-2	GREAT BARRINGTON	H-20	BROCKTON
C-14	ASHBY	E-8	MOUNT TOBY	G-3	MONTEREY	H-21	WHITMAN
C-15	TOWNSEND	E-9	SHUTESBURY	G-4	OTIS	H-22	HANOVER
C-16	PEPPERELL	E-10	QUABBIN RESERVOIR	G-5	BLANDFORD	H-23	DUXBURY
C-17	NASHUA SOUTH	E-11	PETERSHAM	G-6	WORONOCO	H-27	PROVINCETOWN
C-18	LOWELL	E-12	BARRE	G-7	MOUNT TOM	H-28	NORTH TRURO
C-19	LAWRENCE	E-13	WACHUSETT MOUNTAIN	G-8	SPRINGFIELD NORTH		
C-20	SOUTH GROVELAND	E-14	STERLING	G-9	LUDLOW	I-17	PAWTUCKET
C-21	GEORGETOWN	E-15	CLINTON	G-10	PALMER	I-18	ATTLEBORO
C-22	IPSWICH	E-16	HUDSON	G-11	WARREN	I-19	NORTON
C-23	GLOUCESTER	E-17	MAYNARD	G-12	EAST BROOKFIELD	I-20	TAUNTON
C-24	ROCKPORT	E-18	CONCORD	G-13	LEICESTER	I-21	BRIDGEWATER
		E-19	LEXINGTON	G-14	WORCESTER SOUTH	I-22	PLYMPTON
D-2	HANCOCK	E-20	BOSTON NORTH	G-15	GRAFTON	I-23	PLYMOUTH
D-3	CHESHIRE	E-21	LYNN	G-16	MILFORD	I-24	MANOMET
D-4	WINDSOR	E-22	MARBLEHEAD SOUTH	G-17	HOLLISTON	I-28	WELLFLEET
D-5	PLAINFIELD			G-18	MEDFIELD		
D-6	ASHFIELD	F-1	STATE LINE	G-19	NORWOOD	J-18	EAST PROVIDENCE
D-7	SHELBURNE FALLS	F-2	STOCKBRIDGE	G-20	BLUE HILLS	J-19	SOMERSET
D-8	GREENFIELD	F-3	EAST LEE	G-21	WEYMOUTH	J-20	ASSONET
D-9	MILLERS FALLS	F-4	BECKET	G-22	COHASSET	J-21	ASSAWOMPSET POND
D-10	ORANGE	F-5	CHESTER	G-23	SCITUATE	J-22	SNIPATUIT POND

QUAD	NAME
J-23	WAREHAM
J-24	SAGAMORE
J-28	ORLEANS
K-18	BRISTOL
K-19	FALL RIVER
K-20	FALL RIVER EAST
K-21	NEW BEDFORD NORTH
K-22	MARION
K-23	ONSET
K-24	POCASSET
K-25	SANDWICH
K-26	HYANNIS
K-27	DENNIS
K-28	HARWICH
K-29	CHATHAM
L-19	TIVERTON
L-20	WESTPORT

QUAD	NAME
L-21	NEW BEDFORD SOUTH
L-22	SCONTICUT NECK
L-23	WOODS HOLE
L-24	FALMOUTH
L-25	COTUIT
L-28	MONOMOY POINT
M-21	CUTTYHUNK
M-22	NAUSHON ISLAND

QUAD	NAME
M-23	VINEYARD HAVEN
M-24	EDGARTOWN
M-28	GREAT POINT
N-22	SQUIBNOCKET
N-23	TISBURY GREAT POND
N-26	TUCKERNUCK ISLAND
N-27	NANTUCKET
N-28	SIASCONSET

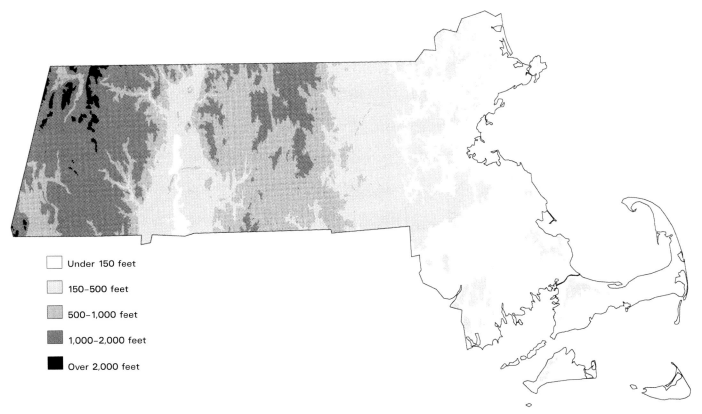

*Figure 5
Massachusetts
Generalized Elevations*

tion with the species distribution maps because it reflects any obvious correlation existing between the distribution of a particular bird species and a specific ecoregion. For example, note the clear affinity of Swainson's Thrush for the Green Mountains/Berkshire Highlands subregion and White-eyed Vireo for the Narragansett/Bristol Lowland.

Major Forest Types of Massachusetts—The major forest types map, especially when used in conjunction with the Ecoregions of Massachusetts map, helps to provide a clear indication of any correlation that may exist between a major forest type and the distribution of a particular species of bird. For example, note the clear affinity of the Golden-crowned Kinglet and Blackburnian Warbler for spruce-fir and northern hardwood forest and the Pine Warbler for Pitch Pine forest.

Massachusetts Generalized Elevations—The map showing generalized elevations is useful for making associations with breeding bird distribution and height above sea level. For example, the relationship between elevation and the breeding distribution of coastal-plain breeding species such as the Piping Plover and several tern species, or highland breeders such as the Blackpoll Warbler and Mourning Warbler, is obvious. However, the correlation between breeding distribution and elevation for the Yellow-bellied Sapsucker and Pine Warbler is considerably more subtle unless the maps are carefully compared. In this example, a relationship between forest type and elevation is also indicated.

Major Drainages of Massachusetts—The major drainages map is self-explanatory. By providing the locations of the principal river systems in Massachusetts, another topographical layer is available for seeking correlations between breeding bird species and Massachusetts waterways. For example, although various habitat factors are involved, the fact remains that the distribution maps for the Grasshopper Sparrow and the Vesper Sparrow show a clear relationship to the agricultural lands associated with the Connecticut River valley.

Massachusetts Average Annual Precipitation 1974–1979—The precipitation map provides a helpful overview of rainfall amounts in Massachusetts during the Atlas period. Undoubtedly, a correlation exists among elevation, forest type, and precipitation, but making the case that rainfall by itself influences breeding bird distribution is not only beyond the scope of this analysis but also highly speculative.

Sum of Birds Surveyed in Each Cell Grid in Massachusetts—This map shows the total number of bird species that were tallied in each block at all levels of confirmation during the 1974 to 1979 period. Since insufficient scientific rigor was applied to the effort factor at the time the Atlas breeding bird data was being collected and analyzed, it is difficult to establish accurate correlations between the numbers indicated on the map and the absolute breeding bird diversity indicated in each Atlas block. Nonetheless, in general, the map does provide a *relative sense* of species richness and species paucity, as well as give a general indication of the general number of breeding bird species to be expected in various areas of Massachusetts.

Pileated Woodpecker

Atlas Methods and Criteria

Atlas Methods

The information that follows is an abbreviated description of the basic techniques and procedures that were used in gathering the information collected by Atlas volunteers in Massachusetts. For a more detailed account of atlas rationales and methods, the reader is referred to the comprehensive *Handbook for Atlasing American Breeding Birds* (Smith 1990), published by the North American Ornithological Atlas Committee.

As described earlier, the Massachusetts Breeding Bird Atlas project achieved almost total state coverage. Although not every block and quad in Massachusetts received equal coverage over the 1974 to 1979 period, at least some coverage was obtained in every block except for several on the borders of the state where the quad maps include only negligible Massachusetts territory. This high degree of coverage was possible because of the dedication of over 600 volunteers who contributed time, expertise, and data to the Atlas project.

Unlike in some other states, no Bay State quads or regions were assigned "priority-block" status. Instead, full state coverage was made a priority, with the result that a nearly complete survey of state habitats was obtained. Throughout the Atlas project, a technique called "block-busting" was employed to provide coverage in underworked blocks. This technique was first initiated by the South Shore Bird Club and was eventually utilized by other groups as well. Block-busting volunteers were organized into teams working in a single block for half a day. This method was usually employed at the height of the breeding season and often made it possible to record a significant percentage of the common breeding species in blocks where previously there were few confirmations.

During the final years of the Atlas project, several observers were contracted by the Massachusetts Audubon Society to cover blocks where there had been little or no previous fieldwork. The transparent overlay entitled Sum of Birds Surveyed in Each Cell Grid in Massachusetts indicates the extent of statewide coverage, although the number of bird species shown in each block on the map has not been normalized to reflect variances in the amount of coverage time per block.

For the purpose of mapping, US Geological Survey topographic quadrangle maps were used to divide the state into 189 manageable survey units. Each quadrangle map was further divided into six blocks. Approximately 75 volunteer coordinators were responsible for obtaining block coverage for as many quads as possible during the course of the five-year Atlas period. Coordinators also served as liaisons between the volunteers and the Atlas coordinator at Massachusetts Audubon by providing block workers with Atlas data cards, answering questions about procedures, and occasionally giving field assistance when it was needed.

In addition to supervising the overall project, the Atlas coordinator maintained regular communication with the quad coordinators and published 15 *Massachusetts Breeding Bird Atlas Newsletter*s over the course of the Atlas period. The newsletters kept volunteers informed about the progress of the project and provided periodic updates on particularly interesting Atlas discoveries at the end of each breeding season.

Printed cards were used for recording field data for each block every year of the five-year Atlas period (see Figure 6). The cards had three columns of boxes next to each breeding species listed on the card. Each time a confirmation was made at the "possible," "probable," or "confirmed" level, the appropriate breeding code was entered on the card. The official American Ornithologists' Union number for each species was also listed on the data cards. The goal of the project after five years was to raise each species recorded in a block to the highest attainable level of confirmation.

Atlas Criteria

A set of standardized criteria was used to establish the breeding status of species recorded during the Atlas period. In order to confirm a species at the "possible," "probable," or "confirmed" level of breeding, a

American Redstart

Ruby-throated Hummingbird

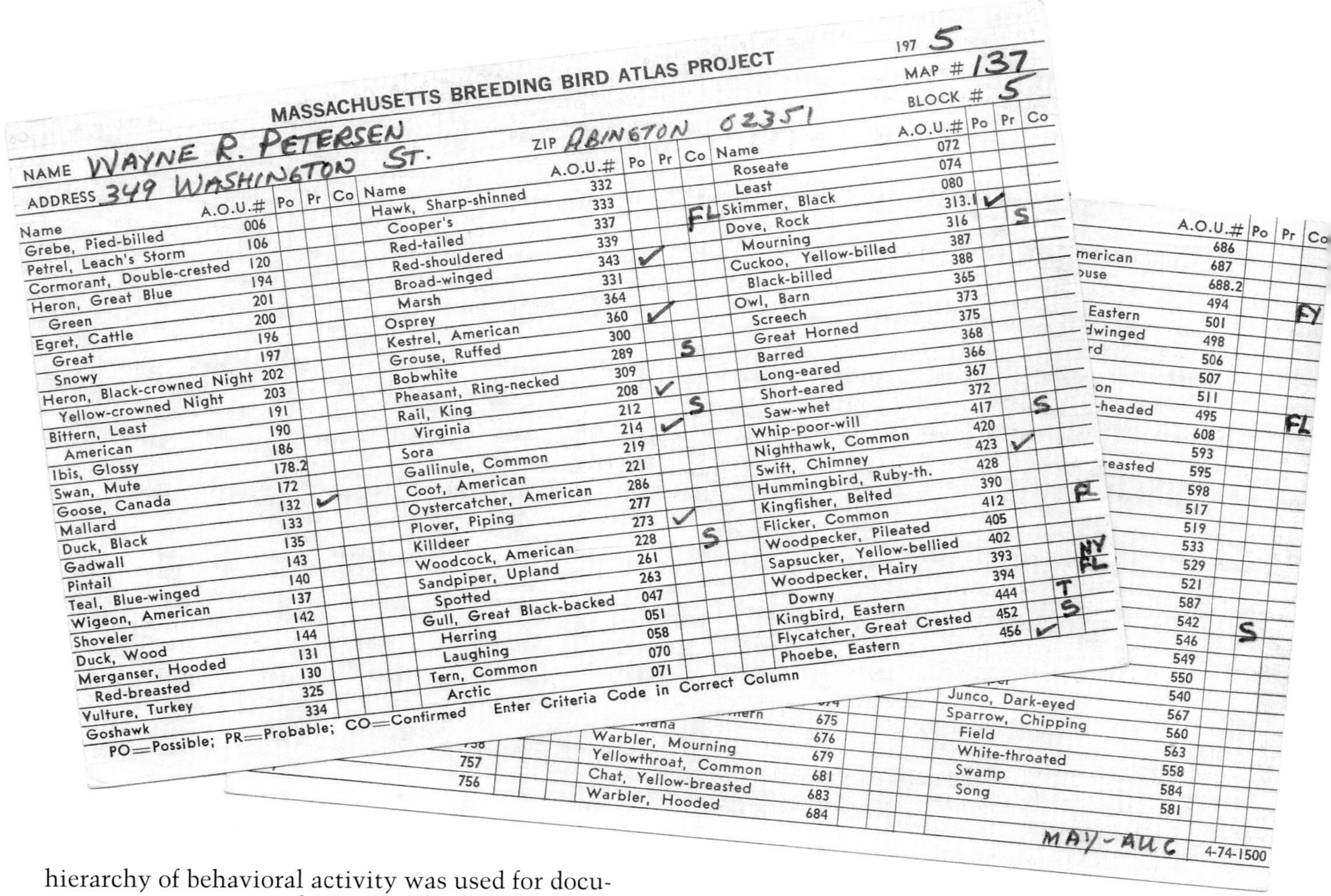

MASSACHUSETTS BREEDING BIRD ATLAS PROJECT

197 5 MAP # 137 BLOCK # 5

NAME WAYNE R. PETERSEN ZIP ABINGTON 02351

ADDRESS 349 WASHINGTON ST.

Name	A.O.U.#	Po	Pr	Co
Grebe, Pied-billed	006			
Petrel, Leach's Storm	106			
Cormorant, Double-crested	120			
Heron, Great Blue	194			
Green	201			
Egret, Cattle	200			
Great	196			
Snowy	197			
Heron, Black-crowned Night	202			
Yellow-crowned Night	203			
Bittern, Least	191			
American	190			
Ibis, Glossy	186			
Swan, Mute	178.2			
Goose, Canada	172			
Mallard	132	✓		
Duck, Black	133			
Gadwall	135			
Pintail	143			
Teal, Blue-winged	140			
Wigeon, American	137			
Shoveler	142			
Duck, Wood	144			
Merganser, Hooded	131			
Red-breasted	130			
Vulture, Turkey	325			
Goshawk	334			
Hawk, Sharp-shinned	332			
Cooper's	333			
Red-tailed	337			FL
Red-shouldered	339			
Broad-winged	343	✓		
Marsh	331			
Osprey	364			
Kestrel, American	360	✓		
Grouse, Ruffed	300			
Bobwhite	289		S	
Pheasant, Ring-necked	309			
Rail, King	208	✓		
Virginia	212		S	
Sora	214	✓		
Gallinule, Common	219			
Coot, American	221			
Oystercatcher, American	286			
Plover, Piping	277			
Killdeer	273	✓		
Woodcock, American	228		S	
Sandpiper, Upland	261			
Spotted	263			
Gull, Great Black-backed	047			
Herring	051			
Laughing	058			
Tern, Common	070			
Arctic	071			
Roseate	072			
Least	074			
Skimmer, Black	080			
Dove, Rock	313.1	✓		
Mourning	316		S	
Cuckoo, Yellow-billed	387			
Black-billed	388			
Owl, Barn	365			
Screech	373			
Great Horned	375			
Barred	368			
Long-eared	366			
Short-eared	367			
Saw-whet	372			
Whip-poor-will	417		S	
Nighthawk, Common	420			
Swift, Chimney	423	✓		
Hummingbird, Ruby-th.	428			
Kingfisher, Belted	390			R
Flicker, Common	412			
Woodpecker, Pileated	405			
Sapsucker, Yellow-bellied	402			NY
Woodpecker, Hairy	393			FL
Downy	394			
Kingbird, Eastern	444		T	
Flycatcher, Great Crested	452		S	
Phoebe, Eastern	456	✓		

PO=Possible; PR=Probable; CO=Confirmed Enter Criteria Code in Correct Column

Name	A.O.U.#	Po	Pr	Co
...merican	686			
...ouse	687			
	688.2			
	494			
... Eastern	501			FY
...dwinged	498			
...rd	506			
	507			
...on	511			
...-headed	495			
	608			FL
	593			
...reasted	595			
	598			
	517			
	519			
	533			
	529			
	521			
	587			
	542			
	546		S	
	549			
	550			
Junco, Dark-eyed	540			
Sparrow, Chipping	567			
Field	560			
White-throated	563			
Swamp	558			
Song	584			
	581			
	757			
	756			
...ern	675			
...ana	676			
Warbler, Mourning	679			
Yellowthroat, Common	681			
Chat, Yellow-breasted	683			
Warbler, Hooded	684			

MAY-AUG 4-74-1500

Figure 6
Massachusetts Breeding Bird Atlas Field Data Cards

hierarchy of behavioral activity was used for documentation purposes. The criteria followed closely those used in *The Atlas of Breeding Birds in Britain and Ireland* (Sharrock 1976), as well as those described in Smith (1990). In some cases, several criteria were applicable in confirming a particular breeding species. Needless to say, some criteria were more frequently used than others, particularly S, B, NB, FL, FY, and NE. A list of the confirmation symbols and the definitions of the accompanying behaviors that were used during the MBBA follows.

Possible Breeding

- ✓ Bird recorded in the breeding season in possible nesting habitat, but no other indication of breeding noted

Probable Breeding

- **S** Singing male present (or breeding calls heard) on more than one date in the same place
- **T** Bird (or pair) apparently holding territory
- **D** Courtship and display; or agitated behavior or anxiety calls from adults, suggesting probable presence of nest or young nearby; brood patch on trapped female or cloacal protuberance on trapped male
- **N** Visiting probable nest site
- **B** Nest building by wrens and woodpeckers

Confirmed Breeding

DD Distraction display or injury feigning; coition

NB Nest building by any species except wrens and woodpeckers

UN Used nest found (use this criterion with caution)

FE Female with egg in oviduct

FL Recently fledged young

FS Adult carrying fecal sac

FY Adult(s) entering or leaving nest site in circumstance indicating occupied nest

NE Nest and eggs or bird setting and not disturbed or eggshells found away from nest

NY Nest with young or downy young of waterfowl, quail, waders, etc.

Data Entry and the Atlas Maps

Two of the more monumental tasks associated with managing the Massachusetts Breeding Bird Atlas project were compiling the breeding-confirmation data at the end of each field season and creating the final Atlas maps. Richard A. Forster, Massachusetts Audubon's Atlas coordinator throughout most of the five-year project, shouldered most of the task of annually updating the breeding confirmation data for every block and every quad in Massachusetts at the end of each field season.

Once the finalized confirmation data for each block and quad was compiled, the information was turned over to David Stemple at the University of Massachusetts Computer Center for computerization. Stemple created a computer mapping program that used symbols to illustrate the level of confirmation for every species recorded during the Atlas period, block by block, on a map of Massachusetts. Without his expertise and tremendous contribution to the project, the *Massachusetts Breeding Bird Atlas* might never have reached fruition.

It is important to point out that the "age of computers" was still in its relative infancy when the original data for this project was being compiled and when the mapping program was being created in the 1970s and early 1980s. Data management during the Atlas period was considerably more tedious and time-consuming than it would be now in the twenty-first century. However, this difficulty was a reality associated with all of the early statewide breeding atlas attempts, a fact that further underscores the diligence of all the volunteers who managed to accurately compile tens of thousands of breeding records using pen, paper, and punch cards rather than Palmtops and PCs.

Prior to their publication in the present format, the original Massachusetts Breeding Bird Atlas project maps and map symbols were variously modified and improved through the efforts of Dorothy Graskamp and Stephen D. McRae, director, Massachusetts Division of Fisheries and Wildlife Geographic Information Systems Program, and of Barry W. Van Dusen, book designer and artist.

White-breasted Nuthatch

Maps and Species Accounts

The data reflected in the maps that follow was collected between 1974 and 1979 and present an accurate snapshot of the distribution of breeding birds in Massachusetts during that decade. No attempt has been made to update the data since this would confound the chief value of any distributional survey, namely to serve as a clearly delineated benchmark with which to compare future surveys using similar methodologies. That some bird species have ceased to nest in the Commonwealth since the 1970s and others have bred for the first time in the same interval, far from making this volume obsolete, demonstrates precisely its value and inherent interest.

It is worth noting here as well that because Massachusetts is small and its birding community large and very knowledgeable, the maps represent more complete coverage of the survey area than has been possible in many larger states or in those where fewer volunteers were available. For species whose status has changed dramatically here in the last 20 years, this has been noted in the appropriate species accounts, and brief accounts of "new" species are described in Appendix 1.

The species accounts represent the labors of 90 authors who were invited to compile relevant facts and add appropriate commentary according to a prescribed format for every bird species that was recorded in Massachusetts at the "possible," "probable," or "confirmed" level during the Atlas period. Authors were selected because they either were particularly knowledgeable about or had a special interest in particular bird species. Despite the fact that multiple authorship inevitably creates significant editorial challenges, the resulting document is undoubtedly far more informative, insightful, and lively than it would have been if it were the product of a single mind.

The content of the individual species accounts, while pleasingly reflective of the writing styles of the different authors, provides information that is consistent in substance and should be of interest to general readers, birders, conservation strategists, and land-use planners. The accounts provide a brief overview of each species' historical status in Massachusetts, along with any notable geographic or ecological association revealed by the Atlas effort.

Much of the general information in the species accounts has been distilled from Edward Howe Forbush's (EHF) *Birds of Massachusetts and Other New England States* (1925, 1927, 1929) and Arthur Cleveland Bent's (ACB) *Life Histories of North American Birds* (1919 to 1968). Five other resources consistently referenced in the species accounts are *American Birds* (AB), *Bird Observer of Eastern Massachusetts* (BOEM) (now *Bird Observer*), the Cornell Nest Record program (CNR), David Kenneth Wetherbee's (DKW) *The Birds and Mammals of Worcester County, Massachusetts* (1945), and *The Chickadee* (TC), a journal published by the Forbush Bird Club.

Besides including information obtained from these resources, the species accounts provide data on habitat preferences, vocalizations, courtship behavior, nest and egg characteristics, breeding chronology, fledgling data, the timing of annual molts, the status of each species after the breeding season, and other valuable or interesting information uncovered by the authors[1]. Each species account is accompanied by an accurate full-color illustration by bird illustrator John Sill or Barry W. Van Dusen.

In summary, this volume is intended to be a comprehensive, useful, and attractive reference work on the breeding birds of Massachusetts. It should be emphasized, however, that it is first and foremost an atlas—a book of maps, painstakingly created out of many thousands of observations made by a small army of dedicated birdwatchers over a five-year period. The data represented in these distribution maps is a significant contribution to the ornithology of the Commonwealth that, we hope, will promote the conservation of the extraordinary bird life of Massachusetts.

Blue Jay

---

[1] Readers seeking additional information about the status, seasonal distribution, and migratory behavior of birds in Massachusetts should refer to *Birds of Massachusetts* by Richard R. Veit and Wayne R. Petersen (1993).

Common Loon
Gavia immer

Egg dates: May 21 to August 16.

Number of broods: one; may re-lay if first attempt fails.

Historically, Common Loons probably nested in suitable locations throughout Massachusetts. However, shooting, loss of undisturbed habitat, and human activities drastically reduced their numbers until, by 1925, Forbush considered them "gone, perhaps never to return" (EHF). Fortunately, northern New England was able to maintain a healthy population of Common Loons, and, when appropriate habitat became available (e.g., the creation of Quabbin Reservoir), they became reestablished as breeding residents by the mid-1970s. Today, a small number of loons also nest in isolated bodies of water in Franklin and Worcester counties. Wintering, and occasionally summering, birds are commonly observed along the coast.

Spring arrival closely follows the melting of ice on the interior lakes and reservoirs used for nesting. These water bodies are generally characterized by having clean, clear water, an abundant food supply, suitable sites for nesting, and freedom from human disturbance. Birds arrive in pairs or pair up shortly after arrival, establish territories of 100 to 200 acres, and begin nest site selection. Nest sites and territories used in previous years are often reoccupied. It is believed that Common Loons pair for life.

Crude nests, often simply shallow scrapes surrounded by sparse vegetative matter, are constructed close to the water's edge, usually on small islands or promontories. These are often protected from wave and wind action, and their location allows for quick access to deep water.

Common Loon vocalizations fall into four basic types: the tremolo, or loon laughter, generally conveys excitement, alarm, or annoyance; the wail, a three-note forlorn-sounding call, is used to make contact with a mate or young; the yodel, given only by males, appears to be associated with territorial defense or other aggressive behavior; and a variety of hoots and mews are used for communication among members of a family or flock.

Following nest construction, one to three (usually two) eggs are laid, and both adults incubate for 28 to 29 days. Nesting birds may flush from the nest or crouch in position when approached or disturbed. Young loons are precocial and generally leave the nest within 24 hours of hatching. Chicks accompany their parents to a nursery area, usually a sheltered bay or cove with adequate food, which is used for approximately two weeks after the eggs hatch. Chicks are dependent on adults for food and protection for at least two months and are fully fledged at ten to twelve weeks of age. Massachusetts records of adults with small young range from June 22 to August 17 (TC).

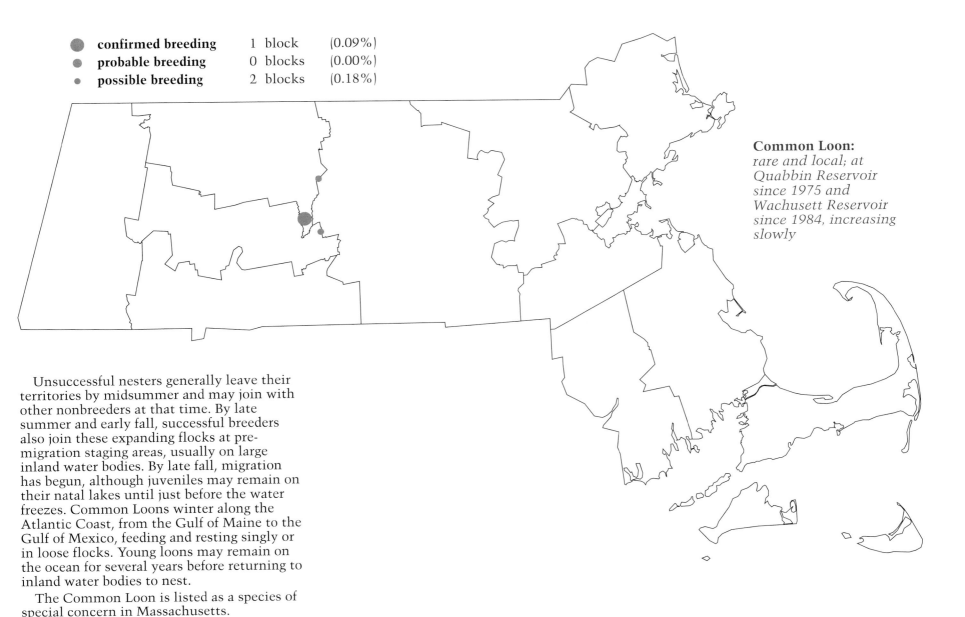

- **confirmed breeding** 1 block (0.09%)
- **probable breeding** 0 blocks (0.00%)
- **possible breeding** 2 blocks (0.18%)

Common Loon: *rare and local; at Quabbin Reservoir since 1975 and Wachusett Reservoir since 1984, increasing slowly*

Unsuccessful nesters generally leave their territories by midsummer and may join with other nonbreeders at that time. By late summer and early fall, successful breeders also join these expanding flocks at pre-migration staging areas, usually on large inland water bodies. By late fall, migration has begun, although juveniles may remain on their natal lakes until just before the water freezes. Common Loons winter along the Atlantic Coast, from the Gulf of Maine to the Gulf of Mexico, feeding and resting singly or in loose flocks. Young loons may remain on the ocean for several years before returning to inland water bodies to nest.

The Common Loon is listed as a species of special concern in Massachusetts.

Paul J. Lyons

Pied-billed Grebe
Podilymbus podiceps

Egg dates: April 23 to July 3.

Number of broods: one; may re-lay if first attempt fails.

Like most marsh-nesting species, the Pied-billed Grebe has declined drastically during this century. It is now a rare and local breeder, with Atlas confirmations only at coastal locations. Historically, Pied-billed Grebes were not known to nest along the coastal plain or on the Islands, but the construction of freshwater impoundments at Parker River and Monomoy national wildlife refuges may have contributed to this change. There were no proven nesting records from the Sudbury River valley, which was formerly a stronghold for the species. Post-Atlas breeding has definitely occurred at three sites in Worcester County.

Resident Pied-billed Grebes appear in March shortly after ice has melted on ponds, and the few migrants that occur have generally departed by the end of April. Residents begin nesting activities in April or May. Preferred habitats have stands of cattails and other emergent vegetation adjacent to areas of open water. Grebes may be difficult to locate in these marshes, but the cuckoolike series of *cuck* and *cow* notes is a sure sign of their presence. Foods include fish, insects, snails, seeds, and soft plants.

Both sexes construct the nest, a well-concealed platform of dead marsh vegetation situated in water 1 to 3 feet in depth and anchored to living plant stalks. The usual clutch of four to seven eggs is incubated, mostly by the female, for slightly more than three weeks. Eggs are covered with nest material whenever the adult leaves them unattended. The distinctively marked black-and-white striped young can swim and make shallow dives soon after hatching. They are often transported on the backs of the adults, loosely held in place by the adult's wings. The parents carry bits of food to the young chicks. Juveniles approach adult size before losing the striped plumage.

A very complete nesting chronology for this species in Massachusetts was recorded in Worcester after the Atlas period. One adult, probably the male, arrived on March 4 but was not paired until April 3. Nesting activity commenced on April 16, and the female was incubating by the end of the month. The two young hatched about May 21 and were

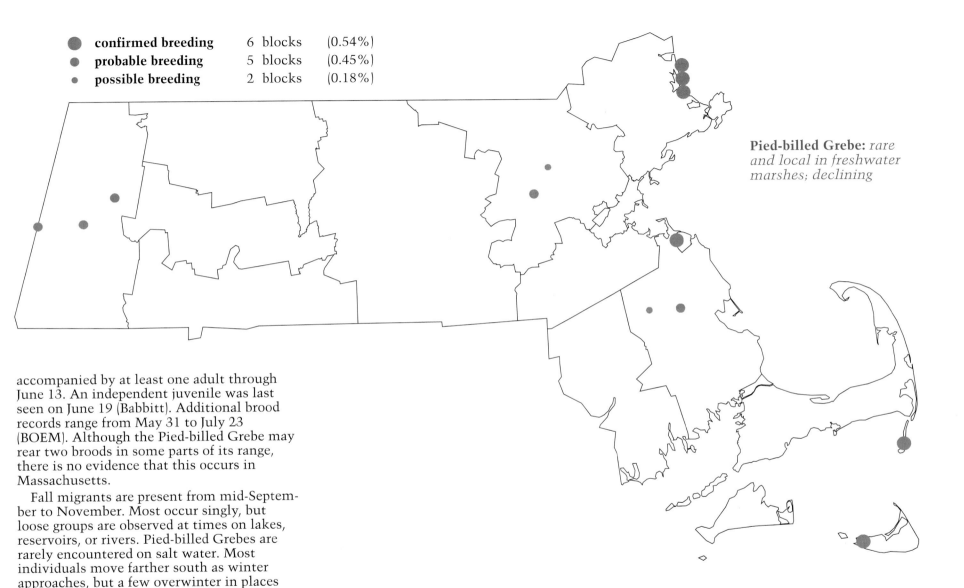

- ● **confirmed breeding** 6 blocks (0.54%)
- ● **probable breeding** 5 blocks (0.45%)
- • **possible breeding** 2 blocks (0.18%)

Pied-billed Grebe: *rare and local in freshwater marshes; declining*

accompanied by at least one adult through June 13. An independent juvenile was last seen on June 19 (Babbitt). Additional brood records range from May 31 to July 23 (BOEM). Although the Pied-billed Grebe may rear two broods in some parts of its range, there is no evidence that this occurs in Massachusetts.

Fall migrants are present from mid-September to November. Most occur singly, but loose groups are observed at times on lakes, reservoirs, or rivers. Pied-billed Grebes are rarely encountered on salt water. Most individuals move farther south as winter approaches, but a few overwinter in places where the water remains open.

The Pied-billed Grebe is listed as a threatened species in Massachusetts.

Allen H. Morgan, Richard A. Forster, and W. Roger Meservey

Leach's Storm-Petrel
Oceanodroma leucorhoa

Egg dates: June 25 to August 3.

Number of broods: one; may re-lay if first attempt fails.

In eastern North America, the only breeding storm-petrel is the Leach's. Even though this species is seldom seen by most birders, it is actually one of the most abundant marine birds in the North Atlantic. Colonies of a few to several thousand pairs are located on islands from the British Isles to Newfoundland and south to Muscongus Bay, Maine (formerly to Casco Bay), with a small disjunct colony on Penikese Island in Buzzards Bay.

Leach's Storm-Petrels were first detected on Penikese Island in 1930. Between 1930 and 1936, only one or two breeding pairs were suspected, but in 1941 State Ornithologist Joseph A. Hagar mapped the locations of 80 nest sites scattered over the island. The first verification of breeding was a nearly fledged chick collected by Hagar on August 24, 1940.

For the next 30 years no published accounts of the petrels on Penikese are available. In 1972, 1 active and 2 inactive burrows were noted; in 1975, 15 to 20 pairs were estimated; and between 1981 and 1985, 5 to 7 pairs were estimated. In July 1984, 5 active burrows were found in a retaining wall, and 21 adults and prebreeding subadults were mist-netted and banded in a two-night period.

In the spring, Leach's Storm-Petrels arrive at their colony on Penikese Island by the last week in April. Farther north, nest burrows are dug into the ground beneath a canopy of spruce or into sod on treeless islands. On Penikese, the traditional nesting site has been in the crevices of a retaining wall built of large cut stones, but other burrows have been found under building foundations, stone walls, and boulders. Although the loose sandy soil on Penikese may be a relatively easy substrate for storm-petrels to dig in, these burrows, except for those in the retaining wall, do not appear to last for as many years as do those constructed in dense sod on the northern nesting grounds. The nest consists of a few dead seed stalks or blades of grass at the end of the underground burrow.

During the spring and summer, shortly after dark, storm-petrels return to the colony from forays at sea, and the colony becomes alive with activity as birds fly irregular courses low overhead, each bird occasionally breaking the silence with a loud chuckle call. In large colonies, individual calls mingle together to form a loud chorus. From their nesting burrows, the birds emit a repetitious purring or churring call that in the spring may continue for much of the night, interrupted occasionally by the chuckle. As the season progresses and incubation duties begin, the nightly activity over the colony is continued by prebreeders returning from sea for the first time at two or three years of age.

Storm-petrels have a highly developed sense of smell, which enables them to follow scent trails on the open ocean. They can also locate their colony and even their individual burrows by scent as they return on dark, foggy nights.

Although Leach's Storm-Petrels are about the size of a robin, characteristics of their life history are similar to those of much larger seabird species. They are long-lived (to at least 29 years), the age at first breeding is great (4 to 5 years), the clutch size is small (one very large white egg), the incubation and fledging periods are long (40 to 50 days and 63 to 70 days, respectively), and they tend to mate for life, although fidelity appears to be directed more toward the burrow than the partner.

- confirmed breeding — 1 block (0.09%)
- probable breeding — 0 blocks (0.00%)
- possible breeding — 0 blocks (0.00%)

Leach's Storm-Petrel: *rare and local at Penikese Island, where population has declined since 1940*

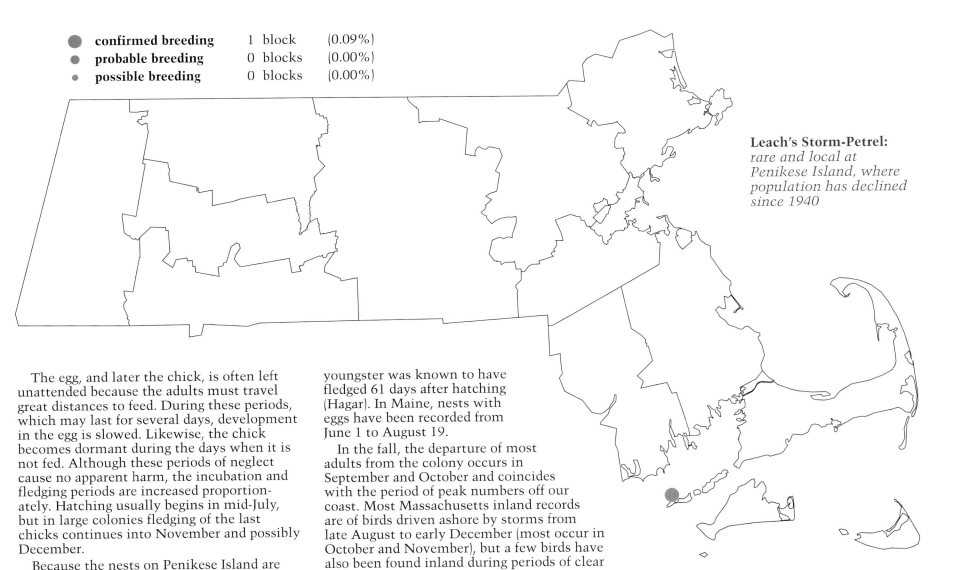

The egg, and later the chick, is often left unattended because the adults must travel great distances to feed. During these periods, which may last for several days, development in the egg is slowed. Likewise, the chick becomes dormant during the days when it is not fed. Although these periods of neglect cause no apparent harm, the incubation and fledging periods are increased proportionately. Hatching usually begins in mid-July, but in large colonies fledging of the last chicks continues into November and possibly December.

Because the nests on Penikese Island are very difficult to reach, there is little specific nesting data for Massachusetts. An egg was discovered in one nest in the rock wall on July 4. A pipped egg found in a nest on August 2 hatched the following day. Allowing for the minimum incubation period, the egg would have been laid about June 25. The youngster was known to have fledged 61 days after hatching (Hagar). In Maine, nests with eggs have been recorded from June 1 to August 19.

In the fall, the departure of most adults from the colony occurs in September and October and coincides with the period of peak numbers off our coast. Most Massachusetts inland records are of birds driven ashore by storms from late August to early December (most occur in October and November), but a few birds have also been found inland during periods of clear weather.

Off our shores, the Leach's Storm-Petrel usually feeds in areas of cold upwellings with high plankton densities near the edge of the continental shelf. Most individuals of the Atlantic winter in tropical seas on both sides of the equator.

The Leach's Storm-Petrel is listed as an endangered species in Massachusetts.

Thomas W. French

Double-crested Cormorant
Phalacrocorax auritus

Egg dates: April 5 to June.

Number of broods: one; may re-lay if first attempt fails.

Cormorants are now familiar sights along the coast and are seen increasingly in small numbers on inland lakes and rivers. The Double-crested Cormorant is a summer resident, replaced in winter by the Great Cormorant. Extirpated in the nineteenth century, Double-crested Cormorants returned to breed in Massachusetts by 1944 and have increased greatly, especially since about 1970. In 1977, 1,760 pairs nested at eleven sites. Subsequent increases have been substantial: in 1984 nearly 5,000 pairs nested at fifteen sites, and in the 1990s approximately 8,000 pairs nested at twenty-five sites. Breeding birds are numerous, apparently because of the presence of preferred nesting sites along the coast between Boston and Cape Ann, and in Buzzards Bay (Weepecket Islands). Nonbreeders are more widely distributed, especially during postbreeding dispersal. Traditional roosting sites are conspicuous, and some of these become breeding colonies after years of use. The increasing numbers of inland breeders are now as close as Lake Champlain so their appearance in Massachusetts may not be far off.

The migratory flights of cormorants can be striking, with many thousands of birds passing by day in a succession of untidy echelons, frequently numbering several hundred birds and sometimes thousands. Most of these movements occur over short periods: in spring many birds arrive in late March and early April, but continue into May; in fall the major departures are early in October, but some flights are reported in September through November. Coastal movements are common along the North Shore of Massachusetts Bay, but in fall many of the migrants fly overland from Boston southwest toward Narragansett Bay.

Breeding colonies vary in size from a handful of pairs to more than a thousand. The preferred sites along the Massachusetts coast are the flatter parts of small, undisturbed islands comprised of bedrock or eroding glacial deposits with boulders on the periphery. Sites with sand or soil are not preferred, and tree nests are very unusual (although reported for one year from House Island). On the Canadian coast and in the interior, many colonies are located in trees and elsewhere occasionally on shipwrecks or abandoned wharves. Thus, it seems likely that, although breeding birds are now absent from the sandy shores of Cape Cod, they may not remain so for long. (By 1994 small numbers nested at four sites [two constructed by people] on the Cape and Islands.)

Double-crested Cormorants are generally found near shallow waters, especially estuaries and bays, and almost never far offshore. They forage by diving from the surface and then swimming underwater with vigorous foot thrusts to catch a variety of fish. A typical dive lasts 30 seconds, and large or awkward prey may be swallowed at the surface. Feeding seems to take up only a minor part of the day, and cormorants spend much time loafing on rocks, pilings, or other nearby perches, often holding their wings out in a characteristic spread-wing posture. There has been some dispute about the function of this behavior, but it commonly is thought to serve in drying the wings. The feeding birds often commute in small parties from the roost (or colony) to feeding areas as far as 10 miles away, quite often inland.

In spring, the newly returned migrants speedily begin to build their bulky nests of seaweed, twigs, and a diverse array of flotsam and jetsam. The nests are generally lined with finer materials, but sometimes newly laid eggs are surrounded by no more than a ring of kelp or rockweed. Nest building may start on top of the remnants of previous years' constructions and continue at least through incubation so that the results are substantial heaps consolidated by large quantities of guano. This last substance combines with abandoned fish and numbers of dead young to give cormorant colonies a powerful reek, detectable far downwind.

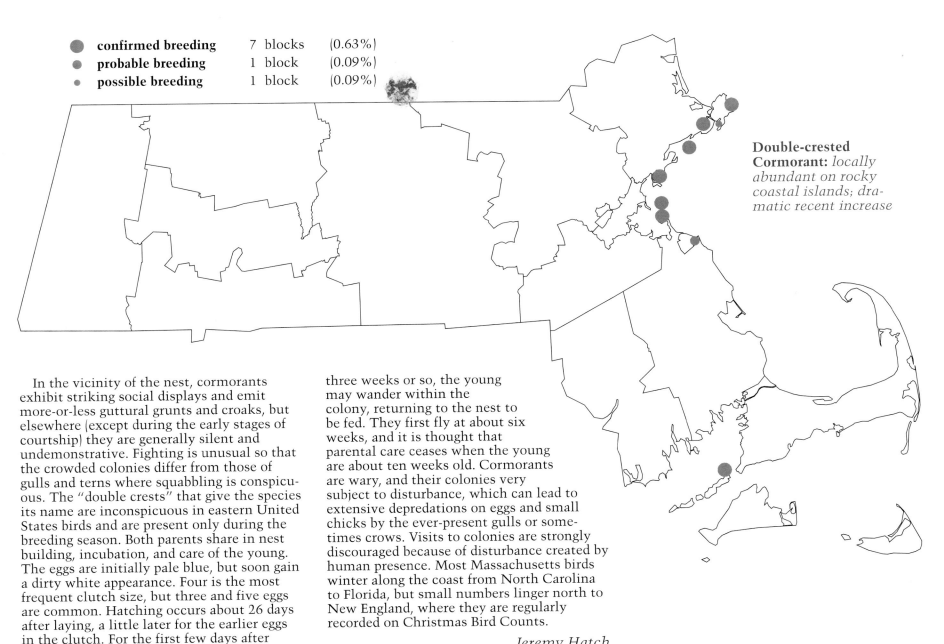

Double-crested Cormorant: *locally abundant on rocky coastal islands; dramatic recent increase*

● **confirmed breeding**	7 blocks	(0.63%)
● **probable breeding**	1 block	(0.09%)
· **possible breeding**	1 block	(0.09%)

   In the vicinity of the nest, cormorants exhibit striking social displays and emit more-or-less guttural grunts and croaks, but elsewhere (except during the early stages of courtship) they are generally silent and undemonstrative. Fighting is unusual so that the crowded colonies differ from those of gulls and terns where squabbling is conspicuous. The "double crests" that give the species its name are inconspicuous in eastern United States birds and are present only during the breeding season. Both parents share in nest building, incubation, and care of the young. The eggs are initially pale blue, but soon gain a dirty white appearance. Four is the most frequent clutch size, but three and five eggs are common. Hatching occurs about 26 days after laying, a little later for the earlier eggs in the clutch. For the first few days after hatching, the chicks are blind, naked, and helpless; then the shiny, nearly black skin is covered by a thick coat of black down. After three weeks or so, the young may wander within the colony, returning to the nest to be fed. They first fly at about six weeks, and it is thought that parental care ceases when the young are about ten weeks old. Cormorants are wary, and their colonies very subject to disturbance, which can lead to extensive depredations on eggs and small chicks by the ever-present gulls or sometimes crows. Visits to colonies are strongly discouraged because of disturbance created by human presence. Most Massachusetts birds winter along the coast from North Carolina to Florida, but small numbers linger north to New England, where they are regularly recorded on Christmas Bird Counts.

Jeremy Hatch

American Bittern
Botaurus lentiginosus

Egg dates: May 5 to June 10.

Number of broods: one.

With the increasing destruction, disappearance, and degradation of wetland habitat, the once common American Bittern has become a local, if not rare, summer resident throughout most of Massachusetts exclusive of Cape Cod, where it no longer breeds at all. Because of its preference for freshwater marshland and moist meadows for nesting, its breeding distribution is defined by areas where these specialized environments currently exist. During the Atlas period, most American Bitterns in the Commonwealth were found nesting in the few remaining extensive marshes and river meadows located in the eastern part of the state, or else in damp hayfields, bogs, and Beaver-created meadows in western Massachusetts.

Bitterns begin to arrive in Massachusetts in early April (earlier in some years) and may be found at breeding localities through late summer; however, they become markedly inconspicuous after their courtship calling ceases in late May. In the spring, the birds are best located by listening for the remarkable courtship "pumping" of the males. The noise produced is a hollow *oonk-ka-choonk* and, depending upon the listener's age and life experience, sounds like the priming of an old pump or the sound of a distant pile driver, the precise effect being a function of one's distance from the calling bird. In addition to producing these unusual sounds, male bitterns display by elevating whitish, rufflike plumes from a point near the shoulder and extending them outward and backward, nearly encircling the hind neck. Much posturing is assumed during such display.

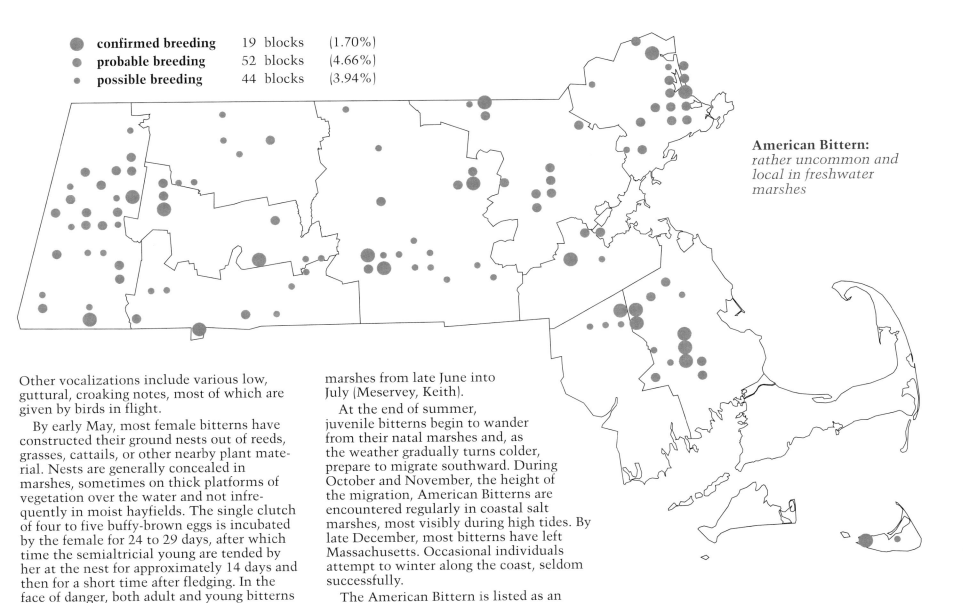

American Bittern: *rather uncommon and local in freshwater marshes*

- confirmed breeding — 19 blocks (1.70%)
- probable breeding — 52 blocks (4.66%)
- possible breeding — 44 blocks (3.94%)

Other vocalizations include various low, guttural, croaking notes, most of which are given by birds in flight.

By early May, most female bitterns have constructed their ground nests out of reeds, grasses, cattails, or other nearby plant material. Nests are generally concealed in marshes, sometimes on thick platforms of vegetation over the water and not infrequently in moist hayfields. The single clutch of four to five buffy-brown eggs is incubated by the female for 24 to 29 days, after which time the semialtricial young are tended by her at the nest for approximately 14 days and then for a short time after fledging. In the face of danger, both adult and young bitterns rely on their cryptic coloration for protection. With necks extended, beaks pointed skyward, and bodies greatly compressed, they resemble just another piece of marsh vegetation or debris. Fledged young have been recorded in the Quaboag River and East Brookfield River marshes from late June into July (Meservey, Keith).

At the end of summer, juvenile bitterns begin to wander from their natal marshes and, as the weather gradually turns colder, prepare to migrate southward. During October and November, the height of the migration, American Bitterns are encountered regularly in coastal salt marshes, most visibly during high tides. By late December, most bitterns have left Massachusetts. Occasional individuals attempt to winter along the coast, seldom successfully.

The American Bittern is listed as an endangered species in Massachusetts.

Wayne R. Petersen

Least Bittern
Ixobrychus exilis

Egg dates: June 1 to June 30.

Number of broods: one; may re-lay if first attempt fails.

The Least Bittern has always been considered a rare and local summer resident in Massachusetts. In earlier years, this status may have had more to do with its secretive and retiring habits than its actual numbers. However, today there is no question that the Least Bittern is scarce, the consequence of the draining and filling of wetlands. Wetland alteration and ecological changes have produced a marked reduction in the extensive areas of cattails in freshwater marshes or broad river meadows that are essential to the species' existence. The birds are distributed sparsely in the eastern portions of the state and are rarely encountered in central and western Massachusetts.

Resident Least Bitterns usually arrive by the middle of May. Migrants are seldom noted, but there are several reports of unseasonal occurrences in coastal areas in late March and April following southerly coastal storms. They are, for the most part, nocturnal migrants. Unlike their larger relative, the American Bittern, Least Bitterns do not engage in an elaborate courtship display. The voice is a series of rapidly repeated coos and is very much reminiscent of one of the calls of the Black-billed Cuckoo. Least Bitterns are most vocal at dawn, but they sometimes also call in the middle of the day.

Nesting normally commences in late May. In a dense stand of cattails, several stalks are bent over to form a loose platform on which the nest is constructed. Additional stalks, grasses, and reeds are then used to complete the nest. There is an early record of a female

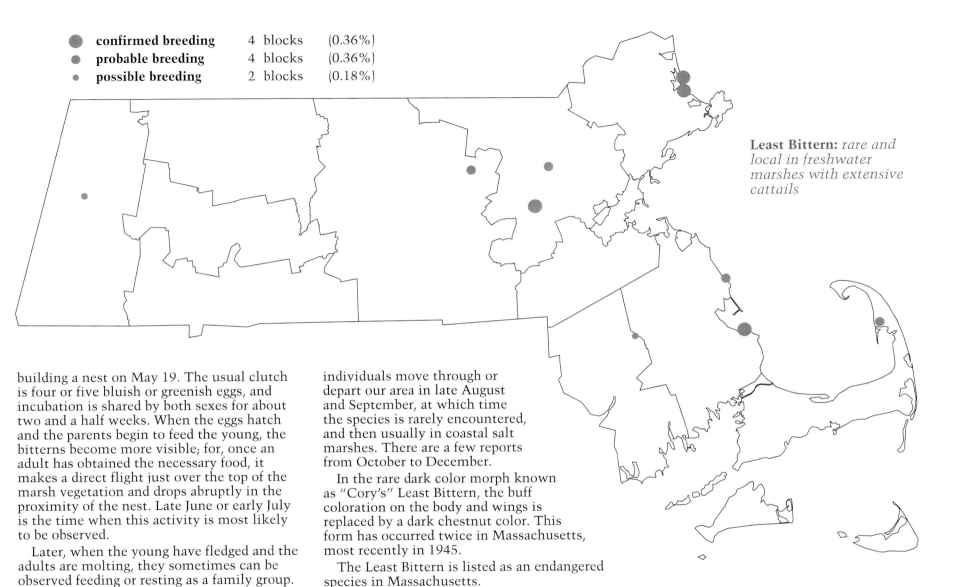

Least Bittern: *rare and local in freshwater marshes with extensive cattails*

- confirmed breeding — 4 blocks (0.36%)
- probable breeding — 4 blocks (0.36%)
- possible breeding — 2 blocks (0.18%)

building a nest on May 19. The usual clutch is four or five bluish or greenish eggs, and incubation is shared by both sexes for about two and a half weeks. When the eggs hatch and the parents begin to feed the young, the bitterns become more visible; for, once an adult has obtained the necessary food, it makes a direct flight just over the top of the marsh vegetation and drops abruptly in the proximity of the nest. Late June or early July is the time when this activity is most likely to be observed.

Later, when the young have fledged and the adults are molting, they sometimes can be observed feeding or resting as a family group. Most sightings of such groups in Massachusetts occur at Plum Island. The favored feeding areas are usually open areas in the marsh, where the birds forage from the edge or by wading in shallow water.

The fall migration is not very apparent, and little is known about it. Presumably, most individuals move through or depart our area in late August and September, at which time the species is rarely encountered, and then usually in coastal salt marshes. There are a few reports from October to December.

In the rare dark color morph known as "Cory's" Least Bittern, the buff coloration on the body and wings is replaced by a dark chestnut color. This form has occurred twice in Massachusetts, most recently in 1945.

The Least Bittern is listed as an endangered species in Massachusetts.

Richard A. Forster

Great Blue Heron
Ardea herodias

Egg dates: April 15 to June 12.

Number of broods: one; may re-lay if first attempt fails.

The first well-documented nesting of the Great Blue Heron in Massachusetts occurred within the Harvard Forest, Petersham, in 1925. This colony of up to 20 nests survived, shrouded in utmost secrecy, until its destruction by the hurricane of 1938. Subsequent to the demise of the Petersham heronry, small colonies likely persisted in remote rural areas prior to the Atlas period, but their presence went largely unnoticed, and they were little known to the general public.

During the Atlas period, 8 nesting sites were found from Westboro and Townsend to Sheffield and Otis in the Berkshires. Some 52 "possible" nesting sites reported in the period appear to be more a reflection of the large number of wandering individuals than actual colonies. Since 1979, the nesting population seems to have grown in size and expanded considerably. In 1984, a total of 229 nests were counted among 23 active heronries that ranged in size from 1 to 48 nests. As of 1984, "confirmed" colonies were distributed quite broadly across the state from a north-south line connecting Dunstable, Maynard, and Norfolk westward, with a distinct center of abundance in Worcester County.

The Great Blue Heron frequents all aquatic environments, including the shallow waters and shores of lakes, ponds, and waterways. It visits coastal beaches and marshes, tidal flats, and sandbars. Shy and wary during the nesting season, it retires into inland backwaters, swamps, and Beaver impoundments, usually away from populous areas. It does not nest communally with other herons in Massachusetts as is frequently the case in the southern United States. In Massachusetts, the recent increase and expansion seems linked to extensive Beaver flowages and associated fisheries that have gradually developed in the state since the return of the Beaver in the mid-1950s.

Nest sites here seem to be predominantly in dead trees drowned by impoundments, especially those created by Beavers. Nests are also found occasionally in living oaks or White Pines. After several years, depositions of excrement tend to burn back leafy vegetation, and trees eventually become sickly. Nests are massive stick platforms 30 to 60 feet above the ground and are repaired and reused annually. Occasionally, a pair of Great Horned Owls will appropriate a platform before spring reoccupation by the herons, but the owls' presence seems to be benign and entirely ignored by the herons.

Spring occupation of heronries begins in mid-March and continues well into April. Nest building and repair commences almost immediately upon arrival. Great Blue Herons engage in highly ritualized breeding behavior. Males initially defend a display site, usually an existing nest. Defended territory is confined to the immediate vicinity of the nest. Territory is used for pair formation and maintenance, copulation, nesting, and rearing of young. Mutual preening and billing by adults are thought to cement pair bonds, which are monogamous for at least seasonal duration. Promiscuous behavior is sometimes observed. First-year birds frequent colonies, sometimes in small visiting parties, but do not normally breed until their second year. Yearlings often occupy sites and build practice nests. These nests are usually flimsy affairs but occasionally are well developed and used as a nest foundation in subsequent seasons.

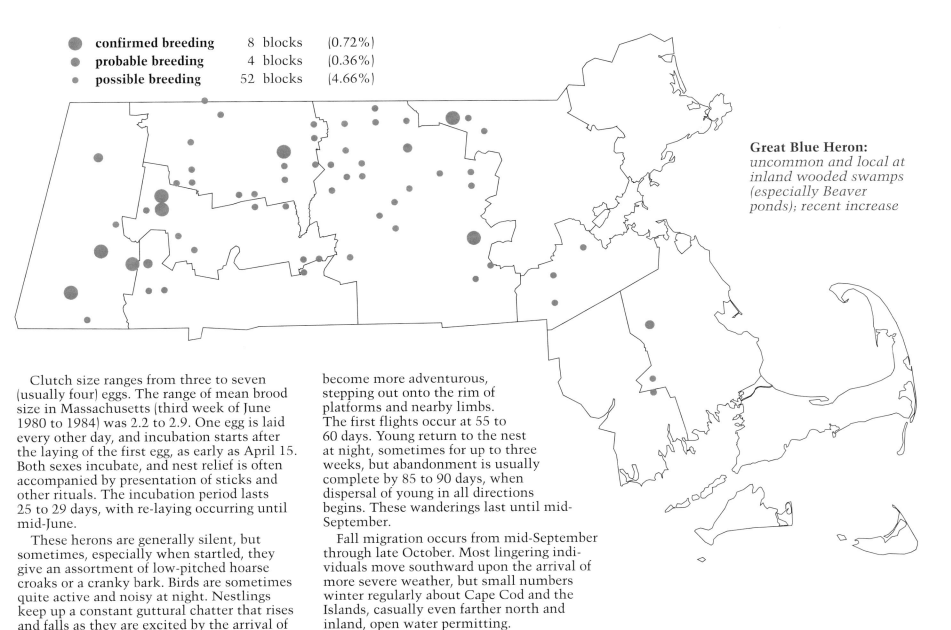

● **confirmed breeding** 8 blocks (0.72%)
● **probable breeding** 4 blocks (0.36%)
· **possible breeding** 52 blocks (4.66%)

Great Blue Heron:
uncommon and local at inland wooded swamps (especially Beaver ponds); recent increase

Clutch size ranges from three to seven (usually four) eggs. The range of mean brood size in Massachusetts (third week of June 1980 to 1984) was 2.2 to 2.9. One egg is laid every other day, and incubation starts after the laying of the first egg, as early as April 15. Both sexes incubate, and nest relief is often accompanied by presentation of sticks and other rituals. The incubation period lasts 25 to 29 days, with re-laying occurring until mid-June.

These herons are generally silent, but sometimes, especially when startled, they give an assortment of low-pitched hoarse croaks or a cranky bark. Birds are sometimes quite active and noisy at night. Nestlings keep up a constant guttural chatter that rises and falls as they are excited by the arrival of adults with food.

For the first two months, the young remain in the nest and are dependent on the parents, both of which tend the young. They gradually become more adventurous, stepping out onto the rim of platforms and nearby limbs. The first flights occur at 55 to 60 days. Young return to the nest at night, sometimes for up to three weeks, but abandonment is usually complete by 85 to 90 days, when dispersal of young in all directions begins. These wanderings last until mid-September.

Fall migration occurs from mid-September through late October. Most lingering individuals move southward upon the arrival of more severe weather, but small numbers winter regularly about Cape Cod and the Islands, casually even farther north and inland, open water permitting.

Bradford G. Blodget

Great Egret
Casmerodius albus

Egg dates: first week of May to third week of June.

Number of broods: one; may re-lay if first attempt fails.

Since the turn of the century, when this species was nearly exterminated in North America by the plume hunters of the millinery trade, the Great Egret has increased steadily in number. During the first half of the century, Great Egrets were occasionally seen in Massachusetts during the late summer and fall and were presumably post-breeding dispersal birds. In 1954, a pair nested in a cedar swamp in South Hanson. Since then they have nested in small numbers at scattered locations along the coast. During the Atlas period, there were breeding records from four sites, with the largest number of birds being 10 pairs at Clark's Island in Plymouth Bay. Smaller numbers were also recorded at House Island in Manchester, Ram Island in Westport, and Sampsons Island on the southern shore of Cape Cod.

Great Egrets begin to arrive in early April, when they can be seen feeding in marshes near their breeding areas. They forage in ponds, tidal inlets, marshes, damp meadows, swamps, and other wet habitats, usually in deeper water than smaller egret species; and they will defend a feeding territory. Great Egrets forage by walking slowly or standing, usually with an upright posture. Occasionally, they will walk quickly or hop between feeding areas, and they may hover and plunge into the water. They feed mostly on fish but also take insects, frogs, salamanders, and small mammals.

In Massachusetts, most Great Egrets are nesting by mid-May. They are colonial breeders, joining mixed-species colonies of Snowy Egrets, Little Blue Herons, Black-crowned Night-Herons, and Glossy Ibises. At Clark's Island, they have nested in Highbush Blueberry and Pitch and White pines, always on the very top of the bush or tree. Males defend a display territory using a "forward threat display" (Palmer 1962) in which all the plumes are erected. The well-developed aigrettes are displayed prominently in both hostile and pair-forming interactions. Aerial chases and advertising "circle flights" (Palmer 1962) are common. Courtship displays include the "stretch display" (Palmer 1962) in which the bird points its bill upward and swings its fully extended neck backward. The "snap display" (Palmer 1962) consists of a downward extension of the head and neck, during which the egret claps its mandibles together. Great Egrets make a variety of harsh single-syllable calls in hostile interactions. The advertising call is a soft *frawnk*. The birds have a repetitive *rrreee* call, which they make during greeting ceremonies at nest relief.

The nest is a loosely woven platform of sticks and twigs up to approximately 3 feet in diameter. Great Egrets have been known to reuse old nests that have remained intact over the winter. The eggs are light bluish

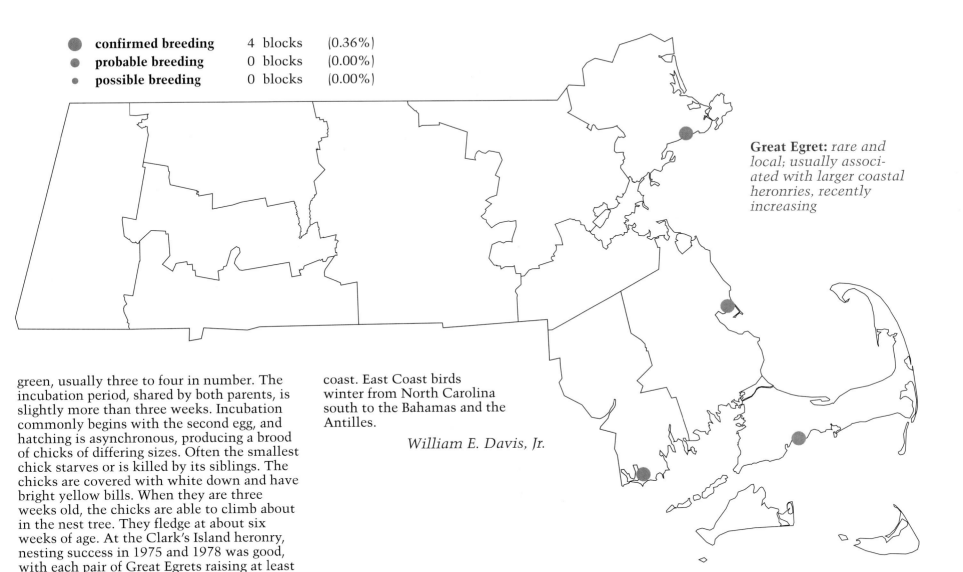

● **confirmed breeding**	4 blocks	(0.36%)
● **probable breeding**	0 blocks	(0.00%)
· **possible breeding**	0 blocks	(0.00%)

Great Egret: *rare and local; usually associated with larger coastal heronries, recently increasing*

green, usually three to four in number. The incubation period, shared by both parents, is slightly more than three weeks. Incubation commonly begins with the second egg, and hatching is asynchronous, producing a brood of chicks of differing sizes. Often the smallest chick starves or is killed by its siblings. The chicks are covered with white down and have bright yellow bills. When they are three weeks old, the chicks are able to climb about in the nest tree. They fledge at about six weeks of age. At the Clark's Island heronry, nesting success in 1975 and 1978 was good, with each pair of Great Egrets raising at least one chick and some rearing two young.

Great Egrets show some northerly postbreeding dispersal and gather in communal nocturnal roosts with other herons in late summer (e.g., Plum Island). They may occur at inland swamps and ponds as well as on the coast. East Coast birds winter from North Carolina south to the Bahamas and the Antilles.

William E. Davis, Jr.

Snowy Egret
Egretta thula

Egg dates: second week of April to June 11.

Number of broods: one.

During the early 1800s, Snowy Egrets nested as far north as New Jersey, but the population was nearly exterminated by the plume hunters of the millinery trade by the turn of the century. Once full protection was afforded, the species managed to recover and expand its breeding range. Snowy Egrets were rare vagrants in Massachusetts until about 1950, when they became regular visitors in small numbers, mostly in late summer and fall.

They were first recorded nesting in July 1955, at Quivett Neck, East Dennis. Since then, they have established about a dozen colonies along the coast, including colonies at Madaket Ditch and Polpis Harbor on Nantucket, several colonies on Chappaquiddick Island, and a colony at Menemsha Pond on Martha's Vineyard. The largest colonies were at House Island, Manchester, with approximately 250 pairs until the colony's collapse in the early 1980s, and at Clark's Island in Plymouth Bay, which supported about 250 pairs from the early 1970s until its final abandonment in the 1980s.

Snowy Egrets begin to arrive in Massachusetts in early April and can be observed foraging in the marshes and ponds near their breeding colonies. Primarily diurnal feeders that frequent a wide variety of open aquatic habitats, they have the most diverse repertoire of foraging behaviors of any egret species thus far studied. Snowy Egrets feed by walking slowly, walking quickly, standing, foot stirring, and hovering, but they very often run actively in pursuit of prey.

They commonly forage in sizable flocks and may follow livestock or other bird species, using them to stir up prey. Snowys have a diverse diet, which consists primarily of small fish, shrimp, snails, aquatic insects, and small frogs.

Breeding activity in Massachusetts usually begins in April and peaks by mid-May. Snowy Egrets are colonial breeders, frequently nesting with other heron species. They are highly social, and a half-dozen or more nests may occupy the same tree. At Clark's Island, they commonly have used Highbush Blueberry, Arrowwood, Red Cedar, and Black Cherry for nesting.

Snowy Egrets are territorial during the breeding season, defending their nest and its vicinity with a variety of aggressive displays, including a "forward display" (Palmer 1962) in which the bird fully erects all of its head, neck, and back plumes and typically emits harsh, low-pitched *aah* calls. In "snap display" (Palmer 1962), a bird extends its neck fully with head level and crest erect and makes snapping sounds with its bill. Fights are common early in the breeding season. Snowys have a variety of vocal displays, including an *aarh* advertising call and an *arg-obble* call. Males attempt to attract females with "stretch display" (Palmer 1962), bending the head over the back with bill pointed upward and pumping the head while uttering

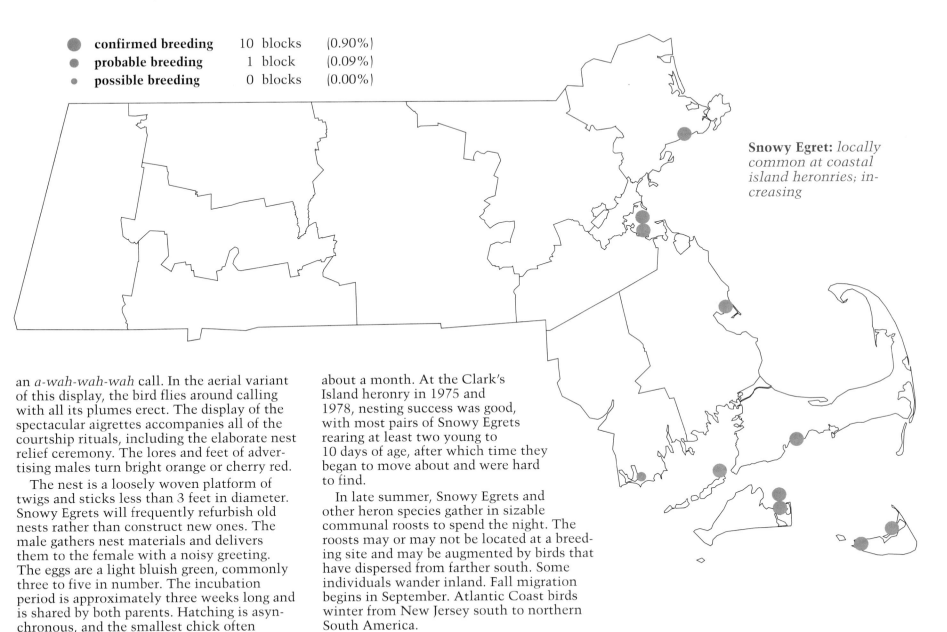

confirmed breeding 10 blocks (0.90%)
probable breeding 1 block (0.09%)
possible breeding 0 blocks (0.00%)

Snowy Egret: *locally common at coastal island heronries; increasing*

an *a-wah-wah-wah* call. In the aerial variant of this display, the bird flies around calling with all its plumes erect. The display of the spectacular aigrettes accompanies all of the courtship rituals, including the elaborate nest relief ceremony. The lores and feet of advertising males turn bright orange or cherry red.

The nest is a loosely woven platform of twigs and sticks less than 3 feet in diameter. Snowy Egrets will frequently refurbish old nests rather than construct new ones. The male gathers nest materials and delivers them to the female with a noisy greeting. The eggs are a light bluish green, commonly three to five in number. The incubation period is approximately three weeks long and is shared by both parents. Hatching is asynchronous, and the smallest chick often starves. Both parents feed the young, and the chicks emit an extremely harsh begging call or shriek. They are capable of climbing about the nest tree after two weeks and can fly after about a month. At the Clark's Island heronry in 1975 and 1978, nesting success was good, with most pairs of Snowy Egrets rearing at least two young to 10 days of age, after which time they began to move about and were hard to find.

In late summer, Snowy Egrets and other heron species gather in sizable communal roosts to spend the night. The roosts may or may not be located at a breeding site and may be augmented by birds that have dispersed from farther south. Some individuals wander inland. Fall migration begins in September. Atlantic Coast birds winter from New Jersey south to northern South America.

William E. Davis, Jr.

Little Blue Heron
Egretta caerulea

Egg dates: first week of May to fourth week of July.

Number of broods: one.

Because the Little Blue Heron lacks fancy aigrettes, it was not subject to wholesale slaughter at the turn of the century; however, because it breeds in colonies with egrets, it nonetheless suffered considerable disturbance. It has been a fairly regular visitor to Massachusetts during the past century, sometimes in substantial numbers. The preponderance of white-plumaged juveniles in the late summer and fall suggests that many are postbreeding dispersal birds. Adults have been seen regularly in spring for the last 50 years, with the first confirmed breeding in 1940 in Marshfield. Little Blue Herons have nested regularly in small numbers in Massachusetts since the formation of large mixed-species heronries in the early 1970s. A maximum of 15 pairs was reported from House Island in Manchester during the mid-1970s, but that colony collapsed and disappeared by 1981. From one to five pairs have regularly nested at Clark's Island in Plymouth Bay from 1975 to 1984 (Andrews 1990). In 1977, six pairs were reported breeding at Big Ram Island in Westport, and five pairs were located at Dead Neck/Sampsons Island in Osterville-Cotuit. It is difficult to census Little Blue Heron nests because the eggs and chicks closely resemble those of the Snowy Egret.

Little Blue Herons begin to arrive in Massachusetts in early April, when they can be observed foraging in coastal marshes, flooded grasslands, and freshwater ponds. The Little Blue Heron uses a wide variety of foraging behaviors, sometimes running actively, but more often walking slowly through shallow water or standing and peering straight down into the water. They often rake the bottom substrate with their feet, presumably to startle hidden prey. Sometimes they also capture prey flushed by other birds or animals. The diet of the Little Blue Heron is diverse, consisting of fish, frogs, insects, and other small animals.

In Massachusetts, Little Blue Herons typically begin nesting in early May. They are colonial breeders, sharing heronries with the much more common Snowy Egret. At Clark's Island they have nested in Highbush Blueberry, Red Cedar, and probably Black Cherry. Breeding activities are highly ritualized. Males defend a territory around the nest and advertise primarily by an elaborate "stretch display" (Palmer 1962), bending the head backward and moving the body up and down in a pumping motion. Aggressive displays include the "upright display" with slightly erected crest and the "forward display" (Palmer 1962) with the feathers of the head, neck, and back raised. Harsh *aarh* calls accompany hostile interactions. The members of a pair have an elaborate series of calls and greeting ceremonies at the nest.

The nest, constructed of small sticks and twigs, is a loosely woven platform less than 3 feet in diameter. Males gather most of the nest materials and pass them to the females, which weave them into the nest. The eggs are a light bluish green, usually three to five in number. The three-week incubation

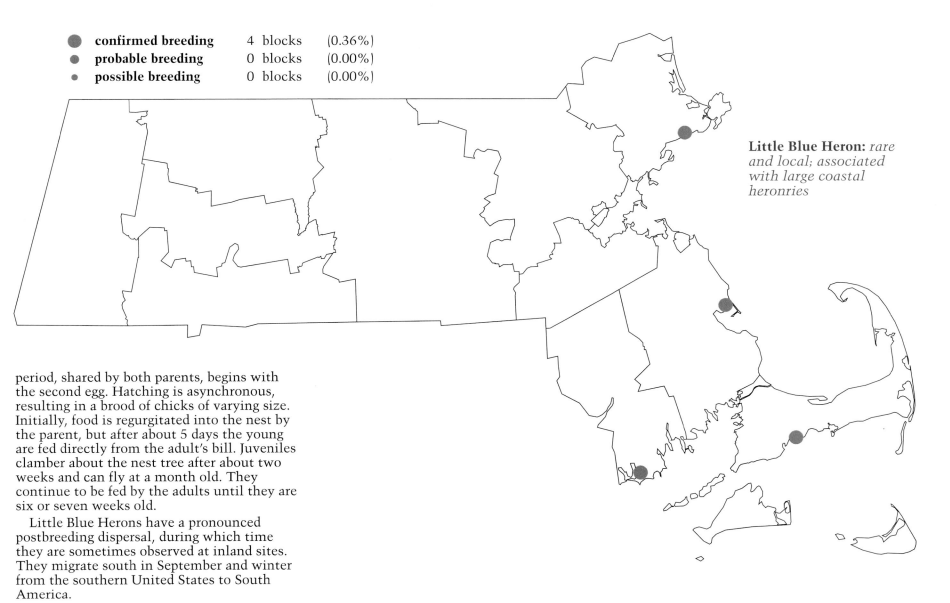

- confirmed breeding — 4 blocks (0.36%)
- probable breeding — 0 blocks (0.00%)
- possible breeding — 0 blocks (0.00%)

Little Blue Heron: *rare and local; associated with large coastal heronries*

period, shared by both parents, begins with the second egg. Hatching is asynchronous, resulting in a brood of chicks of varying size. Initially, food is regurgitated into the nest by the parent, but after about 5 days the young are fed directly from the adult's bill. Juveniles clamber about the nest tree after about two weeks and can fly at a month old. They continue to be fed by the adults until they are six or seven weeks old.

Little Blue Herons have a pronounced postbreeding dispersal, during which time they are sometimes observed at inland sites. They migrate south in September and winter from the southern United States to South America.

William E. Davis, Jr.

Tricolored Heron
Egretta tricolor

Egg dates: not available.

Number of broods: one.

For reasons as yet unexplained, there was a marked influx of southern herons and related species into Massachusetts in the early 1970s. Many of these birds established mixed breeding colonies, most notably at Clark's Island in Plymouth and House Island in Manchester. The Tricolored Heron, formerly known as the Louisiana Heron, was part of this invasion.

The first generally accepted appearance of the Tricolored Heron in the state occurred in 1940 at Ipswich. Since then, reports have increased, and the species has occurred annually since 1960. It is usually restricted to the coast, although there are a few recent records of late summer occurrences in the Connecticut River valley. Breeding was suspected first in the early 1970s, when Tricolored Herons were observed flying in to roost in the evening with other herons at Clark's Island. However, studies of the Clark's Island heronry produced no evidence that tricolors were nesting. Not until 1976 was breeding confirmed in the state, when 3 pairs, each with two young, were located in July on House Island. The young were well feathered but still showed traces of down, especially on the head, and were not yet capable of sustained flight. Breeding has not been recorded in the state since the herons abandoned House Island about 1980.

Most years Tricolored Herons return by late April, but some individuals may appear several weeks earlier. They search for fish, amphibians, crustaceans, gastropods, and insects in marshes, swamps, mudflats, and tidal creeks. Foraging is done in shallow water, where the birds may wade slowly, stand, or run actively in pursuit of prey.

Tricoloreds usually nest in mixed-species heronries. Males choose and defend a nest site and advertise for mates with a series of highly ritualized displays. An "upright display" is used during aggressive encounters, accompanied by low *aah* calls. In "forward display," the plumes are raised. "Circle flight" and "stretch-snap display" (Palmer 1962) are also part of the behavioral repertoire. When meeting at the nest, the members of a pair greet one another with calls and crest raising.

The nest, a platform of sticks 9.5 to 12 inches wide, is lined with twigs and leaves. The usual clutch is three to four greenish blue eggs. Both parents incubate and care for the young. Hatching occurs after about 21 days, and the juveniles are fed by

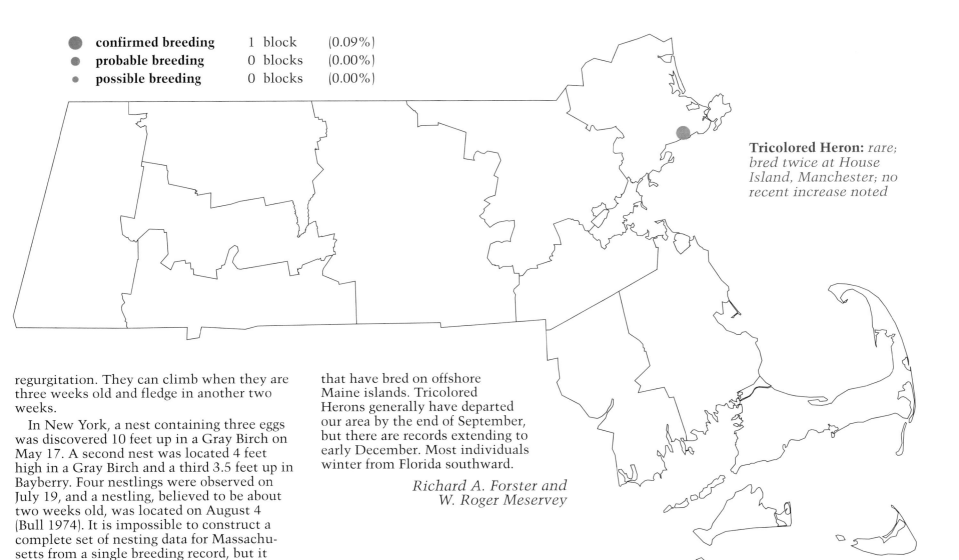

- confirmed breeding   1 block   (0.09%)
- probable breeding   0 blocks   (0.00%)
- possible breeding   0 blocks   (0.00%)

Tricolored Heron: *rare; bred twice at House Island, Manchester; no recent increase noted*

regurgitation. They can climb when they are three weeks old and fledge in another two weeks.

In New York, a nest containing three eggs was discovered 10 feet up in a Gray Birch on May 17. A second nest was located 4 feet high in a Gray Birch and a third 3.5 feet up in Bayberry. Four nestlings were observed on July 19, and a nestling, believed to be about two weeks old, was located on August 4 (Bull 1974). It is impossible to construct a complete set of nesting data for Massachusetts from a single breeding record, but it seems likely that this species would follow a chronology similar to that of the Little Blue Heron.

Unlike most heron species, tricolors do not engage in widespread postbreeding dispersal. Thus, the appearance of birds in late summer roosts, especially at Plum Island, would seem to suggest the possibility of continued local breeding but might also represent migrants that have bred on offshore Maine islands. Tricolored Herons generally have departed our area by the end of September, but there are records extending to early December. Most individuals winter from Florida southward.

*Richard A. Forster and
W. Roger Meservey*

Cattle Egret
Bubulcus ibis

Egg dates: late May to first week of July.

Number of broods: one; may re-lay if first attempt fails.

If reproductive vigor, adaptability, and achievement of transoceanic dispersal are measures of success, the Cattle Egret may be nominated as the world's avian winner. This species probably evolved in the seasonally flooded plains of tropical Africa, where it was forced to feed exclusively on land during the dry season and became a commensal of the African Buffalo. It is theorized that irrigation projects in Africa induced a population explosion beginning early in this century and set the stage for the emigration of "surplus" birds across the Atlantic. The Cattle Egret was first recorded in South America around 1880 and began dispersing toward North America via two routes: north through Central America and from island to island across the Antilles. Cattle Egrets may have reached Florida by 1930, but the first documented North American record was a bird collected in Wayland, Massachusetts, on April 23, 1952. The species is now cosmopolitan between the 45° north latitude and 45° south latitude and is apparently continuing its expansion toward the poles. In eastern North America, it breeds mainly along the coast, north to Maine and Nova Scotia, and is also established at Lake Champlain.

Cattle Egrets increased steadily as annual visitors to Massachusetts beginning in the late 1950s. A total of 155 were seen in the state on May 5 to 17, 1964, and over 100 were seen at single localities in the early 1970s (Veit & Petersen 1993). The first breeding record for Massachusetts was during the Atlas period in July 1974 at House Island, Manchester. The 2 to 4 pairs present in 1974 had increased to 10 pairs by 1976; however, between 1978 and 1980, the House Island colony was abandoned. In 1982, 5 nests were found on Eagle Island, and 2 pairs were seen there in 1984. During the Atlas period, Cattle Egrets were also discovered nesting on Clark's Island in Plymouth Bay, but this site was also abandoned. The breeding population in Massachusetts has yet to exceed a total of 10 pairs at two sites.

Migrants now are recorded regularly throughout the state and have been seen in every month except February. Cattle Egrets return to Massachusetts in April (late March in mild seasons). Males establish territories within mixed colonies containing Snowy Egrets, Black-crowned Night-Herons, and possibly other species. They are very vocal and aggressive and defend territory against conspecifics and other species with a variety of threat displays. The most intense of these is the "full forward threat" in which, with head, chest, and back plumes fully erected, the male crouches and lunges at intruders with bill agape while uttering a rasping *kowwh-kowwh* call. There are also an "upright threat display" (Palmer 1962) and a variety of other calls. The territories defended by Cattle Egrets are said to be notably smaller (and colonies more dense) than those of other heron species.

The beginning of breeding activity is marked by the appearance of intense changes in soft-part colors. Males at first react to females with an aggressive "bill snap display" (Palmer 1962) while extending the head and neck downward and outward. This is followed by a "stretch display," accompanied by an *ow-roo* call. Females carefully watch displaying males and eventually single out a favorite. After pair formation, nest building begins with the males gathering materials and the females placing them. These roles are occasionally reversed. Members of a pair use ritualized greeting ceremonies when they meet at the breeding site. The typical nest has a base of large twigs topped by smaller ones but contains no fine lining materials. The finished structure is a rather crude, shallow cup 8 to 17.5 inches in diameter. It is usually placed in shrubbery or high in small trees. The two to five (usually four or five) pale blue-green eggs are laid at

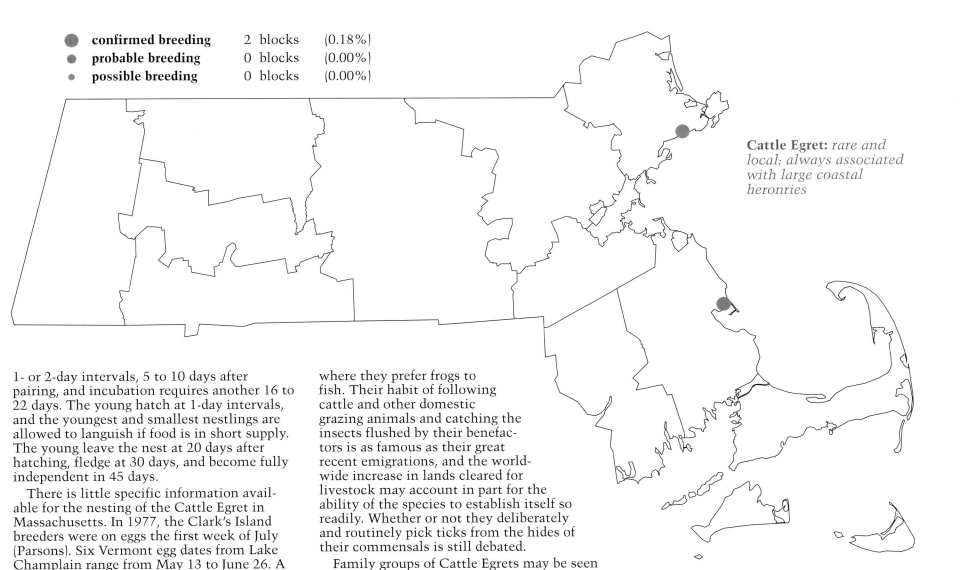

- confirmed breeding 2 blocks (0.18%)
- probable breeding 0 blocks (0.00%)
- possible breeding 0 blocks (0.00%)

Cattle Egret: *rare and local; always associated with large coastal heronries*

1- or 2-day intervals, 5 to 10 days after pairing, and incubation requires another 16 to 22 days. The young hatch at 1-day intervals, and the youngest and smallest nestlings are allowed to languish if food is in short supply. The young leave the nest at 20 days after hatching, fledge at 30 days, and become fully independent in 45 days.

There is little specific information available for the nesting of the Cattle Egret in Massachusetts. In 1977, the Clark's Island breeders were on eggs the first week of July (Parsons). Six Vermont egg dates from Lake Champlain range from May 13 to June 26. A New York nest 5 feet up in Catbrier contained five eggs on June 7 and two nestlings on July 7, and 3 other nests had five young each on June 9.

When not on the breeding islands, Cattle Egrets typically feed on insects (especially grasshoppers) in dry pastures, but they are also very much at home in flooded meadows, where they prefer frogs to fish. Their habit of following cattle and other domestic grazing animals and catching the insects flushed by their benefactors is as famous as their great recent emigrations, and the worldwide increase in lands cleared for livestock may account in part for the ability of the species to establish itself so readily. Whether or not they deliberately and routinely pick ticks from the hides of their commensals is still debated.

Family groups of Cattle Egrets may be seen feeding together in local cow pastures for a brief period in late summer, during which time they return to the nesting island to roost, but the breeding area is deserted by late August. Dispersing and migrating Cattle Egrets occur in variable numbers in fall, mainly in eastern Massachusetts, including as many as 44 counted in Ipswich and South Dartmouth in August 1976 (Veit & Petersen 1993). The wintering grounds of the state's breeders is unknown, but most of the population winters from Florida south to South America.

Christopher W. Leahy

Green Heron
Butorides virescens

Egg dates: early May to early July.

Number of broods: one; may re-lay if first attempt fails.

One of the most cosmopolitan heron species in the world, the Green Heron is widely distributed in Massachusetts, although in most parts of the state it is uncommon. It frequents both inland wetlands and coastal marshes, and breeding was confirmed during the Atlas period from the Berkshires to Cape Cod and the Islands.

In eastern North America, Green Herons are highly migratory, arriving in Massachusetts around the last week of April. While loose breeding colonies are known for this species, most pairs are solitary nesters. These herons are highly vocal, and characteristically, when a bird is surprised at its roost, it will fly off with a loud *skeow* call, leaving a white line of excrement in its wake; hence the popular nickname "Chalk Line." There is another seldom-heard croak made when the birds are very startled as well as an aggressive *raah* call. Green Herons often raise their crests and nervously flick their tails when they are excited.

The male Green Heron uses a series of ritualized displays to defend his nest site and attract a mate. In the "forward display," the mouth lining is shown, while in the "snap display" (Palmer 1962), the bill is clicked and a bobbing motion is performed. Pursuit flights are common.

Nests may be on the ground or close to it but are usually built from 10 to 20 feet high in a bush or small tree on a stream bank or in the midst of a wetland. Some pairs may select an upland site. The heights of 7 Massachusetts nests ranged from 4 to 23 feet. Three were placed in unidentified shrubs, 1 in a lilac, 2 in a Red Maple, and 1 in a White Birch. Nestling dates in Massachusetts are reported for June and July (BOEM, CNR, Anderson, Parsons). Juveniles fledge at about 35 days of age.

The nest is ordinarily a flat, unlined platform of twigs 10 to 12 inches in diameter and is generally neither tightly constructed nor cup shaped so that the eggs often can be seen from below and appear to be in a precarious situation. Up to seven pale green or bluish eggs may be laid at 2-day intervals, with average clutches of four or five. Of 4 Massachusetts nests, 1 contained five eggs and 3 contained four eggs. Incubation is shared by both sexes for 21 to 25 days, and, because it begins with the first egg, hatching is asynchronous. Reported brood size in the Commonwealth ranged from three to five young per nest.

One and a half to two weeks after hatching, the young begin to leave the nest and climb on nearby branches and can scramble out of reach when an intruder climbs the nest tree. They are also good swimmers, a feature that may save their lives if they fall into open water. One other survival trait is the regurgitation of food in self-defense.

Green Herons are known to re-lay if they lose their first clutch of eggs. In New York,

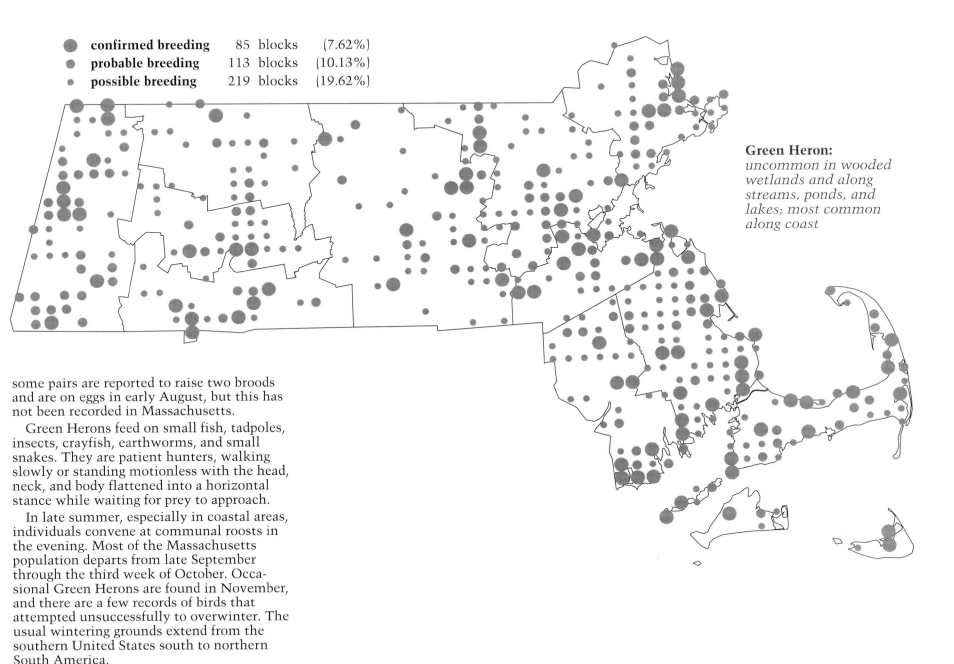

Green Heron: *uncommon in wooded wetlands and along streams, ponds, and lakes; most common along coast*

●	**confirmed breeding**	85 blocks	(7.62%)
●	**probable breeding**	113 blocks	(10.13%)
·	**possible breeding**	219 blocks	(19.62%)

some pairs are reported to raise two broods and are on eggs in early August, but this has not been recorded in Massachusetts.

Green Herons feed on small fish, tadpoles, insects, crayfish, earthworms, and small snakes. They are patient hunters, walking slowly or standing motionless with the head, neck, and body flattened into a horizontal stance while waiting for prey to approach.

In late summer, especially in coastal areas, individuals convene at communal roosts in the evening. Most of the Massachusetts population departs from late September through the third week of October. Occasional Green Herons are found in November, and there are a few records of birds that attempted unsuccessfully to overwinter. The usual wintering grounds extend from the southern United States south to northern South America.

Soheil Zendeh

Black-crowned Night-Heron
Nycticorax nycticorax

Egg dates: first week of April to third week of July.

Number of broods: one; may re-lay if first attempt fails.

One of the best known of its tribe, the Black-crowned Night-Heron is widely distributed in coastal Massachusetts. At one time it was the most abundant heron, but recently its population has suffered a dramatic decline. From 1915 to 1920, there were more than 3,000 birds in three colonies; in 1955, there were 3,600 in ten colonies; in 1975, there were 1,894; in 1976, there were about 1,500 in fifteen colonies; and in 1977, there were 1,958 birds in fourteen colonies. During the 1950s, many heronries were shot out or dynamited because the birds were considered noisy and smelly neighbors. Pollution, DDT contamination, and the steady loss of habitat from coastal development have also contributed to the population decline.

Black-crowned Night-Herons forage in coastal salt marshes, freshwater marshes, ponds, creeks, and tidal flats. They are primarily nocturnal and crepuscular feeders, roosting by day in the vegetation bordering wetlands. Food items include fish, amphibians, insects, crustaceans, and sometimes the young of gulls and terns. The common call of this species is very distinctive and is the origin of the common name of "Quawk." The call may be uttered as the birds fly to and from their roosts or when they are disturbed.

Black-crowned Night-Herons usually arrive in mid-March. They are colonial breeders, often nesting in mixed-species heronries. Males claim and defend small nest territories and use several ritualized displays including "upright display," "forward display," and a "bill snapping display" (Palmer 1962). Courting males land near a female and bow toward her while raising the head plumes and fluffing out the breast, neck, and back feathers. This display may be followed by bill touching.

Nests are usually built in trees or shrubs from 2 to 42 feet above the ground. At Clark's Island, most pairs nested above 10 feet in Red Cedar. The birds' excrement generally kills the vegetation, and eventually the nesting area must be relocated. The nests are crudely

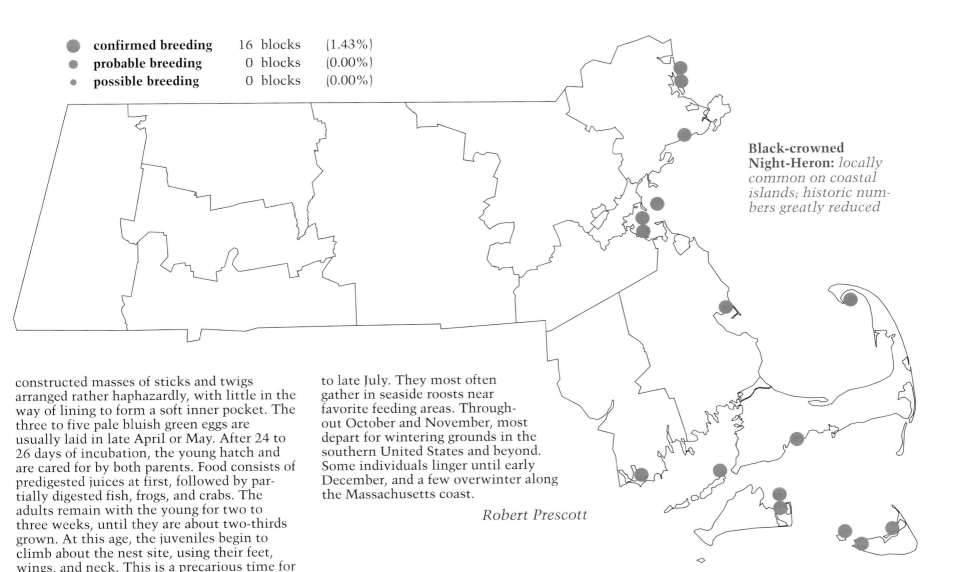

●	**confirmed breeding**	16 blocks	(1.43%)
●	**probable breeding**	0 blocks	(0.00%)
·	**possible breeding**	0 blocks	(0.00%)

Black-crowned Night-Heron: *locally common on coastal islands; historic numbers greatly reduced*

constructed masses of sticks and twigs arranged rather haphazardly, with little in the way of lining to form a soft inner pocket. The three to five pale bluish green eggs are usually laid in late April or May. After 24 to 26 days of incubation, the young hatch and are cared for by both parents. Food consists of predigested juices at first, followed by partially digested fish, frogs, and crabs. The adults remain with the young for two to three weeks, until they are about two-thirds grown. At this age, the juveniles begin to climb about the nest site, using their feet, wings, and neck. This is a precarious time for young herons, and many fall to a premature death. Fledging occurs at from six to seven weeks of age. In 1975 and 1978 at Clark's Island, average pairs reared two or three young to at least 10 days of age. This is considered to be good nesting success.

Adults and juveniles disperse far and wide from the nesting colonies beginning in mid- to late July. They most often gather in seaside roosts near favorite feeding areas. Throughout October and November, most depart for wintering grounds in the southern United States and beyond. Some individuals linger until early December, and a few overwinter along the Massachusetts coast.

Robert Prescott

Yellow-crowned Night-Heron
Nyctanassa violacea

Egg dates: April 30 to June 10.

Number of broods: One; may re-lay if first attempt fails.

The evidence supporting the placement of the Yellow-crowned Night-Heron among the breeding avifauna of Massachusetts is minimal and, in some cases, tenuous. Incubating adults or nests containing young have been recorded on only two occasions during this century. Townsend secured one of four nestlings in Ipswich in 1928, and Hagar documented breeding birds in Marshfield in the late 1940s. Other purported breeding records are at Provincetown in 1891 where adults were present and an immature was collected in July, and at Chatham in 1940 and 1941 where Maclay noted nearly fledged young (Hill 1965). The continued presence of adult birds at the same location for successive years during the nesting season is strongly suggestive of breeding but lacks the necessary evidence for confirmed breeding.

Migrant Yellow-crowned Night-Herons normally appear in the state in mid-May, but there are a number of April reports. Occasionally during spring migration a stray individual may appear in wooded swamps or trees bordering meandering streams at inland locations. Otherwise, they inhabit coastal bays and marshes and are most frequently encountered in southeastern coastal areas, Cape Cod, and the Islands. Throughout most of the twentieth century, the Yellow-crowned Night-Heron has been a very uncommon or rare resident. Adults are often found in potential breeding locations from June through August. The appearance of immature birds, often in the company of adults, is indicative of breeding. However, since herons, most notably young of the year, undertake extensive postbreeding dispersal northward, the origin of these immatures from breeding colonies on Long Island or farther south cannot be discounted.

During the 1960s and 1970s, southern herons spread dramatically northward and established breeding colonies in Massachusetts and Maine, areas where they had not been known to breed or bred only in limited numbers. This group included Great, Snowy, and Cattle egrets, Tricolored and Little Blue herons, and the Glossy Ibis. However, there is no evidence that the Yellow-crowned Night-Heron shared in this range extension since its status remained relatively unchanged.

Yellow-crowned Night-Herons breed widely from the southern United States, the Caribbean, and Middle America south along the coast to central South America. In the United States, they breed at interior locations along the lower Mississippi River drainage, along the Gulf Coast, and north along the Atlantic coast to Massachusetts. They usually nest in small to large mixed-species colonies, but, at the northern periphery of their range, they frequently nest in isolated solitary pairs. On Long Island, Yellow-crowned Night-Herons breed both in mixed colonies with other herons and alone in wooded swamps. Yellow-crowned Night-Herons have yet to be discovered in any of the heronries that have been surveyed periodically since the late 1970s. Hagar discovered them as solitary breeders, and it is likely that this remains the case at present.

Nests are usually located 15 to 45 feet above the ground in a live tree. The nest is a rather large, well-built platform of heavy twigs and is often lined with leaves or grasses. Pair bond formation involves rituals as with other heron species and includes "stretch displays," "circle flights," "billing," and "allopreening" (Palmer 1962).

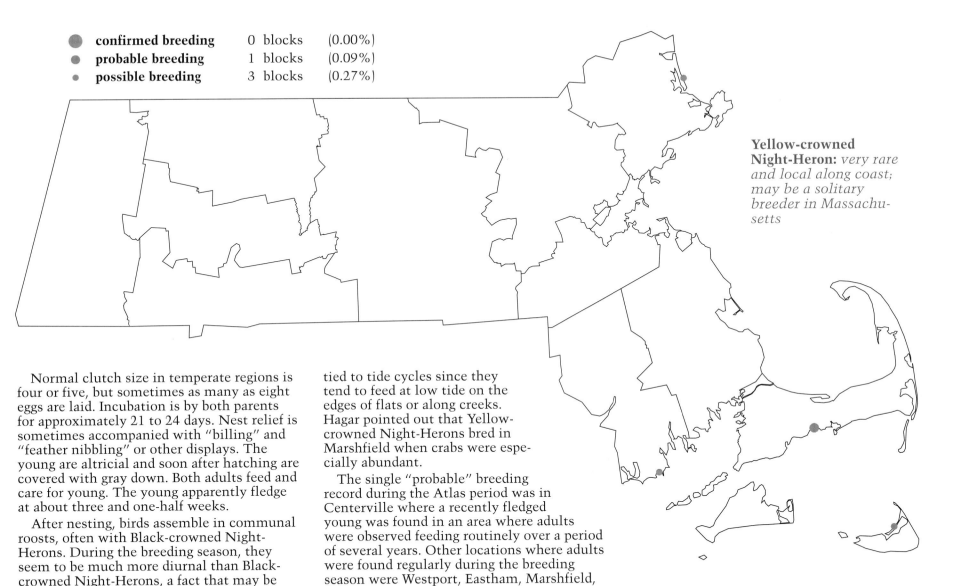

Yellow-crowned Night-Heron: *very rare and local along coast; may be a solitary breeder in Massachusetts*

Normal clutch size in temperate regions is four or five, but sometimes as many as eight eggs are laid. Incubation is by both parents for approximately 21 to 24 days. Nest relief is sometimes accompanied with "billing" and "feather nibbling" or other displays. The young are altricial and soon after hatching are covered with gray down. Both adults feed and care for young. The young apparently fledge at about three and one-half weeks.

After nesting, birds assemble in communal roosts, often with Black-crowned Night-Herons. During the breeding season, they seem to be much more diurnal than Black-crowned Night-Herons, a fact that may be due to a difference in their method of feeding. The heavier, stouter bill of the Yellow-crowned Night-Heron enables them to feed primarily on crustaceans, especially crabs. Black-crowned Night-Herons consume mainly fish. The diurnal feeding habits of Yellow-crowned Night-Herons are closely tied to tide cycles since they tend to feed at low tide on the edges of flats or along creeks. Hagar pointed out that Yellow-crowned Night-Herons bred in Marshfield when crabs were especially abundant.

The single "probable" breeding record during the Atlas period was in Centerville where a recently fledged young was found in an area where adults were observed feeding routinely over a period of several years. Other locations where adults were found regularly during the breeding season were Westport, Eastham, Marshfield, and the Plum Island area. More recent suspected breeding areas are Hingham and Wareham.

Yellow-crowned Night-Herons routinely linger until mid-September, at which time they, along with other herons, undertake an unheralded departure for more southerly climes. Yellow-crowned Night-Herons are rarely encountered in October, and there are only a few winter reports.

Richard A. Forster

Glossy Ibis
Plegadis falcinellus

Egg dates: second week of May to first week of July.

Number of broods: one; may re-lay if first attempt fails.

This small member of the ibis family is thought to be a relatively recent arrival from the Old World. Historically, its center of greatest abundance has been the southeastern coastal states, but the species has a notable habit of wandering great distances and has established nesting records over a wide area of North America. However, its nesting distribution remains erratic, and its status at any given area is often unpredictable from year to year. This pattern seems to hold for Massachusetts as well.

Recorded as an occasional vagrant in Massachusetts as early as 1850, it became regular only after 1947, particularly as a spring visitor, with its periods of occurrence tending gradually to lengthen. It was first discovered nesting in the state at Clark's Island, Plymouth, in 1974. It occurs generally as a rare and irregular nester at a very limited number of coastal stations (four sites from 1974 to 1979). There were 112 pairs at two known sites in 1977 and 27 pairs at four sites in 1984. To date, there have been two important nesting groups, both associated with Black-crowned Night-Heron and Snowy Egret colonies. The first, at Clark's Island, increased rapidly to 66 pairs in 1976 but experienced poor reproductive success and slowly declined to zero in 1983-1984. The other important group developed farther north at House Island, Manchester, at about the same time. This colony peaked at 107 pairs in 1977 and was abandoned by 1982.

Spring arrivals appear in early April. Small flocks of migrants may occur inland as well as along the coast into the third week of May, but inland appearances are highly irregular. Colonies are occupied by late April. As is typical of this gregarious species, breeding almost always occurs in association with other waders, usually on islands overgrown with small trees and dense vegetation. On Clark's Island, nests were located in Arrowwood and Red Cedar. The nests are fashioned from sticks and twigs and are usually 6 to 12 feet above ground. The Glossy Ibis sometimes nests in such great densities with other species of herons that heavy excrement from the colony causes leaf burn and eventual ruination of the habitat. In such instances, the birds abandon the site and move on to other areas.

Save for an occasional series of rasping guttural notes likened to *graa graa graa*, often vocalized when the birds are disturbed, the Glossy Ibis is very seldom heard away from the immediate vicinity of the nest. At the nest, a more varied repertoire of hoarse grunts and cooing contact calls are exchanged

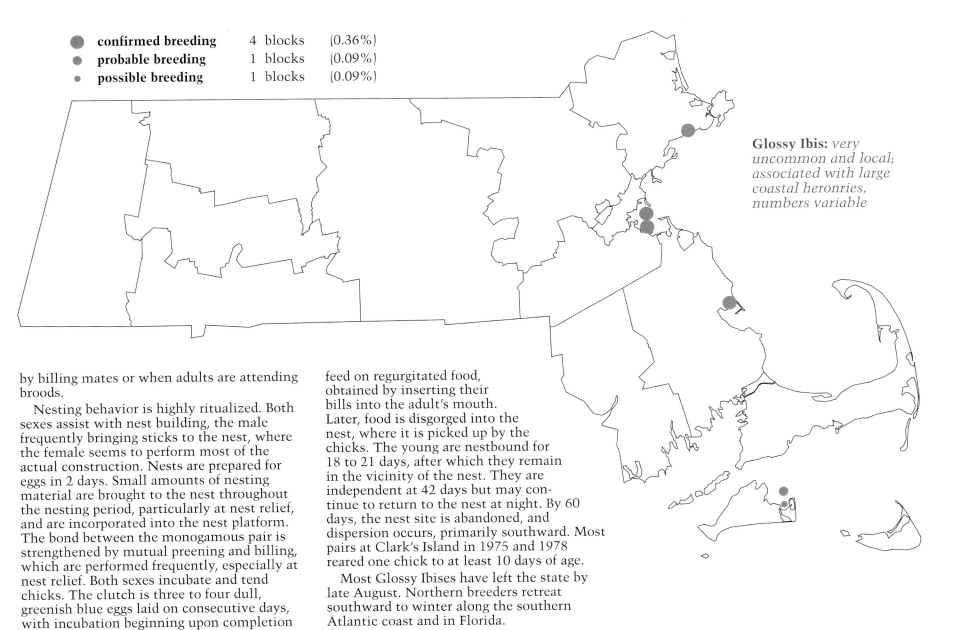

● confirmed breeding	4 blocks	(0.36%)
● probable breeding	1 blocks	(0.09%)
● possible breeding	1 blocks	(0.09%)

Glossy Ibis: *very uncommon and local; associated with large coastal heronries, numbers variable*

by billing mates or when adults are attending broods.

Nesting behavior is highly ritualized. Both sexes assist with nest building, the male frequently bringing sticks to the nest, where the female seems to perform most of the actual construction. Nests are prepared for eggs in 2 days. Small amounts of nesting material are brought to the nest throughout the nesting period, particularly at nest relief, and are incorporated into the nest platform. The bond between the monogamous pair is strengthened by mutual preening and billing, which are performed frequently, especially at nest relief. Both sexes incubate and tend chicks. The clutch is three to four dull, greenish blue eggs laid on consecutive days, with incubation beginning upon completion of the clutch and lasting for approximately 21 days. Hatching is synchronous.

Young hatch from mid-June to early July and are dependent upon the adults. Chicks feed on regurgitated food, obtained by inserting their bills into the adult's mouth. Later, food is disgorged into the nest, where it is picked up by the chicks. The young are nestbound for 18 to 21 days, after which they remain in the vicinity of the nest. They are independent at 42 days but may continue to return to the nest at night. By 60 days, the nest site is abandoned, and dispersion occurs, primarily southward. Most pairs at Clark's Island in 1975 and 1978 reared one chick to at least 10 days of age.

Most Glossy Ibises have left the state by late August. Northern breeders retreat southward to winter along the southern Atlantic coast and in Florida.

Bradford G. Blodget

Turkey Vulture
Cathartes aura

Egg dates: early May to early August.

Number of broods: one.

For the better part of this century, the Turkey Vulture was no more than a rare or casual straggler in the state, with the majority of sightings concentrated in southernmost Berkshire County. The first proven record of breeding occurred in this region at Tyringham, where a nest with two downy young was found on August 29, 1954. Nesting may have taken place as much as a decade or more earlier because an immature with down on the head and neck was found dead at Mount Everett in August 1945.

There has been a pronounced increase in the Turkey Vulture population since the early 1970s, and the species now breeds in northern Vermont and Maine. It is difficult to determine exactly where vultures nest in the state because of the birds' wide-ranging flights in search of carrion and their reclusive nesting habits. These behaviors account for the preponderance of "possible" records on the map. Undoubtedly, vultures are much more common as a breeding species, although not necessarily more widely distributed, than the map would indicate. There were two confirmations during the Atlas period, one in Barre and one in Quabbin. From 1982 to 1987, eight nests were located in eastern Massachusetts in the Blue Hills Reservation in Milton (Smith), and in 1986 breeding was confirmed in Sturbridge.

The Turkey Vulture is most common from New Jersey southward throughout much of temperate and tropical America. Its splayed wing tips and long, two-toned wings held in a pronounced dihedral over the back are characteristic, as is the typical, leisurely, rocking flight as a bird courses over field and treetop. Although it prefers not to fly on cloudy days or during strong winds, the "Buzzard" (as it is known south of the Mason-Dixon line) is fully capable of flying with deep, slow wing beats, but its soaring flight uses less energy. The birds are generally silent, except for occasional hissing and growling sounds used near the nest.

The first spring migrants normally begin appearing in early March, but they may arrive as early as mid-February with unseasonably warm weather. Reports formerly were concentrated in central and western Massachusetts, but migrants are now seen throughout the state, including outer Cape Cod.

Pairing, which may involve dancing antics of groups of birds on the ground, apparently occurs during late April or early May. All known nest sites in Massachusetts have been located in caves or crevices, or on shelves of rocky ledges on hillsides, but in other regions they may be in hollow logs, tree stumps, old buildings, or on the ground. No nest materials are used. Clutch sizes range from one to three (average two), and for 7 Massachusetts nests were as follows: one egg (3 nests), two eggs (3 nests), and three eggs (1 nest) (Smith). The Barre nest contained eggs on June 22 (Cardoza). The elliptical eggs are creamy white and blotched with varying shades of brown. The incubation period is not known precisely but generally is considered to be five to six weeks. Both sexes incubate and feed the young by regurgitation for eight to

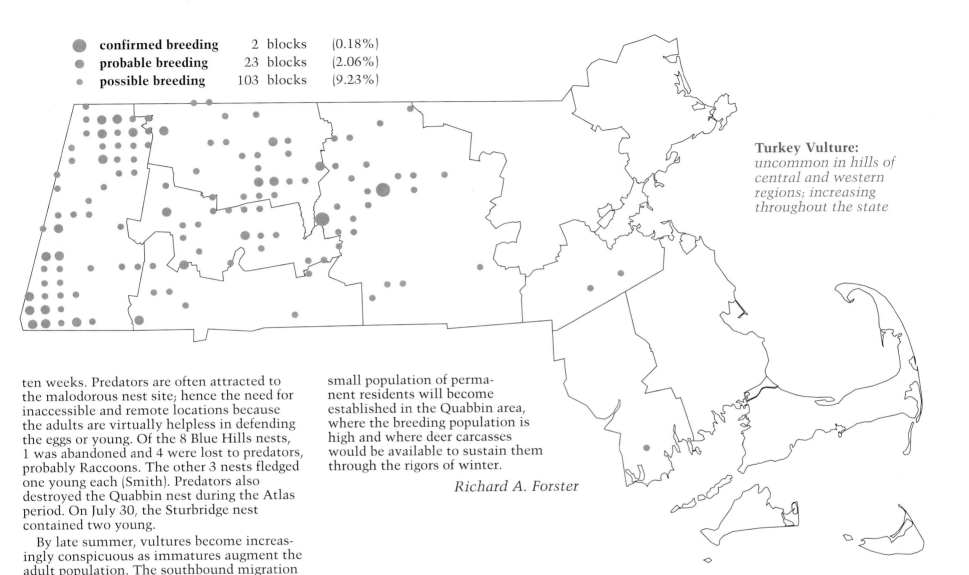

Turkey Vulture: *uncommon in hills of central and western regions; increasing throughout the state*

●	**confirmed breeding**	2 blocks	(0.18%)
●	**probable breeding**	23 blocks	(2.06%)
·	**possible breeding**	103 blocks	(9.23%)

ten weeks. Predators are often attracted to the malodorous nest site; hence the need for inaccessible and remote locations because the adults are virtually helpless in defending the eggs or young. Of the 8 Blue Hills nests, 1 was abandoned and 4 were lost to predators, probably Raccoons. The other 3 nests fledged one young each (Smith). Predators also destroyed the Quabbin nest during the Atlas period. On July 30, the Sturbridge nest contained two young.

By late summer, vultures become increasingly conspicuous as immatures augment the adult population. The southbound migration occurs from mid-September to mid-October, but an increasing number of sightings are made in late fall and early winter. Breeding residents probably winter in the southern United States, although, concurrent with the population increase in the Northeast, winter roosts have become established as far north as Connecticut. It seems probable that a small population of permanent residents will become established in the Quabbin area, where the breeding population is high and where deer carcasses would be available to sustain them through the rigors of winter.

Richard A. Forster

Canada Goose
Branta canadensis

Egg dates: April 1 to May 30.

Number of broods: one; may re-lay if first attempt fails.

Few sounds represent the call of the wild better than the drawn-out honking of Canada Geese as they migrate high overhead to and from their arctic breeding grounds and southern wintering sites. That call loses its evocative charm when you hear it daily as geese move between golf course and reservoir.

The Canada Geese that breed in Massachusetts are relatively new to the state. There is little evidence that geese bred here before the 1900s. Two factors led to the establishment of the present population. Prior to the arrival of European colonists, the forested Northeast provided little in the way of suitable goose nesting habitat, except for some of the offshore islands. As people cleared the forests, built mill and farm ponds, and later made large reservoirs, goose habitat was also created.

Second came the importation of geese. Bird fanciers kept a few Canada Geese in their collections, but wealthy hunters imported birds from Michigan, North Carolina, and other areas for use as live decoys. When this practice was outlawed in 1935, large numbers of geese were liberated, and they joined others that had previously been released or had escaped. These birds were a mixture of races and forms, but a Midwestern subspecies described as the "Giant" Canada Goose predominated. They were large, hardy types able to survive severe winter weather and not prone to migrate. Goose populations started to increase very slowly, but, as the best genetically adapted individuals survived and began to breed, numbers reached nuisance levels in some areas by the 1960s. The Division of Fisheries and Wildlife began a program of transplanting goslings from problem flocks in eastern Massachusetts to goose-free areas of the central and western part of the state. By the time the program ended in 1976, geese were breeding throughout the Commonwealth. Numbers were swelled by an influx of geese from Connecticut and New York, where similar events had occurred. Local Canada Goose populations have continued to increase. The 1993 late summer population estimate was 32,000 resident geese based on collar re-sighting data.

Our Canada Geese are largely suburban birds. Many nest in "wild" areas of the state but move into suburban and rural sections in late summer. While some geese may form pair bonds as yearlings, nesting does not occur until they are two years old. Unlike ducks, geese pair for life, and the gander stays with the female all year. However, if one partner dies, the other will re-mate. Breeding in Massachusetts begins in early April, usually about April 10, but sometimes earlier in interior areas and later along the coast. Nests are established on slightly elevated sites such as Muskrat houses, tussocks, and islets. They begin as scrapes, and the females reach out and pull in surrounding cattails and grasses to add to the base and rim. Down is added as egg laying progresses. Normally, only one pair utilizes a small pond, but, when Canada Geese nest on reservoirs, they frequently use larger islands, and many pairs may establish nests in close proximity. Clutches usually contain five or six white eggs, and incubation lasts 26 to 28 days. Most goslings in Massachusetts hatch in a four-week period from May 3 to May 30, especially after May 10.

The goslings leave the nest shortly after hatching, accompanied by both parents. Particularly aggressive females may adopt other youngsters or entire broods, in which case a single pair may be seen with more

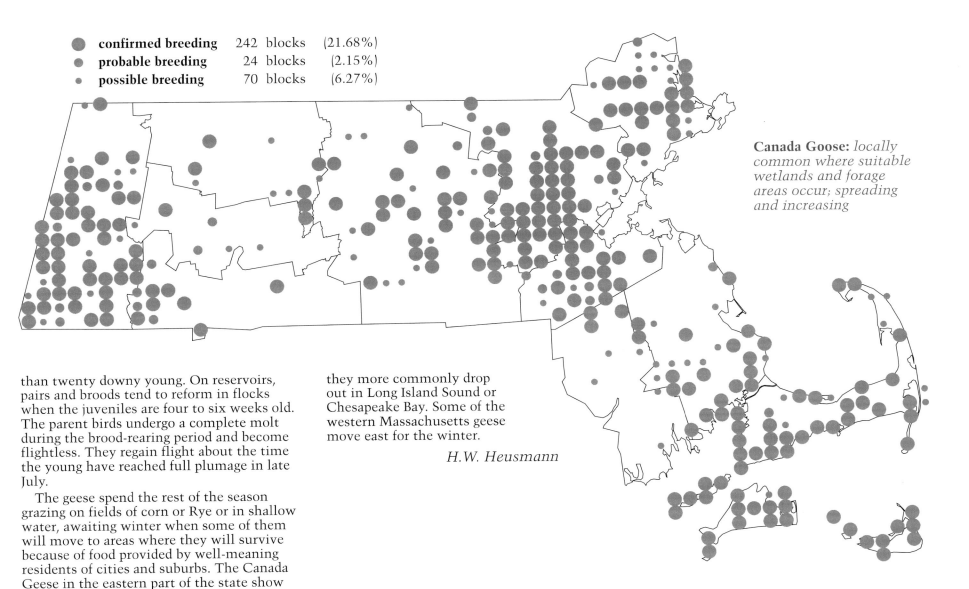

- **confirmed breeding** 242 blocks (21.68%)
- **probable breeding** 24 blocks (2.15%)
- **possible breeding** 70 blocks (6.27%)

Canada Goose: *locally common where suitable wetlands and forage areas occur; spreading and increasing*

than twenty downy young. On reservoirs, pairs and broods tend to reform in flocks when the juveniles are four to six weeks old. The parent birds undergo a complete molt during the brood-rearing period and become flightless. They regain flight about the time the young have reached full plumage in late July.

The geese spend the rest of the season grazing on fields of corn or Rye or in shallow water, awaiting winter when some of them will move to areas where they will survive because of food provided by well-meaning residents of cities and suburbs. The Canada Geese in the eastern part of the state show little tendency to migrate, moving instead from frozen reservoirs to major rivers such as the Sudbury, Concord, and Blackstone, or to coastal waters. Birds in the western part of the state are more likely to join flocks of geese migrating south from Canada and may end up as far south as North Carolina, but they more commonly drop out in Long Island Sound or Chesapeake Bay. Some of the western Massachusetts geese move east for the winter.

H.W. Heusmann

Mute Swan
Cygnus olor

Egg dates: April to third week of May.

Number of broods: one; may re-lay if first attempt fails.

This naturalized resident of the Northeast has established a breeding population, with increasing numbers, primarily along the southeastern coast of Massachusetts. During the nineteenth and early twentieth centuries, Mute Swans from Eurasia were imported by estate and park managers, and it is thought that our population stems from escaped and released birds from that era. Through the 1960s, breeding was irregular, and most swans were considered to be wanderers. By 1974, there were nearly 200 birds reported on Christmas Bird Counts in Massachusetts. During the Atlas period, numbers continued to build.

A post-Atlas survey conducted in 1983 showed a breeding distribution that approximates that of the late 1970s. Nesting Mute Swan pairs were reported from Westport to Dennis along the southeastern coast, from the Islands, and on the South Shore from Plymouth to Cohasset. A few isolated pairs nested on the North Shore. None were discovered breeding inland or on outer Cape Cod. The total for eastern Massachusetts was approximately 135 pairs. The areas with the greatest density included Martha's Vineyard with 35 pairs, Falmouth with 20+ pairs, and Plymouth with 15+ pairs (Clapp, Hall).

Usually beginning in April, a pair of Mute Swans chooses a site at a coastal pond or tidal creek, where they build a large nest of vegetation lined with feathers. They are highly territorial at this time and will defend their nest areas from other swans, geese, ducks, dogs, and even humans, biting and delivering blows with their wings. The usual clutch size is four to six grayish eggs, which the female will incubate while the male remains on guard nearby. Hatching usually takes place within five to six weeks. Young cygnets are precocial, and the first ones to hatch will often join the male on the water while the female attends to the remaining eggs. Broods of three to four small cygnets are commonly reported, generally beginning in mid-May. The young stay with the adults for about fifteen weeks until they are able to fly.

Breeding success appears low, and all data and observations suggest a high mortality of cygnets. In Rhode Island, only about one-third fledge, and the same can be assumed from the 1983 Massachusetts data. There is continued mortality until the birds are old enough to breed. Unmated birds often flock together, creating a situation in which avian disease can produce significant mortality. Those that survive to adulthood have an average life span of six to ten years.

This easily recognized regal bird has developed a human following. Feeding swans,

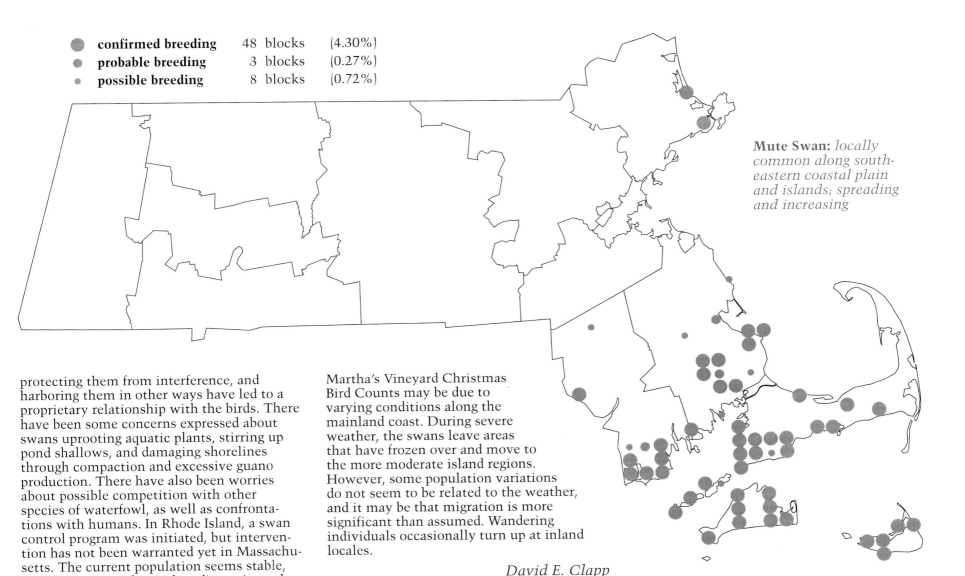

Mute Swan: *locally common along southeastern coastal plain and islands; spreading and increasing*

- confirmed breeding — 48 blocks (4.30%)
- probable breeding — 3 blocks (0.27%)
- possible breeding — 8 blocks (0.72%)

protecting them from interference, and harboring them in other ways have led to a proprietary relationship with the birds. There have been some concerns expressed about swans uprooting aquatic plants, stirring up pond shallows, and damaging shorelines through compaction and excessive guano production. There have also been worries about possible competition with other species of waterfowl, as well as confrontations with humans. In Rhode Island, a swan control program was initiated, but intervention has not been warranted yet in Massachusetts. The current population seems stable, with approximately 135 breeding pairs and a wintering number approaching 600.

Mute Swans are permanent residents of the coastal plain, moving to unfrozen ponds and bays for the winter. Migration is weather dependent and seems to be minimal. The great fluctuation of birds recorded on the Martha's Vineyard Christmas Bird Counts may be due to varying conditions along the mainland coast. During severe weather, the swans leave areas that have frozen over and move to the more moderate island regions. However, some population variations do not seem to be related to the weather, and it may be that migration is more significant than assumed. Wandering individuals occasionally turn up at inland locales.

David E. Clapp

Wood Duck
Aix sponsa

Egg dates: late March to mid-July.

Number of broods: one; may re-lay if first attempt fails.

By the turn of the last century, the colorful Wood Duck was nearly extirpated from Massachusetts as well as throughout most of the Northeast. As settlers cut the forests to clear land for agriculture, the Wood Duck lost the cavity trees it needed for nesting. Popular as table fare, the duck was commercially exploited as well. Populations began to recover after the bird was given complete protection in 1918, and nesting habitat was restored as New England farms reverted to forest. A limiting factor was the paucity of trees old enough to develop sizable nesting cavities, but a massive statewide nest box program in the early 1950s helped to restore the Wood Duck population to previous levels.

However, the extensive use of DDT and other pesticides about this time destroyed populations of insects required by young ducklings, and the population began another decline that lasted until the early 1970s. The combination of a ban on DDT, the maturation of the forest, an increasing Beaver population whose activities created favorable breeding habitat, and carefully controlled hunting seasons has allowed the Wood Duck once again to become a common nester. Populations are densest in the eastern third of the state, excluding Cape Cod and the Islands, where suitable cavity trees are still uncommon.

While a few Wood Ducks overwinter, most birds arrive in Massachusetts in early to mid-March, having left their winter quarters in late February. They can be found on Beaver ponds and river floodplains, along slow-moving streams, and in deep marshes throughout the state. Secretive ducks, they prefer the cover of Buttonbush and emergent vegetation and are not often seen on open water. Males have a distinct whistling note, while the females give a characteristic squealing *wee-ee-k*.

Strictly a cavity nester, the Wood Duck will utilize both living trees and standing snags, requiring at least a 3.5-inch opening and a cavity about 9 inches in diameter. Females bring in no nesting material but rely on the rotted pulp of the cavity, later adding down from their breasts when egg laying advances. They adapt readily to nest boxes supplied with shavings.

Nesting commences the last week of March in southeastern Massachusetts and about the first week of April in the rest of the state. Yearlings begin nests about a month after adult birds, and many do not breed the first year, perhaps because of the scarcity of cavities. Clutch size ranges from seven to fifteen eggs, averaging ten for yearlings and twelve for adults. Larger clutches are the result of other hens laying eggs in a nest, and it is not unusual for a hen to hatch a brood of twenty-four or more ducklings. At times, hens hatch a combination of their own young and those of the Hooded Merganser (see account for that species).

Eggs hatch in 28 to 32 days, depending upon the clutch size and ambient temperatures. Three main hatch times in Massachusetts are May 10 to May 31 (for adult hens), June 1 to June 21 (for yearling hens), and June 22 to July 19 (for renesters). Ducklings leave the nest 24 to 36 hours after hatching, jumping out of the entrance in response to

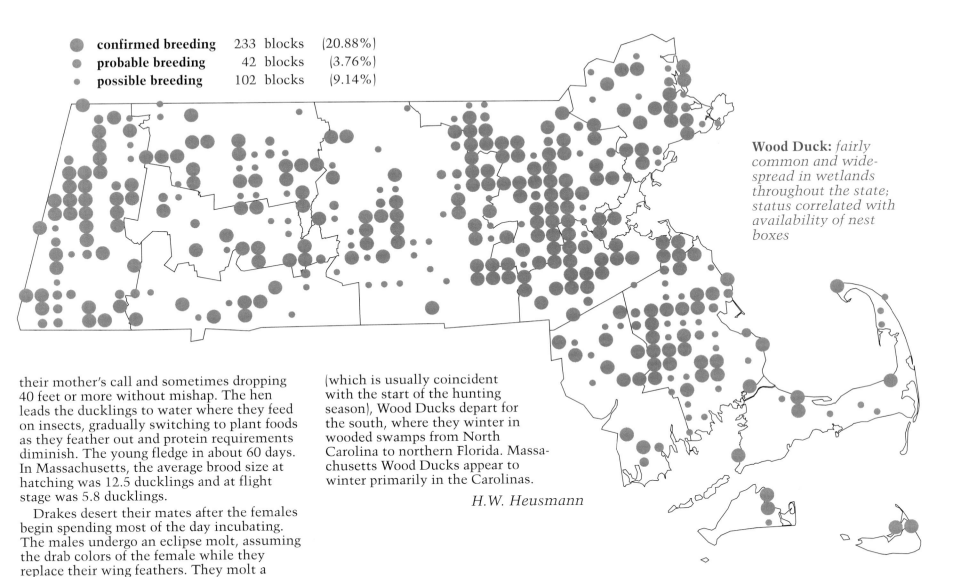

●	**confirmed breeding**	233 blocks	(20.88%)
●	**probable breeding**	42 blocks	(3.76%)
·	**possible breeding**	102 blocks	(9.14%)

Wood Duck: *fairly common and widespread in wetlands throughout the state; status correlated with availability of nest boxes*

their mother's call and sometimes dropping 40 feet or more without mishap. The hen leads the ducklings to water where they feed on insects, gradually switching to plant foods as they feather out and protein requirements diminish. The young fledge in about 60 days. In Massachusetts, the average brood size at hatching was 12.5 ducklings and at flight stage was 5.8 ducklings.

Drakes desert their mates after the females begin spending most of the day incubating. The males undergo an eclipse molt, assuming the drab colors of the female while they replace their wing feathers. They molt a second time to regain their breeding plumage. Some drakes begin leaving the area in August, but most Wood Ducks remain and begin gathering nightly at roosts in September and early October. Often small groups of birds can be observed flying in at dusk to wooded swamps until several hundred have arrived. With the onset of persistent freezing weather (which is usually coincident with the start of the hunting season), Wood Ducks depart for the south, where they winter in wooded swamps from North Carolina to northern Florida. Massachusetts Wood Ducks appear to winter primarily in the Carolinas.

H.W. Heusmann

Gadwall
Anas strepera

Egg dates: late April to late May.

Number of broods: one; may re-lay if first attempt fails.

The Gadwall, a common resident of the prairie pothole region of North America, is a relatively new nesting species in Massachusetts. A small breeding colony was established at Great Meadows National Wildlife Refuge in Concord by introducing flightless young from the Delta Wildlife Research Station in Manitoba, Canada, in 1965. In 1971, the Felix Neck Wildlife Sanctuary imported forty-two Gadwall ducklings from the same source in an effort to establish them on Martha's Vineyard. In 1977, Gadwalls nested in a marsh at Felix Neck and also on nearby Chappaquiddick Island. Each year since then, there have been two to four pairs nesting on Martha's Vineyard.

Possibly due to these introductions, but more likely as a result of natural range expansion, the Gadwall became established in coastal northern Essex County at Parker River National Wildlife Refuge and vicinity. This population has increased gradually. A small group is also present at Monomoy National Wildlife Refuge. The overall number of Gadwalls throughout their normal nesting and wintering ranges has risen considerably, with some tendency toward eastward expansion.

Breeding birds arrive as soon as the ice melts on ponds and marshes, often as early as late February in mild seasons; migrants are occasionally noted away from nesting areas. Gadwalls may delay breeding, possibly due to the retarded growth of vegetation in some cooler areas.

The drake Gadwall is relatively quiet but does produce sounds similar to, but somewhat higher pitched than, those of the Mallard drake. During courtship he makes a burping sound and uses several ritualized displays. The hen utters the typical quack calls. Gadwalls, like most dabbling ducks, prefer to nest on dry, open upland near water but will occasionally travel some distance from it. Small islands are favorite sites. A snug, grass nest is built and lined with down.

Gadwalls lay five to fourteen creamy white eggs, with ten being the normal clutch. The female incubates the eggs for 24 to 27 days, then leads the downy young to water within 24 hours after hatching. On Martha's Vineyard, most broods appear during the first week of June (Ben David). Brood dates at Parker River and Concord range from June 12 to August 16. On Monomoy a female with four young was observed on July 5 (Humphrey). Gadwalls nesting on ponds with sparse cover suffer high brood mortality. The young mature rapidly and are capable of flight as early as 50 days after hatching.

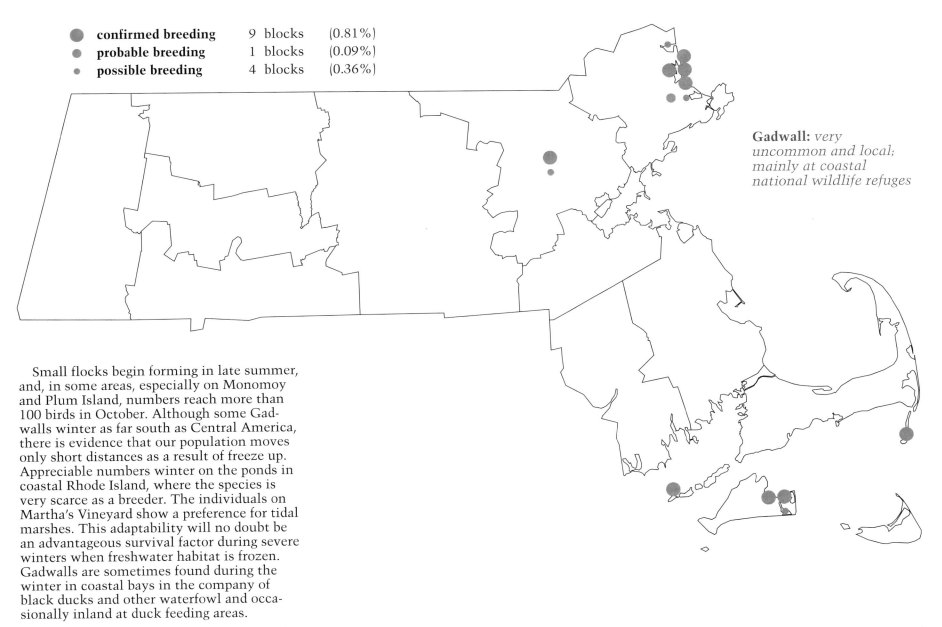

- confirmed breeding — 9 blocks (0.81%)
- probable breeding — 1 blocks (0.09%)
- possible breeding — 4 blocks (0.36%)

Gadwall: *very uncommon and local; mainly at coastal national wildlife refuges*

Small flocks begin forming in late summer, and, in some areas, especially on Monomoy and Plum Island, numbers reach more than 100 birds in October. Although some Gadwalls winter as far south as Central America, there is evidence that our population moves only short distances as a result of freeze up. Appreciable numbers winter on the ponds in coastal Rhode Island, where the species is very scarce as a breeder. The individuals on Martha's Vineyard show a preference for tidal marshes. This adaptability will no doubt be an advantageous survival factor during severe winters when freshwater habitat is frozen. Gadwalls are sometimes found during the winter in coastal bays in the company of black ducks and other waterfowl and occasionally inland at duck feeding areas.

Augustus Ben David II

American Black Duck
Anas rubripes

Egg dates: March 20 to late July.

Number of broods: one; may re-lay if first attempt fails.

Historically, the common dabbling duck of the inland waters and salt marshes of Massachusetts was the American Black Duck. Black ducks were abundant throughout the Northeast, which was outside the traditional breeding range of the Mallard. Popular with waterfowlers for the past 200 years, the wary black ducks withstood even the pressures of market hunting and spring shooting. However, they have not been able to withstand the changes wrought by modern America.

Black ducks still nest widely in the state, but the numbers have been declining throughout their range since the mid-1950s. Much of the early drop was due to loss of both breeding and wintering habitat. In Massachusetts, 50 percent of the shallow freshwater marshes existing in 1951 were gone by 1971, while many deep marshes were dammed to create open ponds. Coastal marshes have not fared any better and have been victim to shortsighted exploitation. The situation has been made more critical by the invasion of undesirable Common Reed into much of the remaining habitat.

Extensive use of pesticides to combat mosquitoes, Spruce Budworms, and Gypsy Moths has occurred throughout the black duck's range, including Massachusetts. Other environmental contaminants are commonplace; the effects of acid rain on waterfowl are yet to be researched fully. Hunting pressure, especially in Canada, increased greatly during the 1970s but has slacked off since then. Urbanization, to which the Mallard readily adapts, is a major threat to this species.

The Mallard currently poses the greatest threat to the American Black Duck. The two species are closely related and have nearly identical calls and displays. They hybridize freely, producing fertile offspring. Over time, the Mallard is threatening to swamp the gene pool of the black duck. The situation has become more critical as Mallard numbers continue to increase (see account for that species). In Massachusetts, Division of Fisheries and Wildlife personnel determined that 15 percent of the black ducks they band along the coast during the winter show signs of hybridization. The actual proportion could be twice that. As wetland habitat in the Northeast continues to succumb to the deleterious effects of urbanization and Mallards proliferate, the future of genetically pure American Black Ducks is in jeopardy.

Breeding from Beaver ponds to salt marshes, the black duck is a territorial bird. The aggressive male will not permit another conspecific pair to nest on a pond he has claimed as his own. The exception is when ducks nest on islands. Perhaps this behavior is related to his drab coloration. For whatever reason, black duck nesting densities are lower than those of other duck species.

Nests are generally located under cover of wooded thickets, conifers, or thick herbaceous growth. In areas subject to flooding, black ducks have adapted to nesting on stumps, snags, or low tree crotches and cavities. Egg laying in Massachusetts begins in late March. Clutches range from seven to twelve eggs and are laid in scrapes lined with vegetation. Most of the dark down is added

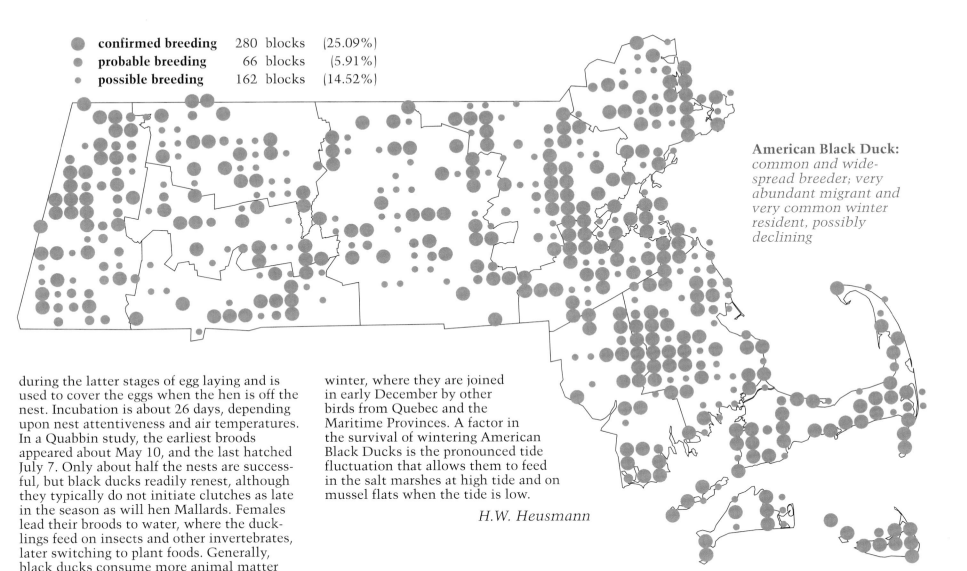

- **confirmed breeding** 280 blocks (25.09%)
- **probable breeding** 66 blocks (5.91%)
- **possible breeding** 162 blocks (14.52%)

American Black Duck: *common and widespread breeder; very abundant migrant and very common winter resident, possibly declining*

during the latter stages of egg laying and is used to cover the eggs when the hen is off the nest. Incubation is about 26 days, depending upon nest attentiveness and air temperatures. In a Quabbin study, the earliest broods appeared about May 10, and the last hatched July 7. Only about half the nests are successful, but black ducks readily renest, although they typically do not initiate clutches as late in the season as will hen Mallards. Females lead their broods to water, where the ducklings feed on insects and other invertebrates, later switching to plant foods. Generally, black ducks consume more animal matter than Mallards. In Quabbin, broods averaged seven ducklings (range four to twelve) on the date they were first seen and averaged five ducklings (range four to twelve) on the last date seen. The young first fly at about 60 days old.

In the fall, few Massachusetts black ducks move south. Most shift to the coast for the winter, where they are joined in early December by other birds from Quebec and the Maritime Provinces. A factor in the survival of wintering American Black Ducks is the pronounced tide fluctuation that allows them to feed in the salt marshes at high tide and on mussel flats when the tide is low.

H.W. Heusmann

Mallard
Anas platyrhynchos

Egg dates: April 1 to late July.

Number of broods: one; may re-lay if first attempt fails.

Just about everybody knows the drake Mallard with his distinctive plumage, but his brown mate frequently goes unnamed, although both sexes are the ducks fed by park visitors. Mallards breed throughout the Northern Hemisphere and are now widespread and common in the Commonwealth. However, accounts by early ornithologists indicate that the Mallard was rare in Massachusetts prior to the 1930s. This population increase has had a negative impact on the American Black Duck (see account for that species).

Mallards, like Canada Geese, were brought here by waterfowlers who used them as live decoys or call ducks until that practice was outlawed in 1935. The liberated birds adapted readily to the lakes and ponds of Massachusetts, especially in areas where they were fed. There is some evidence that Mallards also have expanded their range eastward from the Midwest in response to the forest clearing and pond construction that occurred during the past 200 years. Numbers have risen steadily since 1950, with a sharper increase after 1970.

Pair formation in Mallards is accomplished through a series of ritualized displays. Females quack and make several special vocalizations when they are feeding or communicating with ducklings. Males make similar but coarser sounds. While there is some migration of Mallards into our state in the spring, the bulk of the birds are year-round residents that winter at some 200 sites where people feed them. These birds move into surrounding areas to nest. True cosmopolites, Mallards utilize any number of nesting sites from fields, wooded edges, Muskrat houses, lidless Wood Duck boxes, and tree crotches to grape arbors and cemeteries. The surrounding habitat may be Beaver ponds, deep marshes, meandering streams, agricultural areas, suburbs, city parks, or coastal salt marshes. It is this adaptability that has made the Mallard so successful.

Mallard nesting begins in early April. The hen makes a nest bowl and adds fragments of nearby vegetation, finishing the structure with down as the clutch nears completion.

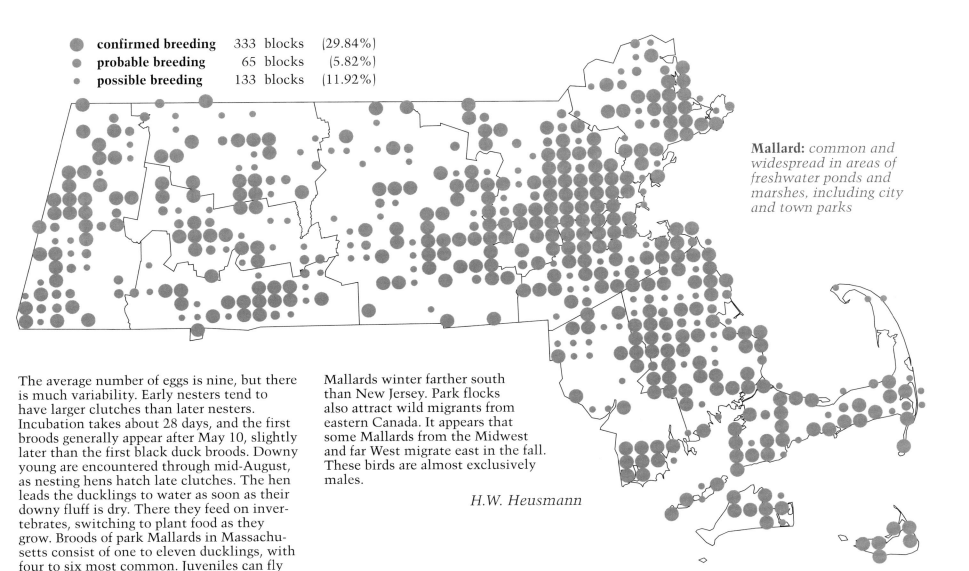

- ● **confirmed breeding** 333 blocks (29.84%)
- ● **probable breeding** 65 blocks (5.82%)
- · **possible breeding** 133 blocks (11.92%)

Mallard: *common and widespread in areas of freshwater ponds and marshes, including city and town parks*

The average number of eggs is nine, but there is much variability. Early nesters tend to have larger clutches than later nesters. Incubation takes about 28 days, and the first broods generally appear after May 10, slightly later than the first black duck broods. Downy young are encountered through mid-August, as nesting hens hatch late clutches. The hen leads the ducklings to water as soon as their downy fluff is dry. There they feed on invertebrates, switching to plant food as they grow. Broods of park Mallards in Massachusetts consist of one to eleven ducklings, with four to six most common. Juveniles can fly from 49 to 60 days after hatching.

In late summer, young and old tend to aggregate on ponds and lakes. In eastern Massachusetts, drakes frequently move into urban parks to undergo their annual molts. Females and juveniles also tend to head for these sites in the fall. A few birds go south, but not very far, and not many Massachusetts Mallards winter farther south than New Jersey. Park flocks also attract wild migrants from eastern Canada. It appears that some Mallards from the Midwest and far West migrate east in the fall. These birds are almost exclusively males.

H.W. Heusmann

Blue-winged Teal
Anas discors

Egg dates: May 4 to late July.

Number of broods: one; may re-lay if first attempt fails.

This small teal characteristically is a breeding bird of the shallow marshes and sloughs of central North America, but it also nests in suitable wetlands to the east from the Maritime Provinces to coastal salt marsh ponds in North Carolina. In Massachusetts, most nesting records in recent years have been from coastal marshes of Essex County or in the Sudbury-Concord River valley, but there have been scattered reports of suspected or confirmed breeding throughout the state, including the ponds on Monomoy Island.

Despite its tame and confiding nature, the Blue-winged Teal remains one of the most abundant of dabbling ducks in North America. The relative success of this species, even in the face of diminishing wetland habitat, has been due both to its ability to breed on small wetlands and to its early autumn migration. It is one of the first duck species to depart in the fall, passing through Florida by early September and thus avoiding most of the hunting season in the United States.

Blue-winged Teal become plentiful during spring migration. Pair formation takes place during this time or on the wintering grounds. It may be late April before pairs reach their breeding grounds, after most other species of ducks have initiated nesting. They select aquatic territories, which the drakes aggressively defend against conspecifics.

During this activity, the normally quiet birds become quite noisy. Males utter excited peeping notes, and the females have a low quack, most frequently given when the nest or brood is disturbed. Hens seek suitable nesting cover, usually areas of low grass or sedge near water but sometimes a distance from it. One egg is laid daily until a clutch of eleven or twelve is reached. Incubation takes 23 to 24 days. Many nests are destroyed by predators. When that happens, renesting usually occurs, but the clutches are smaller. Only one brood is raised annually. As with other ducks, hens do all the incubating and brood rearing. The females lead the ducklings to shallow wetlands with an abundance of

- confirmed breeding   18 blocks   (1.61%)
- probable breeding   11 blocks   (0.99%)
- possible breeding   11 blocks   (0.99%)

Blue-winged Teal: *uncommon and local in eastern marshes and refuge impoundments; rare in central and western areas*

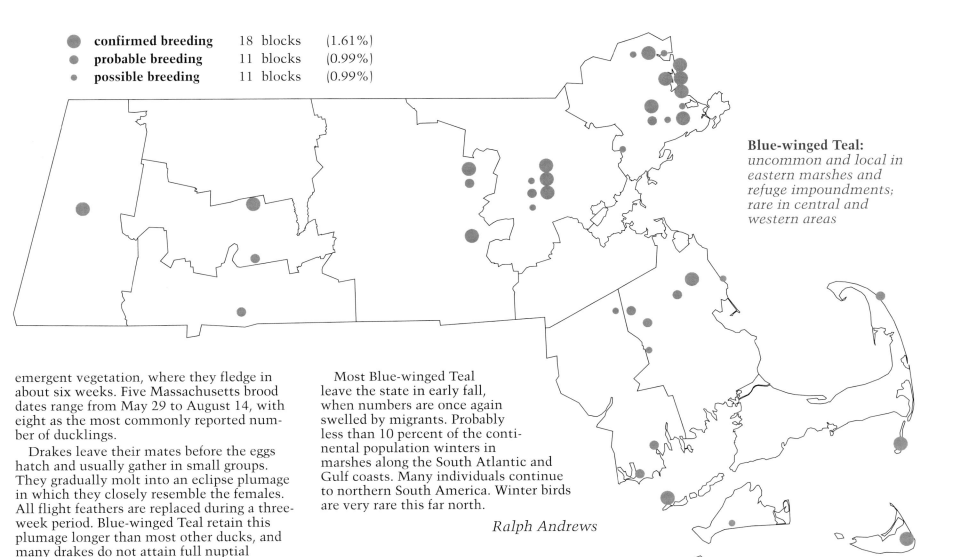

emergent vegetation, where they fledge in about six weeks. Five Massachusetts brood dates range from May 29 to August 14, with eight as the most commonly reported number of ducklings.

Drakes leave their mates before the eggs hatch and usually gather in small groups. They gradually molt into an eclipse plumage in which they closely resemble the females. All flight feathers are replaced during a three-week period. Blue-winged Teal retain this plumage longer than most other ducks, and many drakes do not attain full nuptial plumage until late winter.

The food of this species consists primarily of duckweed and the vegetative parts of aquatic plants and seeds of various sedges, grasses, smartweed, and other wetland plants. In summer larger amounts of insects and mollusks are consumed. Young ducklings feed almost entirely on small invertebrates.

Most Blue-winged Teal leave the state in early fall, when numbers are once again swelled by migrants. Probably less than 10 percent of the continental population winters in marshes along the South Atlantic and Gulf coasts. Many individuals continue to northern South America. Winter birds are very rare this far north.

Ralph Andrews

Northern Shoveler
Anas clypeata

Egg dates: mid-May to early July.

Number of broods: one; may re-lay if first attempt fails.

This medium-sized dabbling duck, commonly called "spoonbill" by waterfowlers, is distinctive in profile due to its large, wide, specialized bill, which has comblike teeth used for sifting small crustaceans and other invertebrates from shallow waters and soft bottoms. However, like most other ducks, shovelers also consume large quantities of small seeds in fall and winter.

This species breeds throughout much of the Northern Hemisphere. In North America, it is primarily a breeding bird of western prairie sloughs and potholes. It has become increasingly common in the Northeast, where small numbers breed in fresh and brackish coastal ponds and shallow impoundments. In Massachusetts, a few pairs have nested in recent years at the Monomoy and Parker River national wildlife refuges. Shovelers are more common during migration, particularly in fall.

Like Blue-winged Teals, Northern Shovelers arrive on their breeding grounds, usually in pairs, after many other ducks have begun nesting. They are less vocal than most duck species. Females utter weak quacks, and males make low grunting notes, particularly during courtship and territorial defense. As with most dabbling ducks, the nest is in a shallow depression in a grassy area, usually close to the wetlands, where shoveler pairs loaf and broods are raised, but may be some distance from the water. One egg is laid each day until a clutch of ten to fifteen is complete. Incubation takes 24 to 25 days. If predators disrupt nests, hens will renest but will lay smaller clutches. Only a single brood is raised. The young soon develop the distinctive bill. They fly from 40 to 60 days after hatching. Females with small young have been recorded as late as mid-July at the Stage Island pool at Parker River National Wildlife Refuge (Gavutis). On Monomoy, a hen with seven young was observed on July 5 (Humphrey).

Drakes have no role in hatching eggs or raising young. They generally desert their mates soon after incubation starts, gather with other drakes of their own and other species on large wetlands with ample emergent cover, and molt into a female-like eclipse plumage that persists for about three months before it is replaced with the colorful nuptial plumage. All flight feathers are

- ● **confirmed breeding** 3 blocks (0.27%)
- ● **probable breeding** 0 blocks (0.00%)
- ● **possible breeding** 0 blocks (0.00%)

Northern Shoveler: *rare and very local at Parker River and Monomoy national wildlife refuges*

molted at one time, and the birds are secretive for the three or four weeks that they are flightless. Hens have their flightless period while they are rearing their broods.

Fall migrants appear as early as late August and reach peak numbers in early October. They can be found in suitable habitats inland as well as on the coast. Most have departed by mid-November for the wintering grounds, which extend from the southern United States south to northern South America. Winter records are rare in Massachusetts.

Ralph Andrews

Northern Pintail
Anas acuta

Egg dates: April 11 to May 22.

Number of broods: one; may re-lay if first attempt fails.

The Northern Pintail is rightfully thought of as a prairie duck. Its greatest numbers occur in the mixed prairie region of Canada and the adjacent tier of northern states, although it breeds throughout much of the Arctic as well. Although there are only a few restricted breeding areas in eastern Canada, there is some evidence that the pintail, like several other duck species, has been extending its range eastward. Here in Massachusetts, pintails nest at the coastal Parker River and Monomoy national wildlife refuges. In the late 1950s, about thirty ducklings were released at the Ipswich River Wildlife Sanctuary in Topsfield, but there is no direct evidence to link those birds to the few scattered breeders now in the Commonwealth.

Pintails are rare to uncommon spring migrants in the state. Most pass through in March and April. Pair formation, accompanied by ritualized displays, occurs during the winter. Pintails tend to nest in open areas of short vegetation, often farther from water than other ducks. The coastal habitat of Massachusetts is a somewhat unusual nesting habitat, but it is not unprecedented because pintails do nest in coastal areas of Alaska.

Hens will use a variety of vegetative covers for nesting and lay an average clutch of seven to eight gray-buff to pale olive eggs. Incubation lasts 22 to 23 days. During the Atlas period, a female with ten young was observed on June 14 at Parker River National Wildlife Refuge. Post-Atlas records from this area are June 13 (a female with eleven young) and

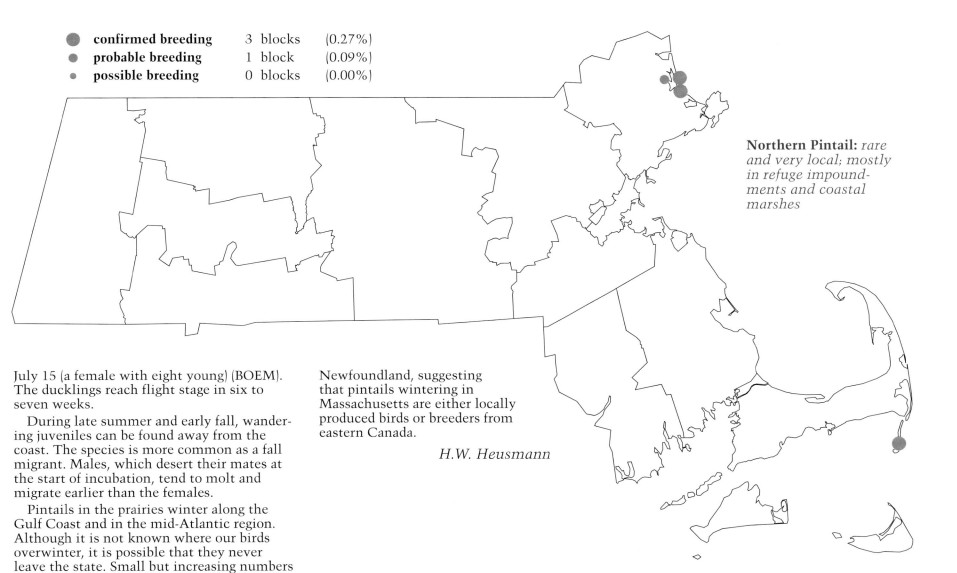

- confirmed breeding 3 blocks (0.27%)
- probable breeding 1 block (0.09%)
- possible breeding 0 blocks (0.00%)

Northern Pintail: *rare and very local; mostly in refuge impoundments and coastal marshes*

July 15 (a female with eight young) (BOEM). The ducklings reach flight stage in six to seven weeks.

During late summer and early fall, wandering juveniles can be found away from the coast. The species is more common as a fall migrant. Males, which desert their mates at the start of incubation, tend to molt and migrate earlier than the females.

Pintails in the prairies winter along the Gulf Coast and in the mid-Atlantic region. Although it is not known where our birds overwinter, it is possible that they never leave the state. Small but increasing numbers of pintails winter on Cape Cod in Barnstable and Yarmouth. A few of these are banded each winter, and some return in subsequent years. Cape birds do not appear to be from the prairies because no banded birds have been reported from that area. However, recoveries have occurred in Ontario, Quebec, and Newfoundland, suggesting that pintails wintering in Massachusetts are either locally produced birds or breeders from eastern Canada.

H.W. Heusmann

Green-winged Teal
Anas crecca

Egg dates: mid-April to late June.

Number of broods: one; may re-lay if first attempt fails.

Although Green-winged Teal are regular breeders and present in Massachusetts in small numbers year-round, they are still considered by most standards to represent a migrant species. Nationally, there has been an upward trend in the number of Green-winged Teal, with the East Coast showing the greatest increase. They breed in small numbers at scattered locations from the outer Cape and Essex County west across the state to Berkshire County. In all seasons, this species shows a predilection for fresh ponds and channels of water with a thick edge of vegetation.

Pair formation begins on the wintering grounds and continues during spring migration, which starts in March. Two or more males may be seen courting a female with vigorous head bobbing and loud whistling notes. The females utter a weak quack. Green-winged Teal prefer to nest in thick, grassy upland areas adjacent to water but will sometimes nest some distance from it. The hen builds a well-concealed nest of soft grasses lined with down in a shallow depression. Usually eight to twelve eggs are laid one per day, with the 21-day incubation period beginning when the last egg is laid. The hen covers the nest when leaving to feed or rest. If the first nest is destroyed, she may renest, although the clutch size is normally smaller. Shortly after hatching, the downy young are led to water by the female. During the rearing process, the brood keeps very close to aquatic cover, seldom venturing out on open water. Four Massachusetts brood dates range from June 7 to July 18. The number of ducklings per brood was four to nine (BOEM, TC). The young fledge at about 44 days old.

The male deserts the female as soon as the clutch is complete. Drakes gather together on a nearby body of water to molt into somber brown eclipse plumage. During this period, they are flightless for about three weeks. They begin to attain their nuptial plumage in late September and October but may have not acquired their full color prior to departure for the wintering grounds.

In late summer, Green-winged Teal gather in sizable flocks that are swelled by migrants from northern breeding areas. Numbers are greatest in October, with many remaining until November. Most depart southward with the advent of freezing weather. The winter

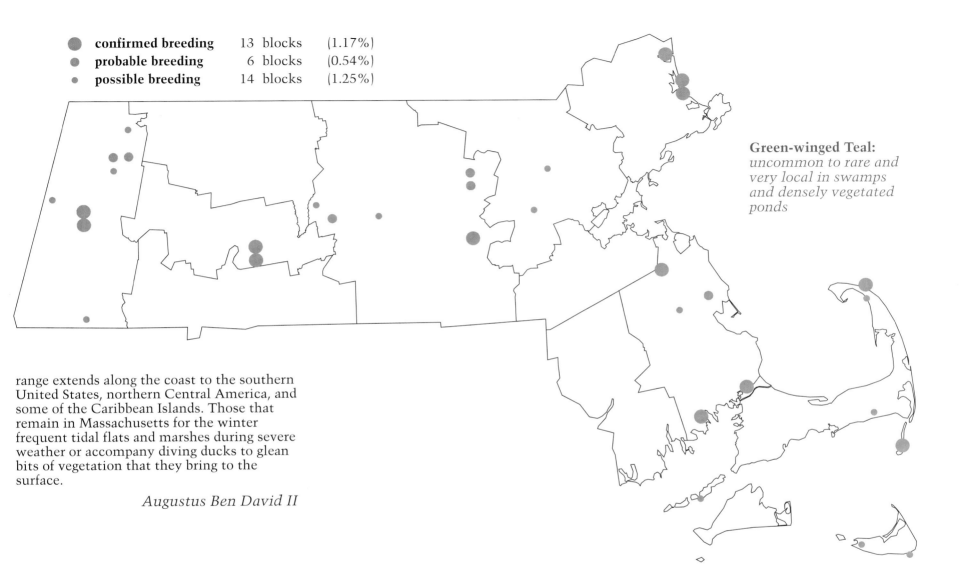

Green-winged Teal: *uncommon to rare and very local in swamps and densely vegetated ponds*

range extends along the coast to the southern United States, northern Central America, and some of the Caribbean Islands. Those that remain in Massachusetts for the winter frequent tidal flats and marshes during severe weather or accompany diving ducks to glean bits of vegetation that they bring to the surface.

Augustus Ben David II

Ring-necked Duck
Aythya collaris

Egg dates: May.

Number of broods: one; may re-lay if first attempt fails.

Prior to 1930, the Ring-necked Duck nested to the north and west of New England and eastern Canada and was seldom observed in Massachusetts, even during migration. Forbush considered it the "rarest duck in the Northeast." Between 1930 and 1935, it became increasingly common as a spring and fall migrant. In 1936, nesting was recorded in Pennsylvania and Maine, and in subsequent years a breeding population was established throughout eastern Canada and northern New England. Ring-necked Ducks became most numerous in Maine, New Brunswick, Ontario, and Quebec. They are best known in Massachusetts as common migrants, but there have been a few breeding records. Nesting occurred in Concord in 1947 and 1951, and during the Atlas period there was a single confirmation from Ashfield.

Inappropriately named for the inconspicuous chestnut-colored band around the neck, the bird should be called "Ring-billed Duck" for the prominent white ring on its bluish gray bill. It is a fairly quiet species. The male utters weak whistling notes, and the female makes faint purring sounds. In display, the drake raises his crown feathers, holds his head erect, and then throws it back. Pair formation may begin in the fall and winter but usually occurs during the spring.

Spring migrants generally appear in Massachusetts during March, with numbers peaking in April. Small groups regularly linger until late May or even early June, but virtually all of these individuals eventually depart. The females tend to return to their natal areas to nest, and pairs tolerate one another so hens may nest in closer proximity than do many other duck species. Preferred nesting habitats are swamps, bogs, or small ponds with marshy borders. The habitat selected by the pair that bred in 1977 during the Atlas period was a Beaver pond surrounded by dead trees and marsh grasses. Other species nesting at the pond included the Canada Goose and Hooded Merganser. Suspicions about the possibility of breeding were raised when two pairs of Ring-necked Ducks were recorded there during May and June of 1976. In 1978, two pairs again were observed on the pond in June, but no nesting activities were discovered. Since that time, the Beavers have disappeared and the water level has become too low for nesting waterfowl.

The nest is a hollow lined with grass, other plant materials, and down. It is always near or surrounded by water and may be concealed in tussocks or clumps of marsh plants. Island nests may have vegetation ramps to make access easier for the hen. In Maine, the laying of eggs usually starts the second week of May and peaks around May 23 (range May 1 to

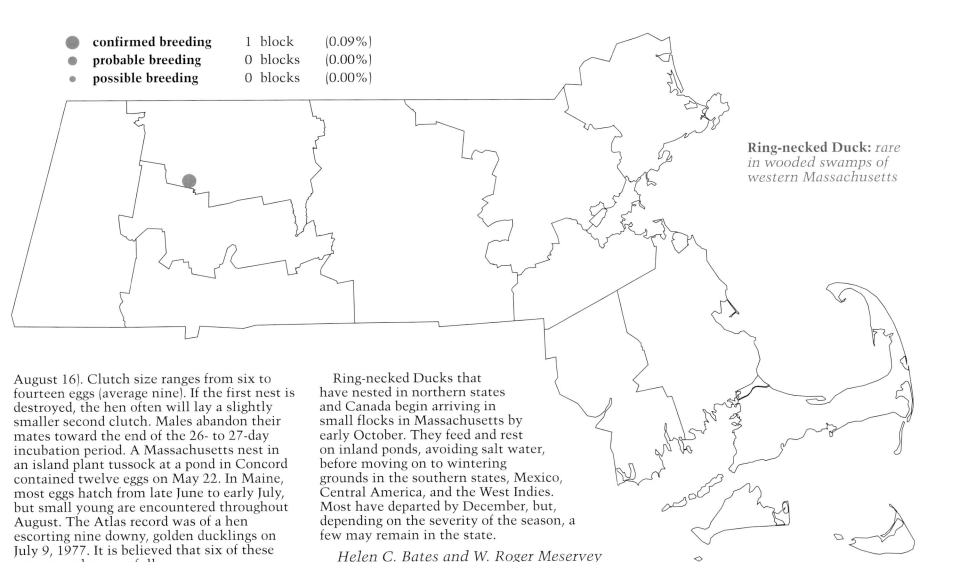

- confirmed breeding 1 block (0.09%)
- probable breeding 0 blocks (0.00%)
- possible breeding 0 blocks (0.00%)

Ring-necked Duck: *rare in wooded swamps of western Massachusetts*

August 16). Clutch size ranges from six to fourteen eggs (average nine). If the first nest is destroyed, the hen often will lay a slightly smaller second clutch. Males abandon their mates toward the end of the 26- to 27-day incubation period. A Massachusetts nest in an island plant tussock at a pond in Concord contained twelve eggs on May 22. In Maine, most eggs hatch from late June to early July, but small young are encountered throughout August. The Atlas record was of a hen escorting nine downy, golden ducklings on July 9, 1977. It is believed that six of these were reared successfully.

The hens brood the young when they are small and usually remain with their ducklings while they molt and replace their flight feathers. The young begin feeding on invertebrates, switching to plant foods as they mature. They fledge at six to eight weeks of age.

Ring-necked Ducks that have nested in northern states and Canada begin arriving in small flocks in Massachusetts by early October. They feed and rest on inland ponds, avoiding salt water, before moving on to wintering grounds in the southern states, Mexico, Central America, and the West Indies. Most have departed by December, but, depending on the severity of the season, a few may remain in the state.

Helen C. Bates and W. Roger Meservey

Common Eider
Somateria mollissima

Egg dates: late April to early July.

Number of broods: one; may re-lay if first attempt fails.

The Common Eider is a circumpolar breeder, nesting along coasts and on offshore islands. In the western North Atlantic, it breeds south regularly as far as the Casco Bay area of Maine. This large robust duck is most familiar as a wintering species in Massachusetts but is also now known to breed in the state in small numbers. There has been a dramatic overall population increase since the 1930s.

In 1975, Philip Stanton of Framingham State College began a program to establish the eider as a nesting species on the Elizabeth Islands. He collected eggs and ducklings from Casco Bay and hand-reared the birds, releasing them at various ages on Penikese Island for four years in succession. Some of these eiders survived and bred in subsequent seasons, producing a small nesting population on the southernmost islands of the Elizabeth chain. In 1984, a nest was discovered at Bird Island in Marion, and this indicated that the new range was expanding.

Small groups of eiders always have been present throughout the nesting season in the vicinity of Boston Harbor, Nantucket, Martha's Vineyard, Cape Cod, and the rocky islands along the North Shore. A female eider with three young seen in Boston Harbor in 1982 almost certainly represents a natural range extension. It is possible that other breeding populations will result from these resident groups.

The heaviest migration in spring occurs during late March and early April. Males utter cooing notes and moans as they posture during courtship, and hens produce hoarse quacking sounds. Common Eiders may nest singly or in dense colonies at variable distances from the shore. Nests are ground depressions lined with some dried vegetation and a large quantity of insulating down pulled from the females' breasts. They are often concealed by clumps of grass, bushes, or rocks. A normal clutch contains three to five olive green eggs, and incubation lasts an average of 28 days. Eiders in Maine begin nesting in late April, and the transplanted Massachusetts birds have adhered to the same schedule. Three Massachusetts nests each contained four eggs when observed on May 13 and May 30 (BOEM). Drakes abandon the hens at the end of laying or during early incubation and gather in their own flocks to molt.

Upon hatching, the brood is led to tidal pools along the ocean edge to begin feeding. Characteristically, Common Eiders merge their broods into large groups called crèches, with a number of females guarding the ducklings and brooding them at night while they are small. Brood dates for the state include July 6 (a female and six young) and July 24 (2 hens with five young) (BOEM). The juveniles require at least eight weeks to fledge. By that time, the adults have molted and renewed their flight feathers.

Common Eiders feed near rocky shores, diving for mollusks, worms, echinoderms, and crustaceans. Blue Mussels and Common Periwinkles are important foods. By late September, adults and immatures are flock-

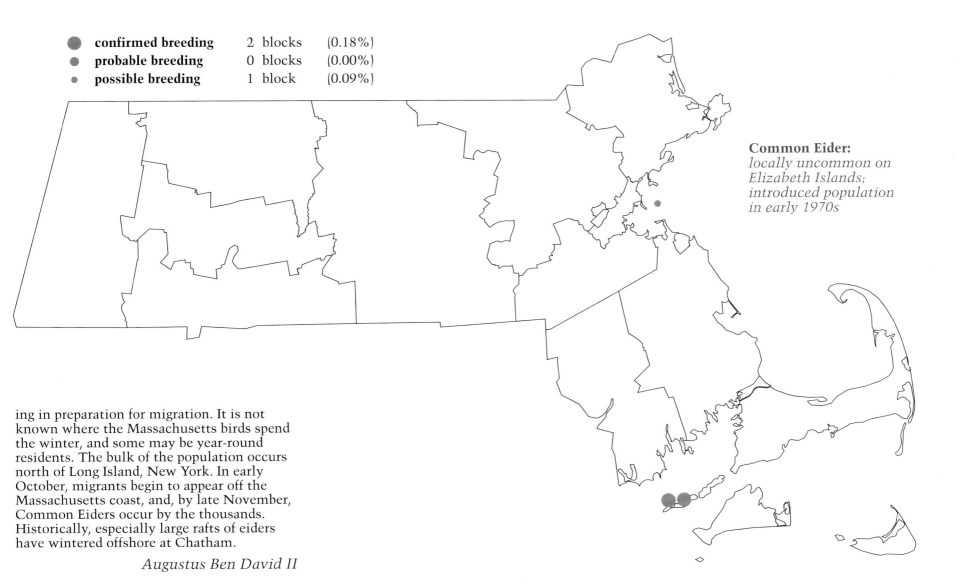

- confirmed breeding 2 blocks (0.18%)
- probable breeding 0 blocks (0.00%)
- possible breeding 1 block (0.09%)

Common Eider:
locally uncommon on Elizabeth Islands; introduced population in early 1970s

ing in preparation for migration. It is not known where the Massachusetts birds spend the winter, and some may be year-round residents. The bulk of the population occurs north of Long Island, New York. In early October, migrants begin to appear off the Massachusetts coast, and, by late November, Common Eiders occur by the thousands. Historically, especially large rafts of eiders have wintered offshore at Chatham.

Augustus Ben David II

Hooded Merganser
Lophodytes cucullatus

Egg dates: mid-March to June 10.

Number of broods: one.

In Massachusetts, Hooded Mergansers have most often been found breeding in the central parts of the Commonwealth, with fewer records from Berkshire, Essex, Bristol, and Plymouth counties, and none from the outer Cape and the Islands. However, they are probably more common than the records would indicate. They are quiet birds, wary and secretive during the breeding season, and their hoarse grunts and chatters are seldom heard. Their preferred habitats are streams, ponds, and freshwater marshes with lots of shrub cover, where they select nest sites in natural tree cavities or Wood Duck nesting boxes. The cautious females use dense cover to keep their broods out of sight, and they doubtless often go undetected even by observant humans.

Hooded Mergansers arrive at their breeding areas as soon as the ice starts to melt. Spring migration peaks in late March, and by late April most northern migrants have passed through. In southeastern Massachusetts, many females start laying eggs in March. They do not bring any nest material into a cavity or box but will rearrange wood chips, decayed wood, leaves, or old nest material present into a shallow, bowl-shaped depression. Eggs (most often ten to twelve) are usually laid at 1- or 2-day intervals, the females covering their eggs with any nest material present before leaving the nest. Often some of the earliest eggs laid in a nest will freeze and crack and will never hatch. Once most of the clutch is laid, the female pulls out down from her breast to cover the eggs, which helps to protect them from freezing before incubation begins and to keep them warm while she is off the nest feeding. Upon her return, she uncovers the eggs by moving the down and reforming the ring around the nest depression.

Male Hooded Mergansers disappear from the breeding area soon after the females start incubation, making the possibility of re-nesting after an initial failure unlikely. Hatching begins after about 32 days of incubation, and the young are usually dry and strong enough to leave the nest within 24 hours of hatching. The female leaves the nest and calls softly until all the young have managed to hop or climb out of the nest cavity and have jumped to the water or ground below.

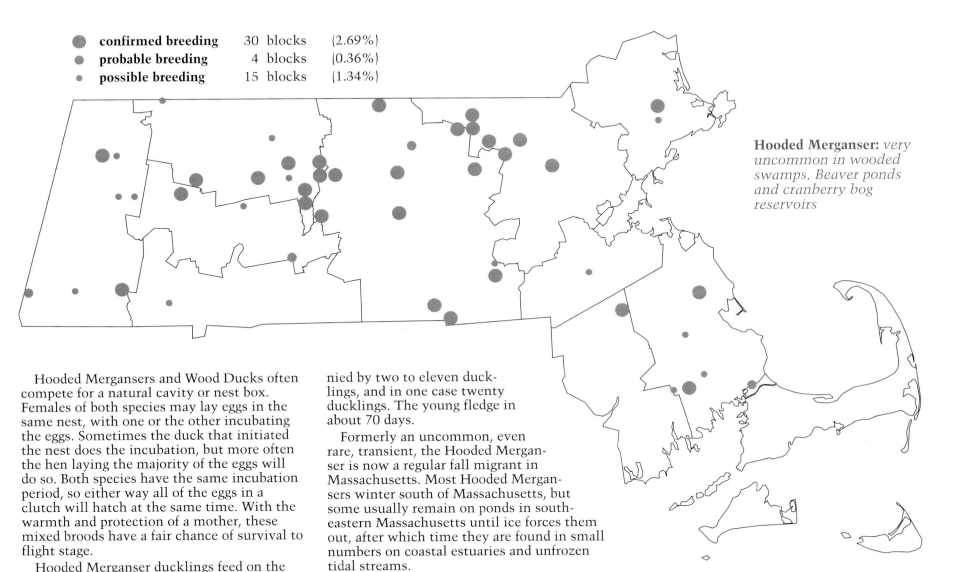

Hooded Merganser: *very uncommon in wooded swamps, Beaver ponds and cranberry bog reservoirs*

● confirmed breeding	30 blocks	(2.69%)
● probable breeding	4 blocks	(0.36%)
· possible breeding	15 blocks	(1.34%)

Hooded Mergansers and Wood Ducks often compete for a natural cavity or nest box. Females of both species may lay eggs in the same nest, with one or the other incubating the eggs. Sometimes the duck that initiated the nest does the incubation, but more often the hen laying the majority of the eggs will do so. Both species have the same incubation period, so either way all of the eggs in a clutch will hatch at the same time. With the warmth and protection of a mother, these mixed broods have a fair chance of survival to flight stage.

Hooded Merganser ducklings feed on the surface of the water or by diving. For the first few weeks of life, they forage for invertebrates such as mayfly and dragonfly larvae. Fish may make up about half the diet of adults, with crustaceans, amphibians, and aquatic insects often making up the other half. Seven Massachusetts brood dates range from May 19 to June 29, with hens accompanied by two to eleven ducklings, and in one case twenty ducklings. The young fledge in about 70 days.

Formerly an uncommon, even rare, transient, the Hooded Merganser is now a regular fall migrant in Massachusetts. Most Hooded Mergansers winter south of Massachusetts, but some usually remain on ponds in southeastern Massachusetts until ice forces them out, after which time they are found in small numbers on coastal estuaries and unfrozen tidal streams.

Richard E. Turner

Common Merganser
Mergus merganser

Egg dates: late April to June.

Number of broods: one.

Believed to have bred widely in the state before 1900, the Common Merganser then became casual or absent as a summer resident. A few pairs now breed locally at Quabbin Reservoir and along the larger swift-moving rivers in the western hill country. Migrant and wintering numbers have increased to common or abundant and have been fairly stable for the last 40 years. Common Mergansers frequent large lakes and ponds until they freeze over and then resort to open water rapids and the mouths of large rivers, or else they move farther south.

Inland peaks of up to 300 occur in March on large rivers as the birds begin moving north. The few that stay to breed seek out wild stretches of the Deerfield, Westfield, and Farmington rivers or large undisturbed bodies of water such as Quabbin Reservoir.

The call note, usually given when the birds are alarmed and about to flush, is a low, hoarse growl or croak. During courtship, soft guttural purrings are made, and the mating dance itself is energetic and elaborate. One or several drakes rush about, raise their salmon-tinted breasts or cock and spread their tails, kick water as they expose their bright red legs, and pursue and frequently strike rivals with their bills. There is much splashing and brief diving. When the drake is inactive, the female sometimes pursues him with head and neck held low in the water.

Nest sites are usually large holes in trees, in banks, or among rocks near or over water, occasionally in dense cover on the ground of small islands, or even in old hawk nests. By mid-May, the single clutch of eggs numbering six to seventeen (usually nine to twelve) is laid in a nest consisting wholly of down if sheltered in a cavity, or of assorted grasses and twigs lined with down and mixed with white feathers when located elsewhere.

After at least a month-long incubation, the young tumble out of the nest to the ground and accompany the female to feed in the shallows along the banks of ponds or in river pools. On June 22, 1978, a family with five small young was seen on the Westfield River in Huntington (Lynes), and on July 1 of the same year, five half-grown young were seen on the Farmington River in Sandisfield. There is a post-Atlas record of four full-grown young on July 26, 1987, on the Westfield River in Cummington (Gagnon). The males do not help in the incubation, and later in eclipse plumage they are found alone in the same habitats. The young reach flight stage in 60 to 70 days. In the early fall, the fully grown young, and perhaps both parents, visit other nearby bodies of water and may appear over a wide area.

The Common Merganser is primarily a fish eater, but insects and vegetation are also consumed. Minnows are the main prey, but this duck will capture any slow-moving warm-water fish. In fact, it plays an important role in controlling fish populations. On the coast, Blue Mussels may be eaten, shell and all. Common Mergansers do feed in brackish water but are seldom found on the open sea or in pure salt water. They are powerful divers and underwater swimmers, using both feet and wings to propel themselves in pursuit of food. As with most birds

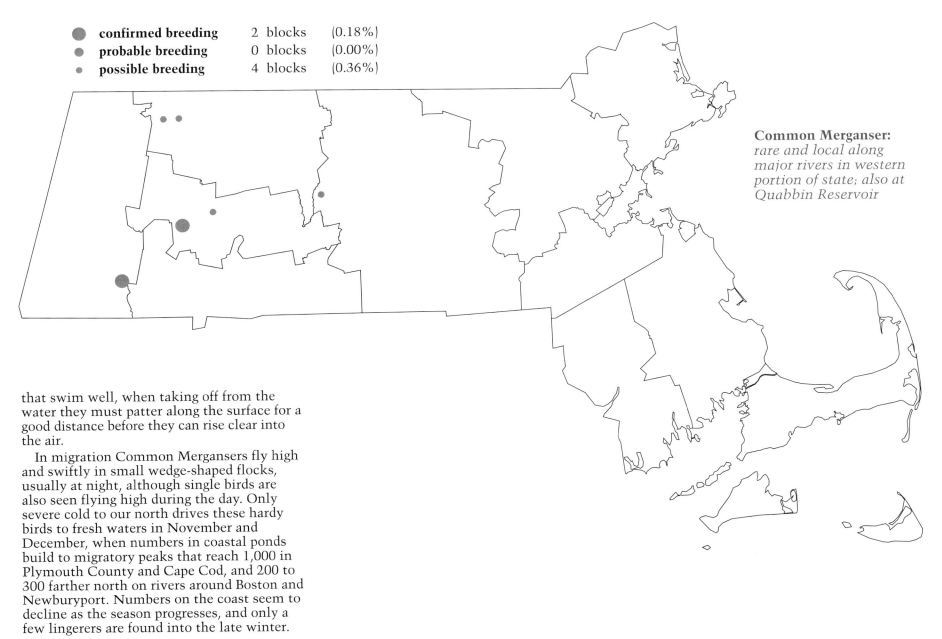

- **confirmed breeding**   2 blocks   (0.18%)
- **probable breeding**   0 blocks   (0.00%)
- **possible breeding**   4 blocks   (0.36%)

Common Merganser: *rare and local along major rivers in western portion of state; also at Quabbin Reservoir*

that swim well, when taking off from the water they must patter along the surface for a good distance before they can rise clear into the air.

In migration Common Mergansers fly high and swiftly in small wedge-shaped flocks, usually at night, although single birds are also seen flying high during the day. Only severe cold to our north drives these hardy birds to fresh waters in November and December, when numbers in coastal ponds build to migratory peaks that reach 1,000 in Plymouth County and Cape Cod, and 200 to 300 farther north on rivers around Boston and Newburyport. Numbers on the coast seem to decline as the season progresses, and only a few lingerers are found into the late winter.

Seth Kellogg

Red-breasted Merganser
Mergus serrator

Egg dates: June to July.

Number of broods: one.

Although the Red-breasted Merganser has been alleged to breed in Massachusetts since 1877, apparently its nest never has been found, and breeding evidence has always been based on the discovery of broods of flightless young or the suggestive behavior of summering adults. Despite its rarity as a state breeding bird, this merganser is recorded consistently in the historical accounts as nesting locally along the southern coast of Massachusetts, primarily at Monomoy Island and on Cape Cod along Nantucket Sound. There is a single breeding record for Essex County in 1916. The Atlas produced two confirmations: one from Duxbury in Plymouth County and one from Monomoy National Wildlife Refuge. Otherwise, the species is abundant as a coastal spring and fall migrant as well as a winter resident. A few nonbreeding individuals can generally be seen during the summer at scattered coastal localities.

By March and April, large numbers of northbound migrants join the wintering birds and gather in spectacular staging aggregations in eastern Cape Cod Bay. Decidedly birds of open ocean or saltwater bays, "Salt-water Sheldrakes" seldom venture far inland, where they occur only as casual migrants on large lakes and reservoirs. In this season, small parties of mixed sexes congregate on quiet waters, where the drakes display with much neck stretching, bobbing, splashing, and calling. Pair bonds are often established before the birds depart from Massachusetts waters. The species apparently breeds in

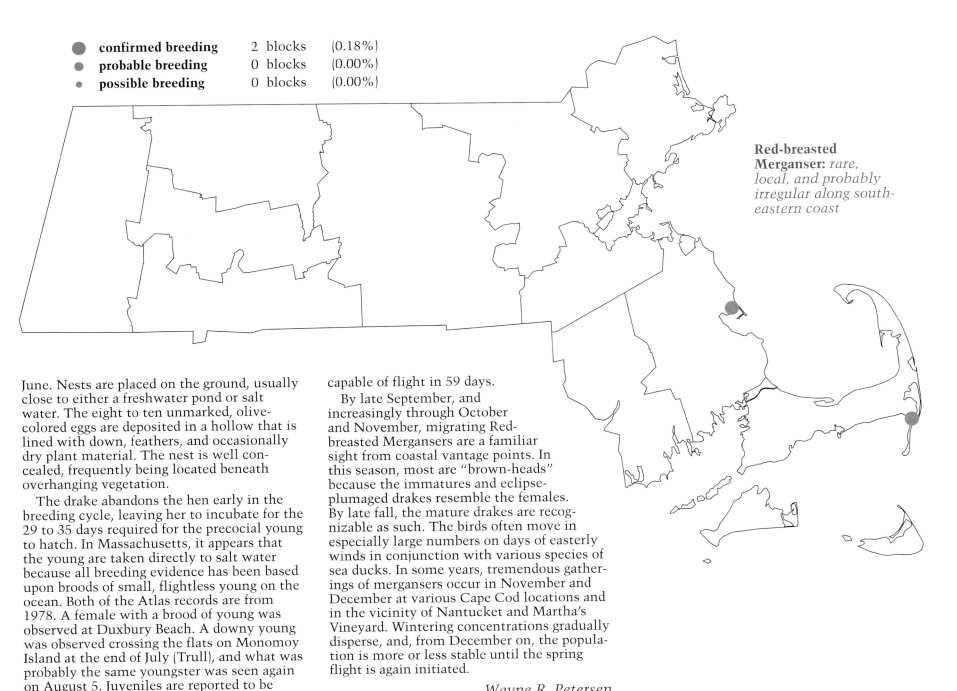

Red-breasted Merganser: *rare, local, and probably irregular along southeastern coast*

- confirmed breeding — 2 blocks (0.18%)
- probable breeding — 0 blocks (0.00%)
- possible breeding — 0 blocks (0.00%)

June. Nests are placed on the ground, usually close to either a freshwater pond or salt water. The eight to ten unmarked, olive-colored eggs are deposited in a hollow that is lined with down, feathers, and occasionally dry plant material. The nest is well concealed, frequently being located beneath overhanging vegetation.

The drake abandons the hen early in the breeding cycle, leaving her to incubate for the 29 to 35 days required for the precocial young to hatch. In Massachusetts, it appears that the young are taken directly to salt water because all breeding evidence has been based upon broods of small, flightless young on the ocean. Both of the Atlas records are from 1978. A female with a brood of young was observed at Duxbury Beach. A downy young was observed crossing the flats on Monomoy Island at the end of July (Trull), and what was probably the same youngster was seen again on August 5. Juveniles are reported to be capable of flight in 59 days.

By late September, and increasingly through October and November, migrating Red-breasted Mergansers are a familiar sight from coastal vantage points. In this season, most are "brown-heads" because the immatures and eclipse-plumaged drakes resemble the females. By late fall, the mature drakes are recognizable as such. The birds often move in especially large numbers on days of easterly winds in conjunction with various species of sea ducks. In some years, tremendous gatherings of mergansers occur in November and December at various Cape Cod locations and in the vicinity of Nantucket and Martha's Vineyard. Wintering concentrations gradually disperse, and, from December on, the population is more or less stable until the spring flight is again initiated.

Wayne R. Petersen

Ruddy Duck
Oxyura jamaicensis

Egg dates: late May to late July.

Number of broods: one; may re-lay if first attempt fails.

This exclusively North American species has an extensive breeding range but is a very uncommon nester in the eastern United States. There are breeding records for Washington County, Maine; Jamaica Bay Refuge and Montezuma National Wildlife Refuge, New York; and Brigantine National Wildlife Refuge, New Jersey. In Massachusetts, there is an old nesting record from Truro. During the Atlas period, Ruddy Ducks bred on the Parker River National Wildlife Refuge, and afterward the species also nested at Monomoy National Wildlife Refuge.

The primary breeding site at the Parker River refuge has been the Stage Island Pool, where a population of Ruddy Ducks has developed since the pool's construction in the 1950s. Generally, a few pairs and a brood or two can be observed there most summers. However, high-water-level management is probably critical to continued successful nesting by Ruddy Ducks. In 1978, possibly in response to declining deep-water emergent plant habitat at Stage Island, breeding activity was also noted for the first time in the refuge's North Pool. Habitat in the North Pool, however, has generally been of lower quality due to lower water levels and less flooding of emergent plants.

Ruddy Ducks are uncommon spring migrants and are generally seen in April, although local breeders may return in March. In summer, the adult male is striking and gaudily colored with a reddish chestnut breast and back and a blue bill. This coloration, coupled with the unusual inflatable air sack in the neck and the curious courtship antics, make the drake somewhat reminiscent of a pouter pigeon or a turkey gobbler. As he raises his peculiar quill-like tail over his back, puffs up, and pumps his bill up and down against his breast, he produces one of several courtship vocalizations, sounding as though he is choking. The drab females are generally silent.

The bulky, basketlike nests will float and are most often constructed about 8 inches above water level in emergent vegetation such as cattails. In the western United States sometimes an old coot nest is used. The six to ten (average eight) rough, granular, thick-shelled, white eggs are substantially larger than a Mallard's. Although the ruddy is one of our smaller ducks, the clutch often weighs two to three times as much as the 1-pound hen, and the females are sometimes parasitic on the nests of other ducks, or even grebes. Hens are very secretive around their nests, usually sliding down a small ramp and swimming away under water without ever being seen. Following a 23- to 26-day incubation period by the female, the extremely large and precocious young hatch and begin diving for food almost immediately, using their big feet most effectively. Unlike the situation with most other duck species, the drake often stays with the family, but the young, which seem to be born "old," are often on their own by the time they are half-grown.

In 1974, a female and five young were seen on July 20, and, in 1975, an adult with four young was observed at Parker River National Wildlife Refuge from July 1 to 26 (BOEM). Young Ruddy Ducks have been recorded in the Stage Island Pool as late as early to mid-August. Estimated production at the Parker

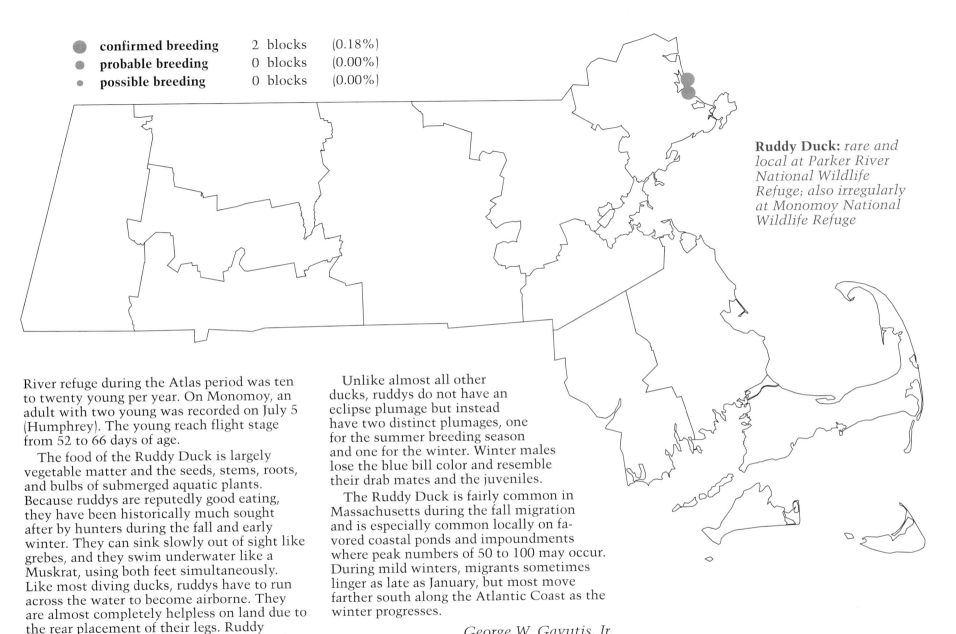

- ● **confirmed breeding**   2 blocks   (0.18%)
- ● **probable breeding**   0 blocks   (0.00%)
- · **possible breeding**   0 blocks   (0.00%)

Ruddy Duck: *rare and local at Parker River National Wildlife Refuge; also irregularly at Monomoy National Wildlife Refuge*

River refuge during the Atlas period was ten to twenty young per year. On Monomoy, an adult with two young was recorded on July 5 (Humphrey). The young reach flight stage from 52 to 66 days of age.

The food of the Ruddy Duck is largely vegetable matter and the seeds, stems, roots, and bulbs of submerged aquatic plants. Because ruddys are reputedly good eating, they have been historically much sought after by hunters during the fall and early winter. They can sink slowly out of sight like grebes, and they swim underwater like a Muskrat, using both feet simultaneously. Like most diving ducks, ruddys have to run across the water to become airborne. They are almost completely helpless on land due to the rear placement of their legs. Ruddy Ducks tend to fly low, even during migration, when they are most often seen traveling near dusk in medium to fairly large groups.

Unlike almost all other ducks, ruddys do not have an eclipse plumage but instead have two distinct plumages, one for the summer breeding season and one for the winter. Winter males lose the blue bill color and resemble their drab mates and the juveniles.

The Ruddy Duck is fairly common in Massachusetts during the fall migration and is especially common locally on favored coastal ponds and impoundments where peak numbers of 50 to 100 may occur. During mild winters, migrants sometimes linger as late as January, but most move farther south along the Atlantic Coast as the winter progresses.

George W. Gavutis, Jr.

Osprey
Pandion haliaetus

Egg dates: April 10 to July 12.

Number of broods: one; may re-lay if first attempt fails.

The North American breeding range of the Osprey, or Fish Hawk, is extensive and includes much of the northern United States and Canada as well as sites in the central and southern states. The species nests at various Atlantic Coast regions from Labrador and Newfoundland to Florida. Ospreys occur statewide as spring and fall migrants, but they have never been common here during the breeding season and until recently have been confined mainly to coastal Bristol County.

According to conservative population estimates from about 1900, there were 60 to 80 pairs of Ospreys in southeastern Massachusetts and eastern Rhode Island. Numbers reached a peak in the early 1940s. After World War II, eastern Osprey populations began a serious, well-documented decline, largely due to reproductive failure induced by organochlorine pesticides such as DDT in the environment. By 1964, only 11 pairs remained in the state. Once these pesticides were banned in 1972, the Osprey began to recover, aided by a program of erecting special poles with platforms to increase suitable nest sites. This latter factor has had a significant impact on the Massachusetts Osprey population, with over 95 percent of the pairs in the state since 1980 using these artificial sites. The result has been an increase in the number of nests in which young are reared as well as in the number of young fledged per nest (Poole 1989).

During the Atlas period, most confirmed nesting records were in Bristol and Dukes counties, with a few confirmations in Plymouth and Barnstable counties. In 1981, there were 45 nests in the Commonwealth. The first recent Nantucket nesting occurred in 1983. By 1985, one hundred forty-five young were produced at 88 active nests. While numbers have climbed appreciably, the range expansion has been much more limited because most young Ospreys return to breed close to their natal areas. It is hoped that eventually the Osprey will recolonize the entire coast and perhaps some inland sites.

Resident Ospreys arrive in the state from late March to early April, with males returning slightly earlier than the females. Males claim their old nest sites or select a location if they are first-time breeders. Favored natural nest sites are tall, living or dead trees in open marshy areas, but Ospreys also adapt readily to structures such as buoys and radio towers in addition to the nest poles. Although the nest sites are variable, they are always located near shallow fresh- or saltwater bodies, where the birds will have a ready supply of fish of various species. Ospreys are not truly colonial, but in favorable areas several pairs may nest amicably in close proximity.

Ospreys pair for life but will choose new mates if one partner fails to return. Courtship is often brief, involving dramatic aerial displays. The female does the nest construction, and the male gathers the necessary sticks, grass, seaweed, and sod. New materials may be added each year, and some nests measure 4 feet across by 3 feet deep, with a shallow central cup. Usually three eggs are laid, but two- or four-egg clutches are not rare. Some pairs will re-lay if their eggs are lost, but in Massachusetts late nesters and renesters generally have completed their clutches by the last week of May. Incubation, which lasts four to five weeks, commences with the first egg and is accomplished mostly by the female. She leaves the nest for brief periods to exercise and eat the food provided by her mate, and the latter may relieve her on the eggs during these times. Peak hatching dates are from late May to early June.

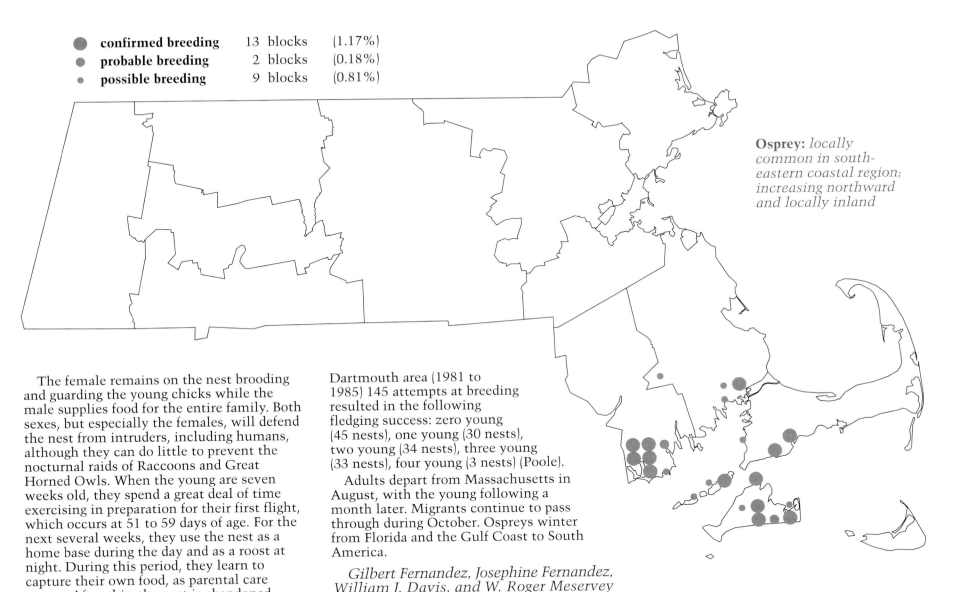

- **confirmed breeding** 13 blocks (1.17%)
- **probable breeding** 2 blocks (0.18%)
- **possible breeding** 9 blocks (0.81%)

Osprey: *locally common in southeastern coastal region; increasing northward and locally inland*

The female remains on the nest brooding and guarding the young chicks while the male supplies food for the entire family. Both sexes, but especially the females, will defend the nest from intruders, including humans, although they can do little to prevent the nocturnal raids of Raccoons and Great Horned Owls. When the young are seven weeks old, they spend a great deal of time exercising in preparation for their first flight, which occurs at 51 to 59 days of age. For the next several weeks, they use the nest as a home base during the day and as a roost at night. During this period, they learn to capture their own food, as parental care wanes. After this, the nest is abandoned, although the juveniles remain in the general vicinity. They are sexually mature at three years of age but may not breed until they are five years old.

Fledging success varies according to nest site quality and food supply. In the Westport-Dartmouth area (1981 to 1985) 145 attempts at breeding resulted in the following fledging success: zero young (45 nests), one young (30 nests), two young (34 nests), three young (33 nests), four young (3 nests) (Poole).

Adults depart from Massachusetts in August, with the young following a month later. Migrants continue to pass through during October. Ospreys winter from Florida and the Gulf Coast to South America.

Gilbert Fernandez, Josephine Fernandez, William J. Davis, and W. Roger Meservey

Northern Harrier
Circus cyaneus

Egg dates: April 26 to June 25.
Number of broods: one.

As with many open-country species, the Northern Harrier, commonly called the "Marsh Hawk," has declined over much of its former range in Massachusetts. Cape Cod and the Islands harbor the state's remaining nesting populations, although there still may be a few breeders in Hampden County. During the past century, development, reforestation, changing land uses, recreational impact, and other types of human disturbance have led to a decrease in the number of harriers in the Commonwealth. These birds are residents of open habitats, including cattail swamps, marshes, bogs, prairies, and agricultural land borders.

Nesting harriers begin arriving on their breeding grounds in late March in Massachusetts. Males are the first to return and establish territories in anticipation of the later arriving females. Northbound transients are observed during April.

Though Northern Harriers utilize territories of open dry grasslands, for the nest site they usually select areas within cattail swamps, marshes, or bogs, preferring wet areas for nesting. Nests are generally located near dense vegetation within the territory. Recent studies on Nantucket indicate that a stand of Highbush Blueberry is preferred. Harriers are sensitive to human disturbance during the early stages of nesting and may abandon a nest if they are bothered at this time.

Harriers are generally silent when they are not near the nest; however, during courtship displays and nest defense, they become quite vocal. During courtship, males, and often females as well, engage in sky-dancing displays, in which they fall from a height into a series of steep undulations. Aerial talon grappling is also a behavior related to territory defense.

The character of the site often determines the size of the nest. In wet areas, harriers may build nests of sticks, weeds, and grasses up to 3 feet across, but average nests are 13 to 20 inches across and 3 to 10 inches deep. Polygyny, the practice of one male mating with two or more females, has often been recorded in the Northern Harrier.

On Nantucket and Tuckernuck islands in 1986, eggs were recorded in nests from May 1 to June 20 and in 1987 from April 26 to June 25. The average clutch size in 1987 was four eggs (range three to five) (Tate, Melvin). A nest on Monomoy contained one egg on May 16 (Humphrey). There are records of up to nine eggs in a clutch, but five is the average number. Eggs are laid a day or more apart, and incubation may commence when the first egg is produced so that hatching is asynchronous. The incubation period ranges from 29 to 39 days. The young remain in the nest or its immediate vicinity until they fledge in approximately 35 to 37 days. Immatures are distinguished easily from the adult female by their fresh-looking plumages with pale feather edgings above and rufous underbody color.

The male supplies the female with food during incubation and the first week of brooding, and transfers can often be observed near the nest site. The female flies to the male and catches the prey that he drops. Both

● **confirmed breeding**	16 blocks	(1.43%)
● **probable breeding**	10 blocks	(0.90%)
· **possible breeding**	10 blocks	(0.90%)

Northern Harrier: *locally uncommon on the Islands, uncommon on outer Cape Cod; rare elsewhere*

sexes join in nest defense and will attack and vocalize when a predator or human intruder approaches too closely.

The fall migration is protracted, with immatures departing as early as late August and some adult males passing through in early November. Peak migration occurs in early October. The wintering grounds extend from the northern states south through Mexico and the West Indies to northern South America. During winter in Massachusetts, Northern Harriers can be found generally along the coast and more uncommonly inland. The number of wintering birds has increased in recent years.

The Northern Harrier is listed as a threatened species in Massachusetts.

Denver W. Holt

Sharp-shinned Hawk
Accipiter striatus

Egg dates: May 3 to June 24.

Number of broods: one; may re-lay if first attempt fails.

The Sharp-shinned Hawk, the most frequently seen accipiter in Massachusetts, is observed primarily during migration and in the winter, when individuals may be seen preying on small birds at bird feeders. Very feisty on migration, the Sharp-shinned Hawk is often seen harassing other hawks or engaging in aerial "dogfights" with others of its kind.

The Sharp-shinned Hawk breeds primarily in the northern and western mountain forests of North America. In the Commonwealth, it is our most secretive and least common nesting accipiter. No "confirmed" breeding sites were found during the Atlas survey, though six "probable" sites were reported. At least 7 nests have been subsequently documented. Although it is by no means common, the Sharp-shinned Hawk is almost certainly more numerous as a breeder than these results indicate, especially in the western half of the state; this hawk's shy, retiring behavior while nesting makes it difficult to find or to identify its breeding sites. The female will often sit quietly on the nest when human intruders approach or will silently vacate the nest area until they leave. Occasionally, some individuals may harass humans within a distance of 150 feet of the nest, though their small size makes them less intimidating than the much larger and more demonstrative Northern Goshawk.

The Sharp-shinned Hawk was apparently a more abundant breeder in Massachusetts during the nineteenth century. Its decline as a breeding species has been attributed in large part to the deforestation that occurred late in that century and early in the twentieth century. The gradual maturation of Massachusetts' forests has not yet precipitated a strong resurgence of the species. It is believed that the Sharp-shinned Hawk population in northeastern North America may have declined during the first half of the 1990s, possibly due to pesticide contamination, although data is incomplete.

The spring migration period extends from mid-March to mid-May, with the largest concentrations, primarily immatures, passing through from late April to early May. Resident birds usually return in late March and April, with the males returning first.

The male establishes the territory, usually in coniferous or mixed coniferous and deciduous woods. They prefer relatively open rather than deep forest, or else large wood lots with scattered secondary growth. Historically, Sharp-shinned Hawks in the state tended to prefer to build their nests in White Pines. Nests found subsequent to the Atlas period indicate a preference for Red Spruce and occasionally White Birch. Nests are usually located 10 to 60 feet high against the trunk or in a notch of the tree, the dense foliage providing protection from predators and the sun. Old nests are rarely reused, but the same stand of trees is often selected. The nests may be up to 2 feet in diameter and are neatly built of sticks, twigs, and occasionally other materials. Both sexes may gather nesting materials, but the female usually does most of the nest building. Birds may breed in immature plumage; nesting pairs are solitary.

Following the rarely observed courtship flights that take place above and beneath the canopy, three to eight eggs are laid, the number depending on the availability of prey.

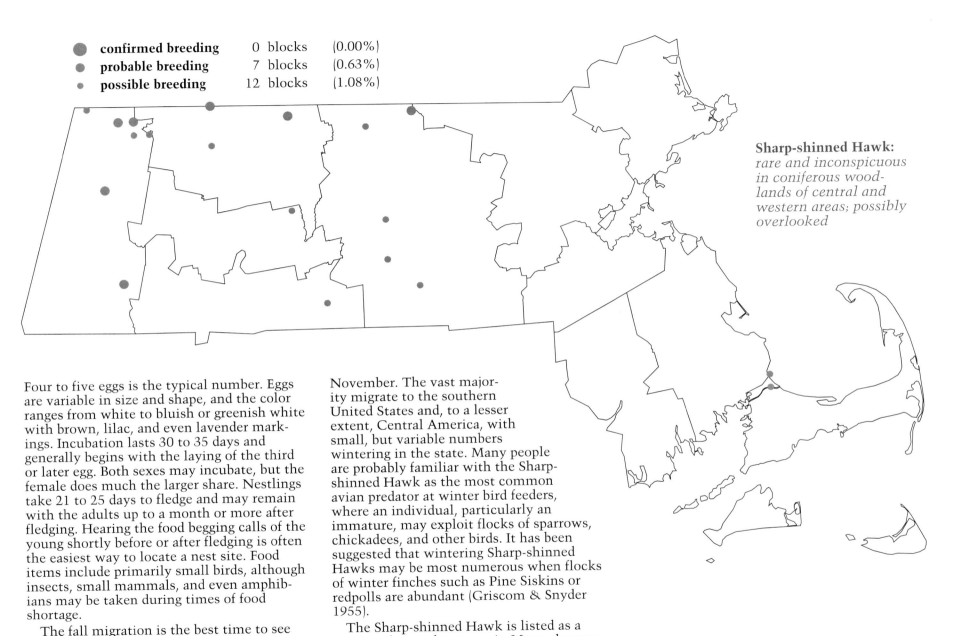

Sharp-shinned Hawk: *rare and inconspicuous in coniferous woodlands of central and western areas; possibly overlooked*

Four to five eggs is the typical number. Eggs are variable in size and shape, and the color ranges from white to bluish or greenish white with brown, lilac, and even lavender markings. Incubation lasts 30 to 35 days and generally begins with the laying of the third or later egg. Both sexes may incubate, but the female does much the larger share. Nestlings take 21 to 25 days to fledge and may remain with the adults up to a month or more after fledging. Hearing the food begging calls of the young shortly before or after fledging is often the easiest way to locate a nest site. Food items include primarily small birds, although insects, small mammals, and even amphibians may be taken during times of food shortage.

The fall migration is the best time to see Sharp-shinned Hawks in the state. The young migrate primarily during September and early October, with smaller numbers of adults predominating from mid-October into November. The vast majority migrate to the southern United States and, to a lesser extent, Central America, with small, but variable numbers wintering in the state. Many people are probably familiar with the Sharp-shinned Hawk as the most common avian predator at winter bird feeders, where an individual, particularly an immature, may exploit flocks of sparrows, chickadees, and other birds. It has been suggested that wintering Sharp-shinned Hawks may be most numerous when flocks of winter finches such as Pine Siskins or redpolls are abundant (Griscom & Snyder 1955).

The Sharp-shinned Hawk is listed as a species of special concern in Massachusetts.

Paul M. Roberts

Cooper's Hawk
Accipiter cooperii

Egg dates: late March to June 11.

Number of broods: one; may re-lay if first attempt fails.

The history of the Cooper's Hawk in Massachusetts has been marked by dramatic changes in its population levels. Formerly present throughout the state, it was a fairly common breeder until the turn of the century, apparently reaching its peak of abundance during the late 1800s. Because the Cooper's Hawk will nest in small groves, forest patches, or partially cut forest as well as in more extensive woodlands, it has adapted readily to the changing conditions associated with the spread of agriculture.

Any species that competes with humans is not regarded favorably, and the Cooper's Hawk is no exception. A predator of game birds, many types of songbirds, and occasionally poultry, it won few, if any, friends, and was inevitably the victim of widespread shooting and trapping until the early part of the twentieth century. After 1900, it began a steady decline, with midcentury habitat destruction and pesticides adding to the bird's problems. The fact that the species diminished from being a common to an uncommon spring and fall migrant was indicative that this trend was occurring throughout the Northeast.

By the 1970s, the status of the Cooper's Hawk was reduced to that of a rare and local breeder. The secretive nature of nesting Cooper's Hawks and the extensive areas of suitable habitat in some regions made censusing difficult. During the survey, breeding was confirmed from only three stations, one each in Middlesex, Worcester, and Franklin counties. The species occurs in greatest numbers in Berkshire County, where it undoubtedly nests, though this is not confirmed. The presence of birds at other locales suggests that the species probably breeds more widely than is generally believed. Since the Atlas survey, the Cooper's Hawk has been "confirmed" at a number of sites and is considered "probable" at several others.

The spring migration period extends from mid-March to mid-May, with resident birds generally arriving prior to or during early April. In Massachusetts, nests are frequently placed in conifers, especially White Pine (2 nests) or Eastern Hemlock (1 nest), but may also be built in oaks or maples, from 20 to 65 feet high. Sometimes the nest of another species will be appropriated or a previous nest will be reoccupied, but most commonly both sexes work in constructing a new nest each year. Small sticks are used for the framework, and a lining of bark, moss, grass, or leaves is added later. From two to six eggs may be produced, with four (1 nest) or five (1 nest) being the usual clutch. Incubation, which lasts from 21 to 24 days, begins with the first egg, resulting in young of different ages being in the nest. Both sexes incubate, but the female does the larger share. At a nest in Stow, five eggs hatched from April 20 to April 27 (Olmstead). Hatching at a Charlton nest occurred around May 24, and

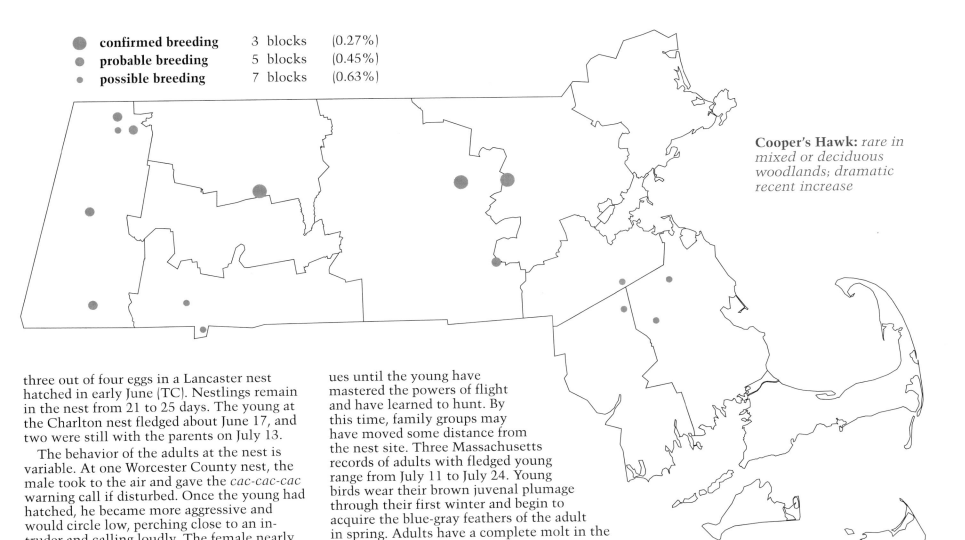

Cooper's Hawk: *rare in mixed or deciduous woodlands; dramatic recent increase*

- ● confirmed breeding — 3 blocks (0.27%)
- ● probable breeding — 5 blocks (0.45%)
- · possible breeding — 7 blocks (0.63%)

three out of four eggs in a Lancaster nest hatched in early June (TC). Nestlings remain in the nest from 21 to 25 days. The young at the Charlton nest fledged about June 17, and two were still with the parents on July 13.

The behavior of the adults at the nest is variable. At one Worcester County nest, the male took to the air and gave the *cac-cac-cac* warning call if disturbed. Once the young had hatched, he became more aggressive and would circle low, perching close to an intruder and calling loudly. The female nearly always disappeared into the woods, but on one occasion she did make a sweeping attack. At a second nest, both birds were much more retiring.

Food items include small mammals, grouse, ducks, poultry, small birds, snakes, frogs, and insects, but young in the nest are fed mostly birds. The young have a high shrill call. Newly fledged birds may return to the nest to be fed, and parental care continues until the young have mastered the powers of flight and have learned to hunt. By this time, family groups may have moved some distance from the nest site. Three Massachusetts records of adults with fledged young range from July 11 to July 24. Young birds wear their brown juvenal plumage through their first winter and begin to acquire the blue-gray feathers of the adult in spring. Adults have a complete molt in the late summer.

The fall migration period lasts from late August to early November. During this time, the resident birds depart and migrants from the north pass southward. Most Cooper's Hawks winter in the southern United States, Mexico, and northern Central America, but a few remain farther north. The species is uncommon but increasingly regular during the winter in Massachusetts.

The Cooper's Hawk is listed as a species of special concern in Massachusetts.

W. Roger Meservey

Northern Goshawk
Accipiter gentilis

Egg dates: March 30 to June 28.

Number of broods: one; may re-lay if first attempt fails.

The Northern Goshawk, one of the most powerful and agile raptors of the Holarctic region, is an uncommon resident of the Alaskan and Canadian boreal and mixed forests and their southward extensions into the United States. It breeds sparsely throughout much of Massachusetts, southeastward to include Plymouth County. The handsome proportions of this bird, as well as its courage and cunning in pursuing prey, have made it a favorite of the world's falconers.

Goshawks usually seek dense forest to breed, preferring remote, secluded stands of mixed conifers and hardwoods. Prior to the 1950s, it was rare to find one nesting in Massachusetts, but, subsequently, Massachusetts forests once again reached sufficient maturity to support breeding pairs. However, in recent years goshawks have been discovered nesting in smaller woodlots and in areas subject to a great deal of human disturbance, especially in eastern and central Massachusetts.

Pairs arrive on their breeding territories by late March. Sometimes a new nest is built, usually in the same woodland occupied previously, or an old nest is refurbished. Rarely is the nest of another species utilized. The structure, located 20 to 70 feet up in a conifer or hardwood, is large (up to 5 feet across) and well constructed of sticks, mostly pine and Eastern Hemlock, broken from living trees. Many of these branches have green needles, and later fresh sprigs may be added. In Massachusetts, White Pine is a preferred nest tree species, along with Eastern Hemlock and Red Oak. A forest pond, stream, or lake often is located nearby because goshawks love water and may spend up to an hour bathing and preening.

The spectacular flight display occurs before and during nest repair or construction, beginning either with the male diving at the female from soaring flight or by a high-speed chase below treetop level. Eventually, the birds fly slower, with deep wing beats and long glides, and gradually get closer together. During the display, both birds are highly vocal. The mating period is prolonged, and the female often initiates copulation by crouching low on a branch with drooped wings while flagging her white undertail coverts.

Goshawks are usually silent except during the reproductive period, when they produce a variety of sounds. Both males and females give the alarm or battle cry, a strident, rapidly repeated *kak-kak-kak-kak* or *kuk-kuk-kuk-kuk*, deeper and hoarser in the female. Pairs converse during courtship flight in high-pitched querulous *kee-a-ah*s. After leaving the nest, the young use a version of this call, a single *kree-ah*. Other calls signify recognition, dismissal, and apprehension.

The Northern Goshawk is the boldest and most reckless of all raptors when defending its nest. Even the vicinity, as far as a half-mile away from the nest, can be dangerous. No intruder is spared; not human, dog, horse, wolf, or bear. When attacking, the goshawk aims for the highest point: humans are struck on the head and animals on the rump.

A clutch of one to five (usually three or four) eggs, depending on the abundance of food, is incubated by the female for 35 to 40 days, during which time she is fed by the

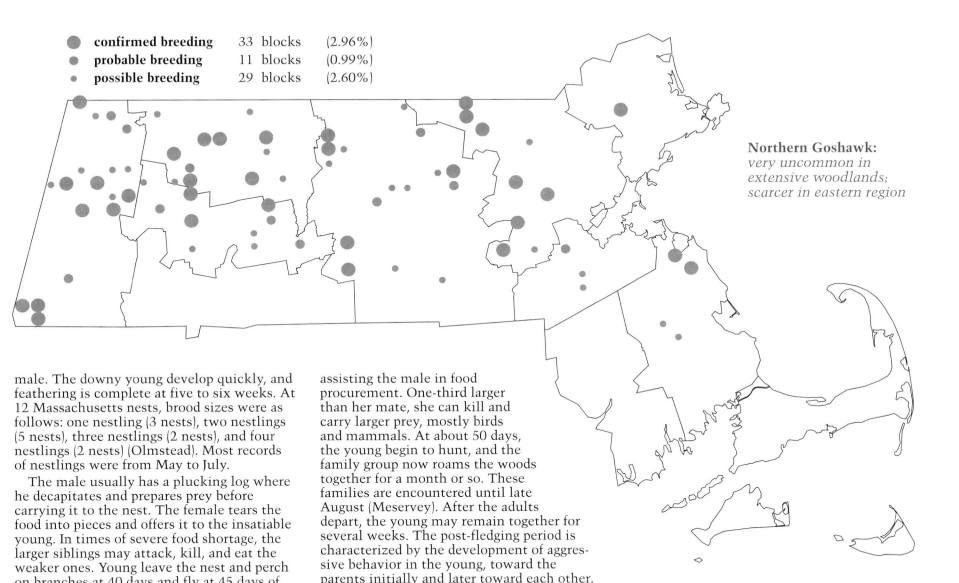

- ● **confirmed breeding** 33 blocks (2.96%)
- ● **probable breeding** 11 blocks (0.99%)
- · **possible breeding** 29 blocks (2.60%)

Northern Goshawk: *very uncommon in extensive woodlands; scarcer in eastern region*

male. The downy young develop quickly, and feathering is complete at five to six weeks. At 12 Massachusetts nests, brood sizes were as follows: one nestling (3 nests), two nestlings (5 nests), three nestlings (2 nests), and four nestlings (2 nests) (Olmstead). Most records of nestlings were from May to July.

The male usually has a plucking log where he decapitates and prepares prey before carrying it to the nest. The female tears the food into pieces and offers it to the insatiable young. In times of severe food shortage, the larger siblings may attack, kill, and eat the weaker ones. Young leave the nest and perch on branches at 40 days and fly at 45 days of age. Approximate fledging dates at 5 Massachusetts nests were June 10, June 14, July 4, July 14, and July 17 (Anderson, Meservey, BOEM).

After fledging, the nest becomes a dinner table for the youngsters, with the female assisting the male in food procurement. One-third larger than her mate, she can kill and carry larger prey, mostly birds and mammals. At about 50 days, the young begin to hunt, and the family group now roams the woods together for a month or so. These families are encountered until late August (Meservey). After the adults depart, the young may remain together for several weeks. The post-fledging period is characterized by the development of aggressive behavior in the young, toward the parents initially and later toward each other.

In the winter, the goshawk exhibits a solitary nature, and there is little association between the members of a pair. Because the Northern Goshawk winters over in its breeding range, it can be seen year-round in this state. Some migrants from farther north are observed each year, and there are infrequent southward irruptions into Massachusetts during winters when prey is scarce to the north.

Nancy Clayton

Red-shouldered Hawk
Buteo lineatus

Egg dates: April 3 to June 5.

Number of broods: one; may re-lay if first attempt fails.

The Red-shouldered Hawk was largely absent from Massachusetts until about 1870. It began increasing rapidly after 1885 and by the early 1900s had become one of the commonest hawks in the state. The population remained high until 1940 but by 1960 had declined to its present level due to a combination of factors including pesticides and habitat destruction. Red-shouldered Hawks nest sparingly throughout the state, except on Cape Cod, Martha's Vineyard, and Nantucket, where they are absent. The species is most numerous in western Massachusetts.

The Red-shouldered Hawk is a bird of lowland forest and swampy areas. The ideal habitat seems to be associated with Beaver ponds and adjacent streams, with mature hardwoods nearby to provide nesting sites. Seclusion from human disturbance is usually a requisite for successful breeding. Sections of southern Berkshire County and Quabbin have an abundance of this type of habitat.

Red-shouldered Hawks are among the first birds to return to our area in spring, arriving about the same time as male Red-winged Blackbirds. In Quabbin, they first appear at territories during early March, when most adults are already paired. The land is usually still snow covered and the lakes and ponds frozen. The birds often seek open water in swamps or inlets of ponds that thaw early. At these locations, they can be seen hunting from low perches near the water's edge. Areas that harbor breeding Wood Frogs, presumed important early-spring food items, are good places to look for one. This buteo has relatively long legs that are used to good advantage in catching amphibians in swampy habitat. More northern nesting individuals pass through Massachusetts, with a peak occurring in late March.

The Red-shouldered Hawk is extremely faithful to its nesting territory and returns year after year to the same woodlot. The nest from the previous season is often repaired and reused, not uncommonly for seven or more years in a row. The string is broken usually when the structure grows too old and rotten to withstand the winter snow and falls to the ground. If the prior year's nesting attempt failed, then a new site generally is selected the next spring. Also, if the eggs or young chicks are lost, the pair will renest nearby the same season.

The nest site is usually in a stand of mature deciduous trees. In Berkshire County, the American Beech is a favored tree, followed by Red Oak, Yellow Birch, and White Birch in that order. Heights of 39 nests ranged from 24 to 60 feet, with an average of 42 feet. In Quabbin, the preferred tree species are Black and Yellow birches. Nest heights of 12 nests ranged from 30 to 48 feet (average 39 feet). Open hardwood forest is favored over mixed hardwood-softwood stands.

Nest building or repairing is a leisurely process for this hawk, sometimes with as much as six weeks passing before eggs are laid. Nests are constructed of sticks in a main crotch of a tree close to the trunk. Sometime during March or early April, old nests will receive a fresh green sprig, usually a bough from an Eastern Hemlock or a pine. More than one alternate nest may be marked this way so that it is not always possible at that point to determine which nest will actually be used.

Each pair of Red-shouldered Hawks requires a territory of about 1 square mile, from which they will evict all conspecific species. Barred Owls and Broad-winged Hawks are tolerated; but the Red-tailed Hawk is antagonistic, and the Great Horned Owl is a serious

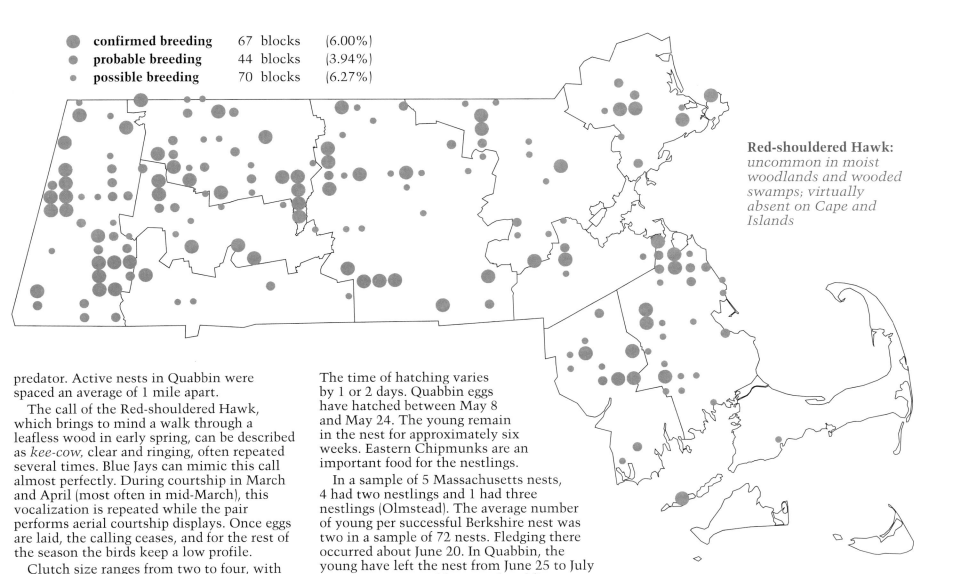

Red-shouldered Hawk: *uncommon in moist woodlands and wooded swamps; virtually absent on Cape and Islands*

predator. Active nests in Quabbin were spaced an average of 1 mile apart.

The call of the Red-shouldered Hawk, which brings to mind a walk through a leafless wood in early spring, can be described as *kee-cow*, clear and ringing, often repeated several times. Blue Jays can mimic this call almost perfectly. During courtship in March and April (most often in mid-March), this vocalization is repeated while the pair performs aerial courtship displays. Once eggs are laid, the calling ceases, and for the rest of the season the birds keep a low profile.

Clutch size ranges from two to four, with an average of three eggs for 9 Quabbin nests. The eggs are the size of those of a domestic hen and are marked with large brown splotches. Some pairs in eastern Massachusetts may have eggs by late March. Most eggs are laid from early to mid-April. Incubation, which starts with the first egg, lasts about 28 days and is accomplished by the female.

The time of hatching varies by 1 or 2 days. Quabbin eggs have hatched between May 8 and May 24. The young remain in the nest for approximately six weeks. Eastern Chipmunks are an important food for the nestlings.

In a sample of 5 Massachusetts nests, 4 had two nestlings and 1 had three nestlings (Olmstead). The average number of young per successful Berkshire nest was two in a sample of 72 nests. Fledging there occurred about June 20. In Quabbin, the young have left the nest from June 25 to July 5. At 5 nests, ten of fourteen hatchlings fledged. Most mortality was due to Great Horned Owl predation and severe weather.

The family stays together well into August, but the birds rarely are seen, and little is known about their habits during this period. Some groups become vocal again by late summer. By September, they have gone their separate ways. Fall migration peaks in late October. The birds winter mainly in the southern states, but an occasional straggler will remain in Massachusetts.

Joseph MacDonald

Broad-winged Hawk
Buteo platypterus

Egg dates: April 28 to June 21.
Number of broods: one.

The Broad-winged Hawk is one of the commonest breeding raptors in the Commonwealth and is found wherever mature forest tracts of sufficient extent are located. It shows a distinct preference for deciduous or mixed woodlands and is equally at home in lowland valleys or hilly uplands. The Broad-winged Hawk was probably widespread in colonial times, but, as the land was cleared for agricultural purposes, it became an increasingly uncommon summer resident. With the decline in agriculture and gradual reforestation of the land, this hawk reoccupied its former territory and is now fairly common throughout much of the state but is scarce on Cape Cod and absent from the offshore islands.

The northward movement of Broad-winged Hawks through Central America can be spectacular because a great number pass through a restricted geographical area in a relatively short period of time. After moving through Texas, the flocks disperse over eastern North America. Normally, first arrivals are expected in Massachusetts in mid-April. The main migration occurs during the latter third of April or the first few days of May in late seasons, but the numbers encountered are never large.

Once a pair is on territory, the breeding cycle begins in earnest. Courtship displays are not spectacular, with circling and flapping just above treetop level constituting the basic activity. During these flights, the pair calls frequently, giving a plaintive, whistled *pe-wee.* The nest is located most frequently in a deciduous tree, although conifers are occasionally chosen. In many regions, the preference for Yellow Birch is so pronounced that the species is referred to as the Birch Hawk. In Massachusetts, nests have been recorded in Yellow Birch, Red Maple, Black Gum, White Pine, and Pitch Pine (CNR, MacDonald). The height of 9 Massachusetts nests ranged from 18 to 60 feet, with an average of 35 feet (CNR, MacDonald).

The nest is an untidy structure of twigs about 1.5 feet in diameter located in a main crotch or on a horizontal limb against the trunk. The inner cup is lined with strips of bark. Green sprigs are added throughout the nesting cycle. Clutches consist of one to three creamy white eggs, marked with blotches of various sizes and shapes. Incuba-

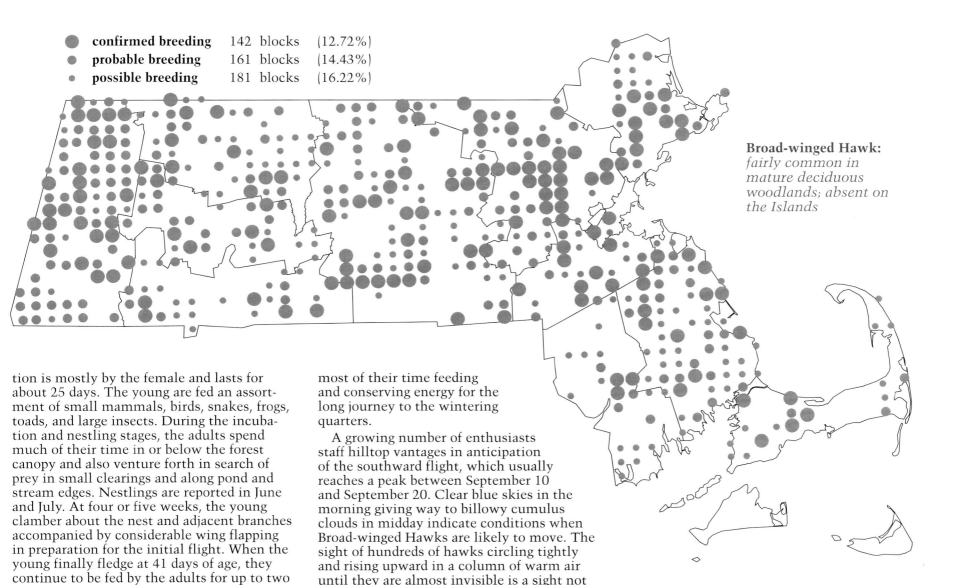

Broad-winged Hawk: *fairly common in mature deciduous woodlands; absent on the Islands*

- ● **confirmed breeding** 142 blocks (12.72%)
- ● **probable breeding** 161 blocks (14.43%)
- · **possible breeding** 181 blocks (16.22%)

tion is mostly by the female and lasts for about 25 days. The young are fed an assortment of small mammals, birds, snakes, frogs, toads, and large insects. During the incubation and nestling stages, the adults spend much of their time in or below the forest canopy and also venture forth in search of prey in small clearings and along pond and stream edges. Nestlings are reported in June and July. At four or five weeks, the young clamber about the nest and adjacent branches accompanied by considerable wing flapping in preparation for the initial flight. When the young finally fledge at 41 days of age, they continue to be fed by the adults for up to two weeks, after which time they are able to seek prey on their own.

From the time the young are fledged until the first migrants begin moving south in the first days of September, Broad-winged Hawks become less conspicuous. During this period, they seldom vocalize and probably spend most of their time feeding and conserving energy for the long journey to the wintering quarters.

A growing number of enthusiasts staff hilltop vantages in anticipation of the southward flight, which usually reaches a peak between September 10 and September 20. Clear blue skies in the morning giving way to billowy cumulus clouds in midday indicate conditions when Broad-winged Hawks are likely to move. The sight of hundreds of hawks circling tightly and rising upward in a column of warm air until they are almost invisible is a sight not soon forgotten. On a good day, thousands can be viewed from favored locations such as Mount Tom or Mount Wachusett. By the end of September, most have departed the state. There are a few reliable sightings for November. With the exception of a small number of individuals found in southern Florida, Broad-winged Hawks winter in the southern portion of Central America and northern South America.

Richard A. Forster

Red-tailed Hawk
Buteo jamaicensis

Egg dates: March 15 to May 6.

Number of broods: one; may re-lay if first attempt fails.

Nearly a century ago, Red-tailed Hawk populations diminished greatly in Massachusetts. This crash was well documented by Brewster (1906) and Griscom (1949) and was due largely to widespread shooting and trapping. Once protection was afforded to birds of prey, Red-tailed Hawks began to recover, and numbers of both residents and migrants climbed. During the late 1940s, populations in many parts of the range began another decline, due to the triple threat of continued direct persecution from people, habitat loss, and pesticides in the food chain.

During the Atlas period, the Red-tailed Hawk was distributed throughout the state and was confirmed in all counties. This species' ability to adapt to a wide range of habitats has been a positive factor, and its preference for open areas with scattered trees and small woodlots allows it to nest in developed areas of eastern Massachusetts. On the map, concentrations of "confirmed" and "probable" breeding, even in suburbs or large cities, should be noted.

Red-tailed Hawks are masters of soaring, and, consequently, they make countless, slow, spiral circles in search of prey, in defense of territory, and during courtship. When courting, the birds sometimes dive and swoop at each other, or even grapple in midair with a grace that seemingly is impossible for so large a bird.

This hawk's call, uttered from a perch as well as on the wing, is a hoarse descending scream. Once recognized, the sound will not be readily forgotten. Blue Jays can mimic the call well. Red-tailed Hawks hunt a variety of mammals, birds, reptiles, and other small animals from flight or a perch.

Migration in spring occurs early, beginning in March and continuing into May. Breeders from farther north pass through, but, as with other raptor species in New England, the spring flight is but a shadow of the autumn migration. Some pairs are year-round residents.

Most pairs in Massachusetts are on their breeding territories by March. Nests are generally placed in trees near the edge of a tract of mature forest. Massachusetts nests have been recorded in White Pine, Black Gum, American Beech, and several species of oaks (CNR, Olmstead). On Cape Cod and the Islands, Pitch Pines may be used. Some pairs show a preference for building in one particular type of tree (Olmstead). The height of 14 Massachusetts nests ranged from 35 to 80 feet, with an average of 53 feet. A nesting site, once selected, will tend to be used year after year, but pairs generally have more than one nest location in their territories. Frequently, a nest built by Red-tailed Hawks will be used in alternate years by the hawks and a pair of Great Horned Owls (Olmstead).

The very large nest, up to 2.5 to 3 feet in outside diameter, is built of sticks and twigs and is decorated regularly with green foliage. When placed in a deciduous tree, the nest is occupied before spring leaves appear. Typical clutches in our part of the continentwide range consist of one to three eggs. Incubation, which is carried out mostly by the female,

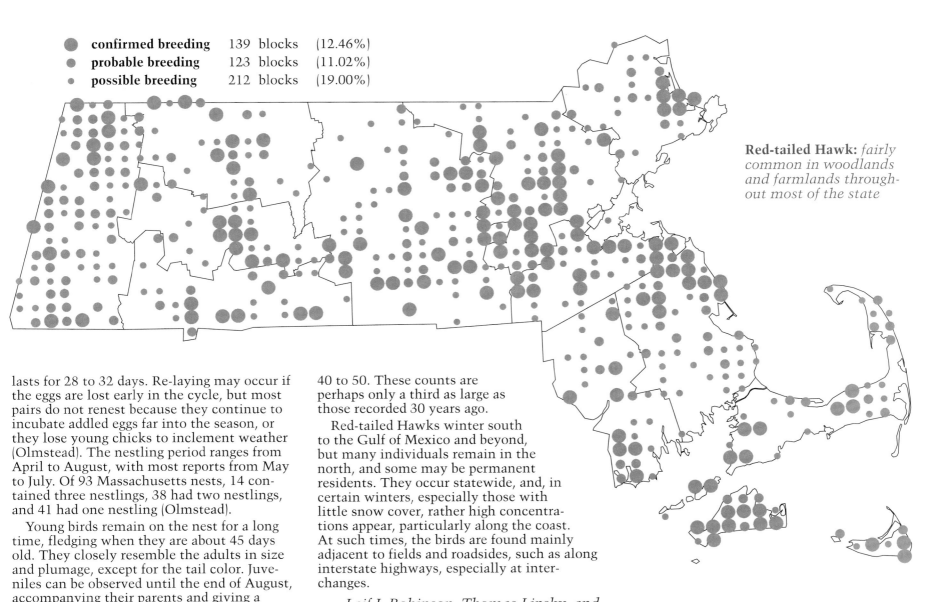

Red-tailed Hawk: *fairly common in woodlands and farmlands throughout most of the state*

- confirmed breeding — 139 blocks (12.46%)
- probable breeding — 123 blocks (11.02%)
- possible breeding — 212 blocks (19.00%)

lasts for 28 to 32 days. Re-laying may occur if the eggs are lost early in the cycle, but most pairs do not renest because they continue to incubate addled eggs far into the season, or they lose young chicks to inclement weather (Olmstead). The nestling period ranges from April to August, with most reports from May to July. Of 93 Massachusetts nests, 14 contained three nestlings, 38 had two nestlings, and 41 had one nestling (Olmstead).

Young birds remain on the nest for a long time, fledging when they are about 45 days old. They closely resemble the adults in size and plumage, except for the tail color. Juveniles can be observed until the end of August, accompanying their parents and giving a shrill food-begging call.

Fall migration runs late, beginning in September and continuing into December. Autumn flights at favored lookouts such as Mount Wachusett in Princeton often yield daily totals of 10 to 20 birds and can exceed 40 to 50. These counts are perhaps only a third as large as those recorded 30 years ago.

Red-tailed Hawks winter south to the Gulf of Mexico and beyond, but many individuals remain in the north, and some may be permanent residents. They occur statewide, and, in certain winters, especially those with little snow cover, rather high concentrations appear, particularly along the coast. At such times, the birds are found mainly adjacent to fields and roadsides, such as along interstate highways, especially at interchanges.

Leif J. Robinson, Thomas Lipsky, and W. Roger Meservey

American Kestrel
Falco sparverius

Egg dates: April 19 to July 5.

Number of broods: one; may re-lay if first attempt fails.

The American Kestrel belongs to a group of small falcons with a worldwide distribution. Our British forefathers mistakenly named it "Sparrow Hawk" after Europe's small accipiter of the same name. In North America, the species nests from the Arctic tree line south across much of Canada and the United States, where it is the commonest falcon. During the Atlas period, kestrels were "confirmed" breeding in all sections of the state. Populations were densest in eastern Massachusetts, including Cape Cod and the Islands, and were sparsest in portions of the hill country in central regions.

The adult male, which has blue wings, two sharp, vertical, black stripes contrasting against a light face, and a bright rufous tail, is certainly the most colorful of North American hawks. The somewhat larger female is not as brilliant as the male and has its upperparts entirely rufous, and barred with black. Kestrels tend to perch high and conspicuously in the open, making them easily visible. They can be tamed readily and will reproduce in confinement. Many falconers have noted that the small males are more docile and tractable than the females.

Spring migration occurs mainly during March and April, and by the latter month local breeders are on their territories at woodland borders, fields, pastures, and the edges of highways. As the breeding season approaches, kestrels abandon their solitary winter habits. Members of a pair often perch side by side, and courtship consists of aerial displays by the male above a perched or flying female. The male ascends on rapidly fluttering wings and then plunges steeply, giving the familiar, repetitive *killy-killy* or *kee-kee* call, which is used not only in courtship but also at other times of excitement. Copulation during this period is frequent and precedes egg laying by several weeks.

Preferring a covered nest site, kestrels choose a hole in a tree, post, or roof. Cavities excavated by flickers may be used, and the birds will readily occupy nest boxes. Because the latter are the most accessible to observe, most Atlas information on nesting comes from pairs using these artificial sites. In one sample of Massachusetts nests, 23 were in boxes and 1 was in a cavity in a maple (CNR). Clutches of three to five (rarely six or seven) eggs are laid. Incubation, mostly by the female, lasts 29 to 30 days. If the first clutch is lost, she often lays a new set of eggs. In

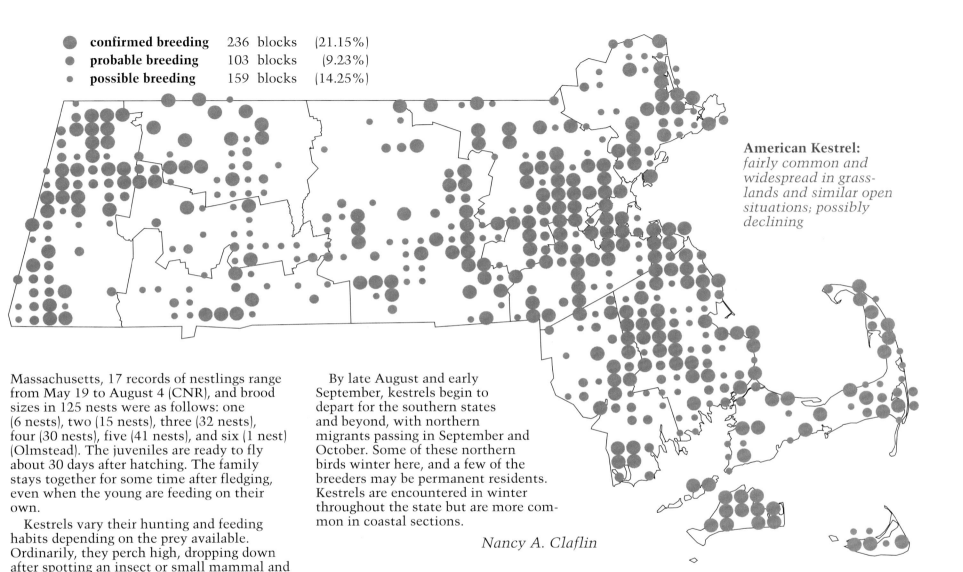

- confirmed breeding 236 blocks (21.15%)
- probable breeding 103 blocks (9.23%)
- possible breeding 159 blocks (14.25%)

American Kestrel: *fairly common and widespread in grasslands and similar open situations; possibly declining*

Massachusetts, 17 records of nestlings range from May 19 to August 4 (CNR), and brood sizes in 125 nests were as follows: one (6 nests), two (15 nests), three (32 nests), four (30 nests), five (41 nests), and six (1 nest) (Olmstead). The juveniles are ready to fly about 30 days after hatching. The family stays together for some time after fledging, even when the young are feeding on their own.

Kestrels vary their hunting and feeding habits depending on the prey available. Ordinarily, they perch high, dropping down after spotting an insect or small mammal and then returning to the same spot, pumping or bobbing the tail upon alighting. At other times, they make forays into the open, heading into the wind with the body tilted diagonally upward and hovering with the tail fanned out. They also pursue small birds, plucking an unsuspecting individual from a perch or capturing it in midair.

By late August and early September, kestrels begin to depart for the southern states and beyond, with northern migrants passing in September and October. Some of these northern birds winter here, and a few of the breeders may be permanent residents. Kestrels are encountered in winter throughout the state but are more common in coastal sections.

Nancy A. Claflin

Ring-necked Pheasant
Phasianus colchicus

Egg dates: April 1 to August 25.

Number of broods: one; may re-lay if first attempt fails.

This exotic, colorful game bird was introduced to Massachusetts in 1894. Thereafter, the population rapidly increased when numerous releases were made in the wild. Widespread favorable farmland habitat resulted in the pheasant's establishment from the Berkshires to Cape Cod and the Islands. Today, despite the dwindling of agricultural land, pheasants continue to thrive under suburban conditions, with substantial numbers occurring in many metropolitan areas. Especially high populations occur in the lower Connecticut River valley, Bristol and Norfolk counties, the Islands, and suburban regions in eastern Massachusetts. The high incidence of feeding stations in the latter contributes greatly to maintaining large numbers of Ring-necked Pheasants. It seems probable, given the decline in suitable pheasant habitat, that it is the survivors of releases made each year from state game farms that maintain the feral population.

The cock Ring-necked Pheasant is vocal principally during the breeding season in late March to early April, when he crows and claps his wings to establish a territory of a few acres. The males may fight fiercely for possession of a territory. Nesting begins in early April. Male pheasants are polygamous and will copulate with several hens, leaving the latter to incubate the eggs and rear the young. In areas with a high population, four or more hens may nest within the territory of a single male. When flushed, males give several loud squawks. Females have a quieter *queep* call.

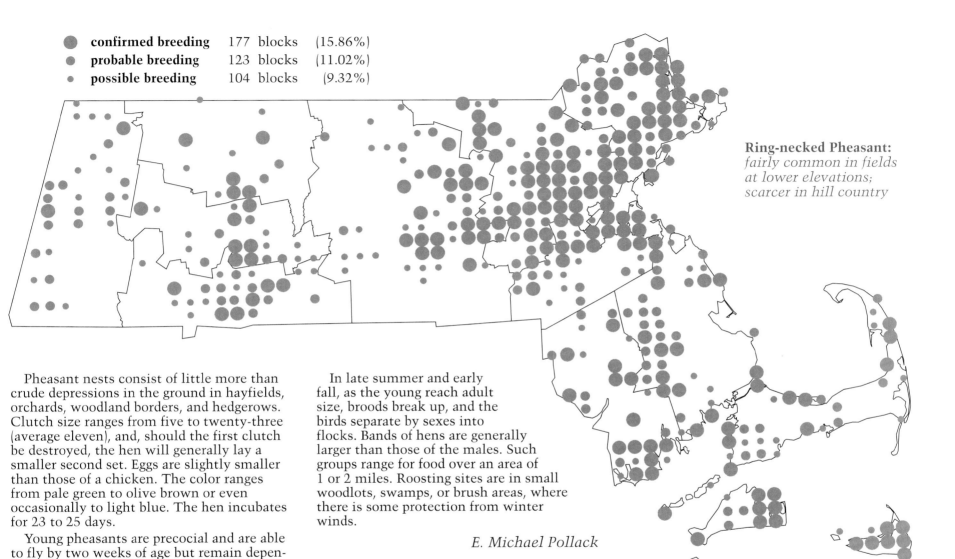

- **confirmed breeding** 177 blocks (15.86%)
- **probable breeding** 123 blocks (11.02%)
- **possible breeding** 104 blocks (9.32%)

Ring-necked Pheasant: *fairly common in fields at lower elevations; scarcer in hill country*

Pheasant nests consist of little more than crude depressions in the ground in hayfields, orchards, woodland borders, and hedgerows. Clutch size ranges from five to twenty-three (average eleven), and, should the first clutch be destroyed, the hen will generally lay a smaller second set. Eggs are slightly smaller than those of a chicken. The color ranges from pale green to olive brown or even occasionally to light blue. The hen incubates for 23 to 25 days.

Young pheasants are precocial and are able to fly by two weeks of age but remain dependent upon the hen for six to eight weeks. For a short time after the chicks hatch, the hen may resort to a feigned broken-wing display to distract enemies, but, as is typical with all ground-nesting species, there may be heavy losses of eggs and young to a host of marauders including domestic cats. The Great Horned Owl is the most significant avian predator on young and adult pheasants.

In late summer and early fall, as the young reach adult size, broods break up, and the birds separate by sexes into flocks. Bands of hens are generally larger than those of the males. Such groups range for food over an area of 1 or 2 miles. Roosting sites are in small woodlots, swamps, or brush areas, where there is some protection from winter winds.

E. Michael Pollack

Ruffed Grouse
Bonasa umbellus

Egg dates: early April to late June.

Number of broods: one; may re-lay if first attempt fails.

The Ruffed Grouse, symbolic of New England, is found throughout Massachusetts, breeding from the western Berkshires to the Pitch Pines of Cape Cod. Populations in Massachusetts were highest during the period between the Civil War and World War II, when much of the state was farmed and the interspersion of woods, fields, and edge made excellent habitat for nesting and feeding. As Massachusetts forests matured, grouse numbers began to decline, and, although the species still occurs throughout the state, the populations of the early 1900s will never return.

The grouse has no equal as an upland game bird. The thunderous takeoff and quick escape enable it to easily outwit or unnerve most hunters. The grouse eats mostly fruits, nuts, berries, and leaves, and much of its life is spent walking rather than flying, so the breast meat is very tender and flavorful.

Grouse are nonmigratory, living out their lives on a few hundred acres. Their unique breeding ritual begins in the spring, with the male selecting and defending one or more drumming sites within his territory. He uses these locations to make the thumping sound so often heard in the spring in New England. Studies in the Quabbin-Hardwick area based on drumming counts showed spring densities of one grouse per 10.1 to 21.6 acres. A cock drums by sitting back on his tail and beating his wings as rapidly as possible for a few seconds. The noise has two very different purposes: advising females of his presence and warning intruding males to stay away. Females crossing this territory are courted actively with much strutting, and, once bred, the hen alone is responsible for ensuring the continuity of the species. The female does all the nest building and incubation, and it takes care of the young.

Vocalization is not the grouse's strong point. That is probably why drumming is used to indicate territory. An alarmed grouse gives a rather sharp *quit-quit* call. A hen may call *crut-crut-car-r-r* to her young and also may squeal when defending them.

The nest is usually in heavy cover in thick woods at the base of a large tree. The eight to sixteen buff-colored eggs are laid in early to mid-April in Massachusetts, with hatching about 23 to 24 days after the last egg is laid. If the first clutch is lost, the hen may lay a slightly smaller second set of eggs. The precocial young begin walking and foraging hours after hatching. At 10 to 12 days of age, they can fly well enough to roost in trees.

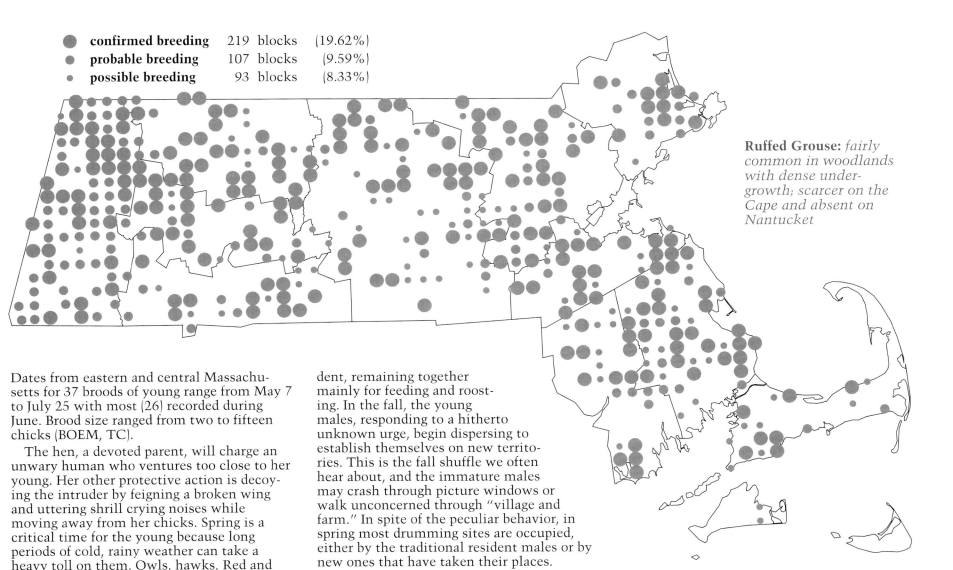

- ● **confirmed breeding** 219 blocks (19.62%)
- ● **probable breeding** 107 blocks (9.59%)
- · **possible breeding** 93 blocks (8.33%)

Ruffed Grouse: *fairly common in woodlands with dense undergrowth; scarcer on the Cape and absent on Nantucket*

Dates from eastern and central Massachusetts for 37 broods of young range from May 7 to July 25 with most (26) recorded during June. Brood size ranged from two to fifteen chicks (BOEM, TC).

The hen, a devoted parent, will charge an unwary human who ventures too close to her young. Her other protective action is decoying the intruder by feigning a broken wing and uttering shrill crying noises while moving away from her chicks. Spring is a critical time for the young because long periods of cold, rainy weather can take a heavy toll on them. Owls, hawks, Red and Gray foxes, and Bobcats all feed on grouse occasionally, but, despite its many enemies, the bird has endured. Ultimately, the presence or absence of suitable habitat will determine the grouse's future. It certainly has the resilience needed to survive.

As the young birds grow and spring eases into summer, they become more independent, remaining together mainly for feeding and roosting. In the fall, the young males, responding to a hitherto unknown urge, begin dispersing to establish themselves on new territories. This is the fall shuffle we often hear about, and the immature males may crash through picture windows or walk unconcerned through "village and farm." In spite of the peculiar behavior, in spring most drumming sites are occupied, either by the traditional resident males or by new ones that have taken their places.

Although winter separates fall shuffle from spring drumming, grouse are well adapted for survival because the scales on their feet grow into "snowshoes." Deep snow poses no obstacle, and, on cold, blustery nights, snow is actually an advantage because grouse will fly right into soft snow to escape overnight cold. Otherwise, thick conifers provide adequate protection. Winter food is mostly buds, with aspen and apple being favorites. When spring returns, the New England hills again echo with the drumroll of the male Ruffed Grouse.

Paul R. Nickerson

Wild Turkey
Meleagris gallopavo

Egg dates: April 24 to May 31.

Number of broods: one; may re-lay if first attempt fails.

The Wild Turkey was extirpated from Massachusetts by 1851 due to widespread clearing of the forests and market hunting. Between 1911 and 1967, about 350 to 380 turkeys were released in Massachusetts during nine restoration attempts. These failed, due principally to the lack of heritable wildness for the pen-raised birds. In 1972 to 1973, 37 wild-trapped New York turkeys were released in Berkshire County. An established population of 4,000 turkeys now occurs west of the Connecticut River, and transplants (162 birds, 1979 to 1984) east of the river are furthering this bird's range expansion.

Turkeys are nonmigratory. Eastern Wild Turkeys may move up to 12 miles between summer and winter habitat; however, movements of less than 5 miles are more typical. Eastern turkeys are commonly associated with mature hardwood forests containing an abundance of mast trees but have adapted to less mature and more diverse mixtures of forest and open habitats. Brood range typically includes grassy clearings and forest openings, usually near a water source.

Turkeys have no song but produce a distinct repertoire of vocalizations including the gobble, yelp, putt, cluck, purr, cackle, and whistle, which variously function in mating, brood assembly, and flock alertness. As daylight hours lengthen, hormonal changes induce gonadal enlargement, gobbling, and display among adult males. Juvenile males are capable of breeding but may not do so due to intimidation by the dominant adults. The males are promiscuous and will breed with several hens. They display weak territoriality, and toms may fight by shoving and spurring each other. Gobbling is high at the onset of breeding, drops off as hens start nesting, and peaks again when most hens are on the nest and males are still sexually active.

Turkeys nest on the ground in a shallow, leaf-lined depression, often in abandoned fields, slash areas, or forest openings. Completed clutches averaged twelve in Massachusetts. Nesting success averages 55 percent. The precocial young are brooded at night by the hen until they can fly to roost at about

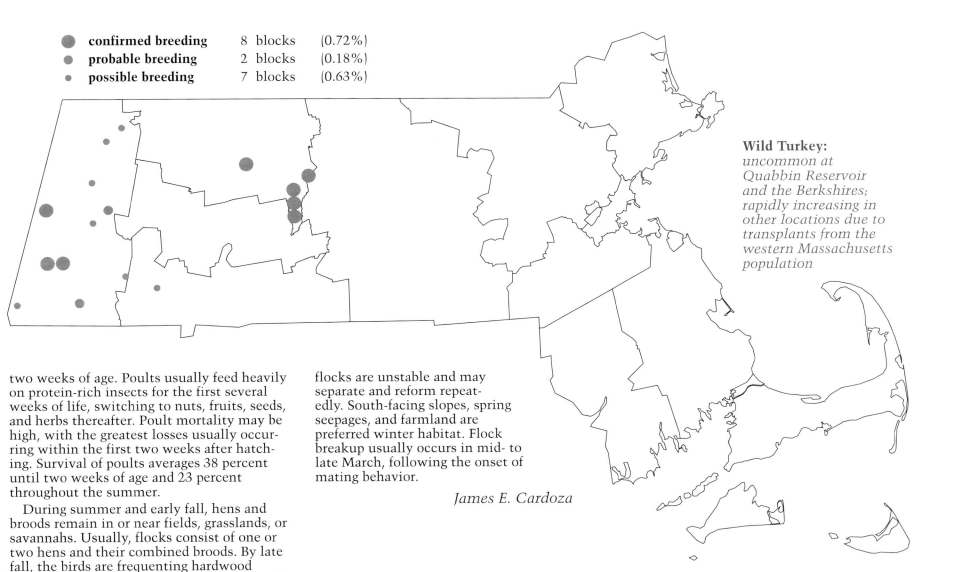

● **confirmed breeding**	8 blocks	(0.72%)
● **probable breeding**	2 blocks	(0.18%)
· **possible breeding**	7 blocks	(0.63%)

Wild Turkey: *uncommon at Quabbin Reservoir and the Berkshires; rapidly increasing in other locations due to transplants from the western Massachusetts population*

two weeks of age. Poults usually feed heavily on protein-rich insects for the first several weeks of life, switching to nuts, fruits, seeds, and herbs thereafter. Poult mortality may be high, with the greatest losses usually occurring within the first two weeks after hatching. Survival of poults averages 38 percent until two weeks of age and 23 percent throughout the summer.

During summer and early fall, hens and broods remain in or near fields, grasslands, or savannahs. Usually, flocks consist of one or two hens and their combined broods. By late fall, the birds are frequenting hardwood ridges, and the young males usually split off to form bachelor flocks. Adult males similarly remain segregated. Depending on habitat, snow depths, and food availability, winter flocks may reach or exceed 100 birds. These consist largely of hens, but males may form temporary attachments. The large flocks are unstable and may separate and reform repeatedly. South-facing slopes, spring seepages, and farmland are preferred winter habitat. Flock breakup usually occurs in mid- to late March, following the onset of mating behavior.

James E. Cardoza

Northern Bobwhite
Colinus virginianus

Egg dates: May 11 to September 14.

Number of broods: one or two.

In areas where bobwhites are common, their loud, cheerful whistle is one of the most distinctive sounds of summer. Although bobwhites were once abundant throughout Massachusetts, by the mid-1800s overhunting and a series of severe winters had resulted in a dramatic decline in the numbers of this appealing species. Attempts to restore the population through an introduction of southern (but less hardy) birds served only to weaken the stock, and a complete recovery never materialized. Today, this popular quail is common only in southeastern portions of the state. Elsewhere, it is a rather rare and sporadic resident whose presence is usually the result of captive-bred birds that have escaped or been released for field trials.

Bobwhites inhabit a variety of open habitats—weedy fields, overgrown pastures, farms, open woodlands, and clearings along power lines. Dense patches of brush and hedgerows within these habitats are important for providing cover and protection from the many predators that find these plump birds to be desirable and vulnerable prey.

The unmistakable call is a robust, whistled *bob-white* or *bob-bob-white,* ascending sharply on the final note. A soft three-note assembly call is used when pairs or covey members have become separated, and a variety of other soft conversational notes also are given. The male performs a courtship display in which he spreads his wings with the tips touching the ground, fans his tail, and bows before the female, turning his head

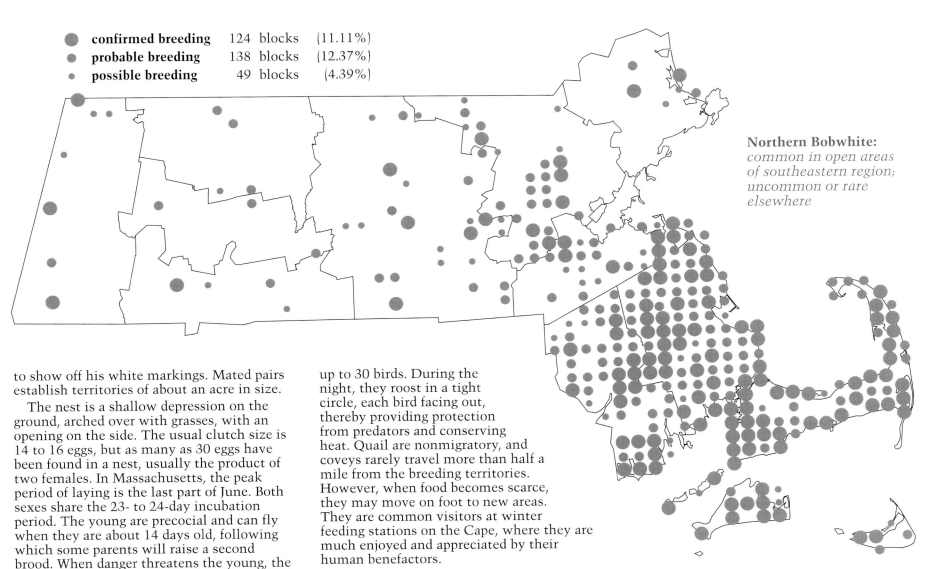

- ● **confirmed breeding**   124 blocks   (11.11%)
- ● **probable breeding**   138 blocks   (12.37%)
- ● **possible breeding**   49 blocks   (4.39%)

Northern Bobwhite: *common in open areas of southeastern region; uncommon or rare elsewhere*

to show off his white markings. Mated pairs establish territories of about an acre in size.

The nest is a shallow depression on the ground, arched over with grasses, with an opening on the side. The usual clutch size is 14 to 16 eggs, but as many as 30 eggs have been found in a nest, usually the product of two females. In Massachusetts, the peak period of laying is the last part of June. Both sexes share the 23- to 24-day incubation period. The young are precocial and can fly when they are about 14 days old, following which some parents will raise a second brood. When danger threatens the young, the adults will use a crippled-wing display in an attempt to lure the predator away. Nonetheless, as with most ground nesters, mortality of both eggs and young is high. The young are full grown at about 49 days of age.

Throughout much of the year, bobwhites are very gregarious, and, when nesting is finished, family groups gather into coveys of up to 30 birds. During the night, they roost in a tight circle, each bird facing out, thereby providing protection from predators and conserving heat. Quail are nonmigratory, and coveys rarely travel more than half a mile from the breeding territories. However, when food becomes scarce, they may move on foot to new areas. They are common visitors at winter feeding stations on the Cape, where they are much enjoyed and appreciated by their human benefactors.

Blair Nikula

Clapper Rail
Rallus longirostris

Egg dates: June 17 to July 3.

Number of broods: one; may re-lay if first attempt fails.

Most literature references cite the Clapper Rail as breeding north to Long Island and Connecticut. The species' occurrence in Massachusetts increased markedly during the early 1950s and caused observers to suspect nesting here in certain favored salt marshes. The Cape's first breeding record came from Barnstable in 1956. After this, nests or broods were found also at Nauset and Yarmouthport. This rails' elusive nature so well preserves the secrets of nesting that Atlas participants were able to make only a single confirmation: at Sippiwisset in Falmouth. Birds calling in May and June were suggestive of breeding at a few other marshes, particularly in the Plum Island region. On Nantucket during 1975 and 1976, birds were present throughout the breeding season at Madaket and Quaise marshes (Andrews). Several other prime areas, particularly on Cape Cod, produced only a surveyor's hunch that rails should be there.

Once incredibly abundant in the south Atlantic coastal marshes, Clapper Rails suffered heavily from gunners and egg collectors around the turn of the century. When relieved by law from such persecution, rails recovered and became plentiful once again. A new threat emerged as salt marshes were filled in and polluted on a wide scale, destroying the birds' year-round habitat, because even during migration they seldom venture far from tidal wetlands.

A healthy marsh supports a good population of fiddler crabs and the mollusks, worms, small fish, insects, and other animals that together form 79 percent of the Clapper Rail diet. Clapper Rails stalk their prey by day in the cover of rank herbage, readily slipping among the grass stems with their narrow bodies. By night at low tide, they emerge to feed on flats and creek beds, striding over the mud on long toes. Capable but reluctant fliers, they prefer to run quickly and silently along a maze of pathways or take to the water and swim away rather than to fly.

Clapper Rails forsake stealth only when they give voice. Calls are usually represented by *clack* or *kik* or *jupe* sounds repeated in a series, beginning loud and fast and ending lower and slower. They vocalize by day as well as at dusk and on moonlit nights and often respond to any sudden loud sound. The

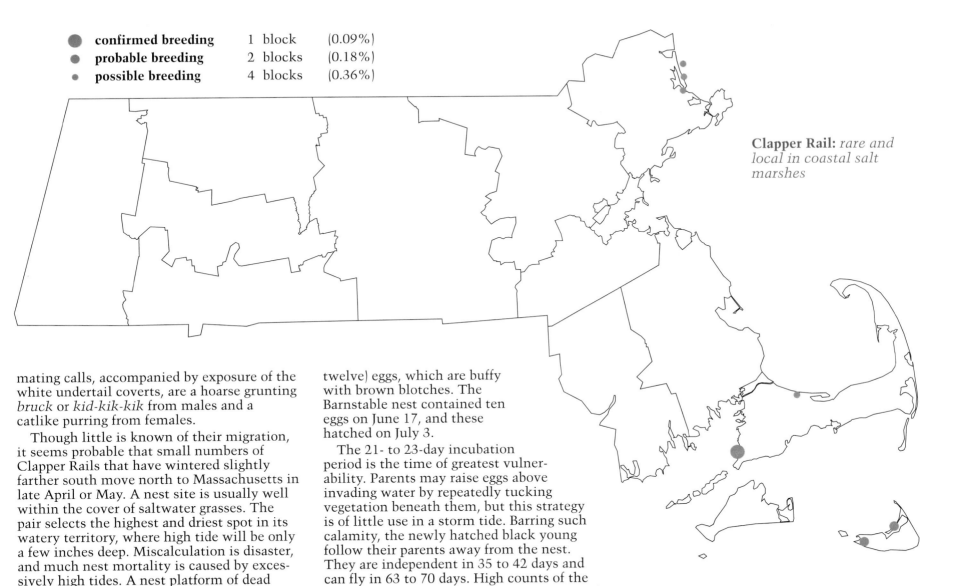

Clapper Rail: *rare and local in coastal salt marshes*

mating calls, accompanied by exposure of the white undertail coverts, are a hoarse grunting *bruck* or *kid-kik-kik* from males and a catlike purring from females.

Though little is known of their migration, it seems probable that small numbers of Clapper Rails that have wintered slightly farther south move north to Massachusetts in late April or May. A nest site is usually well within the cover of saltwater grasses. The pair selects the highest and driest spot in its watery territory, where high tide will be only a few inches deep. Miscalculation is disaster, and much nest mortality is caused by excessively high tides. A nest platform of dead plant material is constructed 8 to 12 inches above the mud in clumps of marsh grass. Growing blades often arch over and conceal the nest. A ramp of dry stalks may connect the nest rim to the ground. Nests measure 8 to 14 inches across and are well cupped to receive up to fourteen (usually eight to twelve) eggs, which are buffy with brown blotches. The Barnstable nest contained ten eggs on June 17, and these hatched on July 3.

The 21- to 23-day incubation period is the time of greatest vulnerability. Parents may raise eggs above invading water by repeatedly tucking vegetation beneath them, but this strategy is of little use in a storm tide. Barring such calamity, the newly hatched black young follow their parents away from the nest. They are independent in 35 to 42 days and can fly in 63 to 70 days. High counts of the year (one to three average; eleven maximum) occur when postbreeding wanderers move into large marshes, such as those at Plum Island, Barnstable, and Wellfleet. On a still October evening, a spontaneous chorus proclaims the gathering of rails. Some may linger into early winter and are fairly regularly reported on Christmas Bird Counts. A few possibly remain year-round, but most move farther south along the Atlantic coast.

Wallace Bailey

King Rail
Rallus elegans

Egg dates: not available.

Number of broods: one; may re-lay if first attempt fails.

Undoubtedly one of the Commonwealth's least known and rarest breeding species, the King Rail was only once confirmed as a breeder in the state during the Atlas project, and to date its nest has never been discovered in Massachusetts. The species has been a presumed breeder since about 1880, when a brood of young was hand-caught in Sandwich. Since that time, several additional sightings of adults with young have occurred, all in eastern Massachusetts, including the fortuitous encounter of an adult crossing a Middleboro roadway with nine chicks on July 17, 1976.

Because the King Rail is at the northeastern limit of its range in Massachusetts, its historic status has been poorly defined. Records would suggest that the species prefers the same extensive freshwater cattail marshes that are occupied by the more familiar Virginia Rail; however, in southeastern Massachusetts there is also a modest scattering of records from grassy river meadows and from marshes with a moderate shrub cover. Perhaps more significantly, the King Rail has been recorded in summer in the salt marshes of the Parker River National Wildlife Refuge at Plum Island, several times in company with Clapper Rails. The interbreeding of these two species has been the subject of considerable scientific speculation, and some authorities have considered the two forms to be conspecific.

King Rails are typically first heard at presumed breeding meadows in May, although specimen evidence would suggest that some undoubtedly arrive earlier. The species' local vocalizing is marked by its irregularity, which is totally different from the situation in the southeastern United States and middle Atlantic coastal areas where calling may be heard frequently almost around the clock early in the breeding season. It is possible that in our area, where the species is at the periphery of its breeding range and its density is low, calling may not be as critical to territory maintenance, even in optimal habitats. In any event, the most frequent call in May is a harsh, grating, and accelerating series of notes that may be likened to *jupe-jupe-jupe-jupe*, etc. This is one of the primary vocalizations of the King Rail, and it seems to correspond to the *wak-wak-wak* call of the smaller Virginia Rail but is given with much more force and volume. Also heard in Massachusetts from time to time in marshes is a loud, two-part call

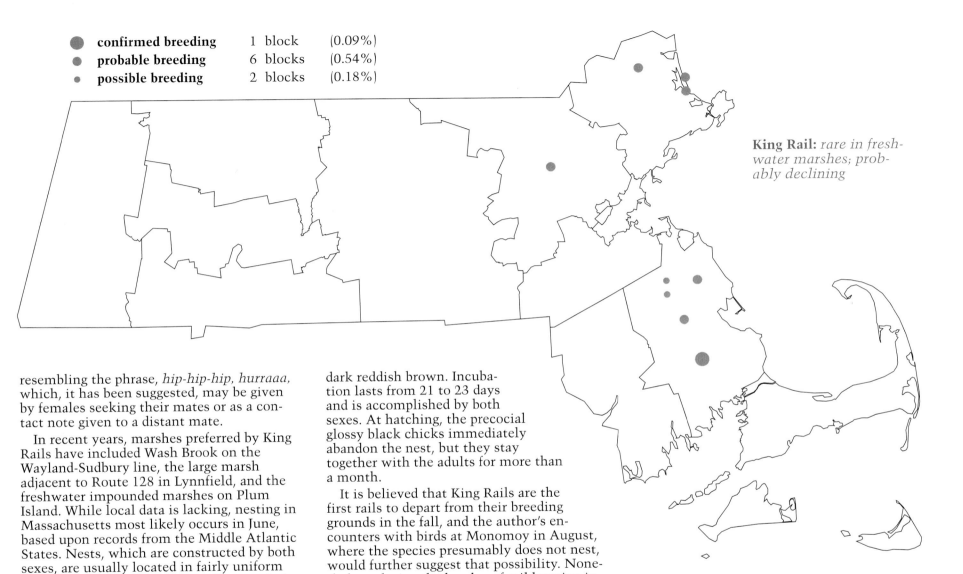

King Rail: *rare in freshwater marshes; probably declining*

- confirmed breeding — 1 block (0.09%)
- probable breeding — 6 blocks (0.54%)
- possible breeding — 2 blocks (0.18%)

resembling the phrase, *hip-hip-hip, hurraaa,* which, it has been suggested, may be given by females seeking their mates or as a contact note given to a distant mate.

In recent years, marshes preferred by King Rails have included Wash Brook on the Wayland-Sudbury line, the large marsh adjacent to Route 128 in Lynnfield, and the freshwater impounded marshes on Plum Island. While local data is lacking, nesting in Massachusetts most likely occurs in June, based upon records from the Middle Atlantic States. Nests, which are constructed by both sexes, are usually located in fairly uniform stands of vegetation and are well concealed, frequently having a cone-shaped or rounded canopy of loosely woven plants overhead. Generally, the nest is on a grassy tussock or amid the stems of some waterside plant. Some nests are actually located over water. Clutches average ten to eleven eggs, which are pale buff in color, sparingly spotted with dark reddish brown. Incubation lasts from 21 to 23 days and is accomplished by both sexes. At hatching, the precocial glossy black chicks immediately abandon the nest, but they stay together with the adults for more than a month.

It is believed that King Rails are the first rails to depart from their breeding grounds in the fall, and the author's encounters with birds at Monomoy in August, where the species presumably does not nest, would further suggest that possibility. Nonetheless, during the heyday of rail hunting in New England, large numbers of birds were shot right through the fall, including many in salt marshes. There is a surprising legacy of winter records for this species, indicating that at least occasionally the King Rail attempts to winter in our area, almost always along the coast. The regular wintering grounds extend south along the mid-Atlantic coast to Georgia and Florida.

The King Rail is listed as a threatened species in Massachusetts.

Wayne R. Petersen

Virginia Rail
Rallus limicola

Egg dates: early May to late July.

Number of broods: one; may re-lay if first attempt fails.

The Virginia Rail is an inconspicuous and somewhat local summer wetland resident whose abundance is largely determined by the availability of suitable breeding habitat. Preferring extensive cattail marshes for nesting, it will also breed along the brushy banks of open river meadows and occasionally around the vegetated margins of undisturbed suburban ponds. The species is notably local as a breeder in central Massachusetts, as well as on the Cape and Islands. South of New England, it is not uncommon as a nesting bird in the upper portions of coastal salt marches, but in our area it is unusual in such situations.

The spring arrival of Virginia Rails from their southeastern wintering grounds generally takes place during the last half of April, with their return apparently contingent on local weather conditions and on appropriate water levels in their chosen breeding marshes. Because of their retiring nature, much of our knowledge of their precise migration timetable must be derived from their vocalizations. These calls, many of which are not well understood, are given irregularly, often at odd hours of the evening. Additionally, it is probable that many individuals may not call immediately upon their arrival, further making it difficult to pin down their migration schedule.

A dawn visit to an appropriate breeding meadow on a mellow and windless May morning is most likely to produce a chorus of territorial male Virginia Rails. In this season, the advertising calls of the males are a metallic-sounding *ki-dik, ki-dik, ki-dik*, with distance distorting the call and making it sound more like *dik, dik, dik*. This vocalization is given with considerable energy, and it often continues uninterrupted for many minutes. A second common call, and one not confined to the breeding grounds, is a piglike grunting sound, which ends with a diminishing quality, much like a bouncing ball coming to a stop. Phonetically, it may be rendered as *wak-wak-wak-wak-wak*. It is likely that

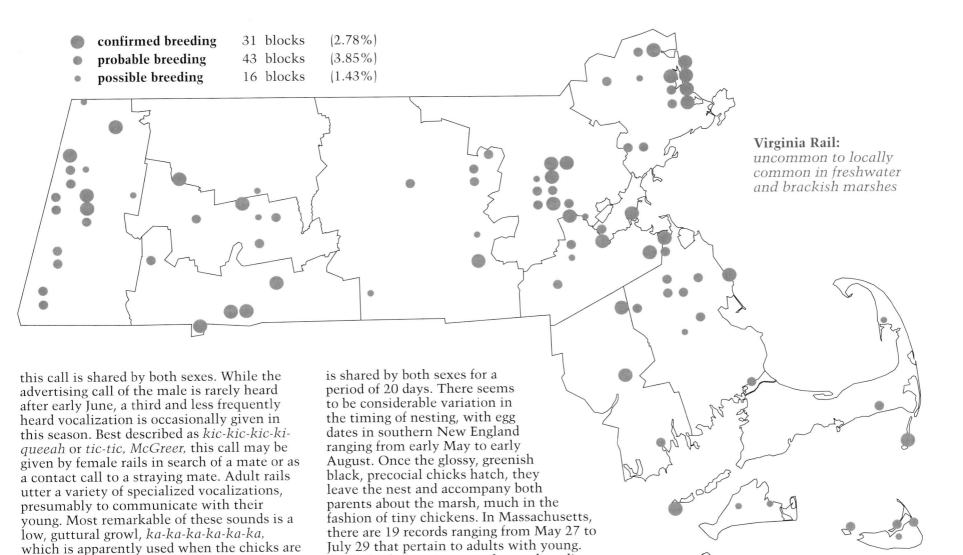

Virginia Rail: *uncommon to locally common in freshwater and brackish marshes*

- confirmed breeding — 31 blocks (2.78%)
- probable breeding — 43 blocks (3.85%)
- possible breeding — 16 blocks (1.43%)

this call is shared by both sexes. While the advertising call of the male is rarely heard after early June, a third and less frequently heard vocalization is occasionally given in this season. Best described as *kic-kic-kic-ki-queeah* or *tic-tic, McGreer*, this call may be given by female rails in search of a mate or as a contact call to a straying mate. Adult rails utter a variety of specialized vocalizations, presumably to communicate with their young. Most remarkable of these sounds is a low, guttural growl, *ka-ka-ka-ka-ka-ka*, which is apparently used when the chicks are in jeopardy.

By May, some Virginia Rails are laying clutches of five to twelve buffy-white eggs, which are sparsely speckled with reddish brown. Their well-concealed nests may be placed on the ground amidst marsh vegetation, in a grass or sedge tussock, or among plants over mud or water. The nest is usually constructed of course grass, dead stalks, cattails, or other plant materials. Incubation is shared by both sexes for a period of 20 days. There seems to be considerable variation in the timing of nesting, with egg dates in southern New England ranging from early May to early August. Once the glossy, greenish black, precocial chicks hatch, they leave the nest and accompany both parents about the marsh, much in the fashion of tiny chickens. In Massachusetts, there are 19 records ranging from May 27 to July 29 that pertain to adults with young. Single birds or pairs were observed tending from one to five chicks (BOEM).

By late summer and early fall, Virginia Rail populations are at their peak in the marshes, and, in this season, a few are hunted legally by sportsmen who are willing to take the trouble to find them. With each autumn cold snap, the numbers lessen as more birds head south. By November, only a hardy few are left to winter, primarily in southeastern Massachusetts, wherever they can find an open spring-fed seep, running brook, or salt marsh along the coast. The main wintering grounds extend from the southern states to northern Central America.

Wayne R. Petersen

Sora
Porzana carolina

Egg dates: mid-May to early July.

Number of broods: one; may re-lay if first attempt fails.

The Sora is North America's most common and widespread rail species. There is hardly a substantial marsh in the country that is not visited by this rail during its autumn migrations, when it is hunted widely as a game bird. In Massachusetts, it is best known as a transient in spring and fall and as a scarce and local breeding resident whose numbers are almost certainly declining as fewer and fewer extensive cattail marshes are available for breeding. Being more selective in its choice of breeding habitat than the Virginia Rail, the Sora is usually found nesting only in relatively extensive and unbroken stands of cattails, preferably with adjacent areas of open water. It seems to require marsh openings and water edges for foraging and generally seems far less tolerant of heavy shrub growth in its habitat than the Virginia Rail, whose overall requirements are otherwise similar. Rare as a breeder in western Massachusetts, the Sora has viable nesting populations in only a handful of marshes in the eastern part of the state.

Soras begin arriving in late April, usually a little after the first Virginia Rails are heard calling. The loquacity of both species is the best indication of their presence, and, because much of their vocalizing occurs at night, special effort is required to assess accurately the local status of either species.

The varied calls of the Sora include a sharp *ker-wee* given by the males and a rapidly descending whinny, which ends more slowly than it begins and has the terminal notes given on a uniform key. The whinny call is probably used by both sexes. Because Soras regularly give these calls on migration, it cannot be assumed that a calling Sora in appropriate habitat in mid-May is necessarily a breeding bird. This vocalizing by migrants possibly has given earlier Massachusetts ornithologists a false impression of the species' abundance as a breeder in the state. As with a number of marsh birds, Soras produce a varied assortment of call notes when they have young in tow. Most typical of these is a sharp *kip*, somewhat reminiscent of the similar but more forceful note of the Common Moorhen.

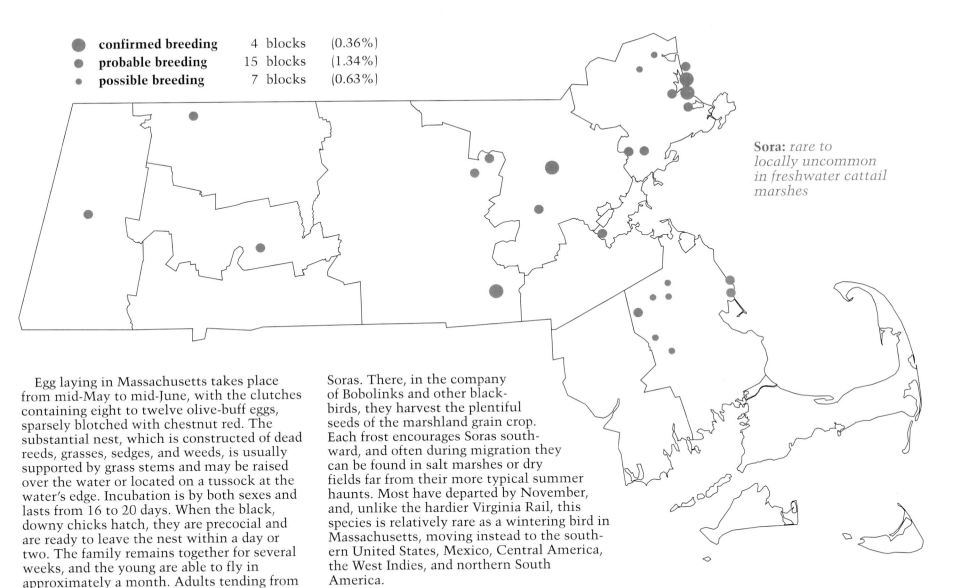

● **confirmed breeding** 4 blocks (0.36%)
● **probable breeding** 15 blocks (1.34%)
• **possible breeding** 7 blocks (0.63%)

Sora: *rare to locally uncommon in freshwater cattail marshes*

Egg laying in Massachusetts takes place from mid-May to mid-June, with the clutches containing eight to twelve olive-buff eggs, sparsely blotched with chestnut red. The substantial nest, which is constructed of dead reeds, grasses, sedges, and weeds, is usually supported by grass stems and may be raised over the water or located on a tussock at the water's edge. Incubation is by both sexes and lasts from 16 to 20 days. When the black, downy chicks hatch, they are precocial and are ready to leave the nest within a day or two. The family remains together for several weeks, and the young are able to fly in approximately a month. Adults tending from four to six chicks have been recorded in Massachusetts from June 24 to July 1, and a juvenile was seen accompanying adults on July 7 (BOEM).

By early fall, marshes and river meadows, especially ones where Wild Rice grows in abundance, sometimes throng with migrant Soras. There, in the company of Bobolinks and other blackbirds, they harvest the plentiful seeds of the marshland grain crop. Each frost encourages Soras southward, and often during migration they can be found in salt marshes or dry fields far from their more typical summer haunts. Most have departed by November, and, unlike the hardier Virginia Rail, this species is relatively rare as a wintering bird in Massachusetts, moving instead to the southern United States, Mexico, Central America, the West Indies, and northern South America.

Wayne R. Petersen

Common Moorhen
Gallinula chloropus

Egg dates: mid-May to early August.

Number of broods: one; may re-lay if first attempt fails.

Formerly known as the Common Gallinule, the moorhen is a widely distributed breeding species in freshwater marshes and weedy ponds in the eastern United States. It is common in the southern portion of its range, but in Massachusetts, where it nears its northeastern breeding limit, it has always been considered rare and local, and there are fewer locations where it now breeds than at any time in history. Certainly, ecological changes and the draining and filling of wetlands have taken their toll on its meager population.

The first Common Moorhens normally arrive in Massachusetts during the later part of April. Unlike their marsh relatives, the rails, they are as likely to be seen as heard because they prefer the edges of open areas in the marsh. Some of their vocalizations are similar to those of the Pied-billed Grebe but can readily be distinguished with experience. The call is loud, harsh, and varied, consisting of a series of clucking sounds. Also, chicken-like grunts and clucks are frequently uttered, and during the nesting season moorhens often make a distinctive explosive croak.

The nest, a low bulky structure with a shallow cup, is constructed of cattails of the previous year's growth. Typically, a runway of rushes or cattails extends from the rim of the nest to the water, allowing easy access to and from the nest. Pairs may construct additional vegetation platforms on which to brood the young. The normal clutch contains seven to twelve buffy eggs marked with irregular spotting. Incubation, which is

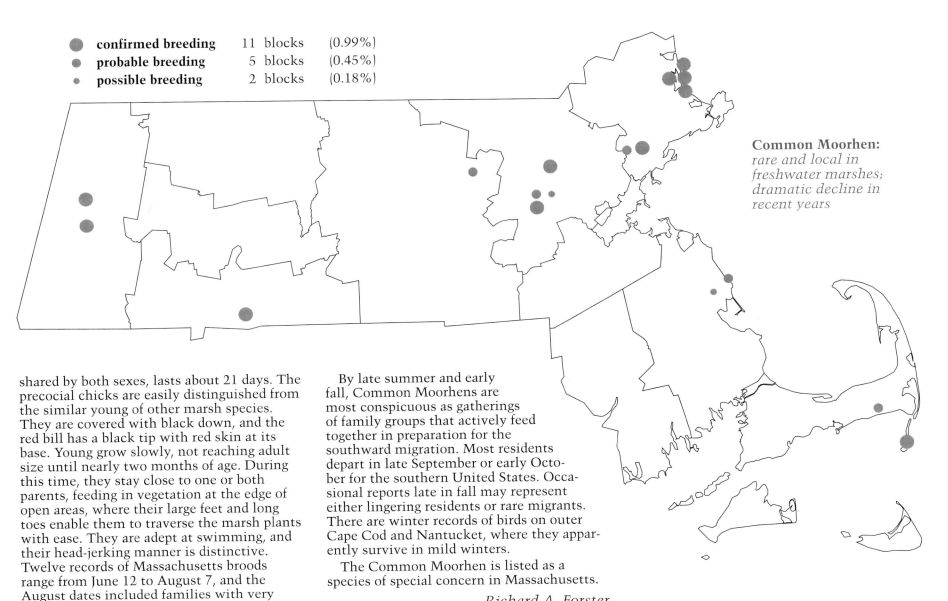

- confirmed breeding   11 blocks   (0.99%)
- probable breeding   5 blocks   (0.45%)
- possible breeding   2 blocks   (0.18%)

Common Moorhen: *rare and local in freshwater marshes; dramatic decline in recent years*

shared by both sexes, lasts about 21 days. The precocial chicks are easily distinguished from the similar young of other marsh species. They are covered with black down, and the red bill has a black tip with red skin at its base. Young grow slowly, not reaching adult size until nearly two months of age. During this time, they stay close to one or both parents, feeding in vegetation at the edge of open areas, where their large feet and long toes enable them to traverse the marsh plants with ease. They are adept at swimming, and their head-jerking manner is distinctive. Twelve records of Massachusetts broods range from June 12 to August 7, and the August dates included families with very small young. The number of young with single birds or pairs ranged from two to nine (BOEM, Petersen). Although two broods are reared in the South, there is no evidence that this occurs here.

By late summer and early fall, Common Moorhens are most conspicuous as gatherings of family groups that actively feed together in preparation for the southward migration. Most residents depart in late September or early October for the southern United States. Occasional reports late in fall may represent either lingering residents or rare migrants. There are winter records of birds on outer Cape Cod and Nantucket, where they apparently survive in mild winters.

The Common Moorhen is listed as a species of special concern in Massachusetts.

Richard A. Forster

American Coot
Fulica americana

Egg dates: early May to early August.

Number of broods: one; may re-lay if first attempt fails.

The American Coot, or Mudhen, is a common breeding species in many freshwater areas, especially in the western United States and Canada. It is also well distributed in the Southeast but is uncommon and local throughout the northeastern United States and eastern Canada. In Massachusetts, there is one old possible breeding record from the western part of the state in Cheshire. In the latter part of the 1960s, American Coots began to breed at Plum Island and at the Great Meadows National Wildlife Refuge in Concord. This represented a range expansion that also occurred during the same period in New York.

During the Atlas project, they were present but not confirmed breeding in Concord, and the small population, at most several pairs, did not increase at Plum Island. Both of these sites are human-made freshwater impounded areas. Perhaps subtle changes in the habitat at these locations prevented the expansion of the population. Breeding is erratic, with nesting attempts not undertaken every year.

The coot is a very uncommon spring migrant, usually arriving in early April. By the middle of May, the few migrants have departed, and only breeding individuals remain. The preferred nesting site is an open body of water with marshlike edges (e.g., cattails and associated plants) rather than the denser marsh areas prefered by its close relative, the Common Moorhen. The voice is a series of grunting *kuk* notes and other clucking noises, again similar to those of the moorhen but sufficiently different to establish the identity of the caller by persons familiar with both species.

The nest is constructed from dried leaves and stems of marsh plants and may be located in the open or concealed in the marsh, sometimes on a natural platform or simply secured to live vegetation. A large clutch of eggs numbering up to a dozen is laid during May. The incubation period lasts about three weeks. The downy chicks are characteristically black in color, with rust on the head and neck, and the bill is red with a black tip. During the early stages of development, the young parade behind the parent. Both sexes share in incubation and care of the young. Juveniles can fly 49 to 56 days after hatching. Four Massachusetts brood dates range from June 2 to August 11. Brood size ranged from one to eight (BOEM, Petersen). It appears that late broods result from re-nesting after the initial attempt has failed.

Little is known about postbreeding events. An immature bird, too young to be a migrant,

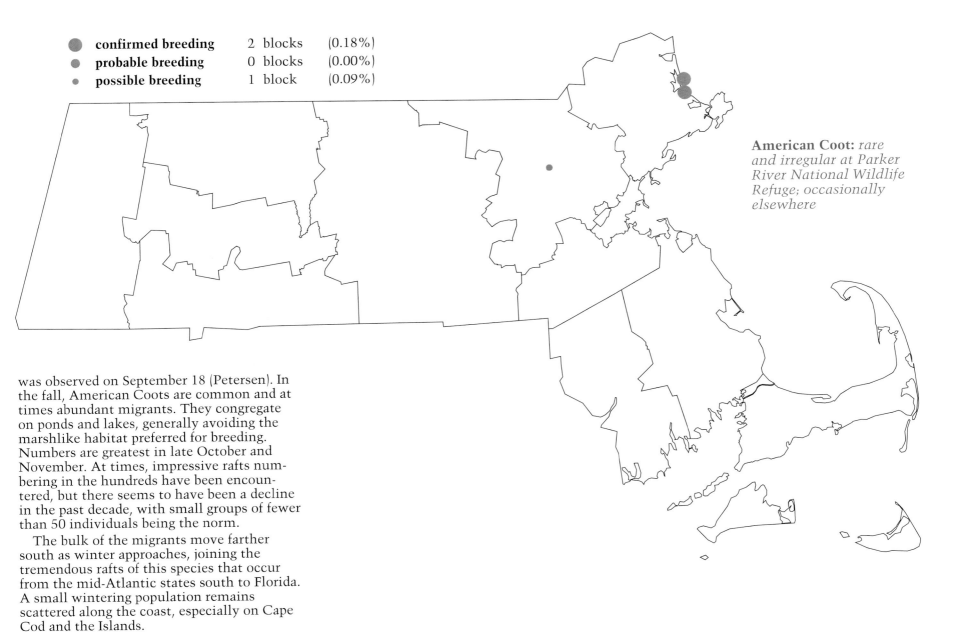

- ● **confirmed breeding**   2 blocks   (0.18%)
- ● **probable breeding**   0 blocks   (0.00%)
- · **possible breeding**   1 block   (0.09%)

American Coot: *rare and irregular at Parker River National Wildlife Refuge; occasionally elsewhere*

was observed on September 18 (Petersen). In the fall, American Coots are common and at times abundant migrants. They congregate on ponds and lakes, generally avoiding the marshlike habitat preferred for breeding. Numbers are greatest in late October and November. At times, impressive rafts numbering in the hundreds have been encountered, but there seems to have been a decline in the past decade, with small groups of fewer than 50 individuals being the norm.

The bulk of the migrants move farther south as winter approaches, joining the tremendous rafts of this species that occur from the mid-Atlantic states south to Florida. A small wintering population remains scattered along the coast, especially on Cape Cod and the Islands.

Richard A. Forster

Piping Plover
Charadrius melodus

Egg dates: mid-April to mid-August.

Number of broods: one; will re-lay numerous times in response to nest loss.

During the Atlas period, nesting Piping Plovers were confirmed in 54 blocks, with the population distributed mainly about Cape Cod and the Islands north to Scituate. Piping Plovers were less frequent about the rocky shores of Buzzards Bay and were almost entirely absent from Massachusetts Bay. A few pairs nested north of Cape Ann. They were unrecorded inland. In January 1986, the Atlantic coastal population was listed as threatened by the United States Fish and Wildlife Service because of the decline of the species in many portions of its range since 1965. Massachusetts had an estimated population of 132 pairs at 43 stations in 1985 and 139 pairs at 49 stations in 1986 (MDFW). Aggressive management, including use of symbolic fencing on nesting grounds, restriction of off-road vehicles where unfledged plover chicks are present on the beaches, and use of specially designed predator exclosures for nests, has resulted in a dramatic increase to about 350 nesting pairs in 1994 (MDFW).

Piping Plovers first appear on Massachusetts beaches in late March, with the bulk of the population arriving in April. The birds occupy wide, barren, sandy beaches, usually preferring areas with scattered grass clumps, and they also use tidal flats for feeding. These plovers are inconspicuous, solitary nesters, often, but not necessarily, found in association with Least Terns, whose spirited group defense undoubtedly confers added protection from predators. The birds themselves have near-perfect protective coloration so that when they are immobile or crouching, they seem to disappear. Except to a trained observer, they would be undetectable were it not for their soft *peep-lo* that gives away their presence.

Courtship is a highly ritualized and protracted affair, with a month or more often passing before the first eggs are laid. Courting males are quite vocal and run around the female, crouching with wings and tail spread and drooping. Piping Plovers are typically monogamous within a nesting season but usually change mates from year to year.

The nest is a mere hollow in the sand, sometimes lined with pebbles or small bits of shell. Several practice scrapes are often fashioned by the male. Nests are normally on very sparsely vegetated beaches, often in the vicinity of blowouts or overwash fans, above the high tide line and usually within sight of water. They may be among small rocks, close to protective clumps of grass or beach detritus, or completely in the open.

Clutches of four eggs are almost invariably produced, with three sometimes appearing in second or subsequent nests. The earliest clutches are usually laid in late April. The eggs and chicks are extremely cryptic. Both sexes share incubation duties for about 27 days. If a nest gets destroyed by a predator or storm, pairs will renest repeatedly—up to five times—sometimes well into July. Usually, a different scrape in the same area is used for each subsequent nest.

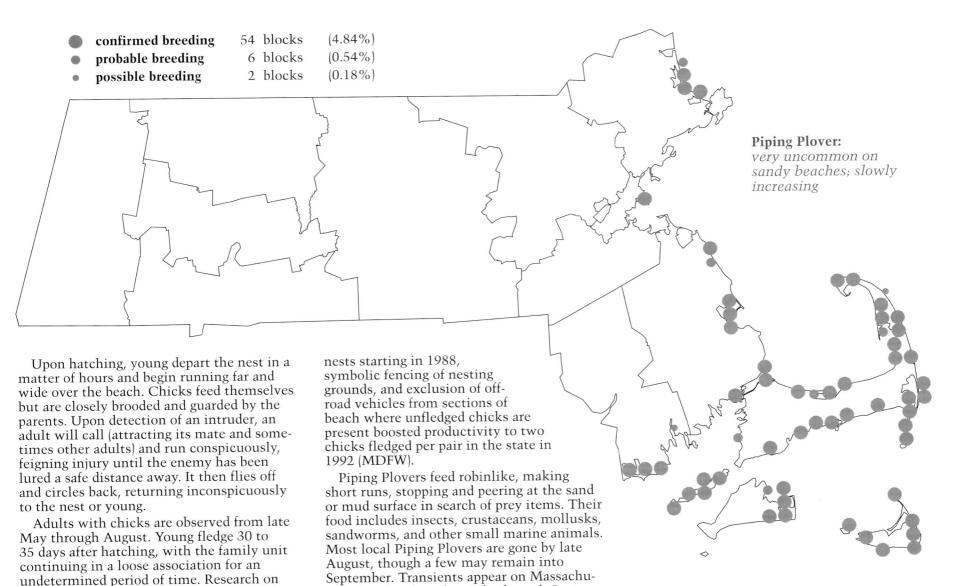

- confirmed breeding — 54 blocks (4.84%)
- probable breeding — 6 blocks (0.54%)
- possible breeding — 2 blocks (0.18%)

Piping Plover: *very uncommon on sandy beaches; slowly increasing*

Upon hatching, young depart the nest in a matter of hours and begin running far and wide over the beach. Chicks feed themselves but are closely brooded and guarded by the parents. Upon detection of an intruder, an adult will call (attracting its mate and sometimes other adults) and run conspicuously, feigning injury until the enemy has been lured a safe distance away. It then flies off and circles back, returning inconspicuously to the nest or young.

Adults with chicks are observed from late May through August. Young fledge 30 to 35 days after hatching, with the family unit continuing in a loose association for an undetermined period of time. Research on outer Cape Cod in 1985 to 1987 showed that breeding success was extremely low due to predation by Red Foxes, Striped Skunks, crows, and gulls, as well as due to human disturbance resulting in mean productivity of less than one chick fledged per pair (MacIvor 1990). Use of special predator exclosures for nests starting in 1988, symbolic fencing of nesting grounds, and exclusion of off-road vehicles from sections of beach where unfledged chicks are present boosted productivity to two chicks fledged per pair in the state in 1992 (MDFW).

Piping Plovers feed robinlike, making short runs, stopping and peering at the sand or mud surface in search of prey items. Their food includes insects, crustaceans, mollusks, sandworms, and other small marine animals. Most local Piping Plovers are gone by late August, though a few may remain into September. Transients appear on Massachusetts beaches from late July through September. The winter range is primarily on the Atlantic and Gulf coasts from North Carolina south to Florida and west to eastern Texas and, less commonly, throughout the Bahamas and Greater Antilles (east to the Virgin Islands).

The Piping Plover is listed as federally threatened in Massachusetts.

Erma J. Fisk and Bradford G. Blodget

Killdeer
Charadrius vociferus

Egg dates: April 15 to July 20.

Number of broods: one or two.

The wanton destruction of birds, especially shorebirds, at the turn of the century nearly extirpated the Killdeer as a breeding species in the state. Its numbers quickly recovered when it was afforded protection, and it soon reoccupied much of its former breeding range. At present, the Killdeer is a widespread but fairly common breeding resident and, although long established on Martha's Vineyard, is rather rare on Cape Cod and Nantucket. Killdeer are scarce in the higher elevations of Worcester and Berkshire counties but do nest there wherever suitable habitat is found.

This species is among the earliest spring migrants to arrive in our area. In normal years, Killdeer appear by the middle of March, but in exceptionally warm years they may return in late February. Early arrivals are found in agricultural fields, congregating around snowless areas where manure has been spread for fertilizer. By mid-April, Killdeer are present in a variety of open areas, where they will establish territories for nesting. These include fields, pastures, airports, golf courses, playgrounds, gravel pits, beach dunes, unpaved driveways, and lawns. In recent years, they have taken to nesting on the flat rooftops of one- or two-story buildings. Despite the questionable suitability of the latter nesting sites, there is evidence that some young are raised to maturity.

The Killdeer derives its name from its familiar call, a strident *killdee, killdee.* When approached in the vicinity of its nest, the bird announces its agitation by uttering a rolling call that often introduces or accompanies its elaborate distraction display. Another vocalization frequently given in the proximity of the nest or chicks is a nervous *dee, dee, dee, dee.*

Killdeer usually lay four eggs in a nest scrape lined with pebbles or grass, as was the case for eight Massachusetts nest records from various sources. If the first nest is lost, the birds generally will lay another set of eggs, and some pairs may rear two broods. Both sexes take an equal share in the 25-day incubation period. Once the eggs hatch, the young almost immediately are capable of moving about and feeding themselves. There is an interesting Massachusetts record of a male taking 2 mates and 1 female deserting her four eggs to help care for the young of the other female (CNR). The young fly about 25 days after hatching.

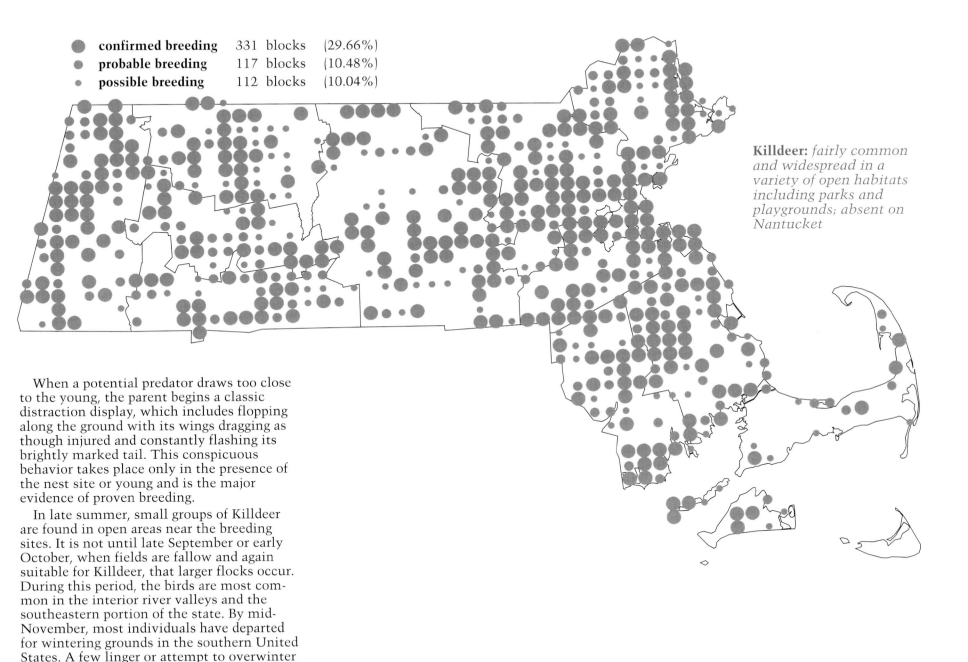

- confirmed breeding — 331 blocks (29.66%)
- probable breeding — 117 blocks (10.48%)
- possible breeding — 112 blocks (10.04%)

Killdeer: *fairly common and widespread in a variety of open habitats including parks and playgrounds; absent on Nantucket*

When a potential predator draws too close to the young, the parent begins a classic distraction display, which includes flopping along the ground with its wings dragging as though injured and constantly flashing its brightly marked tail. This conspicuous behavior takes place only in the presence of the nest site or young and is the major evidence of proven breeding.

In late summer, small groups of Killdeer are found in open areas near the breeding sites. It is not until late September or early October, when fields are fallow and again suitable for Killdeer, that larger flocks occur. During this period, the birds are most common in the interior river valleys and the southeastern portion of the state. By mid-November, most individuals have departed for wintering grounds in the southern United States. A few linger or attempt to overwinter at coastal locations.

Richard A. Forster

American Oystercatcher
Haematopus palliatus

Egg dates: April 22 to July 4.

Number of broods: one; may re-lay if first attempt fails.

One of the state's most spectacular birds, the American Oystercatcher, is a fairly recent addition to Massachusetts' breeding avifauna. However, its appearance is more a matter of reestablishing itself in territory once occupied than a bona fide range extension. Oystercatchers were apparently common in this area during the early 1800s (and Audubon claims to have seen them as far north as the Saint Lawrence River), although it is not known if they nested during that period. In any event, by the mid-1800s they had disappeared from the entire Northeast, due in part to excessive hunting; and the species remained a rare vagrant until the mid-1900s. With protection, oystercatchers gradually began to reclaim some of their former range, and in 1969 a pair nested on Martha's Vineyard, establishing the first confirmed breeding record for the state. By 1984, the Massachusetts population had increased to an estimated 42 pairs.

In Massachusetts, nesting oystercatchers are found almost exclusively on the islands off the southern coast—Monomoy, Nantucket, Martha's Vineyard, and the Elizabeth Islands, where they inhabit sandy beaches and salt marshes adjacent to the tidal flats where they feed. Oystercatchers are among the earliest residents to appear on our coastal beaches, often arriving by mid-March if the weather is mild. Soon after, they begin to establish territories, often with the same pairs occupying the same sites as in previous years. Some pairs defend a single territory that contains both nesting and feeding areas, while others have disjunct nesting and feeding territories that may be separated by as much as several hundred yards.

Oystercatcher vocalizations are complex and varied, particularly during the breeding season. A frequently heard call is a loud, ascending *wheep,* often given in flight. A loud piping note, given singly or in series, is also common. Courtship and territorial displays are elaborate and noisy affairs that can best be described as comical and that involve exaggerated posturing, display flights, and various displacement activities, all accompanied by loud, rapid calls. Birds from neighboring territories as well as nonbreeding immature individuals (up to three or more years old) often become involved in these performances, resulting in chaotic scenes.

The nest, merely a shallow depression in the sand or dry marsh grass, is located along the dune or marsh edge or on an elevated ridge within the marsh. Each year, a few nests that are elevated inadequately are lost to the spring tides, in which case the pair usually renests. Clutch size ranges from two to five eggs, although the majority of nests

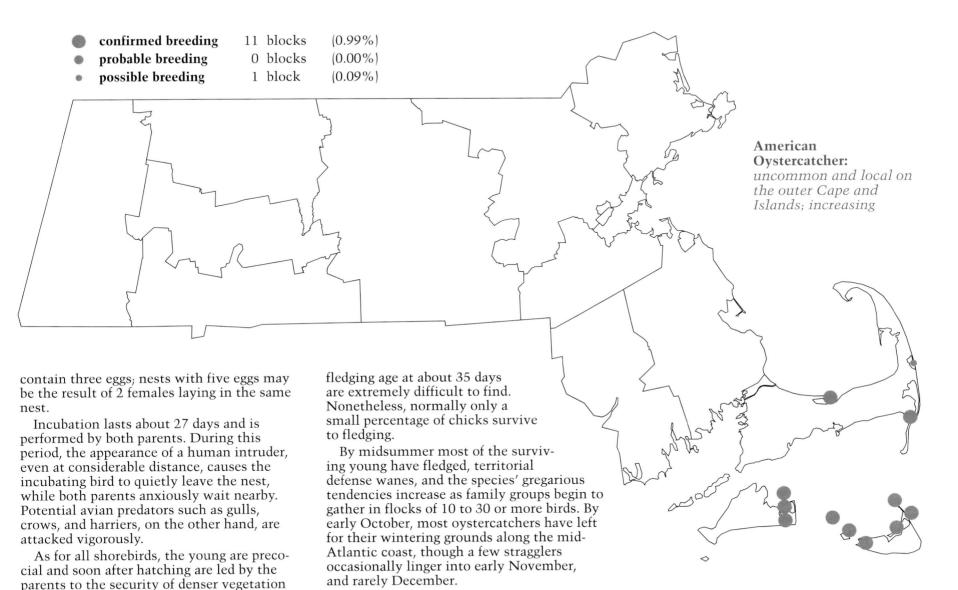

American Oystercatcher: *uncommon and local on the outer Cape and Islands; increasing*

- ● confirmed breeding — 11 blocks (0.99%)
- ● probable breeding — 0 blocks (0.00%)
- • possible breeding — 1 block (0.09%)

contain three eggs; nests with five eggs may be the result of 2 females laying in the same nest.

Incubation lasts about 27 days and is performed by both parents. During this period, the appearance of a human intruder, even at considerable distance, causes the incubating bird to quietly leave the nest, while both parents anxiously wait nearby. Potential avian predators such as gulls, crows, and harriers, on the other hand, are attacked vigorously.

As for all shorebirds, the young are precocial and soon after hatching are led by the parents to the security of denser vegetation near the pair's feeding territory. However, unlike most shorebirds, young oystercatchers are totally dependent upon the adults for food, and not until they are several months old are they able to fully feed themselves. The chicks are remarkably adept at hiding in the marsh grasses, and until they reach fledging age at about 35 days are extremely difficult to find. Nonetheless, normally only a small percentage of chicks survive to fledging.

By midsummer most of the surviving young have fledged, territorial defense wanes, and the species' gregarious tendencies increase as family groups begin to gather in flocks of 10 to 30 or more birds. By early October, most oystercatchers have left for their wintering grounds along the mid-Atlantic coast, though a few stragglers occasionally linger into early November, and rarely December.

Blair Nikula

Willet
Catoptrophorus semipalmatus

Egg dates: late May to mid-June.

Number of broods: one.

Most early accounts of the Willet note its absence as a nesting species in Massachusetts. Once a breeder along the coast from Nova Scotia to Texas, the Willet suffered drastic population declines due to market gunning in the late 1800s and early 1900s. The result was a fragmented breeding range, with the birds largely wiped out north of Virginia. Willets remained rare spring and casual fall migrants in the Commonwealth; however, with full protection, the species has begun to recover. Willets returned to breed in New England in 1971, when nesting was confirmed in southern Maine. The first recent Massachusetts nesting occurred during the Atlas period in 1976.

Migration in the spring may be offshore, with Willets dropping out along the coast at their nesting beaches. The birds begin moving north in March and reach Massachusetts by May 1. They can be seen foraging in and around salt marshes, tidal flats, and sandy beaches for insects, worms, mollusks, fish, seeds, and plant shoots.

Willets are very vocal birds, certainly more so than most other shorebirds. They make intruders aware of their presence with a loud *pill-will-willet*, a repetitive *kip-kip-kip*, or a *whee-wee-wee* while in flight. The vocalizations diminish as the nesting season winds down. Courtship is equally visible and noisy, with the male chasing the female, calling and fluttering his boldly marked wings, and seeming to hang motionless in the air. On the ground, the male holds up his wings in order

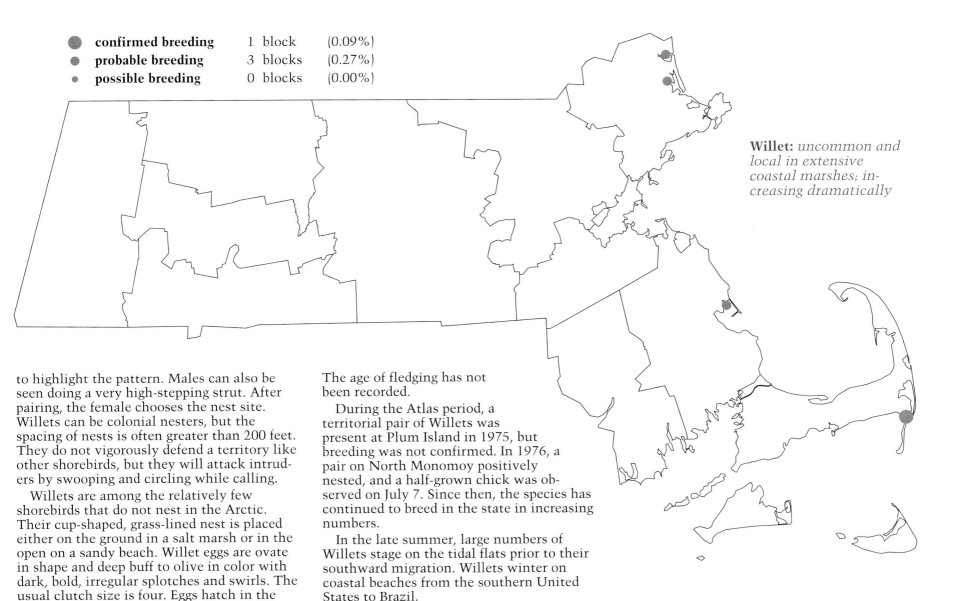

- ● **confirmed breeding** 1 block (0.09%)
- ● **probable breeding** 3 blocks (0.27%)
- • **possible breeding** 0 blocks (0.00%)

Willet: *uncommon and local in extensive coastal marshes; increasing dramatically*

to highlight the pattern. Males can also be seen doing a very high-stepping strut. After pairing, the female chooses the nest site. Willets can be colonial nesters, but the spacing of nests is often greater than 200 feet. They do not vigorously defend a territory like other shorebirds, but they will attack intruders by swooping and circling while calling.

Willets are among the relatively few shorebirds that do not nest in the Arctic. Their cup-shaped, grass-lined nest is placed either on the ground in a salt marsh or in the open on a sandy beach. Willet eggs are ovate in shape and deep buff to olive in color with dark, bold, irregular splotches and swirls. The usual clutch size is four. Eggs hatch in the order in which they are laid after an incubation period of 22 to 29 days. The young have short beaks, are precocial, and are camouflaged in buff to sepia down. Juvenal plumage is attained when they are half-grown. Both parents incubate and help raise the young.

The age of fledging has not been recorded.

During the Atlas period, a territorial pair of Willets was present at Plum Island in 1975, but breeding was not confirmed. In 1976, a pair on North Monomoy positively nested, and a half-grown chick was observed on July 7. Since then, the species has continued to breed in the state in increasing numbers.

In the late summer, large numbers of Willets stage on the tidal flats prior to their southward migration. Willets winter on coastal beaches from the southern United States to Brazil.

Robert Prescott

Spotted Sandpiper
Actitis macularia

Egg dates: May 6 to July 5.

Number of broods: one; may re-lay if first attempt fails.

The familiar Spotted Sandpiper, or Teeter-up, is a common migrant and fairly common breeding species throughout the state. Although it is frequently encountered along the shores of freshwater lakes, ponds, and streams, it is more abundant both as a migrant and as a summer resident along the seacoast.

A few Spotted Sandpipers arrive in late April, but it is not until the early part of May that they become evident on a wide scale. Assemblages of migrants are sometimes encountered along the coast in mid-May. By late May, most of the resident breeders are paired. At this time, when an observer approaches along a shoreline, a Spotted Sandpiper will fly out over the water in distinctive bowed-winged flight, uttering a *peet-weet* call.

In this species, it is the slightly larger female that is more aggressive, defending territory and courting males. When displaying, a female flies up and then glides in to land near a prospective mate. She may strut about like a miniature turkey. Both sexes give a *weet-weet-weet-weet-weet* call during courtship.

In some cases, Spotted Sandpipers form monogamous pair bonds. Studies have shown that females may be polyandrous, mating with two to four males in succession and sharing parental duties with the final mate. It appears that even when the female is present, the male assumes most of the responsibility of incubation and care of the young.

The nest is located on the ground and is generally a hollow, lined sparingly with grasses. Along the coast, the nest may be no more than a shallow depression in the dunes. Although the nest is usually placed near water, Spotted Sandpipers may select a site at the edge of a cultivated field or other area of sparse vegetation some distance from water.

The usual shorebird complement of four eggs (sometimes two, three, or five) is completed by the end of May. Later clutches probably represent attempts at renesting. Incubation lasts 20 to 24 days, and the chicks leave the nest soon after hatching, although they may return there in the evening to be brooded. The cryptically marked young quickly adopt the curious and comical bobbing behavior of the adults. Nine Massachusetts records of an adult with one to four young range from June 1 (chicks several days old) to July 17. The latest date for a downy youngster was July 14 (CNR, TC). The young fledge when they are about 17 days old, at which time the family breaks up.

Spotted Sandpipers are rather inconspicuous in midsummer, at which time the adults molt into a drab plumage that lacks their distinctive black-spotted underparts and

- confirmed breeding   87 blocks   (7.80%)
- probable breeding   79 blocks   (7.08%)
- possible breeding   109 blocks   (9.77%)

Spotted Sandpiper: *common and widespread, usually near open water but occasionally in damp meadows*

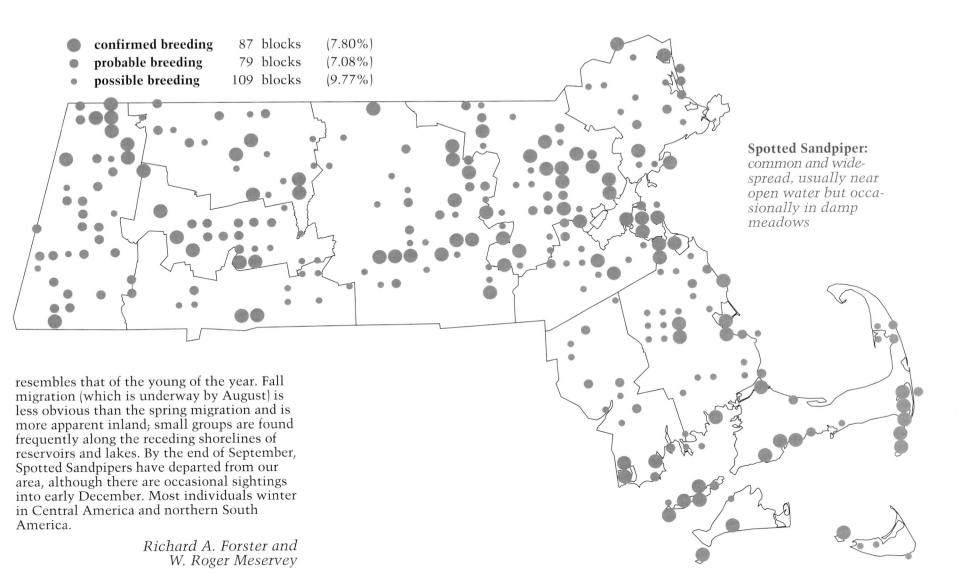

resembles that of the young of the year. Fall migration (which is underway by August) is less obvious than the spring migration and is more apparent inland; small groups are found frequently along the receding shorelines of reservoirs and lakes. By the end of September, Spotted Sandpipers have departed from our area, although there are occasional sightings into early December. Most individuals winter in Central America and northern South America.

*Richard A. Forster and
W. Roger Meservey*

Upland Sandpiper
Bartramia longicauda

Egg dates: May 9 to July 6.

Number of broods: one; may re-lay if first attempt fails.

A bird of the prairies and open grasslands, the Upland Sandpiper was probably uncommon in Massachusetts prior to colonial times when only unforested areas of Cape Cod and the larger islands provided suitable habitat. European settlement created extensive nesting areas through clearing for agriculture and grazing, and by the mid-1800s the Upland Sandpiper had become abundant. Heavy hunting pressure led to severe population reductions, and by the 1920s the species was again a rare breeder in the state.

Although now protected from hunting, the Upland Sandpiper is threatened again, this time by a loss of habitat brought about by development pressure, decline of agricultural lands, and the natural succession of open grasslands to forests. It breeds at fewer than 10 sites in the state; breeding no longer occurs on Nantucket or Martha's Vineyard. Since the Atlas period, nesting has been confirmed at military bases on Cape Cod and in Worcester County.

The Upland Sandpiper returns to Massachusetts during mid- to late April, where its preferred habitat is limited to scattered blocks of pastures and hayfields and the flat open expanses of several of the larger airports. Formerly, the species was an occasional nester in salt marshes. Management to maintain large tracts of open grasslands is essential to the conservation of this species. Lightly to moderately grazed pastures provide excellent habitat.

Courtship takes place both in flight and on the ground. Courting Upland Sandpipers will circle in the air at great heights, calling with a long, drawn-out whistled *whip-whee-ee-you*, then plummet swiftly toward the ground. A *guip-ip-ip-ip* alarm call and *quitty-quitty-quit* are other common vocalizations. In flight, the stiff downward curve of the bowed wings is distinctive. When alighting on fence posts or utility poles and wires, the birds characteristically hold their wings aloft

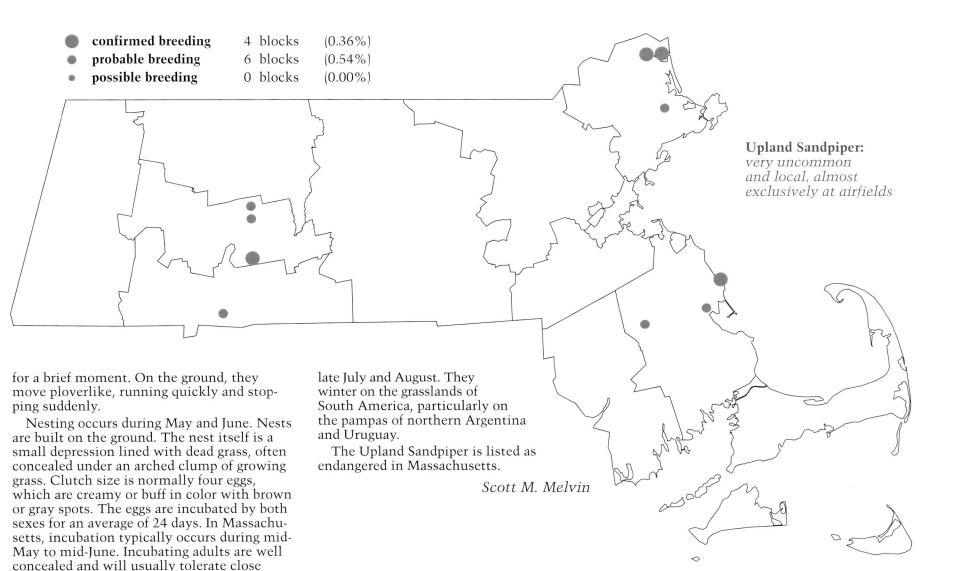

Upland Sandpiper: *very uncommon and local, almost exclusively at airfields*

- ● confirmed breeding — 4 blocks (0.36%)
- ● probable breeding — 6 blocks (0.54%)
- • possible breeding — 0 blocks (0.00%)

for a brief moment. On the ground, they move ploverlike, running quickly and stopping suddenly.

Nesting occurs during May and June. Nests are built on the ground. The nest itself is a small depression lined with dead grass, often concealed under an arched clump of growing grass. Clutch size is normally four eggs, which are creamy or buff in color with brown or gray spots. The eggs are incubated by both sexes for an average of 24 days. In Massachusetts, incubation typically occurs during mid-May to mid-June. Incubating adults are well concealed and will usually tolerate close approach before flushing from the nest. Chicks are precocial and fledge during July, when they are about 30 days old. Both adults and juveniles feed on a variety of insects, weed seeds, and, when available, small grains.

Upland Sandpipers begin to flock together in July and initiate their fall migration during late July and August. They winter on the grasslands of South America, particularly on the pampas of northern Argentina and Uruguay.

The Upland Sandpiper is listed as endangered in Massachusetts.

Scott M. Melvin

Least Sandpiper
Calidris minutilla

Egg dates: not available.

Number of broods: one; may relay if first attempt fails.

Although a common migrant in suitable habitat throughout most of Massachusetts, the Least Sandpiper takes its place among the state's breeding birds on the basis of a single chick, no more than 3 days old, found freshly dead on Monomoy Island on July 12, 1979. The chick was completely downy, smaller than a Ping-Pong ball, white below and dark chestnut brown above, the white-tipped down on the back and crown giving it a spangled appearance. The normal breeding range of the Least Sandpiper stretches from western Alaska across northern (but not high Arctic) Canada east to Newfoundland, with a disjunct population on Sable Island off Nova Scotia and on the Nova Scotia mainland. There is also a recent breeding record from Machias Seal Island in New Brunswick.

As migrants, Least Sandpipers like mudflats, muddy margins of rivers and ponds, and wet agricultural land rather than sandy beaches. Wherever found, Least Sandpipers are the smallest of our peeps and the most mouselike, running about in a crouching posture along the edges of vegetation, often keeping up a constant chittering as they pick at minute organisms. When flushed, they dart off like tiny snipe, uttering a high-pitched *threeep* note. Compared with other native peeps, they are smaller, darker, browner on both back and breast, and the only ones with yellowish legs. Their spring migration is limited mostly to May, peaking during the second week.

No one has ever reported seeing a displaying Least Sandpiper in Massachusetts. However, working back from July 12, it is reasonable to presume that, sometime during the second week of June 1979, a male Least Sandpiper probably engaged in an early-morning aerial display over the Monomoy mudflats. First flying upward at a steep angle, then leveling off to alternate fluttering with a stiff-winged sailing, a female was attracted to this display, which was accompanied by a song of short, high notes. Later, she

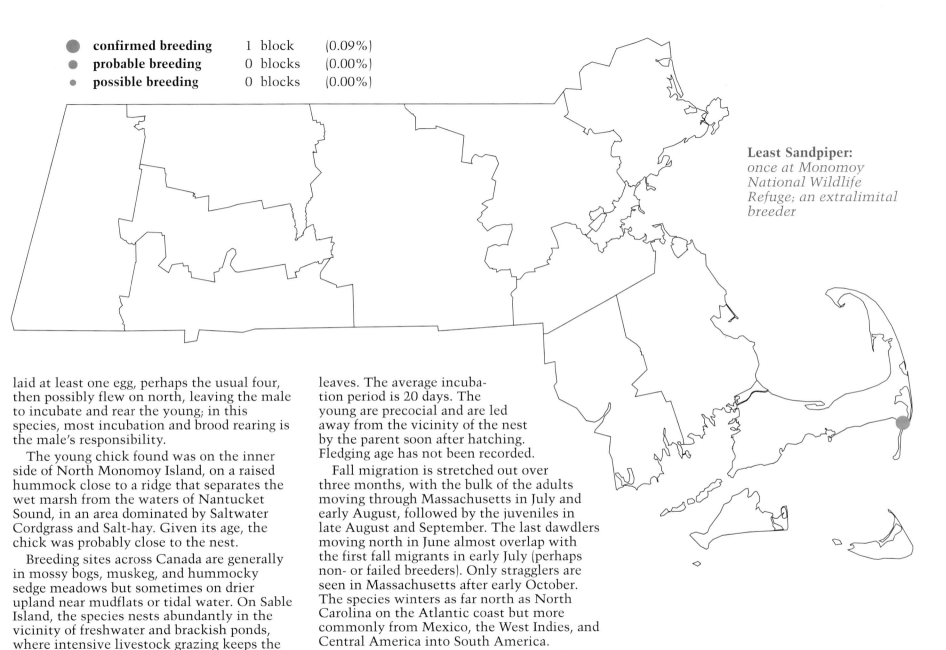

- confirmed breeding 1 block (0.09%)
- probable breeding 0 blocks (0.00%)
- possible breeding 0 blocks (0.00%)

Least Sandpiper: *once at Monomoy National Wildlife Refuge; an extralimital breeder*

laid at least one egg, perhaps the usual four, then possibly flew on north, leaving the male to incubate and rear the young; in this species, most incubation and brood rearing is the male's responsibility.

The young chick found was on the inner side of North Monomoy Island, on a raised hummock close to a ridge that separates the wet marsh from the waters of Nantucket Sound, in an area dominated by Saltwater Cordgrass and Salt-hay. Given its age, the chick was probably close to the nest.

Breeding sites across Canada are generally in mossy bogs, muskeg, and hummocky sedge meadows but sometimes on drier upland near mudflats or tidal water. On Sable Island, the species nests abundantly in the vicinity of freshwater and brackish ponds, where intensive livestock grazing keeps the vegetation short. The nest is a small depression, lined with a bit of grass and a few dead leaves. The average incubation period is 20 days. The young are precocial and are led away from the vicinity of the nest by the parent soon after hatching. Fledging age has not been recorded.

Fall migration is stretched out over three months, with the bulk of the adults moving through Massachusetts in July and early August, followed by the juveniles in late August and September. The last dawdlers moving north in June almost overlap with the first fall migrants in early July (perhaps non- or failed breeders). Only stragglers are seen in Massachusetts after early October. The species winters as far north as North Carolina on the Atlantic coast but more commonly from Mexico, the West Indies, and Central America into South America.

Kathleen S. Anderson

Wilson's Snipe
Gallinago delicata

Egg dates: not available.

Number of broods: one; may re-lay if first attempt fails.

Snipe were hunted heavily year-round in the 1800s but did not suffer extreme declines due to their occurring in loose flocks and their predilection for remote habitats. They can, however, be stressed seriously by droughts on the breeding grounds and freezing spells on their wintering grounds. Currently, the snipe is believed to be common to abundant throughout most of its range. In Massachusetts, the discontinuance of mowing wet meadows after 1910 resulted in a decrease in snipe habitat. Snipe historically bred in low numbers in eastern and central Massachusetts (and rarely in western Massachusetts), and they continue to breed locally in small numbers.

Snipe, like woodcocks, are early migrants. They begin departing the wintering grounds in early or mid-March, first as individuals or small groups, but flocking as the migration peaks in late March or April. Gregarious behavior diminishes en route, and the birds appear to arrive singly at the breeding grounds. Males arrive 10 or more days before the females. Spring migrants pass through Massachusetts between mid-March and mid-May, with resident birds usually at breeding sites by late March or early April.

The primary nesting grounds of the snipe are the peat lands of the boreal forest, composed of bogs, fens, and swamps with an understory of alder, Sweet Gale, and willow interspersed with clumps of larch, spruce, and fir. Snipe secondarily frequent matted areas of rotting vegetation along ponds, creeks, and wet meadows. In winter, they use coastal fresh and brackish marshlands, wet pastures, rice fields, and other sporadically flooded wetlands.

Male snipe begin displaying on arrival on the breeding grounds. The winnowing display, usually manifested at dusk or on moonlit nights (but sometimes during daylight), consists of a rapid high flight followed by a steep dive, during which air vibration around the center tail feathers produces a humming sound. During pair formation, males also fly over a female with arched wings and legs dangling. Displays or calls by intruding males may elicit sparring or other agonistic behavior. Both sexes produce a monotonous *cutacuta...*call or an excited yakking sequence during ground displays. Snipe also voice a hoarse *scaipe* call when flushed and use an abbreviated *cutacuta* call to summon lost chicks.

The nest is usually just above the water level and may be lined thoroughly with grass and sedges. Clutch size is normally four, but three eggs are not uncommon. Yearlings nesting in late summer appear to lay smaller clutches. Peak hatching in Newfoundland is in June. The young leave the nest immediately but are fed and brooded by the parents for about 10 days. Both sexes play a role in tending the young.

Little specific information about the nesting habits of snipe in Massachusetts is

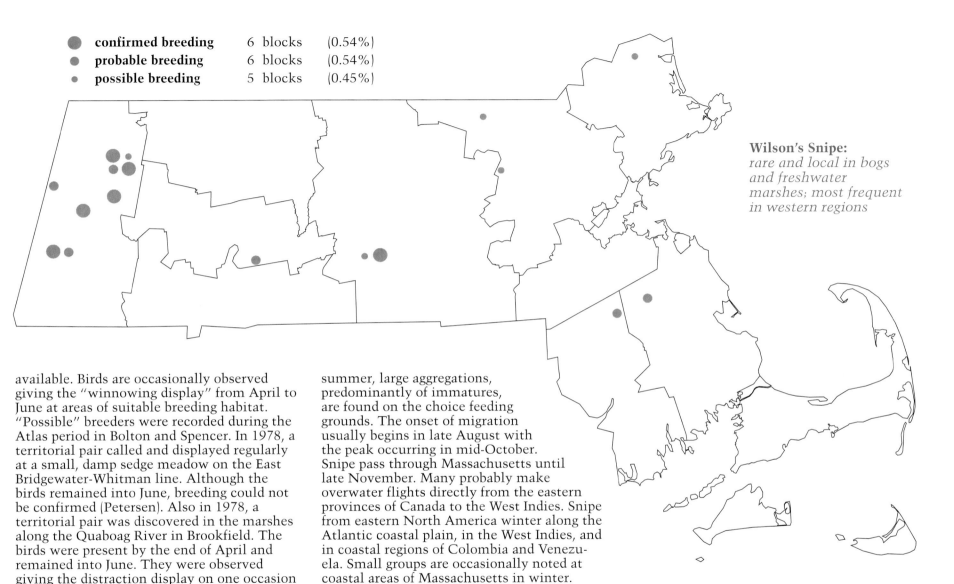

Wilson's Snipe: *rare and local in bogs and freshwater marshes; most frequent in western regions*

- confirmed breeding — 6 blocks (0.54%)
- probable breeding — 6 blocks (0.54%)
- possible breeding — 5 blocks (0.45%)

available. Birds are occasionally observed giving the "winnowing display" from April to June at areas of suitable breeding habitat. "Possible" breeders were recorded during the Atlas period in Bolton and Spencer. In 1978, a territorial pair called and displayed regularly at a small, damp sedge meadow on the East Bridgewater-Whitman line. Although the birds remained into June, breeding could not be confirmed (Petersen). Also in 1978, a territorial pair was discovered in the marshes along the Quaboag River in Brookfield. The birds were present by the end of April and remained into June. They were observed giving the distraction display on one occasion in early June (Meservey). After the Atlas period, a small, downy youngster was seen in Bolton on July 4 (McMenemy).

When the chicks are about six weeks old, they aggregate with others of the same age class while the adults are molting. By late summer, large aggregations, predominantly of immatures, are found on the choice feeding grounds. The onset of migration usually begins in late August with the peak occurring in mid-October. Snipe pass through Massachusetts until late November. Many probably make overwater flights directly from the eastern provinces of Canada to the West Indies. Snipe from eastern North America winter along the Atlantic coastal plain, in the West Indies, and in coastal regions of Colombia and Venezuela. Small groups are occasionally noted at coastal areas of Massachusetts in winter.

James E. Cardoza

American Woodcock
Scolopax minor

Egg dates: March 20 to June 15.

Number of broods: one; may re-lay if first attempt fails.

At the turn of the century, the American Woodcock was a vanishing bird, principally due to unregulated spring shooting and wetland drainage. The enactment of protective legislation resulted in a resurgence of woodcock populations by the 1920s. However, woodcock numbers in the Northeast are once again on the decline because of loss of habitat and lowered survival rates resulting from development and succession of abandoned-field and alder habitats. Annual ground surveys of singing males were instituted in 1953 as a measure of breeding trends. During the Atlas period, woodcocks were "confirmed" in all sections of the state.

Woodcocks are exceptionally early migrants, often arriving in the Northeast when there is still snow on the ground. In most years, they reach central Massachusetts by mid-March, but February records occur frequently. Woodcocks habitually return to the same breeding grounds each year, as demonstrated by research in Quabbin, where 80 percent of 84 banded males returned to within 1 mile of the initial capture site.

Typical woodcock singing grounds are old fields, pastures, or boggy areas. Daytime habitat includes alder swales, patches of second-growth hardwoods, and mixed forest with rich, moist soil. Night roosting occurs in fields and forest openings. On the wintering grounds, woodcocks concentrate in bottomlands, cleared croplands, and pastures.

The evocative courtship activities of the male woodcock consist of an aerial display and flight song, preceded and followed by location calls and occasional strutting on the ground. The singing area is defended aggressively against other males. The auditory element of the display flight is both mechanical and vocal, with a twittering, whistling sound generated by air currents rushing through three notched outer primaries, combined with vocalizations consisting of a series of high-pitched chirps. Ground calls consist of a nasal, buzzy *peent* preceded by a soft *took-oo*. Females may *peent* occasion-

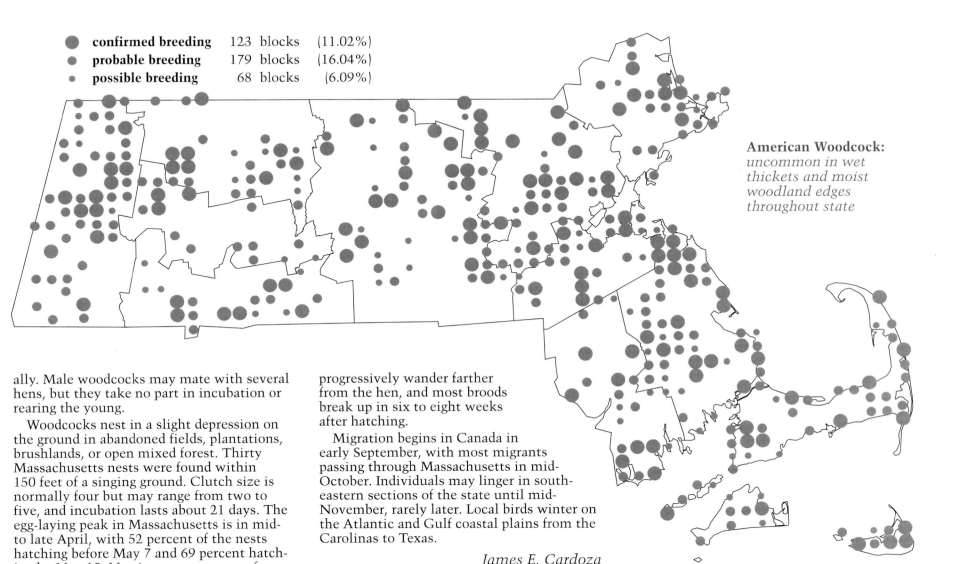

- ● **confirmed breeding** 123 blocks (11.02%)
- ● **probable breeding** 179 blocks (16.04%)
- · **possible breeding** 68 blocks (6.09%)

American Woodcock: *uncommon in wet thickets and moist woodland edges throughout state*

ally. Male woodcocks may mate with several hens, but they take no part in incubation or rearing the young.

Woodcocks nest in a slight depression on the ground in abandoned fields, plantations, brushlands, or open mixed forest. Thirty Massachusetts nests were found within 150 feet of a singing ground. Clutch size is normally four but may range from two to five, and incubation lasts about 21 days. The egg-laying peak in Massachusetts is in mid- to late April, with 52 percent of the nests hatching before May 7 and 69 percent hatching by May 15. Nesting success ranges from 42 percent in Massachusetts to 62 percent in Maine. The young are precocial, although the hen broods them frequently for the first several days and for lesser periods until their primary feathers develop. Weather extremes commonly affect breeding woodcocks by chilling chicks and by depressing earthworm availability. As they mature, the juveniles progressively wander farther from the hen, and most broods break up in six to eight weeks after hatching.

Migration begins in Canada in early September, with most migrants passing through Massachusetts in mid-October. Individuals may linger in southeastern sections of the state until mid-November, rarely later. Local birds winter on the Atlantic and Gulf coastal plains from the Carolinas to Texas.

James E. Cardoza

Wilson's Phalarope
Phalaropus tricolor

Egg dates: June.

Number of broods: one.

The main breeding grounds of the Wilson's Phalarope are the shallow, cattail-rimmed sloughs, marshy farm ponds, and wet grassy meadows of the western plains states and occidental provinces of Canada. There, in prairie pothole country, the ornately plumaged Wilson's Phalarope shares its home with the vast majority of the continent's breeding "puddle ducks."

The coast of Massachusetts is a long way from the prairies, but the pioneering Wilson's Phalarope, the landlubber of the phalarope trio and the only species limited to the New World, has worked its way eastward in recent years, extending its limits and establishing outposts disjunct from its normal range. By 1978, seven pairs were recorded in Quebec about 50 miles northeast of Montreal, and in previous seasons there had been suspected nesting in New Brunswick, upstate New York, and Virginia. The discovery of a pair and nest with three eggs at Plum Island on June 29, 1979, provided the first breeding record for Massachusetts and the East Coast. An unusual aspect at this site was the phalaropes' choice of nesting habitat: a Salt-hay marsh instead of the typical freshwater slough or impoundment. The summer after the Atlas period, a male Wilson's Phalarope was flushed from a nest also containing three eggs in the Monomoy Island marsh on June 7 (Petersen). These two occurrences remain the only "confirmed" evidence of breeding activity in the region, although the continued presence of multiple pairs at Plum Island each summer strongly suggests that they are now nesting regularly there.

Wilson's Phalaropes arrive in spring during May and are seen at coastal locations throughout the summer, particularly at the known breeding sites. An interesting aspect of the breeding biology of all three species of phalaropes is a complete sex-role reversal. The larger, brighter female performs the displays, attracts a mate, and then leaves the rather drably attired male to incubate the eggs and begin tending the young. Both Massachusetts nests consisted of a grass-lined depression well concealed in saltwater grasses. Each contained three eggs, at least at the time of their discoveries, rather than the usual four typical of most shorebirds. Excessive high tides pose a very real danger to salt marsh nests, and the ultimate success of both nests was not known. After a 20- to 21-day incubation period, the precocial, downy young hatch and are soon led away from the nest to avoid detection by predators. Typically, a number of pairs nest in close proximity, forming a loose colony.

If one enters a territory, the male phalarope will fly up out of the marsh, seemingly

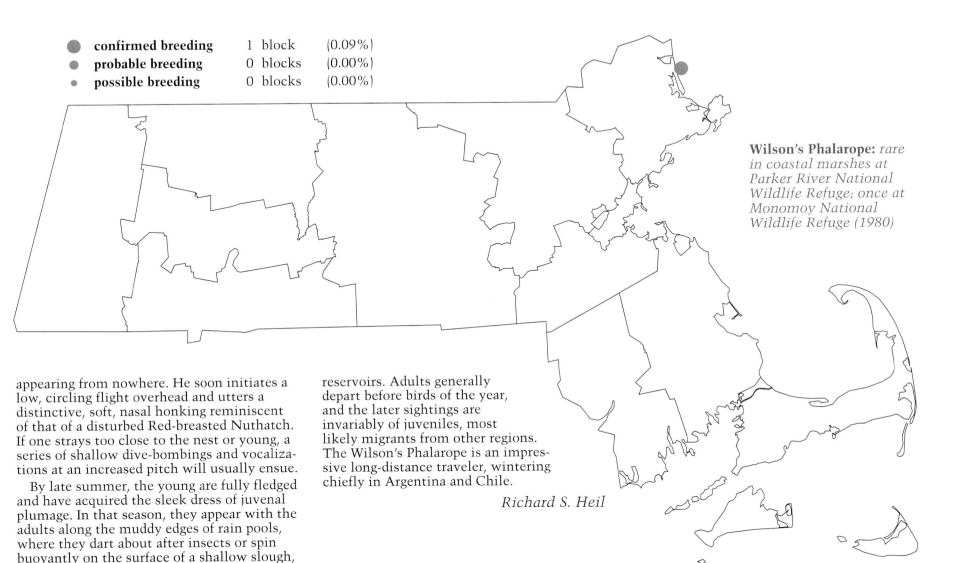

Wilson's Phalarope: *rare in coastal marshes at Parker River National Wildlife Refuge; once at Monomoy National Wildlife Refuge (1980)*

appearing from nowhere. He soon initiates a low, circling flight overhead and utters a distinctive, soft, nasal honking reminiscent of that of a disturbed Red-breasted Nuthatch. If one strays too close to the nest or young, a series of shallow dive-bombings and vocalizations at an increased pitch will usually ensue.

By late summer, the young are fully fledged and have acquired the sleek dress of juvenal plumage. In that season, they appear with the adults along the muddy edges of rain pools, where they dart about after insects or spin buoyantly on the surface of a shallow slough, picking items from the surface. In Massachusetts they are attracted to both salt pannes and muddy tidal flats. From July through September, the small Massachusetts population is usually supplemented by migrants from farther west. These migrant birds typically occur on the coastal plain, seeking out salt marsh pools, shallow farm ponds, and the muddy perimeters of drawn-down reservoirs. Adults generally depart before birds of the year, and the later sightings are invariably of juveniles, most likely migrants from other regions. The Wilson's Phalarope is an impressive long-distance traveler, wintering chiefly in Argentina and Chile.

Richard S. Heil

Laughing Gull
Larus atricilla

Egg dates: May 18 to July 12.
Number of broods: one.

The Laughing Gull can be seen along much of the Massachusetts coastline from April to November but is confined as a breeding bird to one or two of the outermost islands. Until 1972, the only recorded colony was on Muskeget Island. Abundant there in the mid-nineteenth century, Laughing Gulls were almost extirpated by plume hunters in the 1870s but recovered under protection in the 1890s, and increased dramatically to more than 20,000 pairs from 1936 to 1945. The island was then occupied by increasing numbers of Herring Gulls, and Laughing Gulls decreased steadily; the last 120 pairs were washed out by a hurricane in 1972. They then moved to Monomoy, where they prospered and increased to 1,000 pairs by 1981. However, the total Massachusetts population has now decreased to about 500 pairs, and many of the birds have moved to New Island in Orleans.

Laughing Gulls arrive in Massachusetts in early April and occupy the breeding areas in early May. One- and two-year-old birds migrate north and can be seen around the breeding colonies, but most do not nest until they are at least three years old. They forage along the shore and in tide rips, tidal bays, and inlets. Most birds feed within 10 miles of the breeding colony, but birds in breeding plumage can be seen regularly up to 40 miles away in Buzzards Bay and at Provincetown and Plymouth. They feed predominantly on crustaceans and other invertebrates, which they pick from the surface of the water. Some feed on Sand Eels or other fish swimming near the surface, while others scavenge at sewage outlets or around boats. They sometimes catch flying insects in the air. Occasionally, they will chase terns at the breeding colonies and steal fish from them. Laughing Gulls are rarely seen inland, but a few visit ponds on Cape Cod, and a few birds at Monomoy bring freshwater fish to their young. They nest on sand dunes, usually selecting areas of dense Dunegrass or Poison Ivy; occupation by Laughing Gulls seems to promote the spread of Poison Ivy into their nesting areas.

The most characteristic call on the breeding grounds is a maniacal laugh, *ha-ha-ha-ha-ha-ha-haah-haah-haah-haah*, usually given in display to other birds. When disturbed from the nesting area, Laughing Gulls circle or hover overhead with a babble of yelping alarm calls. The flight call is a gooselike *hap-hap.* Laughing Gulls commonly feed in groups, and they rest in flocks on the sand, on jetties, or on the water. During the breeding season, off-duty birds collect in flocks in an open area within or adjacent to the nesting area.

Laughing Gulls usually lay two or three eggs in a nest built from dried vegetation. Most eggs are laid between May 21 and June 10, although a few birds do not lay until late June. Nests are usually well concealed

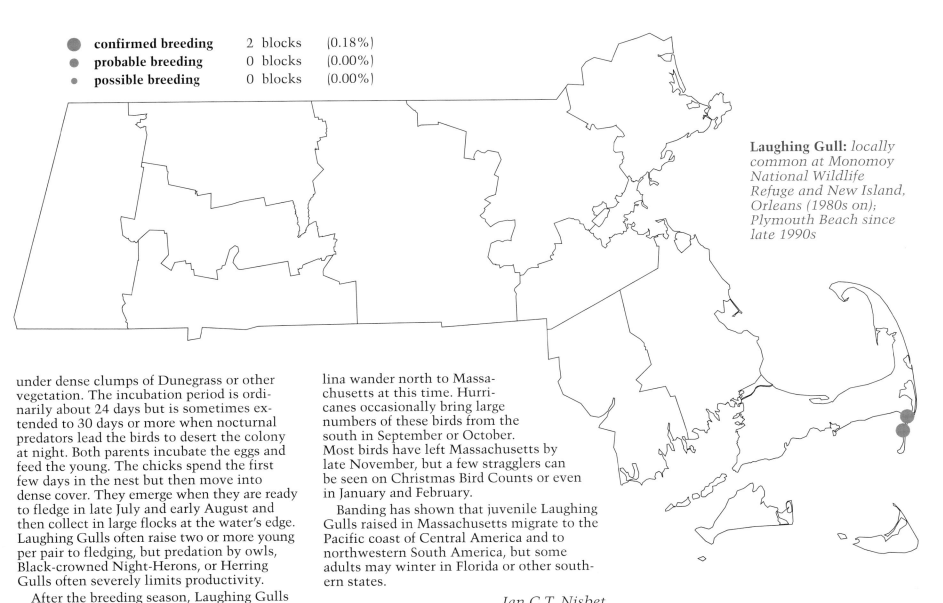

- confirmed breeding — 2 blocks (0.18%)
- probable breeding — 0 blocks (0.00%)
- possible breeding — 0 blocks (0.00%)

Laughing Gull: *locally common at Monomoy National Wildlife Refuge and New Island, Orleans (1980s on); Plymouth Beach since late 1990s*

under dense clumps of Dunegrass or other vegetation. The incubation period is ordinarily about 24 days but is sometimes extended to 30 days or more when nocturnal predators lead the birds to desert the colony at night. Both parents incubate the eggs and feed the young. The chicks spend the first few days in the nest but then move into dense cover. They emerge when they are ready to fledge in late July and early August and then collect in large flocks at the water's edge. Laughing Gulls often raise two or more young per pair to fledging, but predation by owls, Black-crowned Night-Herons, or Herring Gulls often severely limits productivity.

After the breeding season, Laughing Gulls disperse along the coast, and, in August through November, flocks of up to 500 birds can be seen around the Cape and Islands, with smaller numbers northward to Boston Harbor and on the North Shore. Banded birds from New Jersey, Virginia, and North Carolina wander north to Massachusetts at this time. Hurricanes occasionally bring large numbers of these birds from the south in September or October. Most birds have left Massachusetts by late November, but a few stragglers can be seen on Christmas Bird Counts or even in January and February.

Banding has shown that juvenile Laughing Gulls raised in Massachusetts migrate to the Pacific coast of Central America and to northwestern South America, but some adults may winter in Florida or other southern states.

Ian C.T. Nisbet

Herring Gull
Larus argentatus

Egg dates: April 25 to August 8.

Number of broods: one; may re-lay if first attempt fails.

This abundant, large, gray-mantled gull is no doubt the most well-known coastal bird in the northeastern United States. Formerly restricted to breeding sites along the seashore, it now nests on inland lakes and reservoirs given the opportunity. Islands are the preferred nesting domain of this omnivorous larid, and the substrate may be rocky or sandy. A dependable food source is the main influence on its distribution, and it may range from an inland garbage dump or coastal beach to an estuary or the waters surrounding an offshore fishing fleet.

The first Massachusetts Herring Gull nest was discovered on Martha's Vineyard in 1912. Subsequent breeding occurred in 1919 on Skiff's Island, an ephemeral sandbar off southeastern Martha's Vineyard. In 1925, Forbush doubted that the Herring Gull could survive as a breeding species in the face of a growing human population and increased beach recreation. However, the Herring Gull took hold, and the population gradually increased until the late 1940s, when it exploded. The increase began to level off in the mid-1960s, and the numbers have remained static since then. (The estimated Massachusetts population was 36,250 pairs in 1972 and 35,750 pairs in 1984.) The Herring Gull population has stabilized, apparently due at least in part to competition with Great Black-backed Gulls on nesting islands. Laughing Gulls and terns have declined as the more aggressive large gulls crowd them from favored nesting areas and prey on their eggs and young. In addition to its coastal colonies, the Herring Gull was also confirmed nesting inland at the Wachusett Reservoir. Breeding at the latter site was first detected in 1965, when 800 adults and 30 flightless young were estimated on July 25. The colony peaked in 1967, with 500 pairs. In that year, the Metropolitan District Commission initiated a gull-control program, collecting and removing the eggs. This policy has proved to be effective, and only a few pairs have nested there in recent years.

The loud buglelike feeding and territorial calls of this raucous bird are well known. Vocalizations are variable on the feeding grounds, while a low throaty *ow, ow, ow* is the common call of an adult soaring overhead. In mid-April, adult Herring Gulls (age four and older) begin using an extensive repertoire of displays and vocalizations to establish their breeding territories. When pair formation is concluded, nest building begins. The nest is built of dry grass obtained from the immediate vicinity and is constructed by both birds. It may be in the open or concealed neatly in a clump of Dunegrass or other vegetation. As is the case with most colonial seabirds, a strong bond exists between the parents. Appeasement displays are continued

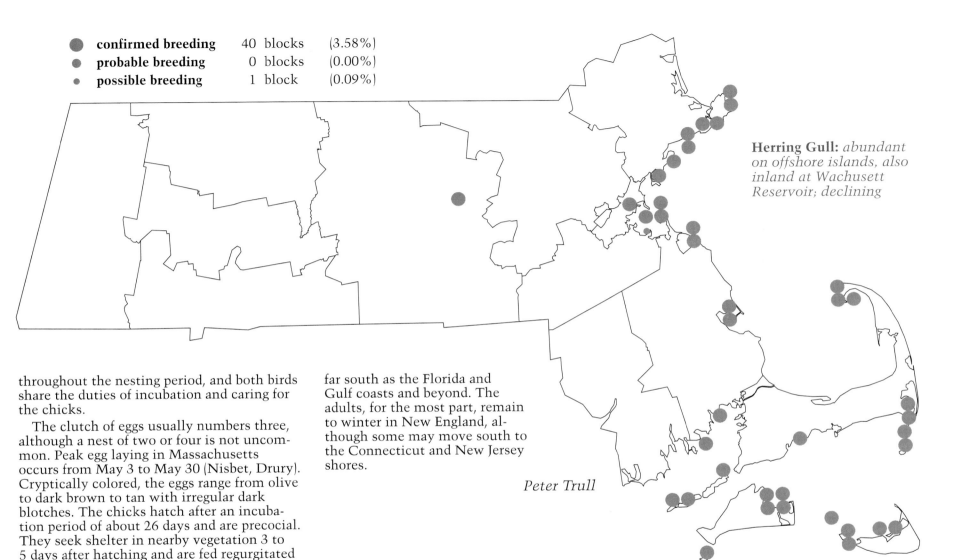

● confirmed breeding	40 blocks	(3.58%)
● probable breeding	0 blocks	(0.00%)
● possible breeding	1 block	(0.09%)

Herring Gull: *abundant on offshore islands, also inland at Wachusett Reservoir; declining*

throughout the nesting period, and both birds share the duties of incubation and caring for the chicks.

The clutch of eggs usually numbers three, although a nest of two or four is not uncommon. Peak egg laying in Massachusetts occurs from May 3 to May 30 (Nisbet, Drury). Cryptically colored, the eggs range from olive to dark brown to tan with irregular dark blotches. The chicks hatch after an incubation period of about 26 days and are precocial. They seek shelter in nearby vegetation 3 to 5 days after hatching and are fed regurgitated fish, crabs, and various bivalves. The average pair raises one or two young to fledging age (Nisbet).

As autumn approaches, the chocolate brown fledglings may be seen accompanying the parents to the feeding areas. Later, juvenile Herring Gulls migrate to wintering grounds that extend from the mid-Atlantic as far south as the Florida and Gulf coasts and beyond. The adults, for the most part, remain to winter in New England, although some may move south to the Connecticut and New Jersey shores.

Peter Trull

Great Black-backed Gull
Larus marinus

Egg dates: April 18 to July 15.

Number of broods: one; may re-lay if first attempt fails.

This large, northern gull has been extending its breeding range southward since the beginning of this century. Nova Scotia was the limit in 1920, but the species continued its southerly expansion, and pioneering Great Black-backed Gulls were found nesting in Massachusetts in Salem in 1931 and on the Weepecket Islands in 1941. As Great Black-backed Gulls continued to forge south, reaching North Carolina in 1985, the population in the Commonwealth grew to 3,000 pairs by 1972 and an estimated 11,000 breeding pairs by 1984. This explosive population growth was the result of protection and abundant food sources, such as town dumps; an increase in offshore fishing; and inappropriate fish handling by processors. All indications are that the breeding population has stabilized or may be declining.

Great Black-backed Gulls can be observed in Massachusetts year-round. As the scientific name implies, this is a bird of the coast and ocean; however, during the nonbreeding season, this gull may be found well inland along with the more numerous Herring Gull. Since the Atlas period, breeding has been confirmed for a few pairs inland at the Wachusett Reservoir. In spring, they hold breeding territories by early April, with courtship and pairing later in the month. Adults are basically permanent residents. Great Black-backed Gulls are usually found nesting in mixed colonies with Herring Gulls, preferably on barrier beaches and rocky islands but also occasionally on the mainland. Aggressive and dominant to other gulls and terns, Great Black-backed Gulls crowd them from desirable nesting areas and prey on their eggs and young.

This gull has a long call that is similar to, but deeper than, that of the Herring Gull. Similarly, its *kyow, kyow* call note is markedly deeper and distinctive and most often heard as it soars overhead. A repeated *kyow, kyow, kyow* is given if an intruder goes too near the nest or young.

The nest is bulky and made of dry grasses and seaweeds. A three-egg clutch is the norm, but two or four eggs are not uncommon. In Massachusetts, the peak laying period is

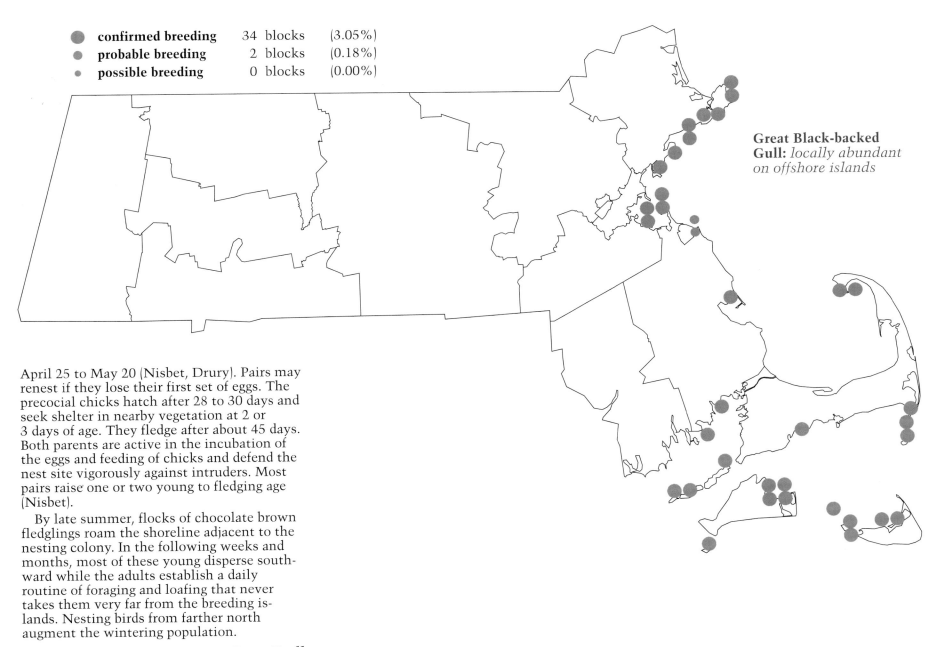

● **confirmed breeding**	34 blocks	(3.05%)
● **probable breeding**	2 blocks	(0.18%)
· **possible breeding**	0 blocks	(0.00%)

Great Black-backed Gull: *locally abundant on offshore islands*

April 25 to May 20 (Nisbet, Drury). Pairs may renest if they lose their first set of eggs. The precocial chicks hatch after 28 to 30 days and seek shelter in nearby vegetation at 2 or 3 days of age. They fledge after about 45 days. Both parents are active in the incubation of the eggs and feeding of chicks and defend the nest site vigorously against intruders. Most pairs raise one or two young to fledging age (Nisbet).

By late summer, flocks of chocolate brown fledglings roam the shoreline adjacent to the nesting colony. In the following weeks and months, most of these young disperse southward while the adults establish a daily routine of foraging and loafing that never takes them very far from the breeding islands. Nesting birds from farther north augment the wintering population.

Peter Trull

Roseate Tern
Sterna dougallii

Egg dates: May 12 to August 18.

Number of broods: one; may re-lay if first attempt fails.

Ever since the first historical records in the 1870s, the state has harbored more than half the Roseate Terns breeding in North America. Roseate Terns in Massachusetts were reduced by plume hunting to about 2,000 pairs in the 1890s, but increased under protection to a peak of about 5,000 pairs in the 1930s. They then decreased again to about 1,600 pairs in 1978 but subsequently have stabilized or slightly increased. Most of the remaining birds nest in one densely packed colony on Bird Island in Marion. The concentration of so much of the population on one small island makes it very vulnerable, and the Roseate Tern has been designated as an endangered species in the state.

Roseate Terns start to arrive on the Cape and Islands in early May. They feed along the coast and in bays and tidal inlets, preferring clear waters where tide rips or predatory fish bring Sand Eels and other bait fish to within 2 feet of the surface. They dive deeply into the water from heights of up to 40 feet. They nest on sandy or rocky islands, where they prefer areas that are more densely vegetated than those selected by other terns.

The Roseate Tern can be distinguished by its sharp, disyllabic call, *chi-vik*, which is uttered during feeding and courtship display. In response to intruders on the breeding colony, it gives a harsh alarm call, *aaach*, like tearing cloth, or a melodious *klew*. Roseate Terns have a spectacular aerial courtship display, in which groups of three to eight birds spiral up to heights of 100 to 600 feet and then descend in pairs in long, weaving glides. They rarely rest on the water but spend much time bathing in small groups in shallow water near shore. Loafing birds collect in flocks on beaches or sandflats, often with Common Terns.

Roseate Terns usually lay one or two eggs in a scrape concealed under a clump of Dunegrass or other plants. Most eggs are laid in late May to mid-June, but some younger birds do not start to lay until late June or July. After the eggs hatch, in 23 to 25 days, the chicks hide in long tunnels under the vegetation and emerge only to be fed. The parents alternate in incubating the eggs, brooding and guarding the chicks, and bringing food. The chicks are

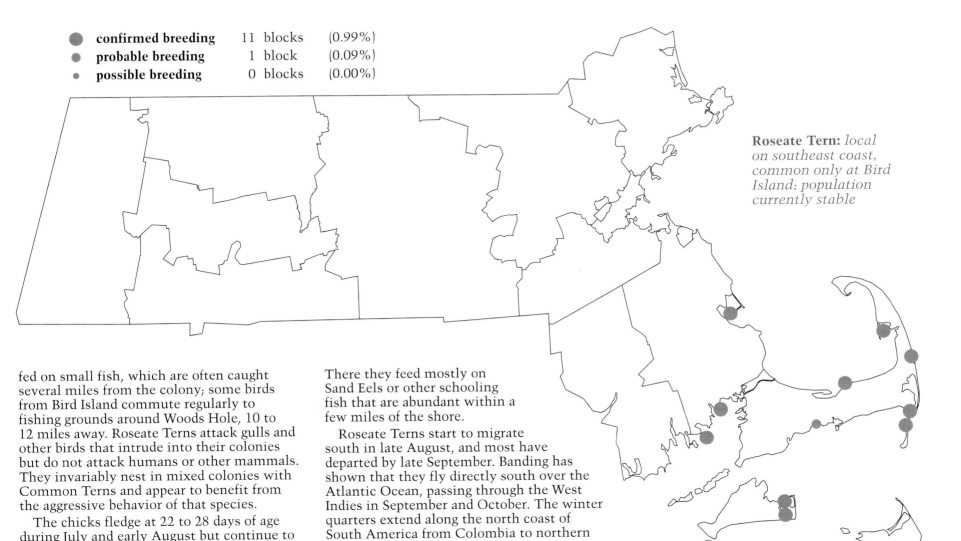

- ● **confirmed breeding**   11 blocks   (0.99%)
- ● **probable breeding**   1 block   (0.09%)
- ● **possible breeding**   0 blocks   (0.00%)

Roseate Tern: *local on southeast coast, common only at Bird Island; population currently stable*

fed on small fish, which are often caught several miles from the colony; some birds from Bird Island commute regularly to fishing grounds around Woods Hole, 10 to 12 miles away. Roseate Terns attack gulls and other birds that intrude into their colonies but do not attack humans or other mammals. They invariably nest in mixed colonies with Common Terns and appear to benefit from the aggressive behavior of that species.

The chicks fledge at 22 to 28 days of age during July and early August but continue to be fed by their parents for at least six weeks until they learn to fish for themselves. If two chicks are raised, the family breaks up at fledging and 1 parent tends each chick. Average fledging success in Massachusetts (late 1960s to early 1980s) was about one chick per pair. After breeding, most birds move to the outer shores of Cape Cod and the Islands, where flocks of up to 5,000 occur at favored places such as Monomoy and Nauset.

There they feed mostly on Sand Eels or other schooling fish that are abundant within a few miles of the shore.

Roseate Terns start to migrate south in late August, and most have departed by late September. Banding has shown that they fly directly south over the Atlantic Ocean, passing through the West Indies in September and October. The winter quarters extend along the north coast of South America from Colombia to northern Brazil. Young birds remain there for about 18 months. They usually migrate north to the breeding area at the age of two, but most do not breed until they are three or four years old. In recent years, Roseate Terns have been killed for human food in their winter quarters, especially on the coast of Guyana. This was probably the main reason for their rapid decline during the 1960s and 1970s, although encroachment by Herring Gulls onto the breeding colonies, predation, and toxic chemicals probably also contributed.

The Roseate Tern is listed as federally endangered in Massachusetts.

Ian C.T. Nisbet

Common Tern
Sterna hirundo

Egg dates: May 4 to August 15.

Number of broods: one; may re-lay if first attempt fails.

The Common Tern breeds on islands and locally on the mainland around the entire coast of Massachusetts and can be seen feeding along most parts of the coast. It has been studied intensively since 1870 and is perhaps the most researched bird in the state. The colony at Muskeget Island was said to contain hundreds of thousands of Common Terns in 1870. However, like most seabirds in the Northeast, Common Terns were almost wiped out by plume hunters and, by 1889, only about 5,000 pairs were left. The population recovered rapidly under protection and reached a peak of 30,000 to 40,000 pairs in the 1920s.

Subsequently, Herring Gulls usurped most of the favored islands for nesting, and Common Terns decreased steadily to about 6,000 pairs in 1977. The population now appears stable at 7,000 pairs, but its status is precarious.

Common Terns start to arrive in late April, and most reach the breeding colonies during May. At this time, small parties of birds migrating to more northern breeding grounds are seen occasionally on lakes and rivers inland. Most birds feed in shallow water close to shore, often in bays, inlets, or tidal creeks. Common Terns characteristically feed in flocks, hovering over schools of fish in shallow water and diving from heights of 10 to 20 feet. Many pick shrimp or insects from the surface of the water. Some birds defend feeding territories along the shore, where they often forage by plunging from perches on docks or boats. Common Terns usually nest on sandy or gravelly islands or on sand dunes on the mainland. They prefer areas that are open enough to provide sitting birds with a clear view but that have enough Dunegrass or other plants to provide shelter for the chicks. A few nest on open sand or on tidal wrack at the edge of salt marshes.

The best-known call is a harsh, descending *kee-arrr*, used when the terns are alarmed by intruders in the colony or when they are threatening other terns. Feeding and quarreling birds utter a sharp *kik, kik,* and birds carrying fish use an advertising call, *keeur, keeri-keeri-keeri*. Male Common Terns carry fish to attract females during aerial courtship display, or later to feed to their mates prior to egg laying. In ground display, assertive birds bend their heads forward, presenting the black cap, while submissive individuals stretch their heads upward and twist their necks to avert the black cap. Common Terns rarely swim except when bathing. Loafing birds collect in flocks on beaches, sand flats, or jetties.

Common Terns lay two or three eggs in a scrape, lined with grasses and shells. Most eggs are laid in the second half of May or early June, but younger birds and birds that have lost early clutches continue to lay until mid-July. Both sexes share the 23- to 27-day incubation period. The young birds are fed in their parents' territories until they can fly at 22 to 28 days old and for up to two weeks thereafter. Family groups then disperse, but the parents feed the chicks for at least six weeks after fledging. Average fledging success in Massachusetts (late 1960s to early 1980s) was one young per pair. From late July until October, Common Terns disperse widely

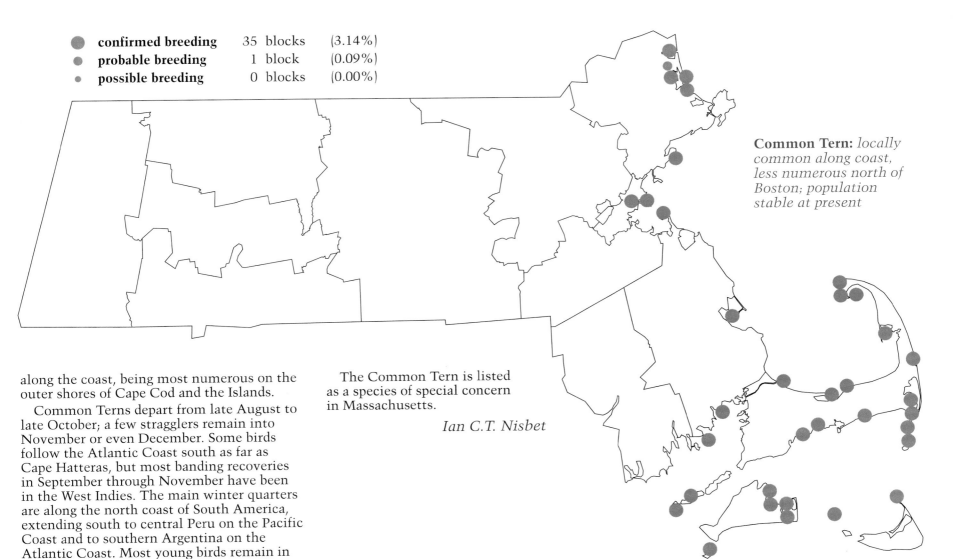

- ● **confirmed breeding** 35 blocks (3.14%)
- ● **probable breeding** 1 block (0.09%)
- · **possible breeding** 0 blocks (0.00%)

Common Tern: *locally common along coast, less numerous north of Boston; population stable at present* along the coast, being most numerous on the outer shores of Cape Cod and the Islands.

Common Terns depart from late August to late October; a few stragglers remain into November or even December. Some birds follow the Atlantic Coast south as far as Cape Hatteras, but most banding recoveries in September through November have been in the West Indies. The main winter quarters are along the north coast of South America, extending south to central Peru on the Pacific Coast and to southern Argentina on the Atlantic Coast. Most young birds remain in the winter quarters for 18 months and do not migrate north until the age of two; most do not breed until age three or four. A few banded birds have been retrapped on nests in Massachusetts at ages 20 to 22. Common Terns have been killed for human food in South America, especially in Guyana, and this may have contributed to their recent decline.

The Common Tern is listed as a species of special concern in Massachusetts.

Ian C.T. Nisbet

Arctic Tern
Sterna paradisaea

Egg dates: May 23 to July 27.
Number of broods: one.

A widespread arctic and subarctic species, the Arctic Tern reaches its extreme southern limit on the coast of Massachusetts, where it is now confined to half a dozen breeding sites on outer beaches from Plymouth Beach, and around Cape Cod, to Nomans Land. It is rarely seen except at the breeding colonies. It has probably never been abundant in the state, but it prospered under protection in the early years of this century and reached a peak of 300 to 400 pairs in the 1940s. Since then it has steadily decreased: only 18 pairs were found in 1985, and it appears headed for extirpation.

Arctic Terns arrive at the breeding colonies in mid- to late May. They commonly catch small schooling fish in open water and are sometimes found farther from land than other terns. However, they also catch fish and invertebrates in shallow water close to shore and may pick up sandworms or other invertebrates from the surface of tidal flats. They prefer to nest on open sandy or gravelly places, usually on the outer edge of colonies of Common Terns but sometimes within colonies of Least Terns.

Arctic Terns are most readily distinguished from Common Terns in mixed colonies by the timbre of their voices. The commonest alarm calls are a descending *ki-yerrr* and a trilling *krrree*; both are higher pitched and have a more squealing quality than those of Common Terns. Birds carrying fish give a high-pitched polysyllabic call, *kittiweewit, kittiweewit*. The aerial display flight is similar to that of Common Terns, although Arctic Terns also have a low-altitude flight display in which two or three birds fly through the colony with wings raised in a *V* over their back. Arctic Terns are lighter on the wing than Common Terns, and they are more prone to hover, both when feeding and when disturbed by intruders near the nest. They can often be seen resting on the sand near the breeding colonies, where their short legs readily distinguish them from accompanying Common Terns.

Arctic Terns usually lay one or two eggs in a scrape on open sand. Most eggs are laid in a short period between May 28 and June 15. Although a few young birds lay in late June or early July, Arctic Terns do not regularly renest if they lose eggs or young. During incubation, scraps of vegetation or shell are added to the nest, but Arctic Terns rarely build nests as substantial as those of Common Terns. After hatching, the chicks are usually moved away from the open nest site to areas with more cover, sometimes outside the area where other terns are nesting. Both sexes share in the 21- to 23-day incubation period and in feeding the young. Arctic Terns vigorously defend their nests and young against humans and other predators. They dive silently on human intruders from behind, pecking them sharply on the head and uttering an angry *kyaaa* as they swoop upward. Despite these active defense measures, the Arctic Terns' habit of nesting on

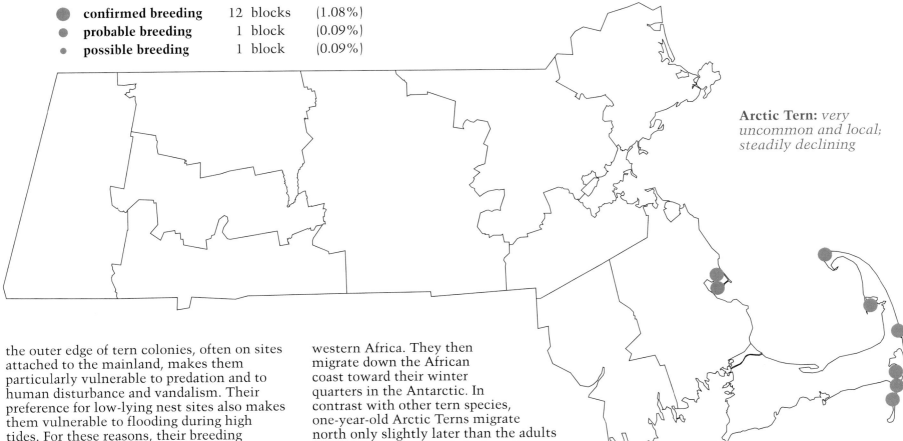

- ● **confirmed breeding**   12 blocks   (1.08%)
- ● **probable breeding**   1 block   (0.09%)
- • **possible breeding**   1 block   (0.09%)

Arctic Tern: *very uncommon and local; steadily declining*

the outer edge of tern colonies, often on sites attached to the mainland, makes them particularly vulnerable to predation and to human disturbance and vandalism. Their preference for low-lying nest sites also makes them vulnerable to flooding during high tides. For these reasons, their breeding success has been very low in Massachusetts in recent decades, and they have failed to benefit greatly from protective measures that have been effective for the other tern species. The average fledging success in Massachusetts from the late 1960s to the early 1980s was .2 young per pair.

Family groups leave the breeding colonies almost as soon as the young can fly, at 23 to 28 days of age, and Arctic Terns are rarely seen in Massachusetts after August 7. Their initial movement on departure is probably toward the northeast, and their autumn migration takes them directly across the Atlantic Ocean to southwestern Europe or western Africa. They then migrate down the African coast toward their winter quarters in the Antarctic. In contrast with other tern species, one-year-old Arctic Terns migrate north only slightly later than the adults and can be seen around the breeding colonies in June and July. These one-year-old birds have a distinctive plumage with a white forehead and underparts, and a black nape, bill, legs, and carpal bar. They were once described as a separate species, *"Sterna portlandica,"* and are still known by that name. They are sometimes present at Monomoy and Nauset in flocks of 100 to 500 birds, far exceeding the local breeding population. Arctic Terns do not breed, however, until they are three or four years old. If they survive to this age, they have an unusually high life expectancy; a number of birds banded in Massachusetts have been found breeding when they were 15 to 20 years old, and one even at age 27.

The Arctic Tern is listed as a species of special concern in Massachusetts.

Ian C.T. Nisbet

Least Tern
Sterna antillarum

Egg dates: May 23 to July 28.

Number of broods: one; may re-lay if first attempt fails.

Although accurate historical data is lacking, the Least Tern is thought to have long been a common to abundant nester in Massachusetts. By the end of the nineteenth century, however, the species had been nearly extirpated from its entire range. After the turn of the century, only a few small pockets of nesting birds survived on remote refuges in southeastern Massachusetts. Under complete legal protection after 1918, the species recovered rapidly. In 1923, Forbush estimated a state population of 300 birds as far north as Duxbury. In an incomplete census in 1935-1936, Hagar raised this estimate to 830 birds.

From 1974 onward, the Massachusetts Audubon Society and the Massachusetts Division of Fisheries and Wildlife have made concerted efforts to track the Least Tern population. During the Atlas period, Least Terns were reported from approximately 54 stations along the Massachusetts coast, primarily at river mouths and along barrier islands and beaches. Nesting occurs mostly from Cape Cod north to Scituate and on the islands of Nantucket and Martha's Vineyard. The species is much less frequent about the predominantly rocky shores of Buzzards Bay and Massachusetts Bay between Cohasset and Rockport and is entirely absent inland.

The mean population, conservatively estimated over the five-year period 1974 to 1979, was 1,300 pairs (range 800 to 1,734). Subsequently, the estimated population trend has been upward, reaching 2,040 pairs at 40 stations in 1980 and 2,415 pairs at 42 stations in 1984. Some of the increase is definitely a reflection of more thorough census coverage. However, aggressive protective programs and public education have likely resulted in a real increase as well. Year-to-year numbers at individual colonies fluctuate widely as a result of habitat changes and human disturbance.

In spring, returning Least Terns are on the Massachusetts beaches by mid-May, though actual commencement of nesting activities is typically delayed until the third week of May and sometimes even to mid-June, on account of weather conditions. The Least Tern is a classic opportunistic nester that capitalizes on ephemeral nesting conditions. The preferred nesting habitat—expansive sandy or pebbly beaches just slightly above the high tide line—is very unstable and subject to dramatic yearly change due to storms. The deposition of dredge spoil upon beaches provides excellent nesting substrate and has been a positive factor for the species. Unlike most terns, Least Terns do not prefer on-island sites, and most colonies are actually situated along mainland or barrier beaches. As a result, they are especially vulnerable to mammalian predator raids and human disturbance.

Courtship involves high-speed ceremonial flights and strutting. Copulation is accompanied by head-turning displays by the male, who holds a fish and presents it to the female immediately after consummation. The birds are highly nervous and defensive of nests and chicks, attacking upon any provocation, with alarmed colonies flying up in group defense. Dubbed in some quarters as "Little Strikers," Least Terns rarely actually strike but instead drop excrement upon intruders with accuracy.

The most frequently given vocalization is a shrill two-note *kidic-kidic* or *kidee-kidee.* Alarmed birds utter a sharp *kip-kip-kip* at varying speeds and intensities, depending

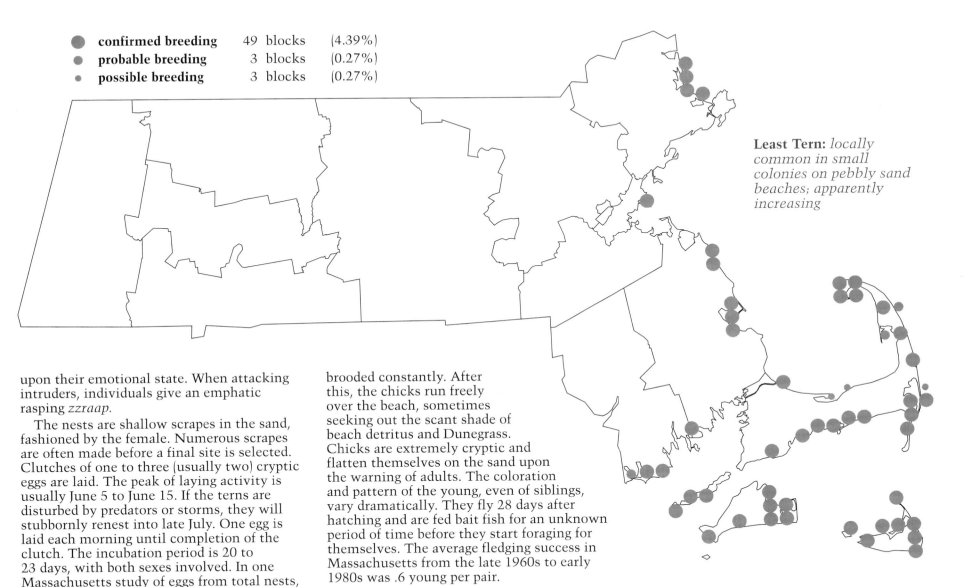

Least Tern: *locally common in small colonies on pebbly sand beaches; apparently increasing*

- ● confirmed breeding — 49 blocks (4.39%)
- ● probable breeding — 3 blocks (0.27%)
- ● possible breeding — 3 blocks (0.27%)

upon their emotional state. When attacking intruders, individuals give an emphatic rasping *zzraap*.

The nests are shallow scrapes in the sand, fashioned by the female. Numerous scrapes are often made before a final site is selected. Clutches of one to three (usually two) cryptic eggs are laid. The peak of laying activity is usually June 5 to June 15. If the terns are disturbed by predators or storms, they will stubbornly renest into late July. One egg is laid each morning until completion of the clutch. The incubation period is 20 to 23 days, with both sexes involved. In one Massachusetts study of eggs from total nests, 60 percent hatched, 27 percent were lost to predators, 11 percent were abandoned, and 2 percent were washed out. The eggs hatch asynchronously, with incubation continuing in most cases until the last egg is hatched.

The precocial chicks remain at the scrape for only about a day, during which they are brooded constantly. After this, the chicks run freely over the beach, sometimes seeking out the scant shade of beach detritus and Dunegrass. Chicks are extremely cryptic and flatten themselves on the sand upon the warning of adults. The coloration and pattern of the young, even of siblings, vary dramatically. They fly 28 days after hatching and are fed bait fish for an unknown period of time before they start foraging for themselves. The average fledging success in Massachusetts from the late 1960s to early 1980s was .6 young per pair.

Departure from Massachusetts occurs very soon after the young are flying, and Least Terns are usually difficult to find after September 1. The duration of the family bond is unknown. The wintering grounds are along the coast of South America from Colombia to eastern Brazil.

The Least Tern is listed as a species of special concern in Massachusetts.

Erma J. Fisk

Black Skimmer
Rynchops niger

Egg dates: June 8 to August 16.

Number of broods: one; may re-lay if first attempt fails.

The Black Skimmer is an erratic visitor and rare breeder on the coast of Massachusetts, most often encountered along sandy beaches or open tidal flats. In eastern North America, the regular breeding range extends from Long Island, New York, southward along the coast to the Gulf of Mexico. Historically, Champlain recorded skimmers in Massachusetts on Cape Cod in 1605. By the late eighteenth century, the species was extirpated from this, its northernmost breeding ground.

Between 1946 and 1984, skimmers were recorded nesting in Massachusetts on 16 occasions. Most nesting attempts (including the 1976 Atlas record) have occurred on Monomoy. Plymouth Beach has had six nestings, and two nests have been recorded at Cotuit. Since 1984, when skimmers were observed breeding on North Monomoy and New Island, the species has nested annually.

Spring migrant skimmers are rarely seen. Most sightings occur from June to September. Barrier beaches are the preferred nesting habitats along the mid-Atlantic and southern coasts of the United States, while coastal dredge spoils along the Gulf of Mexico support large colonies. In our area, skimmers usually nest in the company of Common Terns.

A hollow depression in the sand serves as a nest, and a normal clutch consists of three to five buffy brown blotched eggs. Sets of two or three eggs were encountered most often in Massachusetts between 1946 and 1984. This and the fact that many of these nesting records occurred in July or August are an indication that they represented renesting attempts of failed and displaced southern breeders or late breeding attempts by inexperienced younger birds. In recent years, it appears that a small population of Massachusetts breeders has become established, and egg laying may begin in early June (Hecker, Nisbet). The nest discovered on Monomoy on August 16, 1976, contained two eggs. Other recent Monomoy egg dates include: three eggs hatched either on or before June 30, and two eggs observed on June 20 (Humphrey). A nest on New Island off Orleans contained four eggs on June 8 (Hecker). Both sexes incubate for 22 to 25 days and care for the young, feeding them regurgitated food at first and later bringing small fish. By kicking sand

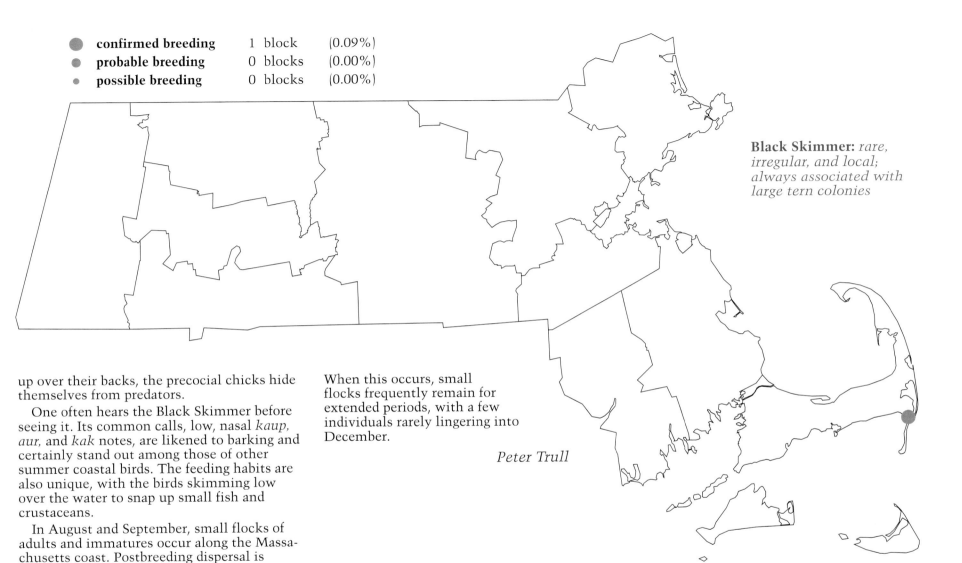

- confirmed breeding 1 block (0.09%)
- probable breeding 0 blocks (0.00%)
- possible breeding 0 blocks (0.00%)

Black Skimmer: *rare, irregular, and local; always associated with large tern colonies*

up over their backs, the precocial chicks hide themselves from predators.

One often hears the Black Skimmer before seeing it. Its common calls, low, nasal *kaup, aur,* and *kak* notes, are likened to barking and certainly stand out among those of other summer coastal birds. The feeding habits are also unique, with the birds skimming low over the water to snap up small fish and crustaceans.

In August and September, small flocks of adults and immatures occur along the Massachusetts coast. Postbreeding dispersal is common among colonial nesters, and skimmers from farther south frequently move northward before beginning their fall migration, which takes them from the Gulf of Mexico to the coasts of South America, where they form flocks of several hundred. Large numbers of skimmers occasionally are driven to our shores after hurricanes or intense storms in late summer or early fall. When this occurs, small flocks frequently remain for extended periods, with a few individuals rarely lingering into December.

Peter Trull

Rock Dove
Columba livia

Egg dates: throughout year; mostly March to June and August to November.

Number of broods: two or more each year, overlapping occasionally; can be up to six per year.

Along with the other two prominent alien imports, the House Sparrow and the European Starling, the Rock Dove is one of Massachusetts' most familiar birds. Despite their many admirable traits, pigeons are often regarded with disdain by ornithologists, as well as by much of the general population.

Rock Doves were originally Palearctic in their distribution. Due to their ease of domestication, they soon became widespread throughout much of Europe. European colonists first introduced them into North America in the early 1600s. These ancestral birds eventually escaped or were released into the wild, resulting in a gradual colonization of the entire eastern seaboard.

Today, Rock Doves can be found almost worldwide, in cities, towns, and agricultural areas. In Massachusetts, they are most numerous around the larger cities, where they can find plenty of suitable nesting sites and cover in the form of buildings and bridges. Away from sprawling urban environs, Rock Doves are unevenly distributed throughout the Commonwealth wherever appropriate habitat exists. Agricultural buildings and bridge abutments frequently offer suitable breeding sites, and highways may provide a network geographically linking concentrated urban populations.

Rock Doves tend to aggregate in large flocks or parties at productive feeding areas. They forage on open ground, primarily in the morning or midafternoon. A variety of foods are eaten. In addition to human refuse and handouts, many weed seeds and fruits are taken. A few invertebrates and even cooked meat may also be consumed.

Rock Dove coloration and body size are extremely variable. Infusions of polymorphic traits still occur as domestic birds find their way into the existing feral populations. The natural coloration of the old-world Rock Dove is an overall blue-gray with a white rump. There are two transverse black bars across the wing and iridescence on the neck and upper breast. All of these features are diagnostic of the ancestral form. A myriad of feral varieties ranging in color from black through reddish brown to white can now be found in Massachusetts. The predominant morph is still very similar to the original import, although albinistic feathering is often present. The female Rock Dove is slightly smaller than the male and has less iridescence on the neck.

Pigeons, though nonmigratory, exhibit a well-documented homing ability. This ability undoubtedly evolved from their need to forage in fields located far away from their ancestral cliff nest sites. One of the most intensively studied of all birds, the Rock Dove has helped to reveal many mysteries of avian navigation. The birds appear to rely on the sun as a compass to find their way home, although on cloudy days they apparently also utilize the earth's magnetic field. Olfactory cues and low-frequency sound may also be involved, but further research is still needed.

Rock Doves typically nest in crevices or on sheltered ledges, preferably in semidarkness. Buildings, bridges, and barns are used as

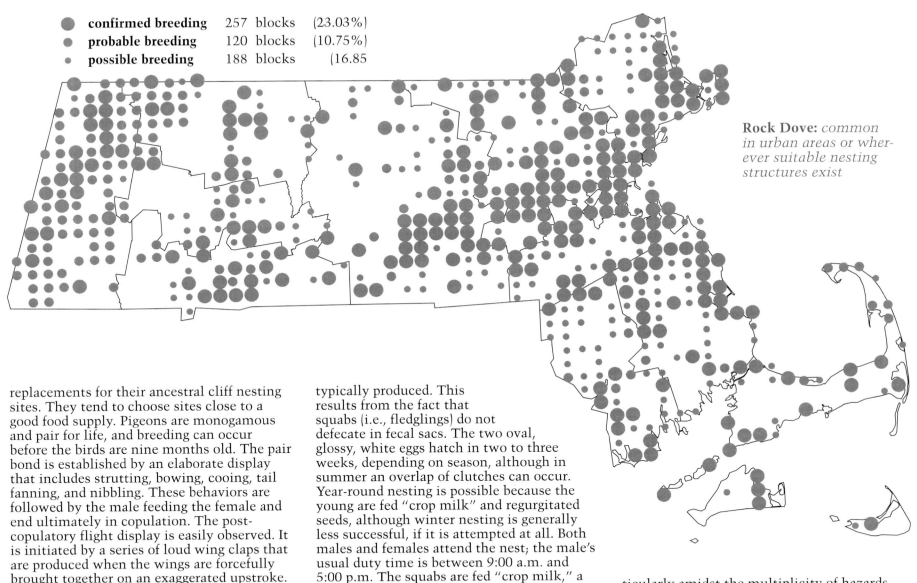

Rock Dove: *common in urban areas or wherever suitable nesting structures exist*

- confirmed breeding — 257 blocks (23.03%)
- probable breeding — 120 blocks (10.75%)
- possible breeding — 188 blocks (16.85%)

replacements for their ancestral cliff nesting sites. They tend to choose sites close to a good food supply. Pigeons are monogamous and pair for life, and breeding can occur before the birds are nine months old. The pair bond is established by an elaborate display that includes strutting, bowing, cooing, tail fanning, and nibbling. These behaviors are followed by the male feeding the female and end ultimately in copulation. The post-copulatory flight display is easily observed. It is initiated by a series of loud wing claps that are produced when the wings are forcefully brought together on an exaggerated upstroke. The male then flies with the wings held in a pronounced V shape between wing claps.

Male Rock Doves usually select the nest site, which is invariably on a flat surface. Because subsequent nesting often occurs in the same nest, a mass of "fecal cement" is typically produced. This results from the fact that squabs (i.e., fledglings) do not defecate in fecal sacs. The two oval, glossy, white eggs hatch in two to three weeks, depending on season, although in summer an overlap of clutches can occur. Year-round nesting is possible because the young are fed "crop milk" and regurgitated seeds, although winter nesting is generally less successful, if it is attempted at all. Both males and females attend the nest; the male's usual duty time is between 9:00 a.m. and 5:00 p.m. The squabs are fed "crop milk," a protein- and fat-rich secretion, three to four times a day for the first few days. By day 7, feedings include mostly seeds and are reduced to twice a day. Little is known about the life of the juveniles after fledging, but presumably this is a period of high mortality, particularly amidst the multiplicity of hazards they face in the urban areas that they prefer.

Thomas Aversa and Steven M. Arena

Mourning Dove
Zenaida macroura

Egg dates: late March to early August.

Number of broods: two.

According to Forbush, the Mourning Dove "had decreased so much in numbers in the early part of the Twentieth Century that Massachusetts led the way in 1908 by giving it perpetual protection under the law, to save it from extirpation." By the 1920s, the species had increased and was an "irregular resident in the eastern counties," and the population has continued to grow. During the Atlas period, the Mourning Dove was a common nester in all sections of the state. It is most abundant in the vicinity of open fields and is absent only from the most heavily forested areas. As the numbers increased, the pressure to allow hunting of Mourning Doves in Massachusetts has also intensified; however, even though the species is now listed as a game bird in over 30 states, the Commonwealth has not agreed to permit shooting of this smaller edition of the Passenger Pigeon.

Although the Mourning Dove is a permanent resident in Massachusetts, the numbers are considerably smaller in winter than they are during and after the breeding season. Migrants generally begin to return in late February or early March and are detected both by the calling of the males and sightings of flocks feeding in plowed fields. The song, given most frequently at dawn or dusk, *coah-coo-coo-coo*, is often mistaken by novices for the soft hooting of an owl. Doves produce several low growling calls, which are heard only at close range, and, like many members of the pigeon family, they have a nest call. When flushed from the ground, trowel-shaped tails flared, they can startle an observer with a twittering made by the wings.

Mourning Doves are believed to mate for life, and the male's courtship ritual is vigorous. With nest feathers ruffled, wings drooping, and tail cocked, he stamps his feet and struts around, much like a turkey, to win the female's attention. In addition, he will fly high into the air, making a loud noise by hitting his wings together over his back. When he reaches a height of 100 feet or so, he returns to earth in a series of circular sweeps on stiffened outspread wings.

Mourning doves build the simplest of nests, untidy but efficient, fairly flat and airy. Both sexes work together to weave course sticks and twigs onto the bough of a conifer, the heavier branches of a deciduous tree, or, rarely, on top of an old nest of another species. Twenty-eight Massachusetts nests were located as follows: White and Red pines (10 nests); Blue Spruce (3 nests); Red Cedar (2 nests); apple, plum, White Oak, Slippery Elm, and Weeping Willow (9 nests); English Ivy (1 nest); vine (1 nest); window ledge (1 nest); old robin nest (1 nest) (CNR). The habitats for 26 of these nests were as follows: wooded area (9 nests), suburban or yard (12 nests), orchard (2 nests), dune thicket (2 nests), tree in field (1 nest) (CNR). Heights ranged from 5 feet to 40 feet, with an average of 15.6 feet (CNR).

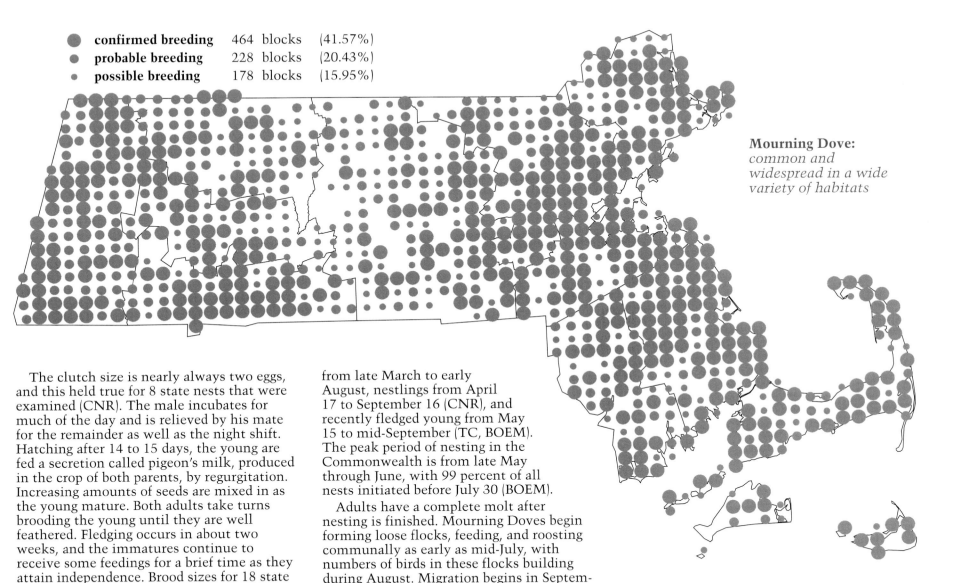

- confirmed breeding 464 blocks (41.57%)
- probable breeding 228 blocks (20.43%)
- possible breeding 178 blocks (15.95%)

Mourning Dove: *common and widespread in a wide variety of habitats*

The clutch size is nearly always two eggs, and this held true for 8 state nests that were examined (CNR). The male incubates for much of the day and is relieved by his mate for the remainder as well as the night shift. Hatching after 14 to 15 days, the young are fed a secretion called pigeon's milk, produced in the crop of both parents, by regurgitation. Increasing amounts of seeds are mixed in as the young mature. Both adults take turns brooding the young until they are well feathered. Fledging occurs in about two weeks, and the immatures continue to receive some feedings for a brief time as they attain independence. Brood sizes for 18 state nests were one young (4 nests), two young (14 nests) (CNR). Once one set of young has fledged, the pair takes a week or more to court, build another nest (occasionally the same nest is used again), and produce a second clutch of eggs. The breeding period is extended, with eggs reported in the state from late March to early August, nestlings from April 17 to September 16 (CNR), and recently fledged young from May 15 to mid-September (TC, BOEM). The peak period of nesting in the Commonwealth is from late May through June, with 99 percent of all nests initiated before July 30 (BOEM).

Adults have a complete molt after nesting is finished. Mourning Doves begin forming loose flocks, feeding, and roosting communally as early as mid-July, with numbers of birds in these flocks building during August. Migration begins in September, and most that are going to leave have departed for the southeastern United States by December. Depending on the severity of the early winter, the numbers overwintering in Massachusetts can vary. Overall, the wintering population seems to have increased, possibly as a result of the growing popularity of winter bird feeding. However, Mourning Doves have some difficulty withstanding a severe New England winter.

Thomas L. Carrolan

Black-billed Cuckoo
Coccyzus erythropthalmus

Egg dates: May 16 to July 5.

Number of broods: one.

In the mixed, mainly deciduous woodlands that cover the greater part of Massachusetts, the Black-billed Cuckoo is a fairly common summer resident. Among the oak, maple, and birch, it lives the life of a hermit, moving furtively among the branches, seldom venturing into the open for more than a few moments. Owing to its retiring nature, the Black-billed Cuckoo is among the least well known of our breeding species; to many it is no more than a voice in the woods.

During the Atlas period, the species was confirmed throughout the state. Many observers have noted that in the years following the Atlas, it has become scarce or absent from many areas formerly frequented. Whether this is part of a cycle or a long-term trend remains to be seen. Like its close relative, the Yellow-billed Cuckoo, it often frequents shrubby tangles and wet thickets, but the Black-billed Cuckoo also inhabits tracts of mature forest.

While occasional individuals arrive in late April, most of the spring migration occurs in May. Sometime during the second week of the month, as the hardwood forest is just attaining full foliage, one may first expect to hear the calls of the Black-billed Cuckoo. From its perch, hidden within the blossoming canopy, the cuckoo issues forth a long series of woody notes, in groups of two to four *coos*, phrase following phrase: *coo-coo-coo, coo-coo-coo, coo-coo-coo, coo-coo-coo, coo-coo, coo-coo-coo*. A long series of single *coo* notes, preceded by a harsher introductory note, also identifies this species. There are a variety of other calls, including some harsh *kuk* notes,

some of which are difficult to distinguish from those of the Yellow-billed Cuckoo.

In parts of their range of the South and West, cuckoos are commonly known as Rain Crows because they are believed to sing more frequently during cloudy periods. In truth, the Black-billed Cuckoo may be heard at almost any hour of the day or night. Well before sunrise on a warm May morning, its calls, with those of the thrushes, initiate the dawn chorus. In the breeding season, there are even observations of this species calling in flight at night.

The elusive cuckoo generally builds its nest at the edge of the woodland, well concealed among the dense cover of a young evergreen or thick shrub. Most nests are placed within 2 to 4 feet of the ground, rarely as high as 6 to 10 feet. Massachusetts nests have been recorded in apple, Red Cedar, barberry, Chinquapin Oak, and honeysuckle (ACB, CNR). Nests are typically rather flimsy affairs, loosely constructed of twigs and lined haphazardly with leaves. There are records of well-made nests, including two from Massachusetts, built more substantially and well lined with leaves and catkins (ACB).

The clutch size of the Black-billed Cuckoo is variable. In an undisturbed nest, two or three eggs may be the rule. Yet, both Black-billed and Yellow-billed cuckoos share a tendency toward brood parasitism, a characteristic of their old-world relatives; and females occasionally drop an egg into any convenient nest, though most often into that of another cuckoo. Nests containing up to eight eggs have been discovered, almost certainly the work of two or more birds. The clutch sizes for 9 Massachusetts nests were two eggs (2 nests), three eggs (5 nests), four eggs (2 nests) (DKW, CNR, Meservey). Massachusetts egg dates have been recorded from mid-May to early July, but in other New England states eggs have been found through the end of August. It is very probable that the situation in Massachusetts is similar, and the recorded egg dates do not represent the full range. Reports of fledged young indicate that eggs are found at least through early August. There is no evidence that more than one brood is raised each season nor that the species renests, although the protracted laying season suggests that this probably does occur.

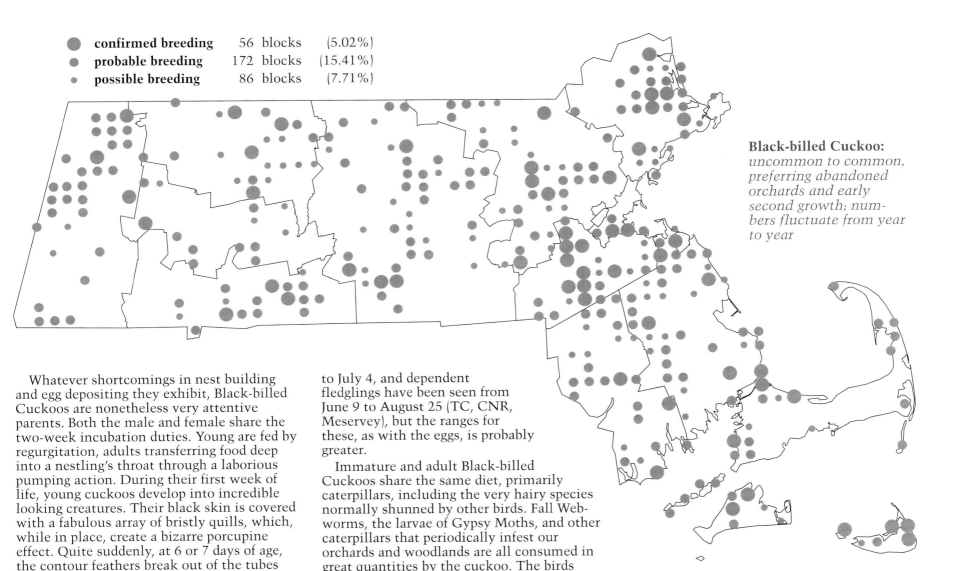

Black-billed Cuckoo: *uncommon to common, preferring abandoned orchards and early second growth; numbers fluctuate from year to year*

- confirmed breeding — 56 blocks (5.02%)
- probable breeding — 172 blocks (15.41%)
- possible breeding — 86 blocks (7.71%)

Whatever shortcomings in nest building and egg depositing they exhibit, Black-billed Cuckoos are nonetheless very attentive parents. Both the male and female share the two-week incubation duties. Young are fed by regurgitation, adults transferring food deep into a nestling's throat through a laborious pumping action. During their first week of life, young cuckoos develop into incredible looking creatures. Their black skin is covered with a fabulous array of bristly quills, which, while in place, create a bizarre porcupine effect. Quite suddenly, at 6 or 7 days of age, the contour feathers break out of the tubes and the young begin to look like real birds. In 8 or 9 days, the young birds abandon the nest and spend their remaining two-week preflight days clambering about among the branches. In late summer, the juveniles are molting into their first winter plumage, which closely resembles that of the adults. Nestlings have been recorded in Massachusetts from May 31 to July 4, and dependent fledglings have been seen from June 9 to August 25 (TC, CNR, Meservey), but the ranges for these, as with the eggs, is probably greater.

Immature and adult Black-billed Cuckoos share the same diet, primarily caterpillars, including the very hairy species normally shunned by other birds. Fall Webworms, the larvae of Gypsy Moths, and other caterpillars that periodically infest our orchards and woodlands are all consumed in great quantities by the cuckoo. The birds appear to follow the infestations to some extent, becoming locally very common in the areas of greatest caterpillar havoc.

In Massachusetts, the autumnal migration of Black-billed Cuckoos stretches through September and October, but the birds become scarce toward the end of the latter month. In this season, cuckoos seem particularly partial to protected stands of saplings and tall shrubs, and can often be sighted upon a visit to a favorite coastal or river valley migrant trap. The species winters in northwestern South America.

Brian E. Cassie

Yellow-billed Cuckoo
Coccyzus americanus

Egg dates: May 20 to July 9.

Number of broods: one.

The Yellow-billed Cuckoo is an uncommon migrant and an uncommon to rare and irregular summer resident. During peak years of Tent Caterpillar, Gypsy Moth, and Fall Webworm infestations, cuckoos become more numerous because as a general rule they are the only birds that will eat large numbers of these caterpillars. Yellow-billed Cuckoos are found in moist thickets, overgrown pastures, orchards, willow and alder groves, and, like the Black-billed Cuckoo, they are regularly encountered in deciduous forests. Yellow-billed Cuckoos are more numerous in eastern Massachusetts and the Connecticut River valley and are rare or absent from much of the hill country of central and western portions of the state.

In spring, the first individuals are usually reported during the first week of May, but the general arrival does not occur until later in the month, and some migration is still continuing in the first week of June. Because cuckoos are skulkers, they are much more often heard than seen. Their voices have ventriloquial qualities, so one may think the bird is far off when it is actually on a limb close by. Knowing the songs of the two cuckoo species may be vital to identification. Whereas the commoner Black-billed Cuckoo's notes are generally a series of *cu-cu-cu* notes, the Yellow-billed Cuckoo typically gives a louder and throatier *kow-kow-kow* series. Two good renditions of the yellow-billed's song are: *kakakakakakaka-ka-ka-ka-ka-ka, kow-kow-kow-kow-kow-kow* and *kuk-kuk-kuk-kuk-kuk, ceaow-ceaow-ceaow-ceaow*, slowing toward the end. There are a variety of other calls, including *coo-coo* notes, some of which are impossible to distinguish from those of the Black-billed Cuckoo.

Yellow-billed Cuckoos will never win a prize for nest building, ranking, in fact, near the bottom of the ladder. The nest is typically a flimsy saucer of twigs, with perhaps a few leaves added for a lining, placed from 2 to 12 feet high in a bush or small tree. Five Massachusetts nests ranged from 2 feet up in low bushes to 12 feet up in an oak sapling in a swampy thicket (ACB). In another sample of state nests, 1 was located at 5 feet in a Red Cedar near a swamp, and the others were all lower in thickets along brooks (ACB).

Two to four bluish green eggs are the rule. When more have been found, the suspicion is that a second female has dumped her eggs in the owner's nest. Eggs may also be left in

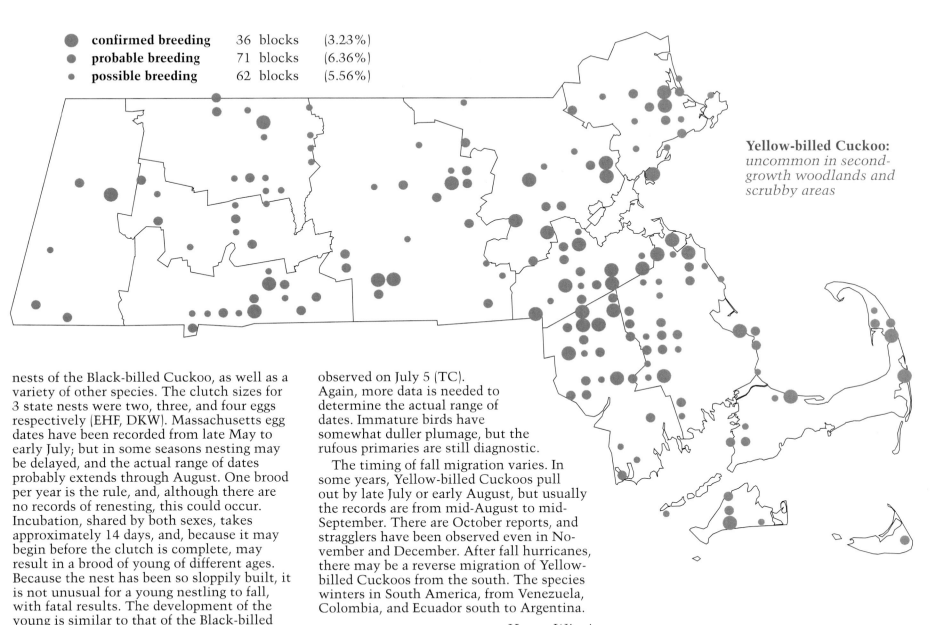

Yellow-billed Cuckoo: *uncommon in second-growth woodlands and scrubby areas*

nests of the Black-billed Cuckoo, as well as a variety of other species. The clutch sizes for 3 state nests were two, three, and four eggs respectively (EHF, DKW). Massachusetts egg dates have been recorded from late May to early July; but in some seasons nesting may be delayed, and the actual range of dates probably extends through August. One brood per year is the rule, and, although there are no records of renesting, this could occur. Incubation, shared by both sexes, takes approximately 14 days, and, because it may begin before the clutch is complete, may result in a brood of young of different ages. Because the nest has been so sloppily built, it is not unusual for a young nestling to fall, with fatal results. The development of the young is similar to that of the Black-billed Cuckoo (see account for that species). Nestlings have been recorded in Massachusetts from June 3 to July 8 (TC, BOEM, Meservey), and an adult with four fledged young was observed on July 5 (TC). Again, more data is needed to determine the actual range of dates. Immature birds have somewhat duller plumage, but the rufous primaries are still diagnostic.

The timing of fall migration varies. In some years, Yellow-billed Cuckoos pull out by late July or early August, but usually the records are from mid-August to mid-September. There are October reports, and stragglers have been observed even in November and December. After fall hurricanes, there may be a reverse migration of Yellow-billed Cuckoos from the south. The species winters in South America, from Venezuela, Colombia, and Ecuador south to Argentina.

Henry Wiggin

Barn Owl
Tyto alba

Egg dates: early April to September 30.
Number of broods: one or two.

The most nocturnal of all owls, this bird in Massachusetts is a rare and local permanent resident in several coastal locales and can be found inland in very limited numbers in the Connecticut River valley. The locations of nests are generally kept a well-guarded secret because disturbing the adults during daylight hours can result in loss of eggs or young and even in death to the flushed adult owl from crows, gulls, or hawks, which are quick to mob or attack a perceived enemy.

The Barn Owl is a permanent resident throughout much of its nearly worldwide range, but in our area some individuals overwinter and others move south to avoid the rigors of that season. Scattered records away from the breeding sites indicate that there are movements in the spring and fall.

The Barn Owl is a bird of open country, agricultural fields, meadows, swamps, marshes, golf courses, or any such habitat that supports a large rodent population that the owls feed upon. As with all owls that hunt by night and roost by day, the majority of these birds go about undetected and only come to the attention of those who stumble upon them or who persistently search them out.

Barn Owls get their name from their habit of nesting in old, mostly run-down buildings, usually barns. In Massachusetts these owls nest under bridges, in holes in cliff faces on Martha's Vineyard, and appropriately in many barns, silos, grain storage buildings, and church belfries. They have also been suspected of breeding in old structures on the Boston Harbor Islands. In 1984, 8 pairs of Barn Owls on Martha's Vineyard produced twenty-eight young birds and in 1985, 9 pairs nested on the island. Augustus Ben David was largely responsible for this increase because he convinced many barn owners to modify their barns by placing an owl nest box near the roof with an opening to the outside. He also developed large boxes that are placed on posts in open country and have been readily used by the owls. During and just prior to the Atlas period, Barn Owls nested in the loft of a cow barn in Sandwich, and in 1983 a pair nested in a church belfry there (Pease). The one inland confirmation occurred in 1979 in a silo in Hadley. These birds returned in 1980 and reared at least three young but abandoned the young in 1981. The nest site has since been destroyed (Allen). Since the Atlas period, breeding has also been confirmed on Nantucket and at Newburyport.

Barn Owls utter some of the most amazing sounds ever heard in these parts in the thick of night. Many listeners have fairly jumped out of their skin at the sound of a loud, blood-curdling, shrieking scream emanating from a flying Barn Owl at close range. The intensity and octave are such that one might believe a violent crime was being committed only a few feet away. There are a variety of other sounds including raspy hisses and a series of bill clicks.

The first eggs are generally laid in early April, earlier in some seasons, and the clutch size varies from four to eleven, with the usual number being six or seven eggs. Incubation lasts about 32 to 34 days and is performed solely by the female. The young hatch asynchronously because incubation begins with the first egg; so a nest usually contains young of different ages. The young fledge at 52 to 56 days of age. Nestlings will often hiss at an intruder and, if older than a couple of weeks, may spread their wings, fall on their backs, and throw their talons up while hissing constantly. They move their monkey-faced heads from side to side and up and down, swaying in a very characteristic fashion. On Martha's Vineyard, the young of the first brood typically fledge during July. It

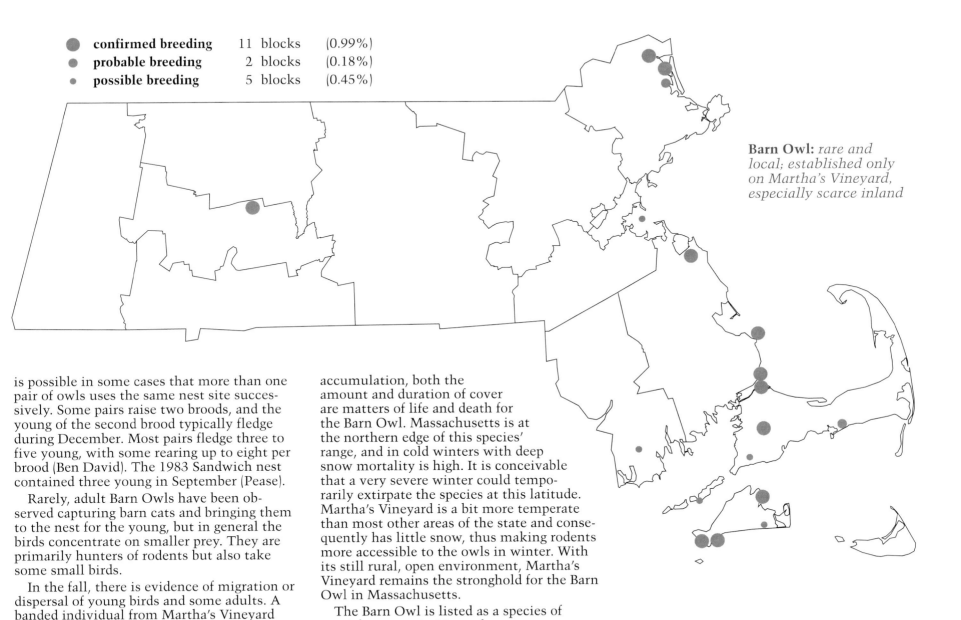

Barn Owl: *rare and local; established only on Martha's Vineyard, especially scarce inland*

- confirmed breeding — 11 blocks (0.99%)
- probable breeding — 2 blocks (0.18%)
- possible breeding — 5 blocks (0.45%)

is possible in some cases that more than one pair of owls uses the same nest site successively. Some pairs raise two broods, and the young of the second brood typically fledge during December. Most pairs fledge three to five young, with some rearing up to eight per brood (Ben David). The 1983 Sandwich nest contained three young in September (Pease).

Rarely, adult Barn Owls have been observed capturing barn cats and bringing them to the nest for the young, but in general the birds concentrate on smaller prey. They are primarily hunters of rodents but also take some small birds.

In the fall, there is evidence of migration or dispersal of young birds and some adults. A banded individual from Martha's Vineyard was recovered from as far away as Maryland. Several resident pairs on Martha's Vineyard are known to leave their nest site in a barn and winter in a dense pine grove a couple of miles away. In winters when there is snow accumulation, both the amount and duration of cover are matters of life and death for the Barn Owl. Massachusetts is at the northern edge of this species' range, and in cold winters with deep snow mortality is high. It is conceivable that a very severe winter could temporarily extirpate the species at this latitude. Martha's Vineyard is a bit more temperate than most other areas of the state and consequently has little snow, thus making rodents more accessible to the owls in winter. With its still rural, open environment, Martha's Vineyard remains the stronghold for the Barn Owl in Massachusetts.

The Barn Owl is listed as a species of special concern in Massachusetts.

E. Vernon Laux, Jr.

Eastern Screech-Owl
Otus asio

Egg dates: March 7 to May 15.

Number of broods: one; may re-lay if first attempt fails.

The Screech-Owl occupies an extensive range in North America east of the Rocky Mountains from southern Canada south to the Gulf Coast and Mexico. It inhabits a variety of habitats, preferring open deciduous forests, small woodlots, parklands, orchards, and even residential areas. In Massachusetts, it is most common and widespread in the lowland, hardwood forests of eastern Massachusetts and the Connecticut River valley. It is decidedly less common or even lacking in the higher elevations of Berkshire and Worcester counties, where coniferous growth begins to predominate, and on outer Cape Cod and Nantucket, where the lack of trees of adequate size is the limiting factor.

Heard throughout the year, screech-owls become most vocal from late winter to early spring, prior to and at the commencement of the breeding season. Typical vocalizations during the nesting period (determined by the response given to imitations of the calls) are low monotone wails, or trills, repeated regularly every few seconds for several minutes' duration. A few whinnies (trills that abruptly rise and slowly descend in pitch) are interspersed among the repeated monotones. Usually, only one member of a pair answers, almost certainly the male. The vocalizations seem to change with the seasons. In winter, responding owls usually produce a long series of descending whinnies, given every few seconds, with only a few of the trills mixed in. A series may last for several minutes and include many repetitions.

Eastern Screech-Owls do not construct a nest. Instead, they select a natural tree cavity, a hole drilled by a woodpecker, or nest box in which to lay their eggs. Nest sites are often near a brook or pond, and the birds will use Wood Duck boxes (Olmstead). Seven Massachusetts nests were located as follows: three natural cavities in apple trees in orchards, three old flicker holes in dead pines or poplar, and one natural cavity in an oak (ACB). Heights ranged from 5 to 35 feet.

The clutch size is usually four or five (range one to eight). In Massachusetts, the peak egg-laying period extends from late March through mid-April. The eggs are laid at 2- or 3-day intervals, and incubation does not always begin with the first egg. Incubation, by the female, has been reported as 21 to 30 days (average 26 days), and the male may roost near her in the cavity. He feeds her during this time and, once the young have hatched, supplies both his brooding mate and the owlets with food for two weeks. Food items include small mammals, insects, earthworms, reptiles, amphibians, fish, and small birds. The young

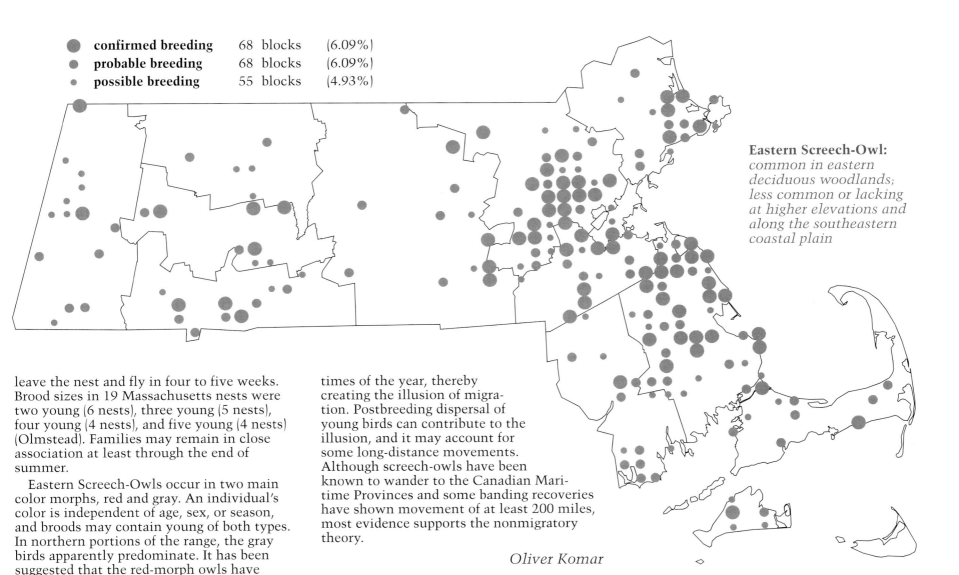

	confirmed breeding	68 blocks	(6.09%)
	probable breeding	68 blocks	(6.09%)
	possible breeding	55 blocks	(4.93%)

Eastern Screech-Owl: *common in eastern deciduous woodlands; less common or lacking at higher elevations and along the southeastern coastal plain*

leave the nest and fly in four to five weeks. Brood sizes in 19 Massachusetts nests were two young (6 nests), three young (5 nests), four young (4 nests), and five young (4 nests) (Olmstead). Families may remain in close association at least through the end of summer.

Eastern Screech-Owls occur in two main color morphs, red and gray. An individual's color is independent of age, sex, or season, and broods may contain young of both types. In northern portions of the range, the gray birds apparently predominate. It has been suggested that the red-morph owls have higher energy demands and are less likely to survive during severe weather conditions.

Seasonal movements of screech-owls are minimal. The species generally is believed to be nonmigratory, except perhaps in the most northerly parts of the range. The size of an owl's territory may be quite large, with the bird occupying various portions at certain times of the year, thereby creating the illusion of migration. Postbreeding dispersal of young birds can contribute to the illusion, and it may account for some long-distance movements. Although screech-owls have been known to wander to the Canadian Maritime Provinces and some banding recoveries have shown movement of at least 200 miles, most evidence supports the nonmigratory theory.

Oliver Komar

Great Horned Owl
Bubo virginianus

Egg dates: February 15 to May 15.

Number of broods: one; may re-lay if first attempt fails.

The extensive range of the Great Horned Owl includes nearly all of North America. It is a common species throughout Massachusetts but is absent from Martha's Vineyard and Nantucket. This owl is definitely much more widespread, especially in Worcester County and surrounding regions, than the Atlas map indicates.

The Great Horned Owl occupies a variety of habitats. It inhabits dense forests, which may be coniferous, deciduous, or mixed, as well as smaller woodlots; and it especially favors woodlands adjacent to marshes, swamps, farmlands, rivers, and ponds. This versatile bird adapts well to the presence of humans, as evidenced by the high number of breeding confirmations from densely settled areas of eastern Massachusetts. For roosting purposes, it has a definite preference for dense stands of evergreens of any kind, which offer good camouflage when an owl sits erect close to the trunk of a tree.

Great Horned Owls become most vocal during late December through January and up to the beginning of nesting in February. Vocalizations are almost as varied as those of the Barred Owl and range from deep booming hoots to whistles, shrieks, screams, and hisses. The resonant hooting call can be heard for a distance of several miles on a still night and is generally composed of five notes that sound like *whooo-whooo-whoooooo-whoo-whooo*.

The distribution of the Great Horned Owl coincides with that of the Red-tailed Hawk, and the owl's nest is often one that has been abandoned by the latter. Many of these nests are used several years in succession, with the hawks and the owls alternating from one nest to another every year or two. In Massachusetts, Great Horned Owls sometimes also use the old nest of another species of hawk or an Osprey, or that of a crow, Great Blue Heron, or squirrel. Hollow tree cavities or crotches where debris has collected are also utilized. The birds have readily accepted open artificial nests constructed for them. Locations of 21 Massachusetts nests were as follows: 13 in White Pine at 40 to 70 feet, 5 in Pitch Pine at 38 to 65 feet, 2 in oak at 25 and 30 feet, and 1 in American Beech at 31 feet (ACB, CNR).

The Great Horned Owl is the earliest nester of our native birds, usually laying its eggs between February 20 and March 25. The clutch almost always consists of two eggs, but the range can be from one to three, and sometimes four. One brood is raised each season, and on rare occasions a second clutch is produced if the first is lost. Incubation usually begins with the first egg and is performed by the female for 28 to 30 days. She will remain on the nest even during severe weather conditions and may be found covered with snow.

The young are brooded nearly continuously for the first two weeks. At three weeks of age they are about half-grown, and the primary feathers are developing. By the sixth week, they are quite active and begin testing their wings and hopping about the nest and nearby

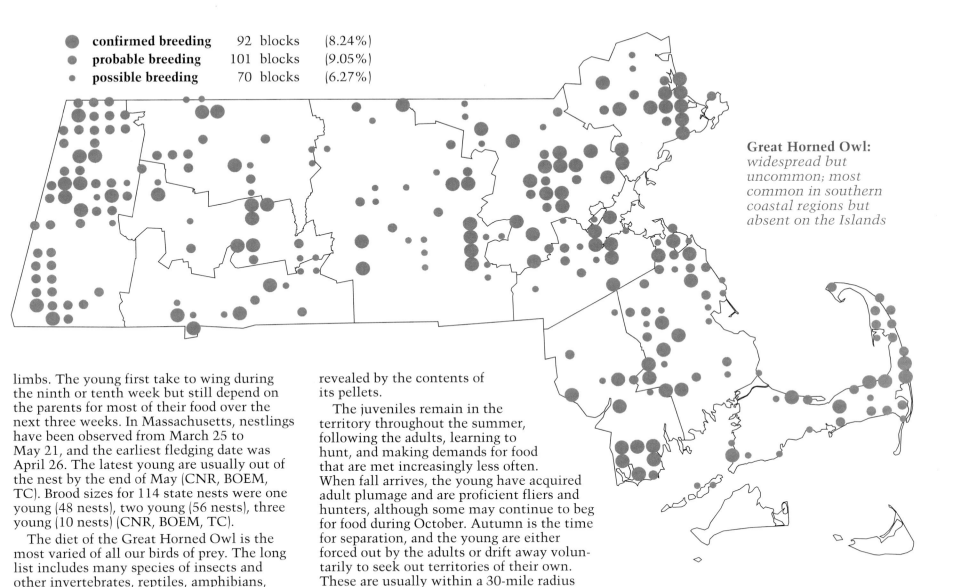

Great Horned Owl: *widespread but uncommon; most common in southern coastal regions but absent on the Islands*

● confirmed breeding	92 blocks	(8.24%)
● probable breeding	101 blocks	(9.05%)
· possible breeding	70 blocks	(6.27%)

limbs. The young first take to wing during the ninth or tenth week but still depend on the parents for most of their food over the next three weeks. In Massachusetts, nestlings have been observed from March 25 to May 21, and the earliest fledging date was April 26. The latest young are usually out of the nest by the end of May (CNR, BOEM, TC). Brood sizes for 114 state nests were one young (48 nests), two young (56 nests), three young (10 nests) (CNR, BOEM, TC).

The diet of the Great Horned Owl is the most varied of all our birds of prey. The long list includes many species of insects and other invertebrates, reptiles, amphibians, fish, birds (from sparrows to geese and including other species of owls and hawks), and mammals (from rats and mice to skunks, foxes, and cats). All undigestible material such as bones, teeth, claws, feathers, and fur are formed into pellets, which are regurgitated on a regular basis. An owl's diet is revealed by the contents of its pellets.

The juveniles remain in the territory throughout the summer, following the adults, learning to hunt, and making demands for food that are met increasingly less often. When fall arrives, the young have acquired adult plumage and are proficient fliers and hunters, although some may continue to beg for food during October. Autumn is the time for separation, and the young are either forced out by the adults or drift away voluntarily to seek out territories of their own. These are usually within a 30-mile radius from where they were reared.

This species is essentially nonmigratory, except during especially severe winters when prey becomes extremely scarce or when immature birds seek out unoccupied territories and occasionally move long distances. One such owl banded in Billerica was recovered two years later in Bingham, Maine, a distance of almost 200 miles.

Michael Olmstead

Barred Owl
Strix varia

Egg dates: February 25 to May 13.

Number of broods: one; may re-lay if first attempt fails.

The Barred Owl is a resident in much of eastern North America from southern Canada to the Gulf Coast. It was undoubtedly scarcer a century ago when the landscape was extensively cleared for agriculture. In Massachusetts, it is now fairly common in the hill country of the western portions of the state, especially in woodlands near extensive swamps, reservoirs, and Beaver ponds. The species nests more sparsely in the Northeast, and, although it is distributed locally in southeastern Red Cedar and Red Maple swamps, it is extremely rare on Cape Cod and the Islands.

One of the most vocal of owls, the Barred Owl can be heard throughout the year, but, as the breeding season approaches, it intensifies its familiar nine-note serenade, *who cooks for you, who cooks for you all?* The birds will respond readily to imitations of the calls, even if these are quite rough. In addition to these standard vocalizations, the owls have other calls and sometimes partake in a fervent caterwauling, a series of wild, almost maniacal, hoots and screams, which, once heard, are unforgettable.

Barred Owls construct no nest; rather, a hollow in a large tree or an old crow, hawk, or squirrel nest is chosen, and the same site is often used for several years in succession. Of 38 Massachusetts nests recorded (ACB), 21 were in White Pine, 6 in deciduous woods, and 11 in mixed forest. The nest sites were as follows: 18 old hawk nests, 15 hollow trees, and 5 squirrel nests. Formerly, when the Red-shouldered Hawk was more common, its old nests were often utilized. However, recent nests have generally been located in tree hollows or on stubs of broken trees.

Two or three (rarely four) pure white eggs are laid. Clutch sizes for 38 Massachusetts nests were two eggs (28 nests), three eggs (8 nests), and four eggs (2 nests) (ACB). Incubation, mostly if not entirely by the female, begins with the first egg and lasts 28 to 33 days. Hatching is asynchronous.

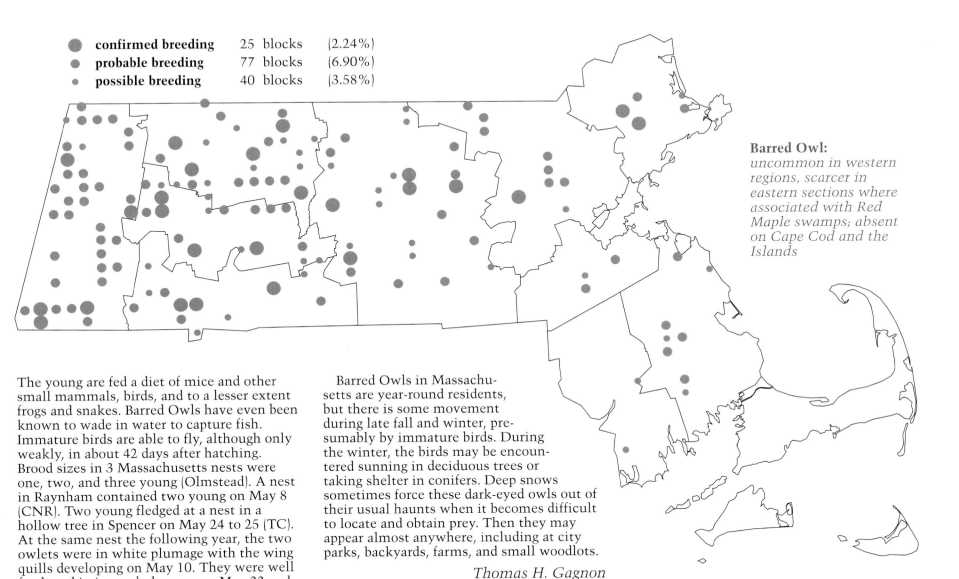

- confirmed breeding — 25 blocks (2.24%)
- probable breeding — 77 blocks (6.90%)
- possible breeding — 40 blocks (3.58%)

Barred Owl: *uncommon in western regions, scarcer in eastern sections where associated with Red Maple swamps; absent on Cape Cod and the Islands*

The young are fed a diet of mice and other small mammals, birds, and to a lesser extent frogs and snakes. Barred Owls have even been known to wade in water to capture fish. Immature birds are able to fly, although only weakly, in about 42 days after hatching. Brood sizes in 3 Massachusetts nests were one, two, and three young (Olmstead). A nest in Raynham contained two young on May 8 (CNR). Two young fledged at a nest in a hollow tree in Spencer on May 24 to 25 (TC). At the same nest the following year, the two owlets were in white plumage with the wing quills developing on May 10. They were well feathered in juvenal plumage on May 22 and fledged from June 1 to 3 (Meservey). The young remain with the adults for most of the summer, begging for food and learning to hunt. At this time, the adults are surprisingly quiet. In the fall, the family group disbands, but mated birds may remain in close proximity.

Barred Owls in Massachusetts are year-round residents, but there is some movement during late fall and winter, presumably by immature birds. During the winter, the birds may be encountered sunning in deciduous trees or taking shelter in conifers. Deep snows sometimes force these dark-eyed owls out of their usual haunts when it becomes difficult to locate and obtain prey. Then they may appear almost anywhere, including at city parks, backyards, farms, and small woodlots.

Thomas H. Gagnon

Long-eared Owl
Asio otus

Egg dates: March 15 to May 14.

Number of broods: one; may re-lay if first attempt fails.

Although the Long-eared Owl is a permanent resident in Massachusetts, it is seldom encountered except during migration and winter, when it is most often seen inconspicuously roosting in a conifer or dense thicket. Formerly more common, the species has always been encountered most frequently in the eastern portions of the state, especially the Southeast. During the Atlas period, confirmations were made only on Martha's Vineyard and Cape Cod.

Spring migration of birds that winter south of Massachusetts occurs mainly during March and April. It is not known to what extent the state population is migratory. Local breeders are on territory by February, but there is a variation in the dates when nesting is initiated. During the courtship period, males offer the female food, but little else appears to be known about courtship activities.

Although infrequently heard in our area, the most familiar call is a mournful hoot uttered singly or in a short series. Near the nest, a variety of vocalizations are heard, including a low-pitched *kitty-kitty-kitty*, cries of distress, yaps, prolonged shrill squeals, whistles, chuckles, snorts, and raspy hisses. Many of these calls are ventriloquial. Bill snapping is also performed.

For nesting, this species generally appropriates an old squirrel, crow, or hawk nest, from 10 to 40 feet high, located in a thicket, grove, or woodlot. Rarely, the nest is in a low shrubby growth or on the ground. Seven southeastern Massachusetts nests were located as follows: 1 in an old squirrel nest at 20 feet in a Pitch Pine in swampy woods and 6 in White Pines in low-growth deciduous woods, mixed forest, or dense pine woods. Of these last 6 nests, 2 were old crow nests and 4 were old hawk nests. Heights ranged from 25 to 40 feet (ACB). During the Atlas period, all of the nests on Martha's Vineyard were in old crow nests in Red Pine plantations (Ben David). The Cape Cod nest (original occupant unknown), constructed of sticks with some oak leaves and pine needles, was located at 40 feet near the top of a Pitch Pine in a pine grove surrounded by swamps (Wood).

The eggs are generally laid in March or April, but a record of six young averaging a week old on March 26 at a nest in Burlington (EHF) indicates that laying may sometimes occur as early as late February. If the birds renest, hatching may not occur until late May. Clutches consist of three to eight (usually four or five) oval glossy-white eggs laid at 48-hour intervals. Clutch sizes in 3 Massachusetts nests were three, four, and five eggs (ACB, DKW). Incubation by the female lasts 26 to 28 days, and the eggs hatch about 2 days apart. Occasionally, the last young hatched are killed and eaten by the older nestlings.

Once the young hatch, the adults boldly defend them with weird, piercing screams, anguished cries, and distraction displays. When an intruder approaches the nest, one or both parents may fly off and plummet to the ground. Immediately, a shrill cry is uttered, accompanied by a great rustling and rattling of feathers, as though the owl were struggling with prey. When the enemy has been decoyed away from the young, the performing bird begins moving away, still holding its wings partially open, as if the imaginary prey were too large to carry off, and eventually flies off to circle back to the young.

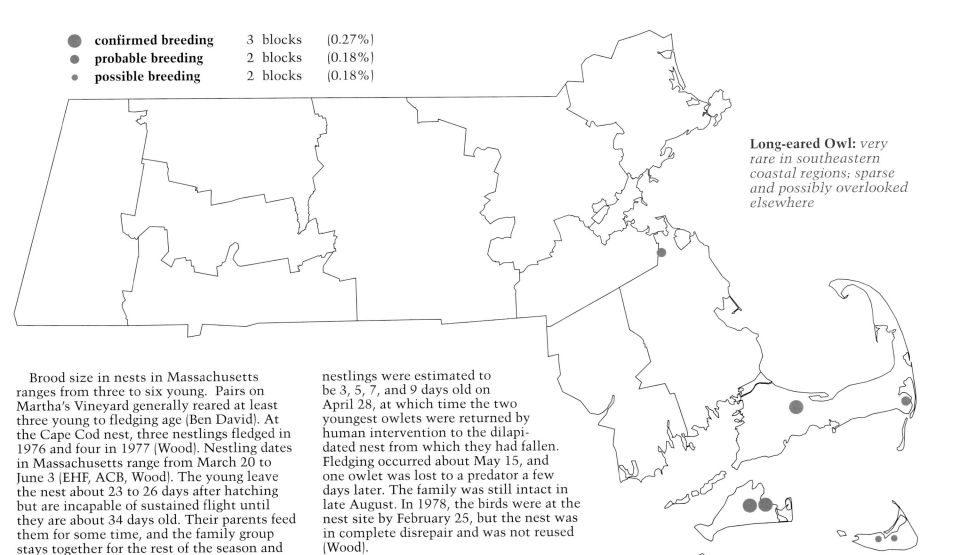

- confirmed breeding — 3 blocks (0.27%)
- probable breeding — 2 blocks (0.18%)
- possible breeding — 2 blocks (0.18%)

Long-eared Owl: *very rare in southeastern coastal regions; sparse and possibly overlooked elsewhere*

Brood size in nests in Massachusetts ranges from three to six young. Pairs on Martha's Vineyard generally reared at least three young to fledging age (Ben David). At the Cape Cod nest, three nestlings fledged in 1976 and four in 1977 (Wood). Nestling dates in Massachusetts range from March 20 to June 3 (EHF, ACB, Wood). The young leave the nest about 23 to 26 days after hatching but are incapable of sustained flight until they are about 34 days old. Their parents feed them for some time, and the family group stays together for the rest of the season and perhaps into the winter.

The most complete recent nesting chronology for Massachusetts is that of the Cape Cod pair. In 1976, the female was discovered on the nest on May 3, three large nestlings were seen on May 28, and these fledged about June 4. In 1977, breeding at this site commenced somewhat earlier, with the female on the nest by March 21. The four nestlings were estimated to be 3, 5, 7, and 9 days old on April 28, at which time the two youngest owlets were returned by human intervention to the dilapidated nest from which they had fallen. Fledging occurred about May 15, and one owlet was lost to a predator a few days later. The family was still intact in late August. In 1978, the birds were at the nest site by February 25, but the nest was in complete disrepair and was not reused (Wood).

Though Long-eared Owls consume many small mammals such as voles, mice, shrews, rats, and squirrels, they have also been known to take birds, including some as large as quail. Small numbers regularly migrate in October and November, and during the winter months they often roost communally along the coast near marshes or streams, or near broad expanses of open fields. These roosts are often in groves of Red or White pine, or stands of Red Cedar, as well as in thickets of low-growing shrubs.

The Long-eared Owl is listed as a species of special concern in Massachusetts.

Helen C. Bates

Short-eared Owl
Asio flammeus

Egg dates: March 15 to July 18.

Number of broods: one.

A bird of open country, the Short-eared Owl requires habitats of grasslands, fields, marshes, and bogs. After much of the land had been cleared for agriculture, this owl was found as an uncommon summer resident throughout the state. More recently, development and changing land use patterns have had a severe impact. The species has declined in numbers, and it is now restricted as a breeder to the islands of southeastern Massachusetts. The nesting population in 1984 was estimated at 15 to 20 pairs, and in 1985 this owl was listed as an endangered species in the state.

Some Short-eared Owls are permanent residents, and others apparently move farther south for the winter. They begin establishing their breeding territories in Massachusetts from mid-March to early April. During courtship, male owls vocalize and sky dance to attract mates. The displaying male often flies to 100 feet or more and gives a soft *hoo-hoo-hoo-hoo-hoo* call 13 to 16 times while drifting on fanned wings and tail. He then drops into a stoop, bringing his wings under the body and clapping them together several times. The bird then loops back up and may begin again, although the wing clapping is not given as frequently as the song, which may be delivered from the ground or a perch as well as in flight. Other vocalizations include a short bark from adults defending a nest and a long, loud, *pssssip* call given by the young when they are begging for food or confronted by an intruder.

Unlike the Northern Harrier, which occupies similar habitats, the Short-eared Owl prefers drier upland locations for nesting. Nest sites are located in open grasslands of tall, dense vegetation, or often under a small shrub. Nests are surrounded on three sides by foliage, with the female facing outward at the open end. The actual nest, small in all respects, is a depression in dried grasses lined sparsely with the owl's downy feathers. Eggs are laid generally from mid-March to early May, but there are later records for Massachusetts. On Nantucket and Tuckernuck islands, nests contained eggs in 1986 from March 15 to July 13 and in 1987 from April 8 to July 18. For both years, the late dates were accounted for by only one nest, possibly second nesting attempts (Tate, Melvin). Although there are literature reports that this species will lay replacement clutches, this appears to be rare in Massachusetts. Typical clutch size is four to seven eggs, but there may be more or fewer. On Nantucket and Tuckernuck islands, clutch sizes were recorded as follows: 1985, six (range three to eight); 1986, eight (range seven to eight); and 1987, seven (range four to nine) (Tate, Melvin). A Monomoy nest contained five eggs (Humphrey). Incubation by the female alone is reported as lasting 26 days.

When an intruder approaches the nest site, the male suddenly appears and hovers overhead, giving the barking call and trying to lead it away by drifting farther off, still

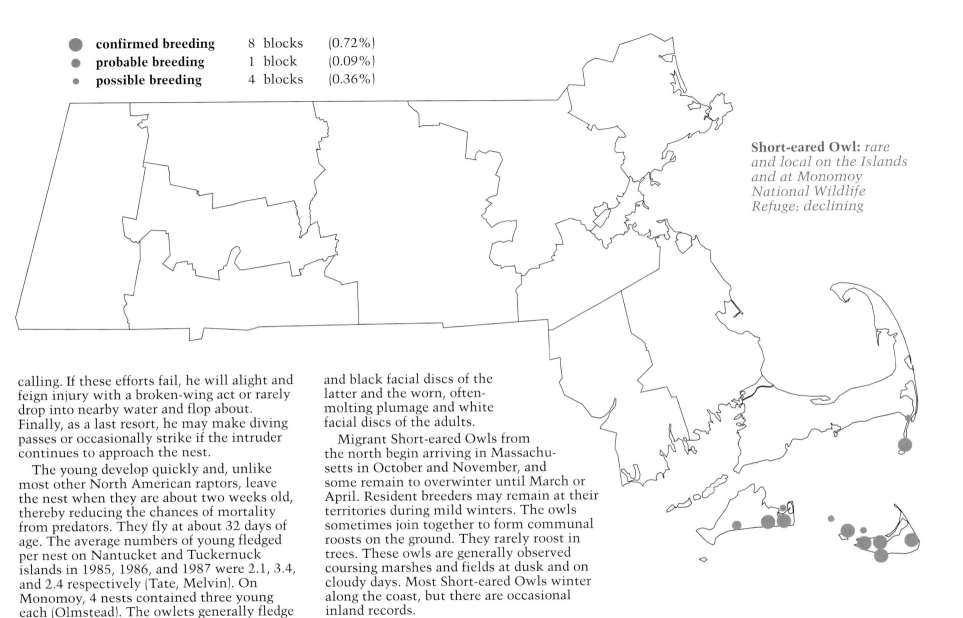

- confirmed breeding — 8 blocks (0.72%)
- probable breeding — 1 block (0.09%)
- possible breeding — 4 blocks (0.36%)

Short-eared Owl: *rare and local on the Islands and at Monomoy National Wildlife Refuge; declining*

calling. If these efforts fail, he will alight and feign injury with a broken-wing act or rarely drop into nearby water and flop about. Finally, as a last resort, he may make diving passes or occasionally strike if the intruder continues to approach the nest.

The young develop quickly and, unlike most other North American raptors, leave the nest when they are about two weeks old, thereby reducing the chances of mortality from predators. They fly at about 32 days of age. The average numbers of young fledged per nest on Nantucket and Tuckernuck islands in 1985, 1986, and 1987 were 2.1, 3.4, and 2.4 respectively (Tate, Melvin). On Monomoy, 4 nests contained three young each (Olmstead). The owlets generally fledge during June, and part of the family may remain together, roosting in close proximity. Individuals observed from this period through fall can be distinguished as adults or hatch-year immatures by the fresh brown plumage and black facial discs of the latter and the worn, often-molting plumage and white facial discs of the adults.

Migrant Short-eared Owls from the north begin arriving in Massachusetts in October and November, and some remain to overwinter until March or April. Resident breeders may remain at their territories during mild winters. The owls sometimes join together to form communal roosts on the ground. They rarely roost in trees. These owls are generally observed coursing marshes and fields at dusk and on cloudy days. Most Short-eared Owls winter along the coast, but there are occasional inland records.

Denver W. Holt

Northern Saw-whet Owl
Aegolius acadicus

Egg dates: late March to July 3.

Number of broods: one; may re-lay if first attempt fails.

Although the Northern Saw-whet Owl is widely distributed, it is seldom seen because of its diminutive size (it is the smallest of the eastern owls) and strictly nocturnal behavior. It is undoubtedly more common than the records would indicate. During the Atlas period, breeding was confirmed only in scattered blocks in eastern Massachusetts, from Essex County and Plymouth County to Cape Cod and Nantucket. The species went unrecorded in most blocks in the interior sections of the state, but it is known to inhabit some of these areas, including northern Worcester County and portions of Franklin County.

This bird's vernacular name derives from one of its calls, a two-syllable vocalization likened to the sound of a file sharpening a large saw. This call is seldom given. Saw-whet owls are most vocal just prior to and early in the breeding season (from mid-March to mid-April) and can be heard at dusk, dawn, and varying intervals throughout the night. At this time, the most commonly recorded call is a short, sweet whistle or toot, repeated endlessly 100 or more times a minute. Other vocalizations include a thrushlike, staccato whistle and various gasps and clacks. Outside this brief period of vocal activity, these owls generally are silent.

The saw-whet owl shows a preference for deep, swampy woodlands, usually in areas of conifers. Its nest is typically in an old woodpecker hole, often that of a flicker. Sometimes a natural cavity is used. The nest is generally unlined, but if there is nesting material present it is most likely left over from a previous avian or mammalian occupant. Locations of 6 Massachusetts nests were as follows: 3 in pine stubs, 1 in a Sycamore stub, 1 in a dead American Chestnut, 1 in a natural hole in an Atlantic White Cedar, and 1 in a natural crevice in a maple stub. Three of the first four cavities were reported to be old flicker holes. Heights ranged from 7 to 25 feet (ACB, Grice 1942, Petersen).

The four to seven eggs are laid at 1- to 3-day intervals, with the young hatching at similar periods and thus varying in size. Clutch sizes for 5 state nests were four eggs (1 nest), five eggs (1 nest), six eggs (2 nests), seven eggs (1 nest) (ACB, Grice 1942). Incubation is reported as taking 21 to 28 days, probably closer to the higher number. The white down of the chicks is replaced with the rich, dark mahogany feathers of the juvenal plumage during the second to fourth week. The young leave the nest at 27 to 34 days of age. Nestlings have been observed in Massachusetts from May 3 to June 15 (Grice 1942, TC, Olmstead). The single nestling was discovered in Templeton, providing a post-Atlas confirmation for Worcester County (TC). Young typically

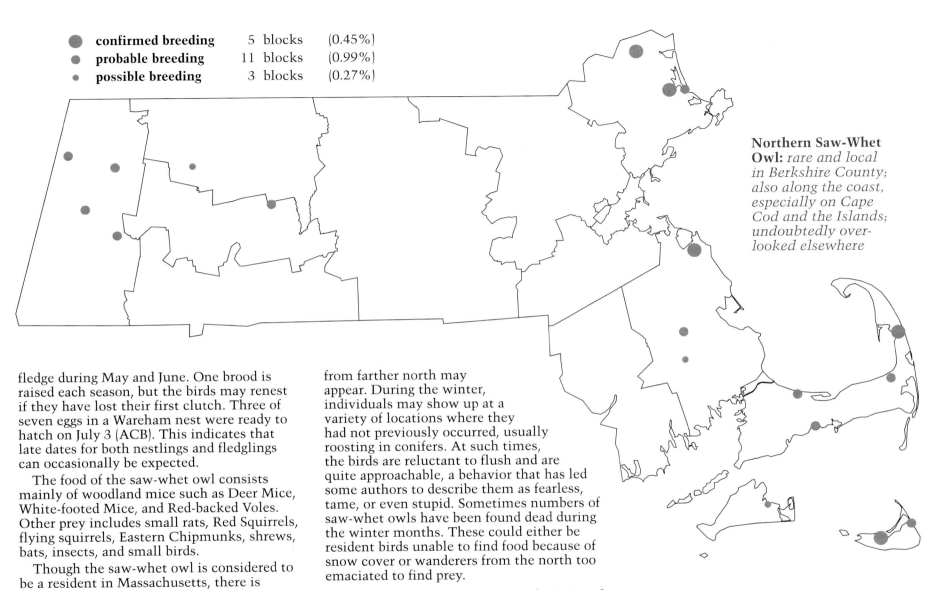

Northern Saw-Whet Owl: *rare and local in Berkshire County; also along the coast, especially on Cape Cod and the Islands; undoubtedly overlooked elsewhere*

confirmed breeding	5 blocks	(0.45%)
probable breeding	11 blocks	(0.99%)
possible breeding	3 blocks	(0.27%)

fledge during May and June. One brood is raised each season, but the birds may renest if they have lost their first clutch. Three of seven eggs in a Wareham nest were ready to hatch on July 3 (ACB). This indicates that late dates for both nestlings and fledglings can occasionally be expected.

The food of the saw-whet owl consists mainly of woodland mice such as Deer Mice, White-footed Mice, and Red-backed Voles. Other prey includes small rats, Red Squirrels, flying squirrels, Eastern Chipmunks, shrews, bats, insects, and small birds.

Though the saw-whet owl is considered to be a resident in Massachusetts, there is evidence that some migration occurs, or at least that the birds may wander widely in the fall. Much of this movement involves a shifting within the extensive breeding range, and it occurs primarily from October through December. There is also evidence of a return from March to May. In severe winters, birds from farther north may appear. During the winter, individuals may show up at a variety of locations where they had not previously occurred, usually roosting in conifers. At such times, the birds are reluctant to flush and are quite approachable, a behavior that has led some authors to describe them as fearless, tame, or even stupid. Sometimes numbers of saw-whet owls have been found dead during the winter months. These could either be resident birds unable to find food because of snow cover or wanderers from the north too emaciated to find prey.

Mark C. Lynch

Common Nighthawk
Chordeiles minor

Egg dates: May 31 to July 25.

Number of broods: one; may re-lay if first attempt fails.

The Common Nighthawk is one of the most widespread breeding birds in North America, though it is now rather local in Massachusetts. This species is more or less confined to the larger cities in the state such as Boston, Cambridge, Worcester, Springfield, and Pittsfield and to other smaller communities where buildings with flat, graveled roofs are present. Before 1900, it was a common summer resident in the pine barrens of the coastal plain and locally inland in burned-over fields and pastures. The extensive reforestation of the vast agricultural areas of the 1800s has undoubtedly contributed to the population decline. There are recent confirmed breeding records for some areas in the Myles Standish State Forest in Plymouth, but increasing human population elsewhere in Plymouth County has substantially decreased the amount of suitable habitat for ground-nesting nighthawks.

The first nighthawks generally arrive by mid-May, but the bulk of the migration occurs during the last week of May and early June. There are a few records for late April and early May, but these are usually storm-driven stragglers. The number of migrants reported in the spring is but a fraction of those of the mass movements that occur in the fall.

To the aware city stroller, the call of the nighthawk is one of the more familiar sounds of summer. It is most often described as an abrupt *peent*, strongly accented at the beginning. This call is uttered only while the nighthawk is in flight and functions as a territorial call. The migrating individuals in flocks seldom call in the fall. Nighthawks also make a resounding *boom* during courtship. The male rises to a considerable height, then dives with wings folded, and ends by swooping sharply upward. During the dive, the stiff outer primaries vibrate, producing the resonant booming sound.

The nighthawk builds no nest but lays two eggs on the ground in sparsely vegetated or burned-over areas or on graveled roofs of flat-topped buildings in cities. Rooftop nesting may be more successful than ground nesting because of fewer predators and a decreased possibility of flooding. Another reason for increased success is that the lights of the city attract a continuous supply of insects for food. The cryptically colored eggs are cream to pale gray and dotted with gray and brown. Incubation, done by the female alone, lasts 19 days on average. The male may feed the incubating female and also helps care for the young. If a potential predator comes in contact with a nesting nighthawk, the parent engages in the classic routine of feigning a broken wing and hissing at the intruder. Nighthawks are known to move both eggs and young in an effort to escape extreme heat or flooding. Young birds usually fly by 21 days after hatching and remain dependent for a short time thereafter.

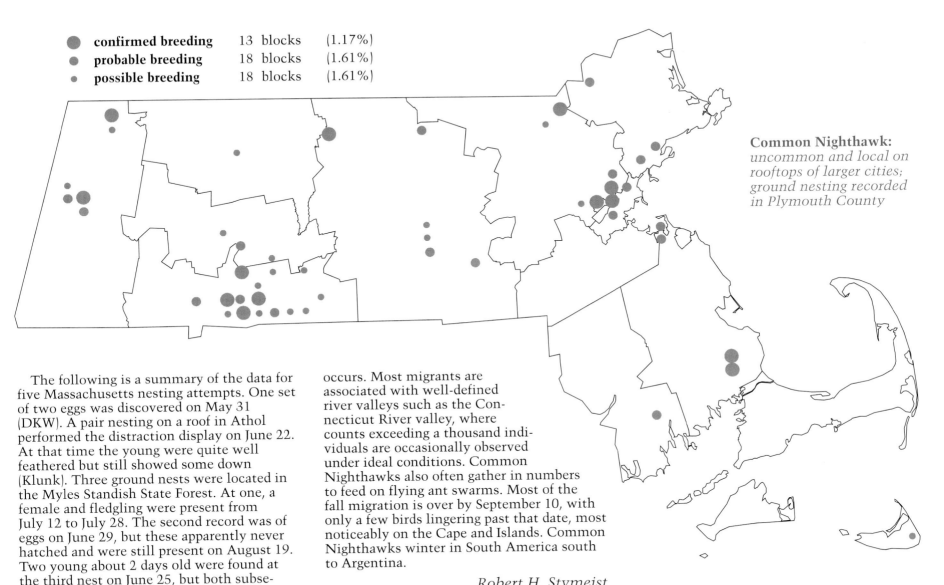

Common Nighthawk: *uncommon and local on rooftops of larger cities; ground nesting recorded in Plymouth County*

The following is a summary of the data for five Massachusetts nesting attempts. One set of two eggs was discovered on May 31 (DKW). A pair nesting on a roof in Athol performed the distraction display on June 22. At that time the young were quite well feathered but still showed some down (Klunk). Three ground nests were located in the Myles Standish State Forest. At one, a female and fledgling were present from July 12 to July 28. The second record was of eggs on June 29, but these apparently never hatched and were still present on August 19. Two young about 2 days old were found at the third nest on June 25, but both subsequently disappeared, presumably lost to a predator. However, this pair renested. On July 5, there was one new egg, and there were two eggs on July 11. On July 28 a 2- or 3-day-old chick was observed (BOEM).

On warm calm nights in late August and early September, a pronounced fall migration occurs. Most migrants are associated with well-defined river valleys such as the Connecticut River valley, where counts exceeding a thousand individuals are occasionally observed under ideal conditions. Common Nighthawks also often gather in numbers to feed on flying ant swarms. Most of the fall migration is over by September 10, with only a few birds lingering past that date, most noticeably on the Cape and Islands. Common Nighthawks winter in South America south to Argentina.

Robert H. Stymeist

Whip-poor-will
Caprimulgus vociferus

Egg dates: May 18 to June 9.

Number of broods: one; may re-lay if first attempt fails.

The loud, persistent call of the Whip-poor-will is familiar to many in Massachusetts, yet few have ever seen this nocturnal recluse. The Atlas results indicate that the species is thinly but widely distributed throughout the state, but it is common only along the southeastern coastal plain of Plymouth County, Cape Cod, and the Islands. In central and western portions of the state, it is very local and basically confined to the lower elevations. The birds inhabit dry woodlands of pine and oak, with little undergrowth and some interspersed clearings, where they hunt for moths and other flying insects; they generally avoid mature forests and higher elevations.

The first *whip*s usually are heard during the last week of April, with a more widespread arrival in the first week of May. The unmistakable call is a ringing, monotonous *whip-poor-will* or *pur-pill-rib* given at a rate of about one per second. A soft *chuck* note preceding each call is audible only at very close range. The birds are most vocal for a couple of hours after sunset and again for two to three hours before dawn. Few birds are as tireless in proclaiming their territories; naturalist William Burroughs once counted 1,088 consecutive calls from one particularly zealous individual.

Whip-poor-wills are nocturnal; consequently, only meager information is available on their behavior and breeding ecology. Territory size has been reported to range from approximately 25 to 77 acres. The male performs a hovering flight display, perhaps

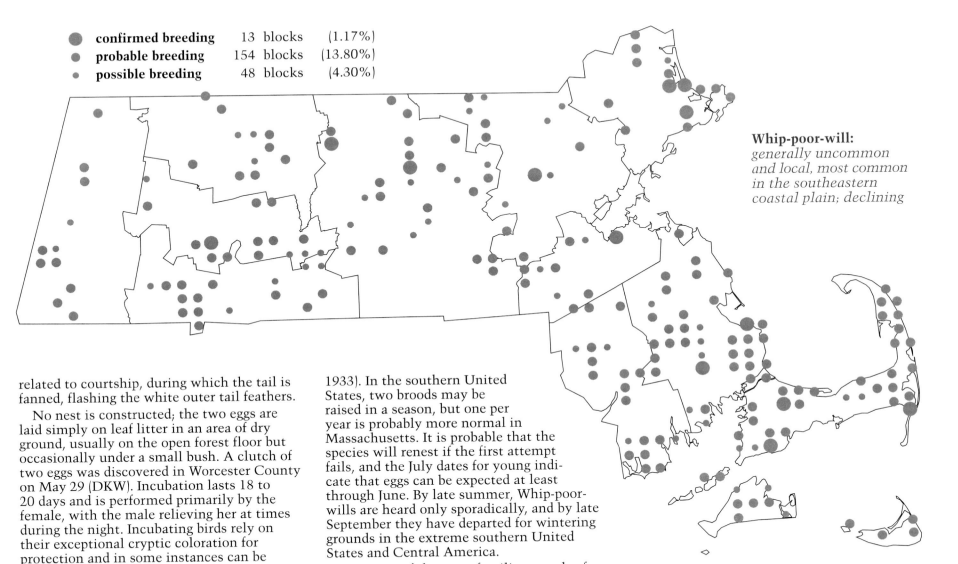

- ● **confirmed breeding** 13 blocks (1.17%)
- ● **probable breeding** 154 blocks (13.80%)
- · **possible breeding** 48 blocks (4.30%)

Whip-poor-will:
generally uncommon and local, most common in the southeastern coastal plain; declining

related to courtship, during which the tail is fanned, flashing the white outer tail feathers.

No nest is constructed; the two eggs are laid simply on leaf litter in an area of dry ground, usually on the open forest floor but occasionally under a small bush. A clutch of two eggs was discovered in Worcester County on May 29 (DKW). Incubation lasts 18 to 20 days and is performed primarily by the female, with the male relieving her at times during the night. Incubating birds rely on their exceptional cryptic coloration for protection and in some instances can be touched without flushing. On other occasions, however, the approach of an intruder elicits a fluttering, feigned, crippled-bird distraction display by the adult.

The cryptically colored young begin to wander short distances from the nest several days after hatching and fledge when they are about 20 days old. Two half-grown young were observed in Pelham on July 31 (Nice 1933). In the southern United States, two broods may be raised in a season, but one per year is probably more normal in Massachusetts. It is probable that the species will renest if the first attempt fails, and the July dates for young indicate that eggs can be expected at least through June. By late summer, Whip-poor-wills are heard only sporadically, and by late September they have departed for wintering grounds in the extreme southern United States and Central America.

Once one of the most familiar sounds of the summer night, the call of the Whip-poor-will is no longer heard in many areas of Massachusetts where it was once common. This decline has continued in the years following the Atlas. Sadly, the species' apparent dependence upon special types of open, scrubby woodlands may soon relegate it to those few areas that remain unfragmented by humanity's inexorable intrusions. Automobiles may take a toll in some areas, and it is believed that the collapse of populations of the large saturniid moths has contributed to the decline of the Whip-poor-will due to the loss of a major food source.

Blair Nikula

Chimney Swift
Chaetura pelagica

Egg dates: May 31 to late July.

Number of broods: one; may re-lay if first attempt fails.

As likely as not, the spring arrival of Chimney Swifts will be announced by a rapid series of loud, harsh *chip*s falling from the sky in late April. A glance skyward will reveal dark, cigar-shaped birds soaring or dashing about on long, sickle-shaped wings. Prior to the arrival of European settlers, Chimney Swifts nested in hollow trees. The number of nests in a given tree was dependent on the size of the cavity. Since then, swifts have adapted to the advances of humankind and in Massachusetts are now reported to nest exclusively in chimneys, although, in other regions, barns, silos, or even the traditional hollow trees may occasionally be used. Swifts are widely distributed in Massachusetts. They are most abundant in river valleys, where smokestacks of abandoned mills provide ample nesting and roosting space, and are decidedly less common on Cape Cod and the Islands than elsewhere in the Commonwealth. During migration, especially in fall, large flocks composed of a hundred or more birds are seen frequently.

More than any other species, the Chimney Swift is an aerial specialist, feeding solely on insects obtained while on the wing. Since swifts are strictly insect eaters, they are best seen in the morning and late in the evening when they course low to the ground. At midday, they are apt to be beyond vision in quest of insects that have ridden warm air currents to great heights. There are also reports of swifts foraging at night. The feet of swifts are poorly developed and serve only to help them cling to vertical surfaces. Most observers will never see one at rest during an entire career of watching. The calls are a series of *chip* or *chit* notes, often run together to produce a shrill chatter.

Occasionally, the first Chimney Swifts appear as early as mid-April, and spring migration continues through mid-May. Breeding normally commences in late May or early June. The only visible sign of nesting activity is called twigging, whereby birds will snap the ends of dead twigs off trees. At the onset of breeding, Chimney Swifts are often seen flying in trios, apparently two males vying for a mate. Copulation, like most activities of swifts, is probably accomplished during flight (Sutton 1928).

The nest is shaped like a half-saucer, constructed solely of twigs and attached to the surface of a chimney with a glutinous saliva. Nests may be located near the top of a chimney or 20 feet or more below the opening. Although firmly attached, they may be washed loose during heavy rainstorms. The normal clutch is four or five (range three to six) white eggs. One Massachusetts nest in the Millbury area contained three eggs on May 31 (DKW), and another in Wendell held four eggs on July 2 (CNR). Both sexes share in incubation, which lasts an average of 20 days. Hatching at the Wendell nest began on July 11 (CNR). As the young mature, they leave the nest and cling to the vertical sides of the chimney. They first fly at 28 to 30 days of age. Various observers in Massachusetts

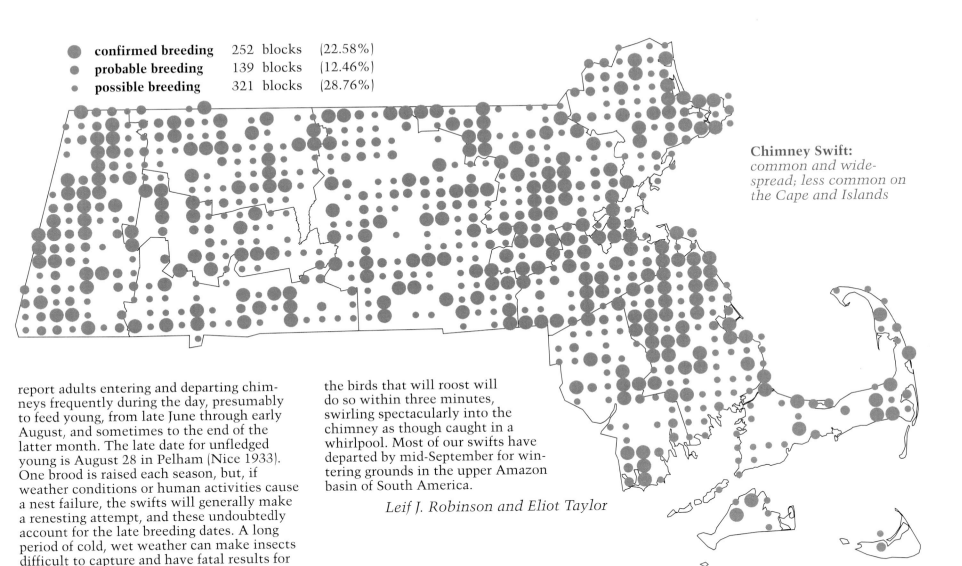

- **confirmed breeding** 252 blocks (22.58%)
- **probable breeding** 139 blocks (12.46%)
- **possible breeding** 321 blocks (28.76%)

Chimney Swift: *common and widespread; less common on the Cape and Islands*

report adults entering and departing chimneys frequently during the day, presumably to feed young, from late June through early August, and sometimes to the end of the latter month. The late date for unfledged young is August 28 in Pelham (Nice 1933). One brood is raised each season, but, if weather conditions or human activities cause a nest failure, the swifts will generally make a renesting attempt, and these undoubtedly account for the late breeding dates. A long period of cold, wet weather can make insects difficult to capture and have fatal results for both old and young swifts.

Autumn migration occurs principally in late August and early September but may continue to mid-October. Numbers of migrants are much greater in fall, and observing several hundred swifts first circling and then entering a roost at dusk can be very engaging. The exercise lasts about half an hour, and during peak activity some 40 percent of all the birds that will roost will do so within three minutes, swirling spectacularly into the chimney as though caught in a whirlpool. Most of our swifts have departed by mid-September for wintering grounds in the upper Amazon basin of South America.

Leif J. Robinson and Eliot Taylor

Ruby-throated Hummingbird
Archilochus colubris

Egg dates: late May to mid-August.

Number of broods: one or two.

An unforgettable sight is the glimpse of a male Ruby-throated Hummingbird on a May morning, especially if he perches briefly, turning his head from side to side, his flame red gorget flashing brilliantly as it catches the light. Declining in numbers over the years in eastern Massachusetts with the maturation of the forests and the disappearance of orchards, this species is now an uncommon but widespread breeder in the state, nesting chiefly in less developed areas. It is most common in the Berkshires and foothills along the Connecticut River valley, with only sparse nesting on the Worcester County plateau. Fair numbers occur along the coastal plain, with the exception of Nantucket.

The males preceding the females, these spritelike birds leave the tropics in late winter, some of them flying 500 miles or more over the Gulf of Mexico, moving northward in pace with the opening flowers and typically arriving in Massachusetts in May. This species is a generally uncommon, although occasionally common, spring migrant, usually seen singly or in a group of two to five, but swarms of fifteen to twenty may gather to feed at a large flowering tree. In spring, these migrants move along the outer coast, skimming over the tops of grasses and shrubs close to the ground, and they are regularly sighted from boats at sea, the tiny birds flying rapidly along, often within the troughs of the waves to avoid strong winds.

Aptly named for the droning sound of its wings and for the male's brilliant throat plumage, the Ruby-throated Hummingbird is one of the smallest members of the largest nonpasserine family. Tiny and beautiful, this species exhibits four prominent hummingbird features: a helicopterlike flight, a very high and variable metabolic rate, an attenuated bill with an extrudable tubular tongue, and a belligerent and aggressive character.

The courtship behavior of this hummingbird emphasizes its most adaptive proficiency—flight, rather than vocalization. In this species, vocalizations are limited to abrupt bursts of rapid chittering and squeaking, more useful in aggression and defense than in attracting a mate. The courtship dance is a pendulumlike flight in which the male swings back and forth in a 180 degree arc, rising 8 to 10 feet above and 5 to 6 feet to the side of a feeding female. Swooping toward her with wings and tail outspread, he creates a loud buzz within inches of her on each downswing. This continues for several seconds at a time and may be repeated every day for as long as a month until the female accepts the male. The male also performs a forward zigzag flight. Copulation occurs only 2 or 3 days before the eggs are laid. No lasting or even temporary pair bond is formed in this species, and the male departs after mating occurs.

The preferred nesting sites are in open woods, occasionally in dense forest or its margins, and in orchards or flower gardens. A variety of different species of trees may be utilized. Two Massachusetts nests were both in pines, and 1 of these was 40 feet up in a White Pine in deep woods (Nice 1933). The female selects the site (usually on a down-sloping, slender branch, often overhanging a stream), 5 to 50 feet above the ground, aggressively defends the area, and quickly (within 1 to 7 days) constructs the nest alone, utilizing bud scales, plant down, lichens, and spider silk. The minute (slightly over 1 inch

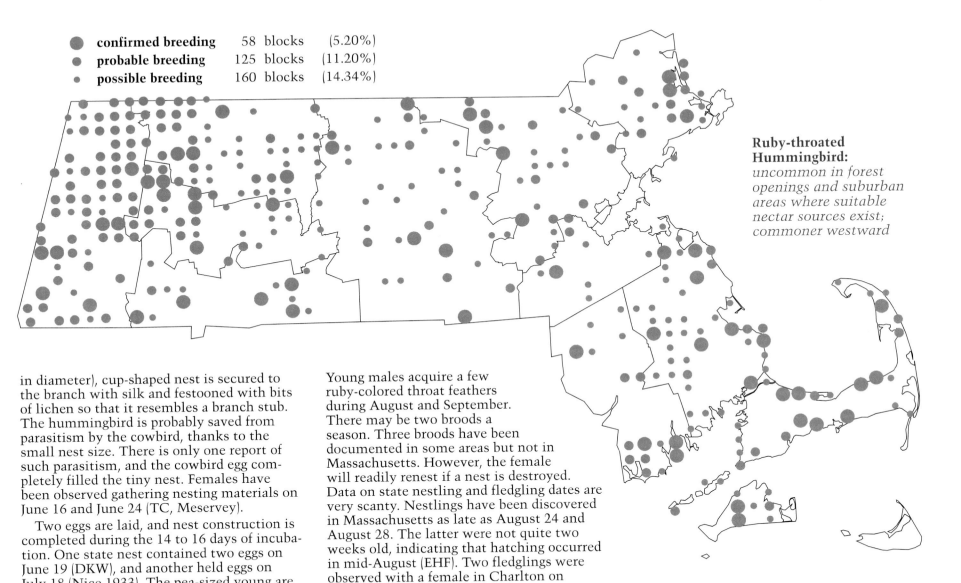

Ruby-throated Hummingbird: *uncommon in forest openings and suburban areas where suitable nectar sources exist; commoner westward*

- confirmed breeding — 58 blocks (5.20%)
- probable breeding — 125 blocks (11.20%)
- possible breeding — 160 blocks (14.34%)

in diameter), cup-shaped nest is secured to the branch with silk and festooned with bits of lichen so that it resembles a branch stub. The hummingbird is probably saved from parasitism by the cowbird, thanks to the small nest size. There is only one report of such parasitism, and the cowbird egg completely filled the tiny nest. Females have been observed gathering nesting materials on June 16 and June 24 (TC, Meservey).

Two eggs are laid, and nest construction is completed during the 14 to 16 days of incubation. One state nest contained two eggs on June 19 (DKW), and another held eggs on July 18 (Nice 1933). The pea-sized young are fed initially with nectar and tiny insects squirted into their mouths, but, as the bills of the young lengthen, the mother inserts the full length of her beak into an infant bird's gullet and regurgitates the food. The young fledge in about three weeks, by which time they are fully feathered in juvenal plumage.

Young males acquire a few ruby-colored throat feathers during August and September. There may be two broods a season. Three broods have been documented in some areas but not in Massachusetts. However, the female will readily renest if a nest is destroyed. Data on state nestling and fledgling dates are very scanty. Nestlings have been discovered in Massachusetts as late as August 24 and August 28. The latter were not quite two weeks old, indicating that hatching occurred in mid-August (EHF). Two fledglings were observed with a female in Charlton on July 31 and August 1 (Meservey).

Between late August and late September, particularly to avoid the first frosts, the Ruby-throated Hummingbirds depart from Massachusetts on a southwesterly course to return to their tropical wintering grounds. There are a number of late records, even into November, but these reports may refer to reverse migrants. The wintering grounds extend from the southern United States through Mexico to Central America.

Dorothy Rodwell Arvidson

Belted Kingfisher
Ceryle alcyon

Egg dates: May 11 to June 6.

Number of broods: one; may re-lay if first attempt fails.

The Belted Kingfisher is an uncommon but widespread and familiar species in Massachusetts. It is most prevalent at lower elevations near salt- or freshwater where there are suitable banks or cliffs for nest burrows. It is decidedly less common in the higher elevations of the central and western portions of the state and is notably scarce on Nantucket due to a lack of nest sites.

Hovering kingfishers, poised over the water with spread tail in a nearly vertical position, are a familiar sight. In normal flight, a series of five to six rapid wing beats is followed by a long glide. Kingfishers feed by plunge diving, plummeting down into the water from 20 to 40 feet in the air or from a perch when they spot their prey, remaining immersed for only a few seconds. Sometimes they will also dive below water to escape predators. After a plunge, the bird returns to its perch, stuns its prey by slapping it against the branch, flips it in the air, catches it headfirst, and then swallows it. Fish are the main food, but crayfish, crabs, insects, and tadpoles also are taken. Sometimes a large fish will protrude from the bill until swallowing can be completed after digestion has begun. If disturbed, kingfishers fly to the limits of their territory and then circle back.

The voice of the Belted Kingfisher is a clattering, dry rattle, aptly likened to the sound of a heavy fisherman's reel. During courtship, the birds produce a mewing sound and also keep up a continual, prolonged version of the rattle call. In this species, it is the female that is more brightly colored, sporting a chestnut band across the breast.

Migrants begin appearing in the interior in March as the ice breaks up. Others arrive in April, and there is still some movement in early May. The habitat is along all types of waterways: lakes, ponds, streams, wooded creeks, rivers, and marine bays and estuaries. The birds adopt favorite perches, near or over the water, to which they return repeatedly. They may roost in evergreens.

Nest sites are burrows near the top of a bank that is relatively free of vegetation. These may be natural bluffs but very often are sand and gravel pits shared with a colony of Bank Swallows. Nests are found rarely in the top of a hollow stump or in a tree cavity, but this has not been documented in Massachusetts. Ten recent nest sites in central Massachusetts were all in gravel banks at a height of 10 to 40 feet (Coyle, Meservey). Two Essex County nests were in sand banks at 7 feet and 20 feet. The latter, in Rowley, was 50 feet from a previous nest hole then being used by Northern Rough-winged Swallows (CNR).

Both sexes excavate the tunnel, which is generally from 3 to 10 feet, sometimes as much as 15 feet, long and which slopes upward to the nest chamber. The latter is often littered with excrement, fish scales, and regurgitated pellets, to which the birds may add a few sticks, leaves, or dry grass. Burrow construction takes from 3 days to three weeks. The male may dig one or two shorter tunnels nearby for roosting. The Belted Kingfisher is highly protective of the nesting area and will attack birds many times its size if the territory is invaded. On smaller inland watercourses, the birds are very territorial, but near the coast several pairs may nest in close proximity. Three pairs were observed to have nest burrows on a small island near Woods Hole (EHF).

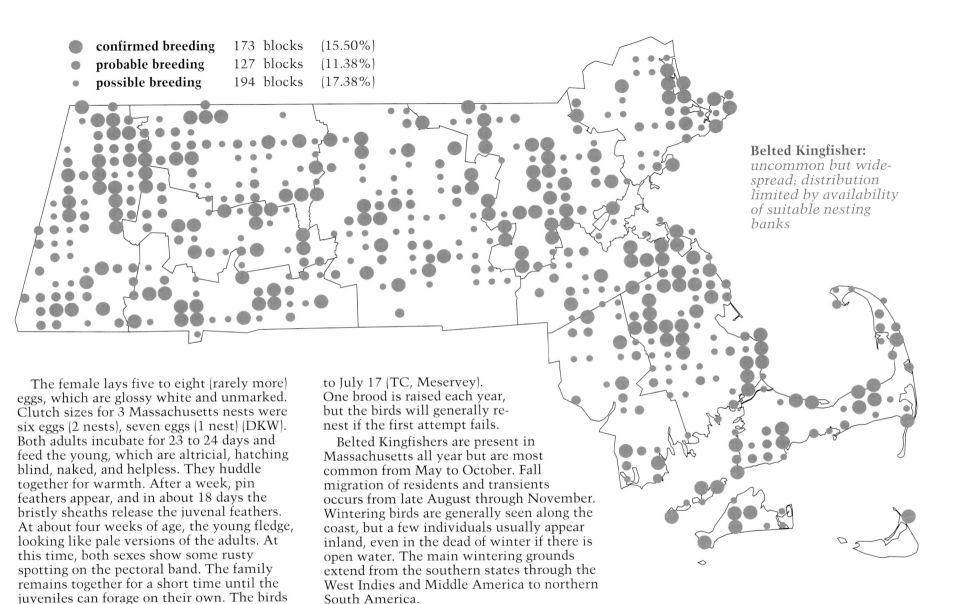

- **confirmed breeding** 173 blocks (15.50%)
- **probable breeding** 127 blocks (11.38%)
- **possible breeding** 194 blocks (17.38%)

Belted Kingfisher: *uncommon but widespread; distribution limited by availability of suitable nesting banks*

The female lays five to eight (rarely more) eggs, which are glossy white and unmarked. Clutch sizes for 3 Massachusetts nests were six eggs (2 nests), seven eggs (1 nest) (DKW). Both adults incubate for 23 to 24 days and feed the young, which are altricial, hatching blind, naked, and helpless. They huddle together for warmth. After a week, pin feathers appear, and in about 18 days the bristly sheaths release the juvenal feathers. At about four weeks of age, the young fledge, looking like pale versions of the adults. At this time, both sexes show some rusty spotting on the pectoral band. The family remains together for a short time until the juveniles can forage on their own. The birds then disperse and are generally encountered as solitary individuals.

In Massachusetts, adults have been recorded feeding nestlings from June 13 to July 1 (CNR, Meservey). Pairs with two to four fledged young were observed from July 3 to July 17 (TC, Meservey). One brood is raised each year, but the birds will generally re-nest if the first attempt fails.

Belted Kingfishers are present in Massachusetts all year but are most common from May to October. Fall migration of residents and transients occurs from late August through November. Wintering birds are generally seen along the coast, but a few individuals usually appear inland, even in the dead of winter if there is open water. The main wintering grounds extend from the southern states through the West Indies and Middle America to northern South America.

Patricia Noyes Fox

Red-headed Woodpecker
Melanerpes erythrocephalus

Egg dates: May to June 17.

Number of broods: one; may re-lay if first attempt fails.

The status of this striking woodpecker varies from a local permanent resident to a very uncommon visitor. Believed formerly to have been more common, the Red-headed Woodpecker apparently declined in the Northeast around the turn of the century. A description of its distribution today in Massachusetts matches that given by Forbush. At best, it is a rare and irregular breeder and an occasional migrant and wintering bird. The species may turn up in any section of the state, but there are generally more records from the western portions.

Open farm country with scattered oak or beech groves is the preferred habitat. The spread of Dutch Elm Disease and subsequent die-off of large elms provided ideal nest locations. However, by the mid-1970s many of these dead giants had fallen to chainsaws or storms and high winds. In other portions of the range, and to a lesser extent in Massachusetts, competition with starlings for nest sites has limited or reduced the populations. In the Commonwealth, Red-headed Woodpeckers will return to the same breeding site for several years in succession, with adults sometimes wintering in the area. Then, for reasons unknown, they will abandon the location.

Spring migrants arrive here in late April and early May, and nesting commences in late May or early June. Nest sites are usually in dead trees adjacent to large fields or farmland, but telephone poles and old fence posts are sometimes used. A post-Atlas nest in Westboro was in a dead tree 20 feet out in the water of an open swampy impoundment (McMenemy). The nesting cavity is 8 to 24 inches deep and may be located from 5 to 80 feet above the ground. Clutches of four to seven pure white eggs are incubated by both sexes for 12 to 14 days. Limited data on nestlings and fledglings indicate that nests with eggs can be found from May to the end of June. If the first attempt fails, the birds appear to be single-brooded in our area. The young fledge at 27 to 30 days of age. At the Westboro nest, a single nestling was still being fed on July 26 and had fledged by August 12 (McMenemy). During the Atlas period, pairs, each with three fledged young, were observed in Lynn and Pittsfield in early July.

The young remain together throughout the summer before migrating. In their brown-and-white juvenal plumage, they appear quite different from their parents. The bright adult plumage, identical for both sexes, is acquired through a gradual fall-winter molt and is renewed each year after the breeding season.

These birds are quite noisy, regularly producing a variety of loud, scold notes. One common call is similar to that of the Great Crested Flycatcher: a loud *churr-churr*. Other listeners liken some raspy notes to those of the Gray Treefrog. Another sound has been described as *quee-o, quee-o*. In spite of being so vocal and showy, these woodpeckers can be surprisingly inconspicuous when nesting.

Like several related species, the Red-headed Woodpecker often will store nuts and acorns, hiding them in holes and crevices. During summer, much of their diet consists of insects, especially grasshoppers and beetles.

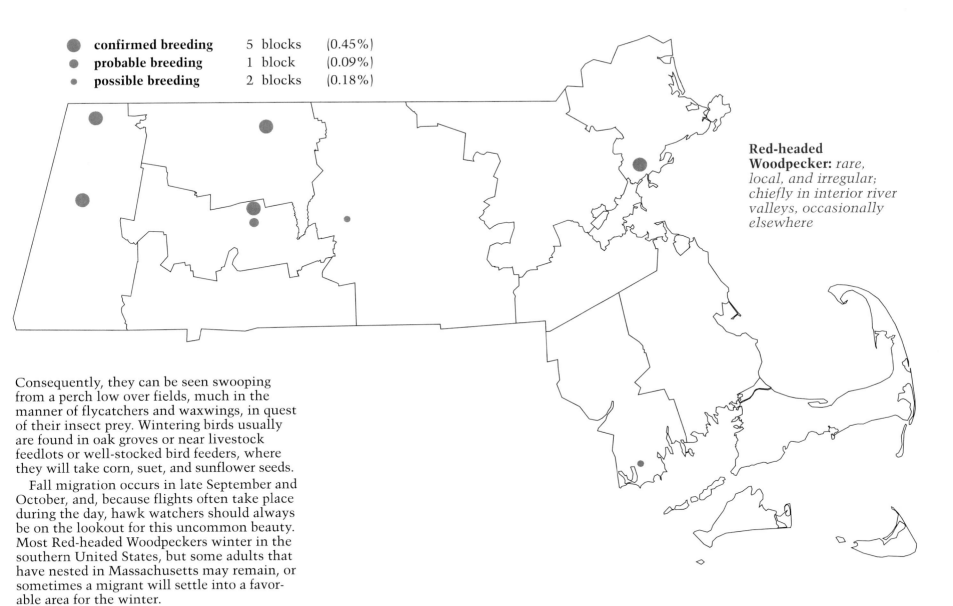

- ● **confirmed breeding**   5 blocks   (0.45%)
- ● **probable breeding**   1 block   (0.09%)
- · **possible breeding**   2 blocks   (0.18%)

Red-headed Woodpecker: *rare, local, and irregular; chiefly in interior river valleys, occasionally elsewhere*

Consequently, they can be seen swooping from a perch low over fields, much in the manner of flycatchers and waxwings, in quest of their insect prey. Wintering birds usually are found in oak groves or near livestock feedlots or well-stocked bird feeders, where they will take corn, suet, and sunflower seeds.

Fall migration occurs in late September and October, and, because flights often take place during the day, hawk watchers should always be on the lookout for this uncommon beauty. Most Red-headed Woodpeckers winter in the southern United States, but some adults that have nested in Massachusetts may remain, or sometimes a migrant will settle into a favorable area for the winter.

Thomas H. Gagnon

Red-bellied Woodpecker
Melanerpes carolinus

Egg dates: April to June.

Number of broods: one or two; may re-lay if first attempt fails.

The Red-bellied Woodpecker is gradually becoming a resident of Massachusetts woodlands, but it has not yet become as thoroughly established as other southern birds such as the Northern Cardinal, Tufted Titmouse, and Northern Mockingbird. Although occasional vagrants have appeared since the mid-1800s, the species has been gradually expanding its range in the northeastern states. Red-bellied Woodpeckers increased in southeastern New York during the mid-1960s and were first recorded nesting in Connecticut in 1971.

The growing number of sightings in Massachusetts led observers to suspect that the birds were breeding, and the first state confirmations occurred during the Atlas period in 1977 in Natick and North Attleboro and then in Westfield and Southwick the following year. Although the species is still a fairly rare and local resident, the number of reports has continued to increase, especially during the winter when the birds visit feeding stations.

The Red-bellied Woodpecker is a resident species, but some individuals may move long distances, and there is some postbreeding dispersal. Males tend to remain on or near the nesting territory year-round while the females move to a new area to establish a separate winter territory. The Red-bellied Woodpecker is often found in river bottomland forest, but it frequents a variety of other habitats including pine barrens, upland forests, farmland, small woodlots, and even shade trees in towns and city parks.

This woodpecker is a noisy species with a variety of vocalizations. Common calls include a series of *chuf-chuf-chuf* or *cherr-cherr-cherr* sounds, a rolling *qui-er-r-r-r-r*, and a *crirrk*, the latter similar to one of the flicker's notes. Both sexes also communicate by tapping and drumming on a trunk or limb to indicate territory.

In Massachusetts, nesting generally begins in April or May. Both sexes work on excavating a nesting cavity in a tree, stump, telephone pole, or fence post. Although dead snags or live trees with softwood (such as Cottonwood) are preferred, a variety of tree species may be utilized. The Southwick nest was in a live Sugar Maple in heavy forest (Kellogg). It takes an average of 7 to 10 days for the birds to complete the cavity. Sometimes an old nest is reoccupied or the abandoned hole of another species of woodpecker is used. Typical nesting cavities have an opening 2 inches in diameter, are 10 to 12 inches deep, and are located from 5 to 70 feet above the ground.

Three to eight (average four to five) white eggs are laid, and a lost clutch usually results in another nesting attempt. Older sources give the incubation period as 14 days, but 11.5 to 12 days is now believed to be typical. Both sexes share incubation, the male by night and the female doing the larger share during the day. The altricial and naked nestlings are fed by both parents and are brooded for at least one week. At 24 to 26 days of age, they have acquired their juvenal plumage, which somewhat resembles that of the adult female, and are ready to fledge. The young remain close to the nesting

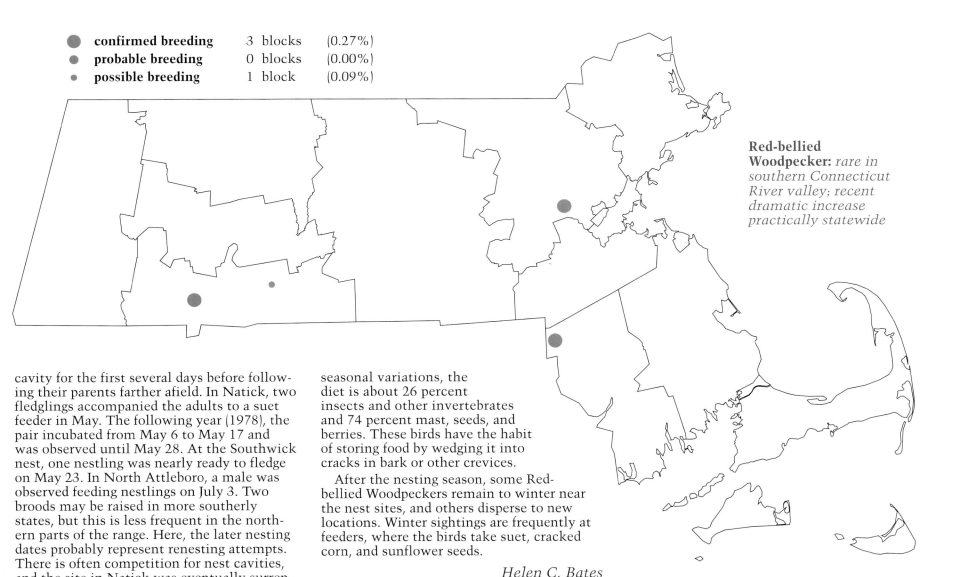

- ● **confirmed breeding**   3 blocks   (0.27%)
- ● **probable breeding**   0 blocks   (0.00%)
- ● **possible breeding**   1 block   (0.09%)

Red-bellied Woodpecker: *rare in southern Connecticut River valley; recent dramatic increase practically statewide*

cavity for the first several days before following their parents farther afield. In Natick, two fledglings accompanied the adults to a suet feeder in May. The following year (1978), the pair incubated from May 6 to May 17 and was observed until May 28. At the Southwick nest, one nestling was nearly ready to fledge on May 23. In North Attleboro, a male was observed feeding nestlings on July 3. Two broods may be raised in more southerly states, but this is less frequent in the northern parts of the range. Here, the later nesting dates probably represent renesting attempts. There is often competition for nest cavities, and the site in Natick was eventually surrendered to invading starlings. The immatures are independent at about 42 days of age. They gradually attain adult plumage. The adults have a complete molt at the end of the summer.

Red-bellied Woodpeckers forage in trees and on the ground. Although there are seasonal variations, the diet is about 26 percent insects and other invertebrates and 74 percent mast, seeds, and berries. These birds have the habit of storing food by wedging it into cracks in bark or other crevices.

After the nesting season, some Red-bellied Woodpeckers remain to winter near the nest sites, and others disperse to new locations. Winter sightings are frequently at feeders, where the birds take suet, cracked corn, and sunflower seeds.

Helen C. Bates

Yellow-bellied Sapsucker
Sphyrapicus varius

Egg dates: May to mid-June (estimated).

Number of broods: one; may re-lay if first attempt fails.

The mystery of the Yellow-bellied Sapsucker's rarity during migration remains unchanged over the years. Its scarcity, particularly in spring, is remarkable, for it is a very common resident of the northern woods. As a breeder in Massachusetts, it has continued to increase long after the hill country reverted to forest. The sapsucker was still considered a rare and local summer resident in the higher hills in the mid-1950s and was said to be steadily decreasing. Curiously, in spite of continual lumbering, this species is currently a very common breeding bird in any forest where birch, beech, and maple predominate and oaks do not, at elevations as low as 500 feet but generally above 600 feet. The Berkshires are a stronghold, and east of the Connecticut River the population falls off dramatically, and the birds are surprisingly scarce in appropriate habitat in Worcester County.

The sapsucker arrives on its local breeding grounds from early to mid-April, with migrants continuing to pass until mid-May. At this time, a few can be seen west and north of Boston or in the Connecticut River valley. It occurs rarely in spring in southeastern Massachusetts or on the Cape and Islands but is regular in small numbers there in the fall. Migrants favor orchards, suburban groves, and cemeteries because the birds seem to avoid the natural oak forests of these lowland regions. In their native hills, they are attracted to Beaver swamps and other wetlands but can be found as well on steep hillsides. They prefer more open woods, such as those recently lumbered.

The common call note heard on the breeding territory, and occasionally on migration, is an emphatic, prolonged squeal or mewing note, repeated several times. It is fairly high-pitched and slightly burry. The sapsucker is a noisy bird at breeding time and has a number of less common calls, some resembling those of the Blue Jay or Red-shouldered Hawk. A variety of low sounds are made in the vicinity of the nest and mate. Both sexes drum, giving a short roll followed by several distinct measured taps, slower toward the end. These are often repeated, with distant males responding to each other.

A nesting pair generally chisels a cavity out of a dead or dying tree but may select a live one, taking care to leave the access hole as small as possible. It may take a week or more to complete the gourd-shaped excavation, which averages 10 to 14 inches deep and is located 8 to 40 feet above the ground. Birch and aspen are preferred tree species. Three Massachusetts nests from the Huntington-Chester area were located as follows: 1 at 20 feet in a dead ash, 1 at 24 feet in a dead snag, and 1 at 25 feet in a dead White Pine (Blodget). A nest in Warwick was situated at a height of 9 feet (Coyle). Surprisingly little has been recorded about the nesting habits of this species in Massachusetts. From what is known about spring arrival dates and nestling and fledgling dates, it appears that nests with eggs can be found from May into June. The four to seven white eggs are incubated for about 14 days in typical woodpecker fashion by both sexes, with the male on duty at night. The young birds remain in the nest for 26 to 29 days and hiss loudly when disturbed. The parents are kept busy feeding the hungry

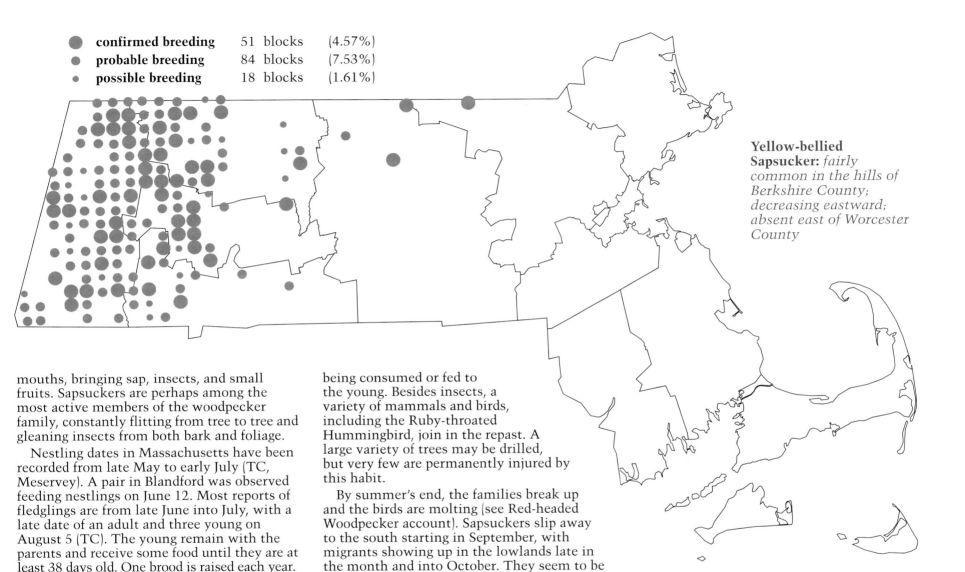

Yellow-bellied Sapsucker: *fairly common in the hills of Berkshire County; decreasing eastward; absent east of Worcester County*

mouths, bringing sap, insects, and small fruits. Sapsuckers are perhaps among the most active members of the woodpecker family, constantly flitting from tree to tree and gleaning insects from both bark and foliage.

Nestling dates in Massachusetts have been recorded from late May to early July (TC, Meservey). A pair in Blandford was observed feeding nestlings on June 12. Most reports of fledglings are from late June into July, with a late date of an adult and three young on August 5 (TC). The young remain with the parents and receive some food until they are at least 38 days old. One brood is raised each year.

The name sapsucker comes from the birds' penchant for drilling a series of small holes in sap-bearing trees and feeding regularly on the flow from the openings. They may use several trees close together in this way, creating a favorite "orchard," which they visit constantly. Insects, captured at the holes or elsewhere, may be dipped in the sap before being consumed or fed to the young. Besides insects, a variety of mammals and birds, including the Ruby-throated Hummingbird, join in the repast. A large variety of trees may be drilled, but very few are permanently injured by this habit.

By summer's end, the families break up and the birds are molting (see Red-headed Woodpecker account). Sapsuckers slip away to the south starting in September, with migrants showing up in the lowlands late in the month and into October. They seem to be as quiet and surreptitious in this season as they are noisy and energetic when nesting. Unlike our other migratory woodpecker, the flicker, this species does not travel in flocks, although two or three might be found together. By November, most have gone south, but occasional individuals are found during the winter, and these may sometimes survive in mild seasons. Yellow-bellied Sapsuckers winter from the southern United States south through Central America to central Panama and on some of the Caribbean Islands.

Seth Kellogg

Downy Woodpecker
Picoides pubescens

Egg dates: May 4 to June 21.

Number of broods: one; may re-lay if first attempt fails.

The Downy Woodpecker, named for the soft appearance of its feathers, is a common bird resident throughout the state, and it is familiar to most people. It nests in a wide variety of wooded areas: deciduous or mixed forest, second-growth scrub, orchards, parks, small woodlots, and urban gardens. This woodpecker is the smallest of those in the United States and Canada, yet its penetrating whinny call can be heard a quarter-mile away.

Drumming and drilling, important activities for most woodpeckers during courtship and nesting, are performed by both sexes in this species. Territorial drumming by Downy Woodpeckers can be heard in Massachusetts throughout the winter. When courtship begins in April and May, or sometimes as early as March, the drumming and the shrill whinnying calls increase in frequency and serve to bring the sexes together. A *tchick* note is also distinctive, sharp, and very brief. Downy Woodpeckers have been observed inspecting trees, apparently to locate the most resonant drumming spot, to which they then return to continue tapping.

Courtship behavior is variable but usually consists of much active chasing and flitting from tree to tree accompanied by loud calls and squeaking, an exaggerated bounding flight, a display of the white spots on the coverts and flight feathers by wing spreading, and elevation of the male's red patch. Often, two males will contest for attention by pursuing the female until one gains a position facing her with wings spread in a preliminary maneuver. Then the other male flies between them and interrupts the ceremony. At times, there may be a very quiet and inactive competition, with the participants perched on separate trees within sight and sound of each other. They remain completely motionless for minutes at a time until two fly off together, a silent selection somehow having been made. Courtship lasts a week or two, and, once mating has occurred, a somewhat prolonged search for a suitable nest site follows, with the partners attracting each other to chosen locations by tapping or drumming. The pair bond may last several years.

A dead branch or stump is usually chosen for nesting, but occasionally a wooden building attracts them. They will chip a hole through the wood and if not driven away by an irate householder will enter to nest between the inner and outer walls on a joist or other horizontal support. Although Downy Woodpeckers may be frustrated occasionally in their drilling by a hard knot, they can excavate easily in solid wood and have been known to chip cement used to plug a former nest hole. Nest cavities are located at heights of between 5 and 50 feet. Four Massachusetts nests found in dead or dying elm and maple trees were between 5 and 20 feet high (CNR). Within a week, a pair excavates a nesting cavity, 8 to 12 inches in depth. The round entrance hole is 1.25 inches in diameter.

A Downy Woodpecker female lays five to seven eggs on a few wood chips left in the bottom of the cavity. Of 3 Massachusetts nests examined, 2 contained four eggs and a third had five eggs (DKW). Both sexes share in the 12-day incubation duties and in feeding the altricial young. The nestlings are fed frequently, at first every two or three minutes, and in 4 or 5 days the parents bring entire insects to the young. In juvenal plumage, the head of the young male is marked

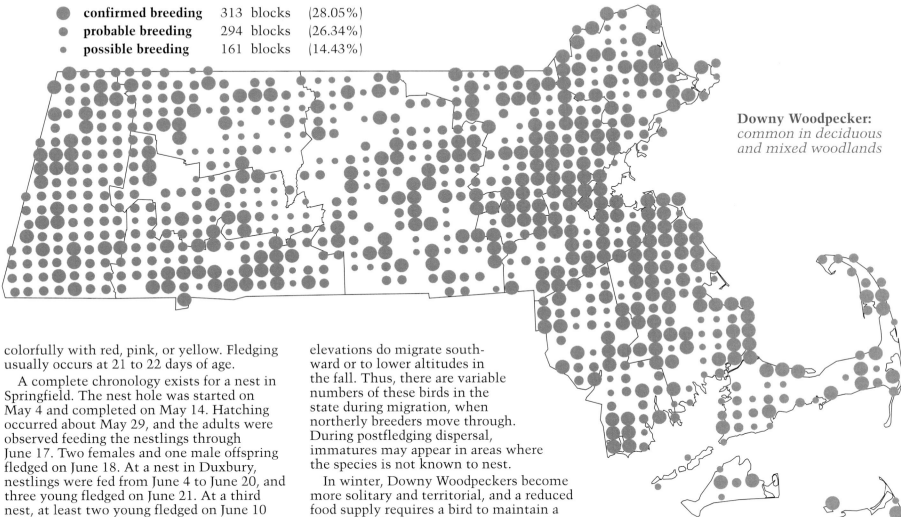

Downy Woodpecker: *common in deciduous and mixed woodlands*

- confirmed breeding — 313 blocks (28.05%)
- probable breeding — 294 blocks (26.34%)
- possible breeding — 161 blocks (14.43%)

colorfully with red, pink, or yellow. Fledging usually occurs at 21 to 22 days of age.

A complete chronology exists for a nest in Springfield. The nest hole was started on May 4 and completed on May 14. Hatching occurred about May 29, and the adults were observed feeding the nestlings through June 17. Two females and one male offspring fledged on June 18. At a nest in Duxbury, nestlings were fed from June 4 to June 20, and three young fledged on June 21. At a third nest, at least two young fledged on June 10 (CNR). The immature birds receive some food from the parents until they are 41 days old. By the end of the summer, the family groups break up, and, as the young attain adult plumage, the mature birds undergo a complete molt.

Throughout their breeding range, Downy Woodpeckers are found year-round, but more northerly populations and those at higher elevations do migrate southward or to lower altitudes in the fall. Thus, there are variable numbers of these birds in the state during migration, when northerly breeders move through. During postfledging dispersal, immatures may appear in areas where the species is not known to nest.

In winter, Downy Woodpeckers become more solitary and territorial, and a reduced food supply requires a bird to maintain a fairly large feeding area to survive. Both sexes excavate roosting holes in decaying trees or find an abandoned nest hole, hollow tree, or birdhouse in which to escape the elements at night. In this season, they become regular visitors to feeders for suet, corn, and peanut butter. Downy Woodpeckers feed primarily on insects and larvae but also eat spiders, snails, and a variety of vegetal materials. The effect of their drillings is primarily beneficial to the tree due to the removal of insect larvae, and little damage is done.

Dorothy Rodwell Arvidson

Hairy Woodpecker
Picoides villosus

Egg dates: April 22 to June 5.

Number of broods: one; may re-lay if first attempt fails.

Together with its close relative, the Downy Woodpecker, this species has an extensive range, occupying virtually all of forested North America south through the highlands of Central America to western Panama. It is everywhere less common than the Downy Woodpecker, but like the latter inhabits mainly deciduous forests and, in the Northeast, seems attracted to wet woodlands. The Hairy Woodpecker will also nest in more open situations such as orchards and shade trees. The species is found throughout the state, even breeding on parts of the Cape and Martha's Vineyard, but not on Nantucket.

During the winter, the female hairy remains on or near the breeding territory while the male, distinguished by his red nape, may be more wide ranging. Pair formation often begins in late winter but becomes accelerated and most obvious in April, at which time vocalizations accompany much of the daily activity. The common call note, *keek*, is louder and sharper than the similar note of the Downy Woodpecker. Another call is a loud, prolonged rattle resembling that of the Belted Kingfisher, but it is of shorter duration and is usually preceded by a single introductory note.

As the time of breeding approaches, drumming becomes a familiar ritual. This is usually performed on a resonant dead branch or tree trunk, and the rolls are shorter and louder than those of the downy. A courting male may pursue a female from tree to tree. Both members of a pair dodge about the trunk and branches and engage in posturing together, frequently spreading the wings while holding the bill on an axis parallel with the body. These activities are accompanied by muted, flickerlike *wicka* calls.

Either live or dead trees are chosen for the nest hole, with excavation by both sexes taking up to three weeks if a live tree is selected. Twelve nests in southeastern Massachusetts were located as follows: 6 in dry upland woods, 2 in maple swamps, 3 in apple orchards, and 1 in a tree in a swampy meadow. Nest trees were maple (3), apple (3), chestnut (2), poplar (2), dead oak (1), and dead beech (1). The heights ranged from 5 to 30 feet above the ground (ACB). Forbush mentions a nest 12 feet up in a live oak. In southern Worcester County, 2 nests were found in live Red Maples in swamps, 1 in a poplar in young second-growth forest, and 1 in a dead branch of a Red Oak (Meservey). Typical nesting cavities are 10 to 15 inches deep and have a circular or slightly elongated entrance hole about 1.5 inches in diameter.

The three to six (average four) white eggs are laid on a bed of wood chips at the bottom of the cavity. The clutch sizes for 6 Massachusetts nests were as follows: two eggs (1 nest), four eggs (5 nests) (DKW). Incubation, by both sexes during the day and the male at night, lasts about 12 days. During the early nestling stage, the young are fed by regurgitation, but later whole grubs or beetles

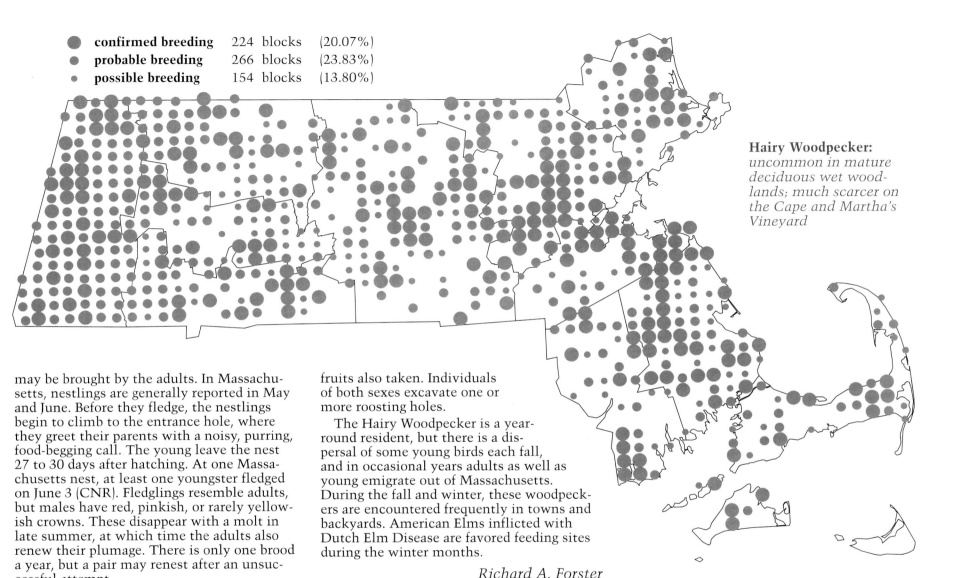

- **confirmed breeding** 224 blocks (20.07%)
- **probable breeding** 266 blocks (23.83%)
- **possible breeding** 154 blocks (13.80%)

Hairy Woodpecker: *uncommon in mature deciduous wet woodlands; much scarcer on the Cape and Martha's Vineyard*

may be brought by the adults. In Massachusetts, nestlings are generally reported in May and June. Before they fledge, the nestlings begin to climb to the entrance hole, where they greet their parents with a noisy, purring, food-begging call. The young leave the nest 27 to 30 days after hatching. At one Massachusetts nest, at least one youngster fledged on June 3 (CNR). Fledglings resemble adults, but males have red, pinkish, or rarely yellowish crowns. These disappear with a molt in late summer, at which time the adults also renew their plumage. There is only one brood a year, but a pair may renest after an unsuccessful attempt.

Juveniles receive some food from the adults until they are 44 days old. The late date for an immature accompanying an adult to a suet feeder is July 21 (Ober), but the birds may roam through the woods in a loose family group until late August. Insects make up the bulk of the diet, with some seeds and fruits also taken. Individuals of both sexes excavate one or more roosting holes.

The Hairy Woodpecker is a year-round resident, but there is a dispersal of some young birds each fall, and in occasional years adults as well as young emigrate out of Massachusetts. During the fall and winter, these woodpeckers are encountered frequently in towns and backyards. American Elms inflicted with Dutch Elm Disease are favored feeding sites during the winter months.

Richard A. Forster

Northern Flicker
Colaptes auratus

Egg dates: April 29 to June 20.

Number of broods: one; may re-lay if first attempt fails.

The Northern Flicker has one of the most extensive ranges of any North American bird. It is a common breeder in all sections of Massachusetts from the Berkshires to Cape Cod and the Islands. Although it can be found nesting in regions of extensive deciduous or mixed forest, the flicker prefers areas where stands of trees are interspersed with open habitats. It has adapted well to civilization, breeding readily in orchards, shade trees, woodlots, and city parks and seeking much of its food on lawns, fields, pastures, and golf courses.

Spring migration occurs mainly during April. At this season, most of the resident breeding birds arrive at the same time that migrants are passing northward. The birds announce their arrival with a loud, ringing *wick-wick-wick-wick* call. Other common vocalizations include a *klee-yer* or *ti-err* call and a *wicker-wicker-wicker* or *yucka-yucka-yucka* series, used frequently at or near the nest site. Both sexes also produce a loud drumming sound by pecking rapidly at a tree limb, the walls of the nest cavity, or even a metal roof, pipe, or insulation box.

One or more courting males may follow a female from tree to tree, creeping and dodging around the trunk and branches. During display, in addition to vocalizing, they bow, nod, turn the head from side to side, and open the wings to expose the yellow linings (hence the former name of Yellow-shafted Flicker).

Flickers nest in cavities that they excavate in live or dead trees, stumps, or posts. Both sexes work for one or two weeks at the task of drilling a new chamber, although old sites may be reused once any debris that has accumulated has been cleared out. Typical cavities are 2 to 4 inches in diameter, 10 to 36 inches deep, and are located from 2 to 60 feet above the ground. Of 6 Massachusetts nests, 1 was in a dead trunk, 1 in a decaying oak, and the others in deciduous trees (maple, elm, and Red Oak). Heights ranged from 7 to 30 feet (CNR). Flickers will also use nest boxes that have been designed for them. Occasionally, the birds will nest in unusual sites. There are historical records of eggs placed on hay in a barn in Lynnfield (the birds first having drilled a hole through the wall to enter) and a set of five eggs was discovered on the ground at the edge of a sand road in a Pitch Pine forest on Cape Cod (ACB).

The three to ten (usually at least five) white eggs are laid on a bed of wood chips that have been left in the bottom of the cavity. Clutch sizes in 4 Massachusetts nests were four, five, six, and seven, respectively (CNR). In another sample of state nests, 5 contained six eggs and 1 contained eight eggs (DKW). If robbed of eggs before her clutch is complete, the female flicker will continue to lay, and there are records of over seventy eggs being produced by a single bird. When a complete clutch is lost, the bird will produce a new set after digging deeper in the cavity or moving to a new site. Squirrels may prey on eggs and small nestlings, and starlings are serious and persistent competitors for nest cavities. Although two broods are reared in portions of the range, this does not appear to be the usual situation in Massachusetts.

Both sexes share in the incubation duties, alternating at intervals during the day. The male remains in the cavity, covering the eggs at night, and the female retires to a separate hole. Hatching occurs after 11 to 12 days, and the nestlings, initially naked and totally helpless, sprawl at the bottom of the cavity for another 11 days. By 17 days of age, they can cling to the sides of the hollow and at three weeks are able to climb up to the entrance hole to be fed. Initially, the youngsters are brooded by the adults on the same schedule as that followed during incubation, with the off-duty bird returning to feed the

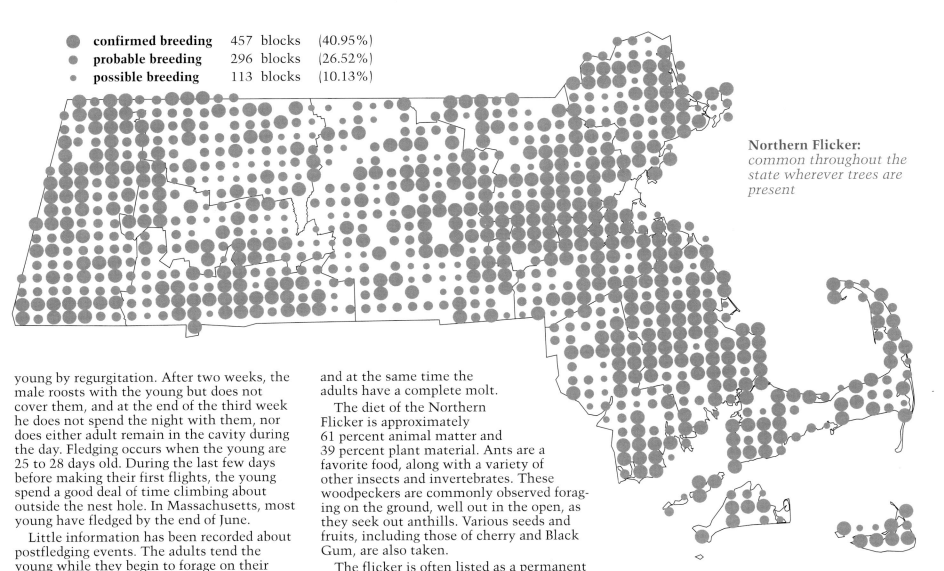

- confirmed breeding 457 blocks (40.95%)
- probable breeding 296 blocks (26.52%)
- possible breeding 113 blocks (10.13%)

Northern Flicker: *common throughout the state wherever trees are present*

young by regurgitation. After two weeks, the male roosts with the young but does not cover them, and at the end of the third week he does not spend the night with them, nor does either adult remain in the cavity during the day. Fledging occurs when the young are 25 to 28 days old. During the last few days before making their first flights, the young spend a good deal of time climbing about outside the nest hole. In Massachusetts, most young have fledged by the end of June.

Little information has been recorded about postfledging events. The adults tend the young while they begin to forage on their own. Old and young generally spend the night in cavities, but they may cling to tree trunks or the sides of buildings. When the young leave the nest, they are garbed in a juvenal plumage similar to that of the adults, except at this stage both sexes have dark mustachioed markings. They attain the adult plumage during a molt from June to October, and at the same time the adults have a complete molt.

The diet of the Northern Flicker is approximately 61 percent animal matter and 39 percent plant material. Ants are a favorite food, along with a variety of other insects and invertebrates. These woodpeckers are commonly observed foraging on the ground, well out in the open, as they seek out anthills. Various seeds and fruits, including those of cherry and Black Gum, are also taken.

The flicker is often listed as a permanent resident because some individuals are present at all times of the year. However, most birds move to the southern United States for the winter. Fall migration peaks from late September through October. During this time, the residents depart and large numbers of northern migrants pass through. Those that remain in Massachusetts are found most commonly in the southeastern sections, although there are regular records from interior regions as well. During severe weather, flickers seek shelter in thickets and stands of conifers.

W. Roger Meservey

Pileated Woodpecker
Dryocopus pileatus

Egg dates: mid-April to early June.

Number of broods: one; may re-lay if first attempt fails.

This spectacular, nearly crow-sized woodpecker has made a comeback since the early 1900s, when its numbers were reduced due to gunning and the clearing of much of the primeval forest for lumber and agriculture. It has been able to coexist with humans and adapt to the secondary forest conditions that have evolved with the abandonment of farmland. The Log-cock or Cock of the Woods is a denizen of extensive forests and is therefore most common in the western third of Massachusetts. Forbush recorded the species as breeding east to Middlesex County. During the Atlas period, confirmations were also made in Essex and Norfolk counties. The pileated is rare in any season in the southeastern portions of the state including Cape Cod and the Islands.

Pileated Woodpeckers mate for life and maintain close pair bonds throughout the year. Because of this, there often is not a great deal of displaying in the spring. However, some individuals perform antics on the ground, bowing, scraping, and sidestepping in the manner of flickers. At times, a pair will engage in bill pecking, hopping up and down the trunk of a tree and occasionally elevating their crests and partly spreading their wings.

These woodpeckers are as wary as grouse, keen of sight and hearing, and often difficult to see in spite of their relatively large size. The common call notes so closely resemble those of the flicker that they are often overlooked. The Pileated Woodpecker's louder and irregular *kik-kik-kikkik-kik-kik* can be compared to the flicker's *wick-wick-wick-wick. Cuk* is another frequent vocalization and appears to have several functions, such as registering excitement, locating the mate, or indicating roosting locations. Drumming occurs in every month of the year but is most frequent in the spring.

Large, rectangular, chiseled-out carvings in the trees indicate the birds' presence. They dig into the heart rot where carpenter ants have their galleries and extract the prey with their tongues. At other times, they will strip away bark and often descend to feed on fallen logs and the ground. It has been postulated that their skill of "sounding" tree trunks enables them to locate their insect prey. Their diet varies with the seasons. Ants (primarily large, black ants), a variety of beetles, and a few caterpillars make up the majority of the animal matter. The vegetable menu (17 percent to 27 percent) consists of wild berries, fruit, and mast.

A pair generally excavates a new nesting hole each spring. Some may begin working in February (Gagnon) and others during March and April, with as much as a month needed to complete the cavity. The birds return year after year to the same location and often to the same tree. Nest sites may be deep in the woods or on the edge, sometimes surprisingly close to areas of human activity. A variety of tree species, live or dead, may be utilized, but the dead portions of deciduous trees with a diameter of 15 to 20 inches are preferred.

Four Massachusetts nests were located as follows: 1 in a live Red Pine (Gagnon), 1 in a dead White Ash, and 2 in dead White Pines (McMenemy). The height of the entrance ranges from 12 to 85 feet, with an average of 45 feet above the ground. Heights of 3 Massachusetts nests ranged from 40 to 60 feet (McMenemy). The opening to the cavity ranges from 3.5 to 4.5 inches and the depth 10 to 30 inches. In this state as elsewhere, the birds may excavate in utility poles and cause damage.

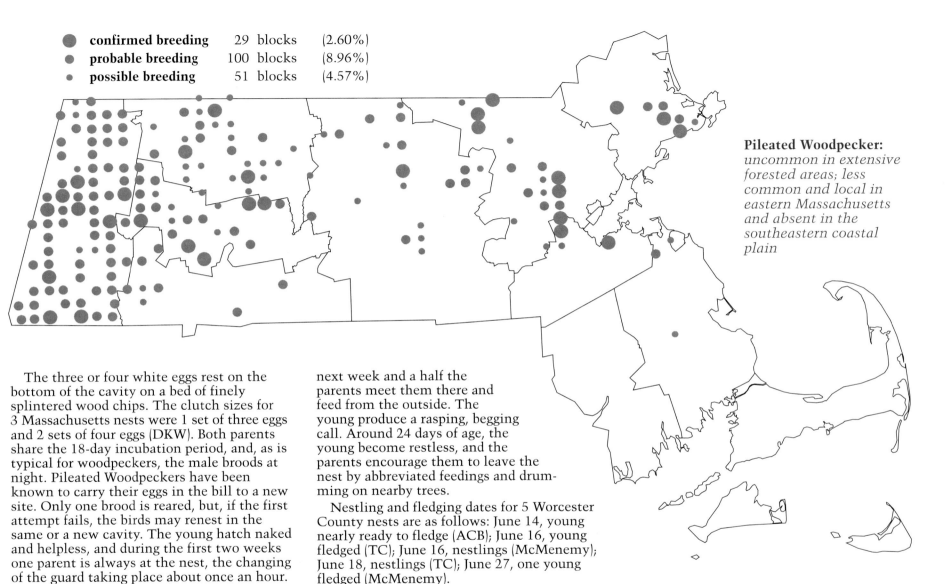

Pileated Woodpecker: *uncommon in extensive forested areas; less common and local in eastern Massachusetts and absent in the southeastern coastal plain*

- confirmed breeding — 29 blocks (2.60%)
- probable breeding — 100 blocks (8.96%)
- possible breeding — 51 blocks (4.57%)

The three or four white eggs rest on the bottom of the cavity on a bed of finely splintered wood chips. The clutch sizes for 3 Massachusetts nests were 1 set of three eggs and 2 sets of four eggs (DKW). Both parents share the 18-day incubation period, and, as is typical for woodpeckers, the male broods at night. Pileated Woodpeckers have been known to carry their eggs in the bill to a new site. Only one brood is reared, but, if the first attempt fails, the birds may renest in the same or a new cavity. The young hatch naked and helpless, and during the first two weeks one parent is always at the nest, the changing of the guard taking place about once an hour. The returning bird feeds the young by an interesting routine, pivoting down into the cavity with the tail braced against the hole. It then inserts its bill into the throat of a youngster and vigorously regurgitates the food. At about two weeks of age, the young can climb to the entrance hole, and for the next week and a half the parents meet them there and feed from the outside. The young produce a rasping, begging call. Around 24 days of age, the young become restless, and the parents encourage them to leave the nest by abbreviated feedings and drumming on nearby trees.

Nestling and fledging dates for 5 Worcester County nests are as follows: June 14, young nearly ready to fledge (ACB); June 16, young fledged (TC); June 16, nestlings (McMenemy); June 18, nestlings (TC); June 27, one young fledged (McMenemy).

The family group will stay together into the fall season, roosting at night in separate holes or together in one large cavity. A pair once established is not likely to leave its area during the changing seasons, but it requires a large territory of several square miles. The young disperse in the fall and winter, and some wander widely, appearing in areas far distant from the breeding grounds. The successful young learn to avoid large accipiters and raccoons, which appear to be their chief enemies.

Winthrop W. Harrington

Olive-sided Flycatcher
Contopus borealis

Egg dates: June 8 to July 10.

Number of broods: one; may re-lay if first attempt fails.

A denizen of northern boreal forests, the Olive-sided Flycatcher is a local summer resident confined almost entirely to the higher elevations of Berkshire County. At the turn of the century, it nested rather sparingly in the Berkshires and northern Worcester County and inexplicably also nested in the pine barrens of southeastern Massachusetts (Wareham, Carver, Plymouth) and Cape Cod (Bourne, Falmouth, Mashpee, Cotuit). Brewster noted that the Olive-sided Flycatcher bred in low numbers within 20 miles to the north and west of Boston but by 1900 had almost entirely deserted the region. The southeastern breeders disappeared as well. During the Atlas period, it was confirmed at only one location (Mount Watatic) east of the Connecticut River.

The Olive-sided Flycatcher, like most flycatchers, is a notably late spring migrant. Seldom is an individual noted before mid-May, and most migrants pass through in late May and early June. Despite its rarity, a nesting Olive-sided Flycatcher seldom goes undetected. The loud, haunting, whistled song consists of three notes, well described phonetically as *hip-three-cheers*, with the introductory note often inaudible from a distance. The song is usually given from an exposed snag of a dead tree with a commanding view of the surrounding woodland. From its prominent perch, the bird will sally forth in quest of insects and then return repeatedly to the same location. During courtship, the male actively pursues the female.

The preferred nesting site is a swamp or bog in spruce forests. At the highest elevations, such as Mount Greylock and Mount Watatic, it nests in spruces unassociated with water. Mated pairs begin nest building in June. The nest is situated in a conifer, usually a spruce, near the end of a horizontal branch and is often hidden in a cluster of needles. Heights range from 5 feet to more than 50 feet. The habits of the species in western Massachusetts are little known. However, information does exist for 15 nests from the former breeding grounds in eastern parts of the Commonwealth. Of these nests, 2 were in Red Cedar from 14 to 50 feet high, 1 in an apple (an unusual situation) at 12 feet, and 12 in Pitch Pines from 7 to 20 feet up and 3 to 5 feet out from the trunk, generally in a twig cluster. Most were in the vicinity of water (ACB).

The nest is constructed of twigs, mosses, and rootlets and is lined with finer materials. It is remarkably flat, thereby adding to the

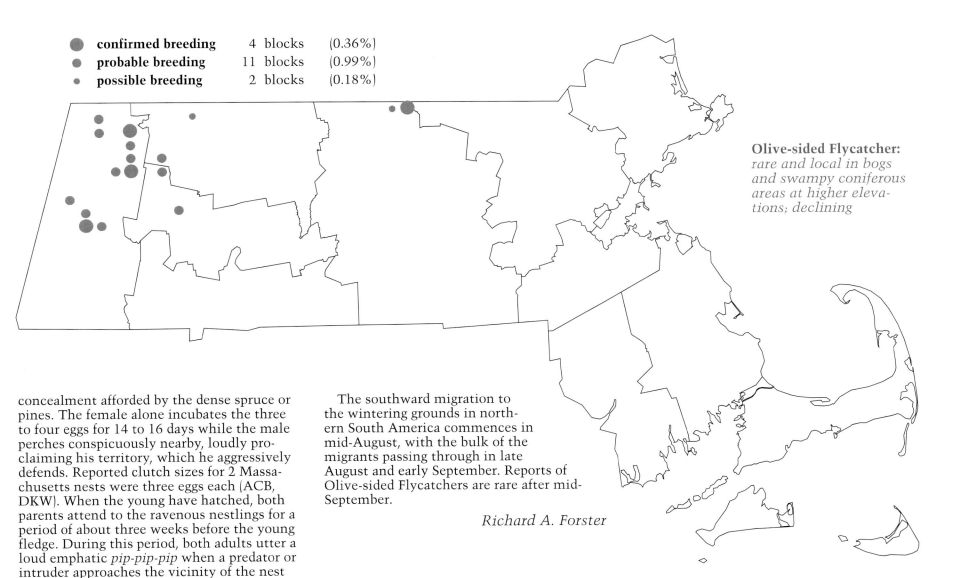

	confirmed breeding	4 blocks	(0.36%)
	probable breeding	11 blocks	(0.99%)
	possible breeding	2 blocks	(0.18%)

Olive-sided Flycatcher: *rare and local in bogs and swampy coniferous areas at higher elevations; declining*

concealment afforded by the dense spruce or pines. The female alone incubates the three to four eggs for 14 to 16 days while the male perches conspicuously nearby, loudly proclaiming his territory, which he aggressively defends. Reported clutch sizes for 2 Massachusetts nests were three eggs each (ACB, DKW). When the young have hatched, both parents attend to the ravenous nestlings for a period of about three weeks before the young fledge. During this period, both adults utter a loud emphatic *pip-pip-pip* when a predator or intruder approaches the vicinity of the nest or fledglings. After undergoing a postjuvenile molt, the young essentially resemble the adults, but the distinctive white tufts at the base of the tail are often less conspicuous. Nestling and fledgling data is lacking for Massachusetts. The few confirmations were mostly of adults carrying food.

The southward migration to the wintering grounds in northern South America commences in mid-August, with the bulk of the migrants passing through in late August and early September. Reports of Olive-sided Flycatchers are rare after mid-September.

Richard A. Forster

Eastern Wood-Pewee
Contopus virens

Egg dates: June 8 to July 25.
Number of broods: one.

The Eastern Wood-Pewee is the common forest flycatcher of Massachusetts. It breeds in deciduous and mixed woodland throughout the state and in the Pitch Pine barrens of the coastal plain. It also nests in orchards, parks, and other situations in which there is an abundance of mature trees. Its status has apparently remained constant throughout the historical period; though, of course, the total population has fluctuated in response to the clearing of land for agriculture through the midnineteenth century and the subsequent reforestation. Bent, writing in the early 1940s, stated that it had decreased markedly in eastern Massachusetts "during the past 20 years," but this observation is not substantiated elsewhere.

Pewees arrive on their breeding grounds with the first major songbird wave in mid-May, and migration continues throughout the month, typically peaking in the third week. Despite their general abundance as breeders, they do not occur in large numbers at coastal migrant traps; the maximum number reported together on spring migration is 10.

A whistled, ascending *pee-a-wee* uttered from amidst concealing foliage is a quintessential sound of the woodlands of eastern North America. In fact, in summer a pewee's presence is far more likely to be detected by its song than by glimpsing the drab little form of the bird itself. The quality has been described as both plaintive and cheerful, depending on the mood of the listener. An alternate song is a down-slurred *pee-ur*. Pewees also make chittering sounds near the nest. The standard call is a soft *chip* without any striking quality. Pewees have also been seen to engage in ecstasy or emotion-release songs, often accompanied by aerobatics. Begging young produce a squeaking call.

The pewee's courtship ritual is not very elaborate or distinctive. Audubon described a butterflylike nuptial flight, and other observers report increased agitation and aggression during the courtship period, with flight pursuit of females by males that are accompanied by chattering call notes.

This flycatcher typically nests at moderate height, 15 to 50 feet, in the shade of the forest canopy. The nest is a notably shallow, thick-walled cup, about 3 inches in diameter, placed on a horizontal limb. It is constructed of bark strips and other plant fibers and lined

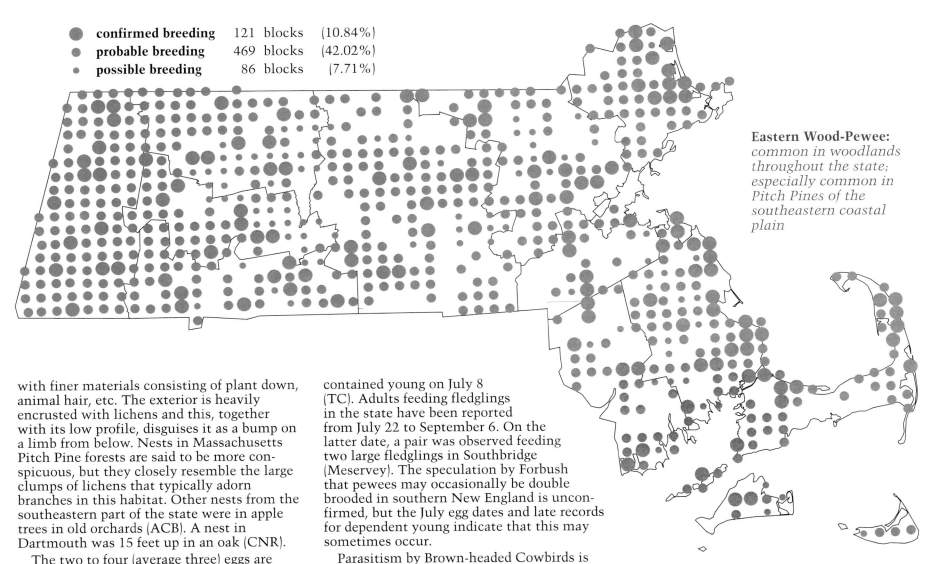

● confirmed breeding	121	blocks	(10.84%)
● probable breeding	469	blocks	(42.02%)
· possible breeding	86	blocks	(7.71%)

Eastern Wood-Pewee: *common in woodlands throughout the state; especially common in Pitch Pines of the southeastern coastal plain*

with finer materials consisting of plant down, animal hair, etc. The exterior is heavily encrusted with lichens and this, together with its low profile, disguises it as a bump on a limb from below. Nests in Massachusetts Pitch Pine forests are said to be more conspicuous, but they closely resemble the large clumps of lichens that typically adorn branches in this habitat. Other nests from the southeastern part of the state were in apple trees in old orchards (ACB). A nest in Dartmouth was 15 feet up in an oak (CNR).

The two to four (average three) eggs are cream to white, usually wreathed at the larger end with lavender to brown blotches and speckles. Clutch sizes for 3 Massachusetts nests were three eggs (2 nests) and four eggs (1 nest) (DKW, CNR). Incubation, exclusively by the female, takes 12 to 13 days. The young leave the nest 15 to 18 days after hatching. A nest in Gardner contained young on July 8 (TC). Adults feeding fledglings in the state have been reported from July 22 to September 6. On the latter date, a pair was observed feeding two large fledglings in Southbridge (Meservey). The speculation by Forbush that pewees may occasionally be double brooded in southern New England is unconfirmed, but the July egg dates and late records for dependent young indicate that this may sometimes occur.

Parasitism by Brown-headed Cowbirds is recorded but is apparently infrequent. The fall migration of the Eastern Wood-Pewee in Massachusetts occurs principally between late August and early October. As in the spring, they are thinly scattered, with numbers seen seldom exceeding 15.

The species winters from Nicaragua south through Colombia and Venezuela to Bolivia and western Brazil. In the tropics, it inhabits semiopen country, including cleared agricultural lands and shaded plantations, from the lowlands to about 5,000 feet.

Christopher W. Leahy

Acadian Flycatcher
Empidonax virescens

Egg dates: June 15 to July 8.

Number of broods: one; may re-lay if first attempt fails.

The Acadian Flycatcher is a southern *Empidonax*, whose range historically came no closer than Long Island, New York. Periodically, breeding birds have invaded southern Connecticut, tantalizing observers in Massachusetts. A nest, eggs, and one adult were collected in Hyde Park in June 1888, but it was another 80 years before even singing summer males were confirmed in the state. Each May in the early 1970s, Manomet Center for Conservation Sciences staff began to band several Acadian Flycatchers, and June singing birds were soon found south and southwest of Boston. Finally, in 1977, two pairs and 1 nest were found in Middleboro, and in 1980 another nesting pair was discovered in Granville, on the Connecticut border in the eastern Berkshires. Now up to 10 pairs breed in Granville, and two or three singing males have also been found for several summers in the hills west of Quabbin. In 1987, the first Worcester County breeding occurred in Royalston.

Spring arrivals can be found during the third week in May, both in southeastern and western Massachusetts. Migration continues into June, and many singing males seem to disappear after failing to attract mates. Actual breeding remains very localized in dense, well-watered woodlands, especially along small streams, and in the west is strongly associated with Eastern Hemlocks. The fact that there is much appropriate habitat in the western hills and that this area is close to the base population in Connecticut accounts for the breeding population that has become established there in the years following the Atlas period. The nesting birds in eastern Massachusetts have all been in deciduous woods, with Red Maple predominating; and no recurring pairs at any one site have been identified in that region.

The birds seem to prefer steep hillsides or ravines, but the nest itself is built near or over the slow-moving portions of the stream itself. There, in the shade of the lower branches of the larger trees, the bird goes about its fly-catching business, the male sneezing out his short emphatic song through most of the summer. The song is basically a two-syllable *peet-seet*, the first part rising and slightly drawn out and the second louder, short, and sharply falling. The only variety seems to be in the degree of harshness in the notes, with the first syllable tending generally to be somewhat less harsh. It resembles the single, almost burry-sounding phrases in one of the Red-eyed Vireo's songs. The call note is a nondescript *peep*. On the breeding grounds, the bird also utters a fast series of soft melodious notes, almost a trill, fluttering its wings as it does so, much like a young bird begging for food. This sound is also occasionally heard on migration.

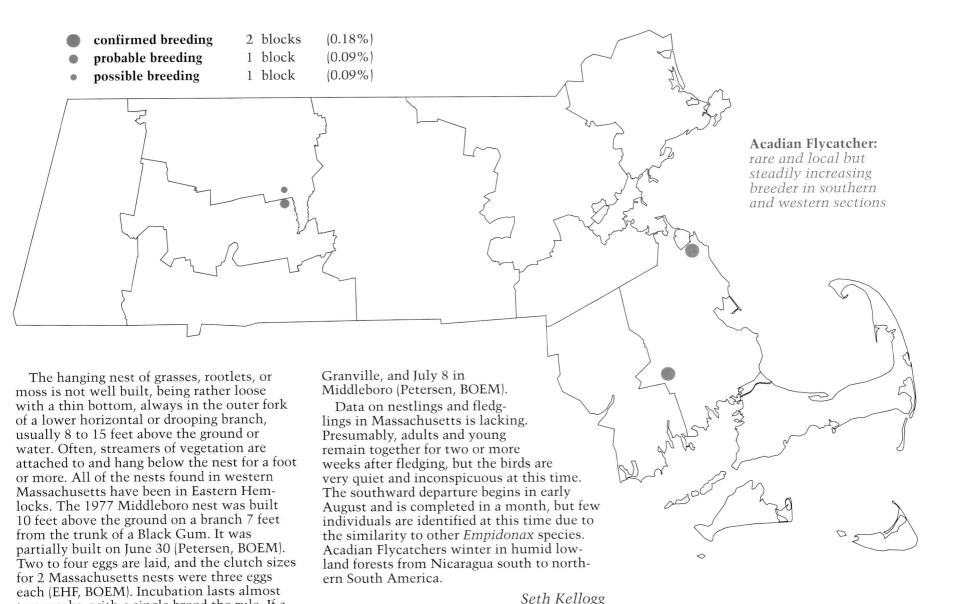

- ● **confirmed breeding** 2 blocks (0.18%)
- ● **probable breeding** 1 block (0.09%)
- · **possible breeding** 1 block (0.09%)

Acadian Flycatcher: *rare and local but steadily increasing breeder in southern and western sections*

The hanging nest of grasses, rootlets, or moss is not well built, being rather loose with a thin bottom, always in the outer fork of a lower horizontal or drooping branch, usually 8 to 15 feet above the ground or water. Often, streamers of vegetation are attached to and hang below the nest for a foot or more. All of the nests found in western Massachusetts have been in Eastern Hemlocks. The 1977 Middleboro nest was built 10 feet above the ground on a branch 7 feet from the trunk of a Black Gum. It was partially built on June 30 (Petersen, BOEM). Two to four eggs are laid, and the clutch sizes for 2 Massachusetts nests were three eggs each (EHF, BOEM). Incubation lasts almost two weeks, with a single brood the rule. If a cowbird disturbs the nest before the eggs are laid, the birds may build again. Acadian Flycatchers have been observed incubating on June 20 in Royalston (TC), June 28 in Granville, and July 8 in Middleboro (Petersen, BOEM).

Data on nestlings and fledglings in Massachusetts is lacking. Presumably, adults and young remain together for two or more weeks after fledging, but the birds are very quiet and inconspicuous at this time. The southward departure begins in early August and is completed in a month, but few individuals are identified at this time due to the similarity to other *Empidonax* species. Acadian Flycatchers winter in humid lowland forests from Nicaragua south to northern South America.

Seth Kellogg

Alder Flycatcher
Empidonax alnorum

Egg dates: June 19 to July 22.

Number of broods: one; may re-lay if first attempt fails.

The Alder Flycatcher is a fairly common and widespread summer resident in Berkshire County. It nests eastward into the northern and central Massachusetts uplands, where it may be locally common. It breeds much less frequently in the Connecticut River valley lowlands and probably very rarely east of Worcester County, where it appears to be largely replaced by the Willow Flycatcher. Historically, this species nested widely over eastern Massachusetts, but, faced with increasing competition from the more aggressive Willow Flycatcher, it has retreated into the more upland and western reaches of the state.

The historic trends of Alder and Willow flycatchers in Massachusetts are somewhat obscure, owing to the fact that until 1973 the two were treated as song-types of the Traill's Flycatcher. The Alder (*fee-bee-o* or *burr-ree-ah* song type) was historically the principal form in Massachusetts, with the Willow (*fitz-bew* or *witch-brew* song type) generally more southern and western in its distribution.

However, Stein (1958), in a classic study, brought some order to a chaotic situation, demonstrating that the Willow Flycatcher appeared to be invading traditional Alder Flycatcher range and ultimately predominating by active displacement in many areas in central New York State, especially at lower elevations. By the early 1960s, Willow Flycatchers were being widely recognized in Massachusetts, and they rapidly expanded their range eastward to the coast, displacing the Alder Flycatcher at many sites along the way and occupying new areas as well.

Although both species have similar advertising songs, a trained ear readily differentiates the songs. There is no overlap of pattern, and neither species will respond to song tapes of the other. Alder and Willow flycatchers typify congeners that are similarly colored and patterned and that rely upon sound instead of visual stimuli for species differentiation. The Alder Flycatcher's buzzy song, *fee-bee-o*, with the second syllable highest and accented, is the only way this species can be positively identified in the field. It has several other vocalizations, including a rasping *queeoo*, delivered in the same manner as the song, and a *pip* call note.

Alder Flycatchers arrive in the third week of May. In addition to its distinctive song, its habitat preference and nest site and structure have been shown to be specifically distinct from that of the Willow Flycatcher. Pure stands of alder are sometimes occupied, as are brushy areas of alder intermixed variably with dogwood and Arrowwood. The thickets used are almost always near water—along the edges of streams, bogs, swamps, and Beaver flowages—and are sometimes at the edges of mature forest.

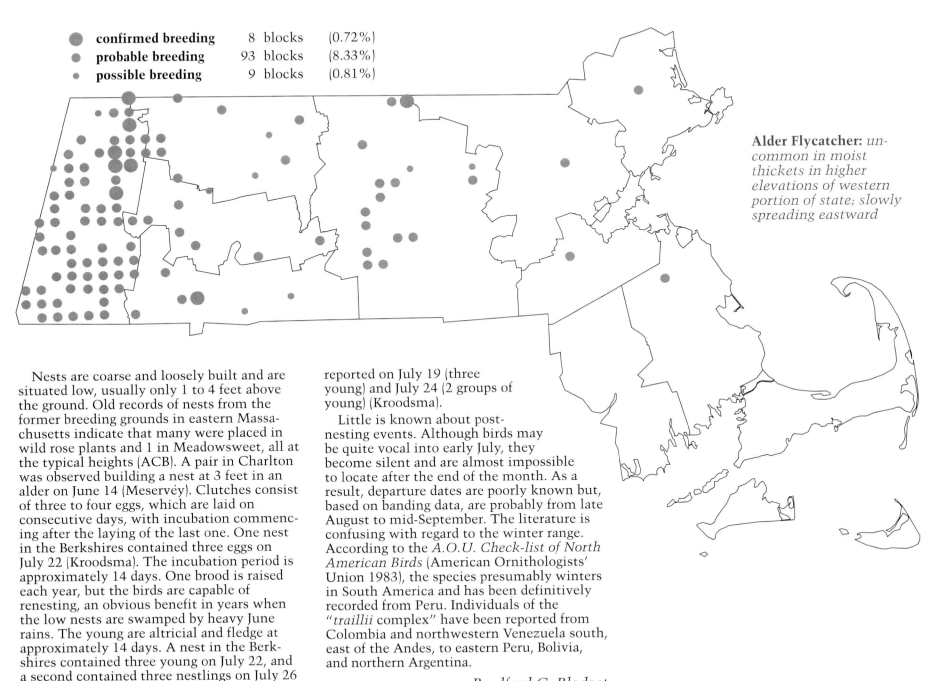

Alder Flycatcher: *uncommon in moist thickets in higher elevations of western portion of state; slowly spreading eastward*

● confirmed breeding	8 blocks	(0.72%)
● probable breeding	93 blocks	(8.33%)
• possible breeding	9 blocks	(0.81%)

Nests are coarse and loosely built and are situated low, usually only 1 to 4 feet above the ground. Old records of nests from the former breeding grounds in eastern Massachusetts indicate that many were placed in wild rose plants and 1 in Meadowsweet, all at the typical heights (ACB). A pair in Charlton was observed building a nest at 3 feet in an alder on June 14 (Meservey). Clutches consist of three to four eggs, which are laid on consecutive days, with incubation commencing after the laying of the last one. One nest in the Berkshires contained three eggs on July 22 (Kroodsma). The incubation period is approximately 14 days. One brood is raised each year, but the birds are capable of renesting, an obvious benefit in years when the low nests are swamped by heavy June rains. The young are altricial and fledge at approximately 14 days. A nest in the Berkshires contained three young on July 22, and a second contained three nestlings on July 26 (Kroodsma). Fledglings in the Berkshires were reported on July 19 (three young) and July 24 (2 groups of young) (Kroodsma).

Little is known about post-nesting events. Although birds may be quite vocal into early July, they become silent and are almost impossible to locate after the end of the month. As a result, departure dates are poorly known but, based on banding data, are probably from late August to mid-September. The literature is confusing with regard to the winter range. According to the *A.O.U. Check-list of North American Birds* (American Ornithologists' Union 1983), the species presumably winters in South America and has been definitively recorded from Peru. Individuals of the "*traillii* complex" have been reported from Colombia and northwestern Venezuela south, east of the Andes, to eastern Peru, Bolivia, and northern Argentina.

Bradford G. Blodget

Willow Flycatcher
Empidonax traillii

Egg dates: June 27 to July 22.

Number of broods: one; may re-lay if first attempt fails.

The Willow Flycatcher was historically more western and southern in distribution than the very closely related Alder Flycatcher (see discussion under that species). Stein (1958) demonstrated that a substantial area of sympatry existed and that the Willow Flycatcher appeared to be aggressively displacing the Alder Flycatcher in many areas, particularly in lowland country in central New York State. This area of sympatry expanded markedly into the Northeast in the 1960s. Willow Flycatchers, once treated as a western song type of the Traill's Flycatcher, were apparently recognized in Massachusetts at least as far back as the early 1930s, but they were not commonly encountered.

In the 1960s, Willow Flycatchers began to turn up more frequently, and range expansion occurred rapidly eastward across the state. Atlas data reveal that the Willow Flycatcher has virtually replaced the Alder Flycatcher in many of its traditional haunts, particularly throughout eastern Massachusetts and in the Connecticut River valley. The former is now a widespread and locally common nester on most of the mainland, probably even on lower Cape Cod. It is particularly well established in Essex and Middlesex counties and in the lower Connecticut River valley. It nests somewhat more sparingly over the central and western uplands, with noticeable concentrations in broad river valleys of the interior. It is very thinly distributed in the southeastern part of the state and absent from the Islands as a confirmed nester.

The sneezing *fitz-bew* advertising song is the surest indication of the Willow Flycatcher's presence. The birds also give a single *creet* and a *whit* call note. Arrival in Massachusetts occurs very close to that of the Alder Flycatcher, about the third week of May.

There is clearly some overlap with the Alder Flycatcher in habitat preference, with the Willow Flycatcher favoring the same alder-dogwood-viburnum thicket associations, but usually in more low open country

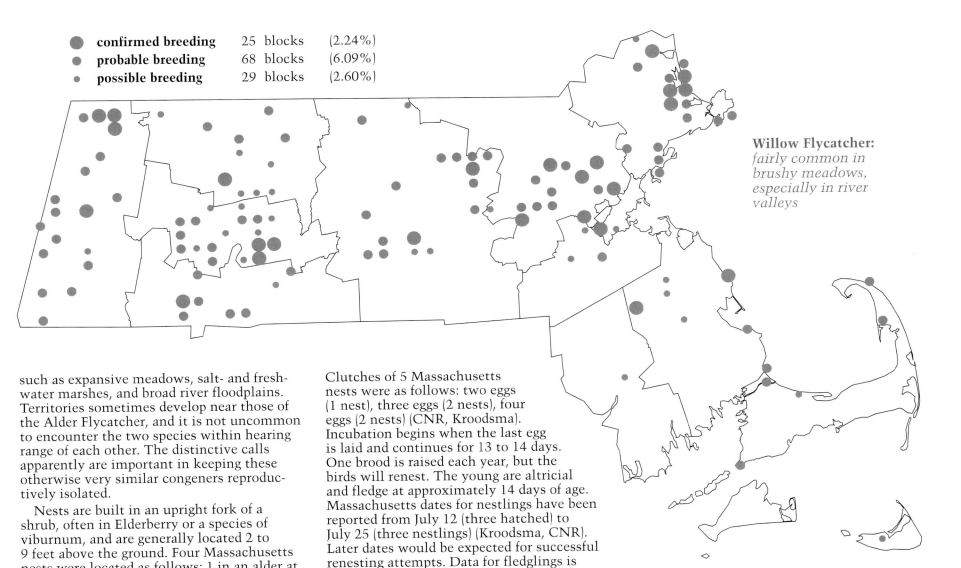

Willow Flycatcher: *fairly common in brushy meadows, especially in river valleys*

● confirmed breeding	25 blocks	(2.24%)
● probable breeding	68 blocks	(6.09%)
• possible breeding	29 blocks	(2.60%)

such as expansive meadows, salt- and freshwater marshes, and broad river floodplains. Territories sometimes develop near those of the Alder Flycatcher, and it is not uncommon to encounter the two species within hearing range of each other. The distinctive calls apparently are important in keeping these otherwise very similar congeners reproductively isolated.

Nests are built in an upright fork of a shrub, often in Elderberry or a species of viburnum, and are generally located 2 to 9 feet above the ground. Four Massachusetts nests were located as follows: 1 in an alder at 6 feet, 1 in a Red Maple at 7 feet, 2 in a Box Elder at 4 and 5 feet, respectively (CNR, Meservey). Nests are neatly and tightly constructed, frequently containing cottony tufts of cattail, fine grasses, and the silky material of aspens and willows.

Clutches consist of three to four eggs, laid at 24-hour intervals until the set is complete. Clutches of 5 Massachusetts nests were as follows: two eggs (1 nest), three eggs (2 nests), four eggs (2 nests) (CNR, Kroodsma). Incubation begins when the last egg is laid and continues for 13 to 14 days. One brood is raised each year, but the birds will renest. The young are altricial and fledge at approximately 14 days of age. Massachusetts dates for nestlings have been reported from July 12 (three hatched) to July 25 (three nestlings) (Kroodsma, CNR). Later dates would be expected for successful renesting attempts. Data for fledglings is lacking. Virtually nothing is known of post-fledging events. After early July, singing abruptly ceases, and the birds once more join the ranks of anonymous *Empidonax*. Departure is presumably in early to mid-September. The wintering grounds are in Middle America from Vera Cruz and Oaxaca south to Panama and possibly farther south (see Alder Flycatcher account).

Bradford G. Blodget

Least Flycatcher
Empidonax minimus

Egg dates: May 20 to July 1.

Number of broods: one or two; may re-lay if first attempt fails.

The Least Flycatcher, or Chebec, is the most widely distributed and familiar *Empidonax* flycatcher in Massachusetts. During this century, it has declined slowly with the reversion of agricultural lands to deep forest, but it is still a widespread and common nesting species across the state east to and including most of Worcester County. However, it has disappeared from much of its former range in eastern Massachusetts, where it was still common into the 1950s. This change has been variously attributed to the decline and disappearance of orchards, agricultural lands, and other favored habitats; increased habitat fragmentation; and intensive insecticide spraying programs that have been widely carried on in rapidly suburbanizing eastern Massachusetts. A handful may rarely nest on Cape Cod west of Barnstable, but otherwise it is absent from Cape Cod and the Islands.

General arrival from the south occurs in the first week of May. Males at this time are notably active and pugnacious, earning the early appellation, Little Feathered Warrior. Rival males and other species are driven from territories, and females are chased about in noisy pursuits. Territories are well established by the third week of May. The advertising call in this species may serve as a form of territorial display. This call, a vigorous *che-bec*, accompanied by a jerk of the head and tail, is often uttered incessantly through the hottest days of summer. Males often call, as in other tyrannid flycatchers, at a noticeably quickened tempo, sometimes up to 60 times per minute, as day is breaking. At times, other notes are interspersed among the *che-bec*s. A guttural *wheu-wheu-wheu* series may be given near the nest. The common call note of both sexes is a simple *whit*.

Preferred habitat includes relatively open deciduous woodlands and wooded swamps where trees are separated by slight spaces. Where trees are not separated, birds will accept scattered forest openings but will not tolerate too many limbs beneath the canopy, which should be at least 30 percent open and contain scattered dead branches and twigs for perches. The open undercanopy provides air space for foraging birds engaged in hunting insects on the wing. In other areas, birds seem to accept old neglected orchards, large shade trees about farms and gardens, advanced second growth, the edges of forests, and not-too-thickly wooded swamps.

Nests are placed 8 to 40 (usually less than 25) feet above the ground in small hardwoods in the crotch of a horizontal forking branch. Bent reported that two-thirds of the nests he discovered in eastern Massachusetts were in

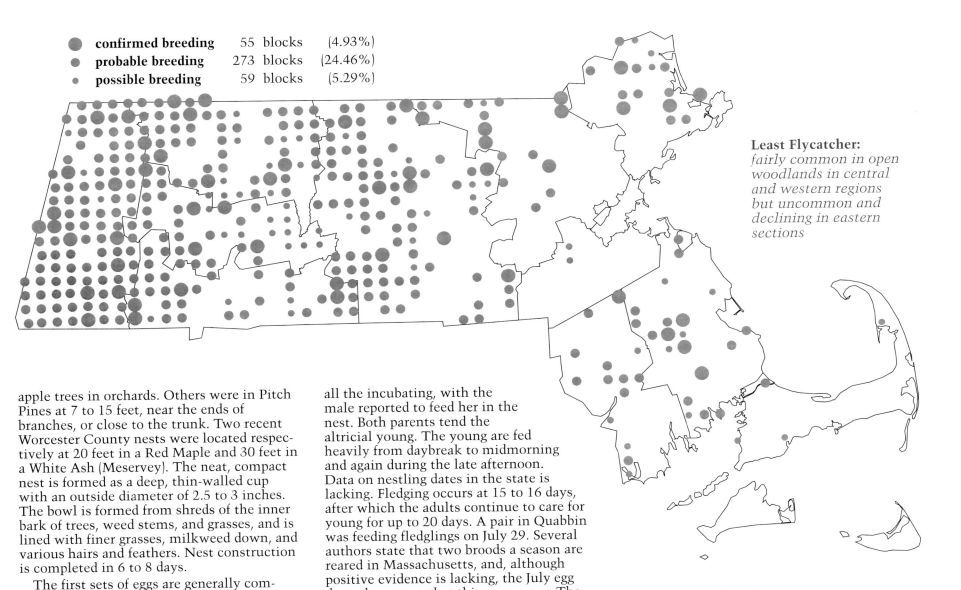

Least Flycatcher: *fairly common in open woodlands in central and western regions but uncommon and declining in eastern sections*

- confirmed breeding — 55 blocks (4.93%)
- probable breeding — 273 blocks (24.46%)
- possible breeding — 59 blocks (5.29%)

apple trees in orchards. Others were in Pitch Pines at 7 to 15 feet, near the ends of branches, or close to the trunk. Two recent Worcester County nests were located respectively at 20 feet in a Red Maple and 30 feet in a White Ash (Meservey). The neat, compact nest is formed as a deep, thin-walled cup with an outside diameter of 2.5 to 3 inches. The bowl is formed from shreds of the inner bark of trees, weed stems, and grasses, and is lined with finer grasses, milkweed down, and various hairs and feathers. Nest construction is completed in 6 to 8 days.

The first sets of eggs are generally completed in the first week of June. Clutches of three to four (rarely five to six) creamy white, unmarked eggs are laid, with eggs deposited in the nest on successive days. Clutch sizes for 4 Massachusetts nests were three eggs (1 nest), four eggs (3 nests) (DKW). The incubation period is 14 days. The female does all the incubating, with the male reported to feed her in the nest. Both parents tend the altricial young. The young are fed heavily from daybreak to midmorning and again during the late afternoon. Data on nestling dates in the state is lacking. Fledging occurs at 15 to 16 days, after which the adults continue to care for young for up to 20 days. A pair in Quabbin was feeding fledglings on July 29. Several authors state that two broods a season are reared in Massachusetts, and, although positive evidence is lacking, the July egg dates do suggest that this may occur. The birds will renest if the first attempt fails.

Least Flycatchers scatter and roam the woodlands in multispecies assemblages in August. Local birds begin to depart in the last week of August, and most are gone by mid-September. Migration is southwestward to Texas and south to the principal wintering areas in southern Sonora and Tamaulipas south along both slopes of Middle America to Honduras and northern Nicaragua.

Bradford G. Blodget

Eastern Phoebe
Sayornis phoebe

Egg dates: April 24 to August 15.
Number of broods: two.

The Eastern Phoebe ranges widely throughout the eastern United States and southern Canada and during the breeding season is widespread throughout Massachusetts, although it is less common on Cape Cod and the Islands. The bird is most often seen perched on small branches from which it makes quick flights to pick off flying insects, followed by a return to the same spot. When perched, phoebes are characterized by the constant downward flipping of their tail.

Phoebes appear in a wide variety of habitats from open country, woods, and farmland to suburbia. Proximity to running water is preferred but not essential. Phoebes are among the earliest passerine migrants, appearing in their breeding areas in late March and early April. From the time the male arrives until the female returns about two weeks later, he sings constantly, both to attract a mate and to establish a nesting territory of several acres in extent. This area is large enough to preclude much conflict with neighbors. If necessary, both the male and female will defend by song and chase, but seldom is there physical contact.

During courtship, the female phoebe frequently appears quite aggressive to the male, keeping him at a distance. The song is a rasping vocalization of the *fee-bee* name, the second syllable alternately rising and falling. After pairing and the initiation of nesting, the male's singing gradually diminishes. Preparation for a second nesting includes renewed singing. The female can sing but seldom does. Both sexes use sharp *chip* notes to keep in contact.

Following courtship, the female builds a cup-shaped nest of mud and moss and lines it with grass, hair, and feathers. She builds on a flat surface that ranges from a rock shelf under a cliff overhang to beams inside buildings or eaves, above lights or shutters, under bridges, or in culverts. The birds frequently nest very close to human activity and seem not to be particularly disturbed by it. In Massachusetts, nests may be completed as early as April 2 (BOEM). The locations of 45 state nests were as follows: bridge (22 nests), house (7 nests), barn (4 nests), shed (3 nests), garage (2 nests), unspecified building (4 nests), rock ledge (3 nests) (CNR, Meservey). Heights of 20 of these nests ranged from 4 to 10 feet, with an average of nearly 7 feet (CNR).

There are usually five eggs (range three to eight) in a clutch. Clutch sizes for 26 Massachusetts nests were as follows: two eggs, possibly an incomplete clutch (1 nest); three eggs (2 nests); four eggs (5 nests); five eggs (17 nests); six eggs (1 nest) (CNR, DKW). The eggs are either completely white or occasionally have a few light brown spots at the large end. Incubation is solely by the female, and she leaves the nest only for short periods. During such an absence, the nest may be parasitized by a cowbird. After an average incubation of 16 days, the nestlings remain in the nest for 15 to 18 days. Nestlings have

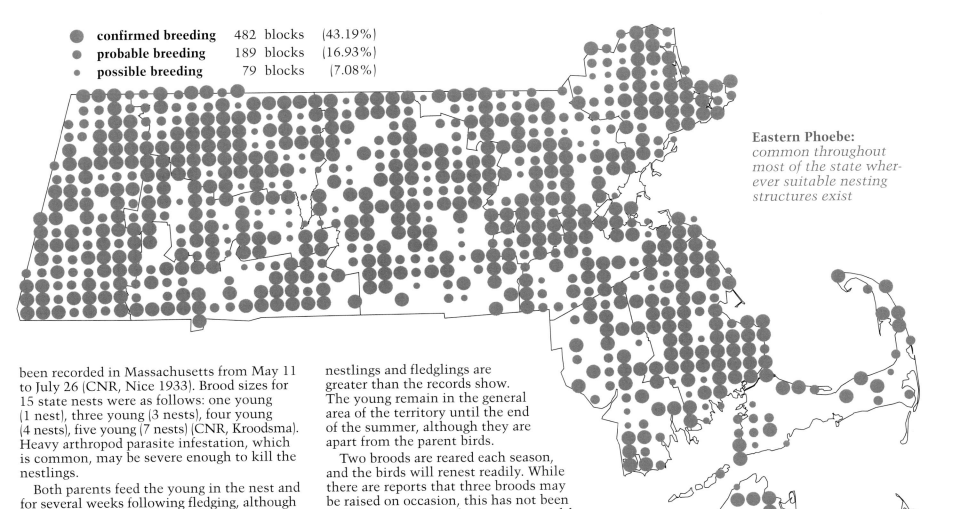

Eastern Phoebe: *common throughout most of the state wherever suitable nesting structures exist*

been recorded in Massachusetts from May 11 to July 26 (CNR, Nice 1933). Brood sizes for 15 state nests were as follows: one young (1 nest), three young (3 nests), four young (4 nests), five young (7 nests) (CNR, Kroodsma). Heavy arthropod parasite infestation, which is common, may be severe enough to kill the nestlings.

Both parents feed the young in the nest and for several weeks following fledging, although the young start some independent feeding soon after leaving the nest. If the female starts to lay a second clutch of eggs within a few days after the fledging of the first brood, the male alone feeds the immatures until they are independent. The diet is primarily a variety of insects captured on the wing or plucked from vegetation or the water's surface while the bird hovers. Fledged young have been observed in the state from late May until early August (CNR, Meservey), but the egg dates indicate that the ranges for both nestlings and fledglings are greater than the records show. The young remain in the general area of the territory until the end of the summer, although they are apart from the parent birds.

Two broods are reared each season, and the birds will renest readily. While there are reports that three broods may be raised on occasion, this has not been well documented in the Commonwealth. Nest failures due to weather, predation, insect parasites, and human disturbance are common. The late egg dates probably represent renesting attempts. The outcomes of 20 state nesting attempts were as follows: young fledged (6 nests), failed (7 nests), unknown outcome (7 nests) (CNR).

Phoebes linger in the fall, departing during late September to mid-October in their fall migration to the southeastern United States and Mexico. After this, only stragglers are reported. During their winter stay in the south, the birds remain solitary and mostly silent.

Mary Baird

Great Crested Flycatcher
Myiarchus crinitus

Egg dates: May 28 to June 26.

Number of broods: one.

At the turn of the century, the Great Crested Flycatcher was an uncommon to rare local summer resident. This is not surprising because this flycatcher is a bird of the forest, and much of Massachusetts was open farmland in the nineteenth century. Agriculture went into a serious decline in the mid-1800s, at which time forest cover once again became a dominant feature of the landscape. As a result, the Great Crested Flycatcher is now a fairly common breeding bird throughout the state wherever deciduous woodlands occur. It is less common on Cape Cod and the Islands and in the higher elevations of the state where coniferous woodlands predominate.

Early individuals are usually seen in late April, but it is not until early May that a widespread arrival is noticed. Migrants pass through until late May, by which time residents have established their territories. The Great Crested Flycatcher is very vocal, emitting a loud, drawn-out *wheep* from an exposed perch in the woods. Often this call is abbreviated and uttered in rapid succession, possibly a sign of agitation. During courtship, the male actively pursues the female.

The Great Crested Flycatcher is unique among Massachusetts flycatchers in that it builds its nest in a hole, either a natural cavity, an old woodpecker hole, or even a nest box. Old records of nests from southeastern Massachusetts indicate that nearly all were found in orchards in cavities in the trunk or main branches of apple trees from 6 to 11 feet in height. One, located in a dead stub in swampy woods, was open at the top and 7 feet aboveground (ACB). More recent records include 6 in nest boxes in mixed forest from 6 to 14 feet (CNR); 1 in an open-ended, hollow, horizontal branch of a Black Cherry at 10 feet (Meservey); 1 at 25 feet in an old woodpecker hole in a dead oak (Meservey). The cavity, which may be 2 to 3 feet deep, is filled with old leaves, bark, and grasses, with a cuplike hollow fashioned near the top. A curious aspect of this structure, and one that is surrounded by much folklore, is that it frequently contains a shed snake-

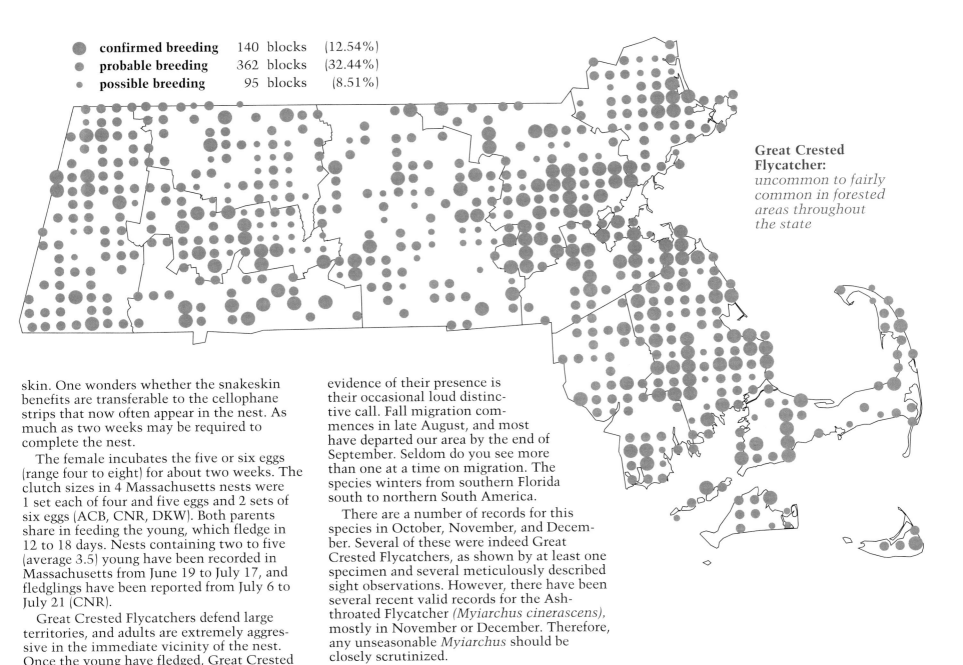

- **confirmed breeding** 140 blocks (12.54%)
- **probable breeding** 362 blocks (32.44%)
- **possible breeding** 95 blocks (8.51%)

Great Crested Flycatcher: *uncommon to fairly common in forested areas throughout the state*

skin. One wonders whether the snakeskin benefits are transferable to the cellophane strips that now often appear in the nest. As much as two weeks may be required to complete the nest.

The female incubates the five or six eggs (range four to eight) for about two weeks. The clutch sizes in 4 Massachusetts nests were 1 set each of four and five eggs and 2 sets of six eggs (ACB, CNR, DKW). Both parents share in feeding the young, which fledge in 12 to 18 days. Nests containing two to five (average 3.5) young have been recorded in Massachusetts from June 19 to July 17, and fledglings have been reported from July 6 to July 21 (CNR).

Great Crested Flycatchers defend large territories, and adults are extremely aggressive in the immediate vicinity of the nest. Once the young have fledged, Great Crested Flycatchers become inconspicuous as they feed primarily in the canopy. The only evidence of their presence is their occasional loud distinctive call. Fall migration commences in late August, and most have departed our area by the end of September. Seldom do you see more than one at a time on migration. The species winters from southern Florida south to northern South America.

There are a number of records for this species in October, November, and December. Several of these were indeed Great Crested Flycatchers, as shown by at least one specimen and several meticulously described sight observations. However, there have been several recent valid records for the Ash-throated Flycatcher *(Myiarchus cinerascens)*, mostly in November or December. Therefore, any unseasonable *Myiarchus* should be closely scrutinized.

Richard A. Forster

Eastern Kingbird
Tyrannus tyrannus

Egg dates: May 30 to July 17.

Number of broods: one; may re-lay if first attempt fails.

The Eastern Kingbird is a common breeding bird throughout the state, and during the Atlas period it was confirmed in all regions, including Cape Cod and the Islands. Perched conspicuously in the open on an exposed branch, fence, or telephone wire or darting out in flight with rapid, fluttering wing beats, the Eastern Kingbird is a familiar sight in all types of open country. Its intolerance of large birds, and sometimes of smaller species as well, along with its aerial attacks on crows, hawks, and vultures, has rightfully earned it its present common name, as well as its former appellation—Tyrant Flycatcher.

Kingbirds may begin to arrive in Massachusetts in late April, but there are no appreciable numbers until early May, with migration continuing through the middle of the month. Fairly large nesting territories are established in orchards, farmlands, old overgrown pastures, wood borders, and suburbs, and any such area with a stream or pond included is especially favored. While the birds may occasionally be found in open woods, especially those that have been partially logged, they do not occur in dense forest.

Courting males perform twisting, tumbling flights with their characteristic short, quick wing beats and vocalize steadily. At times, particularly dawn and dusk, the birds may ascend to considerable heights. The common calls are various shrill *tzi* and *tzee* notes, which may be given singly or in a series. The male's song is actually a series of such calls uttered in an alternating pattern: *tzi-tzee-tzi-tzee-tzi-tzee*. In addition, there are also several *kip* and *kipper* notes, again given in the same varying sequences.

Nests are generally placed from 2 to 60 feet high in a shrub or tree, often near or over water. Massachusetts nests have been recorded in elm, Norway Spruce, honeysuckle, Beach Plum, crabapple, apple, pear, alder, Buttonbush, Red Maple, and White Ash (CNR, Petersen). Heights ranged from 4 to 40 feet. Both sexes gather straw, twigs, grass, and even string and strips of cloth for the construction of a bulky cup nest, which appears somewhat rough and ragged on the outside but has a neat interior lined with fine grasses, rootlets, hair, and plant down. The earliest date for nest building in the state was May 10 (CNR).

Typical clutches consist of three or four (rarely five) eggs, which are white with irregular brownish blotches. The clutch sizes for 6 Massachusetts nests were three eggs (3 nests) and four eggs (3 nests) (DKW, CNR). The female does most of the incubating for 12 to 14 days, while the ever-watchful male remains perched nearby. Young remain in the nest for about two weeks and have been reported in Massachusetts from June 24 to

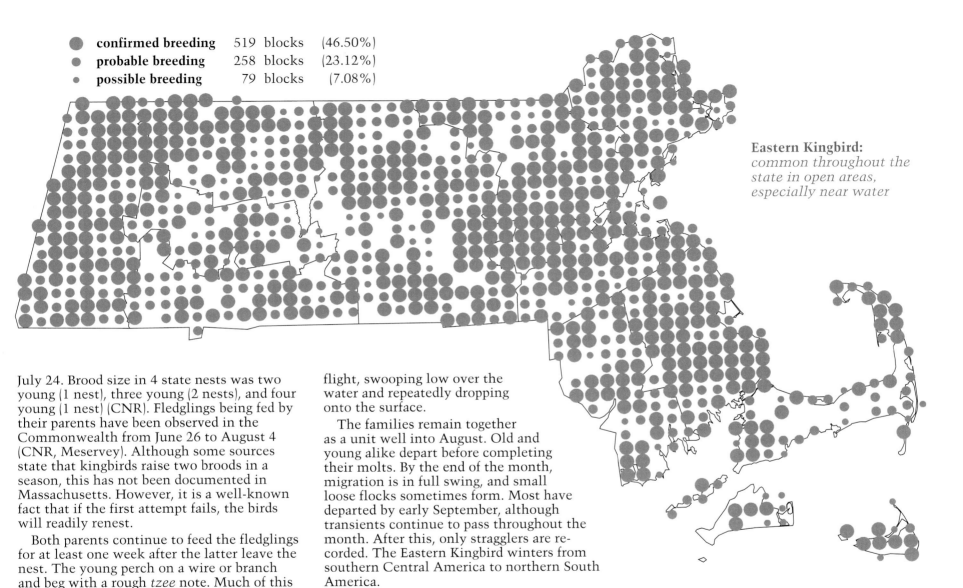

- confirmed breeding 519 blocks (46.50%)
- probable breeding 258 blocks (23.12%)
- possible breeding 79 blocks (7.08%)

Eastern Kingbird:
common throughout the state in open areas, especially near water

July 24. Brood size in 4 state nests was two young (1 nest), three young (2 nests), and four young (1 nest) (CNR). Fledglings being fed by their parents have been observed in the Commonwealth from June 26 to August 4 (CNR, Meservey). Although some sources state that kingbirds raise two broods in a season, this has not been documented in Massachusetts. However, it is a well-known fact that if the first attempt fails, the birds will readily renest.

Both parents continue to feed the fledglings for at least one week after the latter leave the nest. The young perch on a wire or branch and beg with a rough *tzee* note. Much of this kingbird's diet consists of insects caught on the wing or picked from the ground or water's edge in hovering flight. Formerly labeled the Bee Martin and persecuted by humans, the Eastern Kingbird in fact does not eat a significant number of honeybees. Small wild fruits are also taken. At times, the birds bathe in flight, swooping low over the water and repeatedly dropping onto the surface.

The families remain together as a unit well into August. Old and young alike depart before completing their molts. By the end of the month, migration is in full swing, and small loose flocks sometimes form. Most have departed by early September, although transients continue to pass throughout the month. After this, only stragglers are recorded. The Eastern Kingbird winters from southern Central America to northern South America.

W. Roger Meservey

White-eyed Vireo
Vireo griseus

Egg dates: May 22 to June 18.

Number of broods: one.

Massachusetts is at the extreme northeastern limit of the White-eyed Vireo's breeding range, with the consequence that its population levels have fluctuated markedly in the last hundred years. Forbush plotted summer records in Massachusetts with the clear implication that until around 1880, White-eyed Vireos probably bred in all counties of the state except Dukes County. A rapid decline followed, and by 1955 the species was recorded only from Westport, where it was a rare straggler. The Atlas map still shows no breeding records from Berkshire County east to Middlesex and Norfolk counties. Compared with those of the late 1800s, populations of White-eyed Vireos are much reduced in Essex and Plymouth counties and remain as scarce as ever on Cape Cod and the Islands. The majority of confirmed breeding records are confined to Bristol County and the Elizabeth Islands. Thus, in the mid-1900s there was only a partial recovery of the 1800s range. Curiously, Manomet Center for Conservation Sciences' spring and fall migration records showed no significant increases from 1971 to 1984, a period when other "southern" birds such as the Tufted Titmouse and Northern Cardinal were clearly increasing.

First spring migrants are seen in late April; movement peaks in mid-May, and birds are in breeding habitat before June. Migrants may be found in any fairly dense cover, usually close to the ground. White-eyed Vireos are persistent singers; indeed, the many "probable" records on the map emphasize that this species is more often heard than seen, even on nesting territory. Songs consist of five to seven loud notes, the first and last often a clear, emphatic *chick*, the middle notes slurred together and delivered rapidly. Individuals may have several songs or seasonal variations plus a variety of calls described as short ticks, whistles, and harsh mewing notes.

Nest sites are generally in moist areas or close to water, in thickets, shrubs, hedgerows, tangles of vines, or briers. For an apt description, one can hardly improve on Forbush's, "among the umbrageous foliage of bosky thickets." The rather bulky nests are suspended in typical vireo fashion from a forked branch 3 to 7 feet above the ground. An outer construction of leaves, bark, and coarse material is lined with fine fibers, grass, or hair. In eastern Massachusetts, nests have been found in brier thickets in swampy areas, but also occasionally in upland pastures in barberry and other bushes. Other state nests include one found in Rehoboth, also on dry ground, at 20 inches in a cherry sapling growing in a thicket among Arbor Vitae and one at 3 feet in a clump of Arrowwood (ACB).

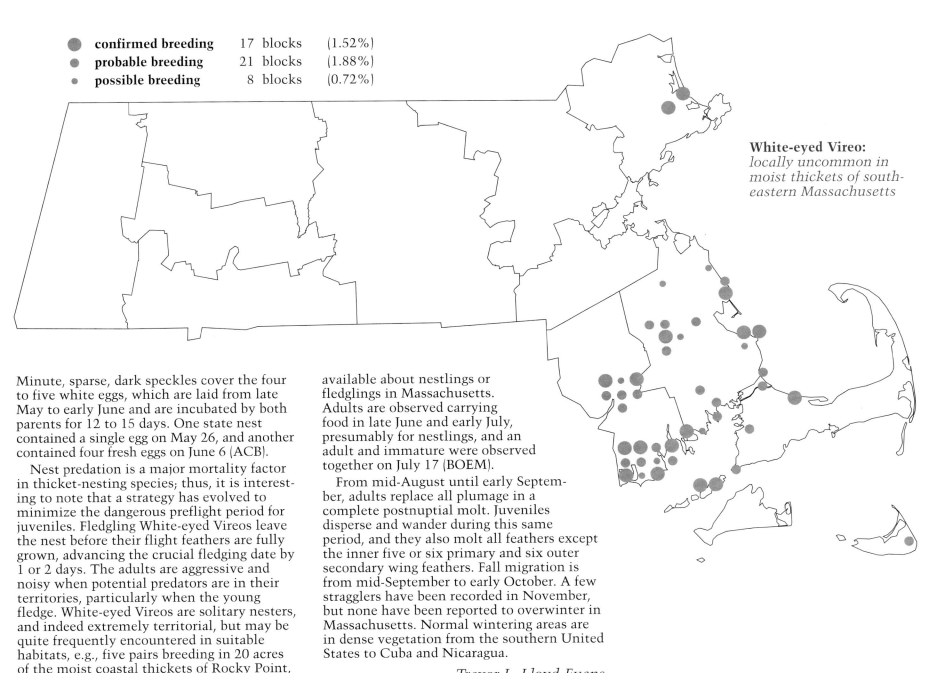

- **confirmed breeding** — 17 blocks (1.52%)
- **probable breeding** — 21 blocks (1.88%)
- **possible breeding** — 8 blocks (0.72%)

White-eyed Vireo: *locally uncommon in moist thickets of southeastern Massachusetts*

Minute, sparse, dark speckles cover the four to five white eggs, which are laid from late May to early June and are incubated by both parents for 12 to 15 days. One state nest contained a single egg on May 26, and another contained four fresh eggs on June 6 (ACB).

Nest predation is a major mortality factor in thicket-nesting species; thus, it is interesting to note that a strategy has evolved to minimize the dangerous preflight period for juveniles. Fledgling White-eyed Vireos leave the nest before their flight feathers are fully grown, advancing the crucial fledging date by 1 or 2 days. The adults are aggressive and noisy when potential predators are in their territories, particularly when the young fledge. White-eyed Vireos are solitary nesters, and indeed extremely territorial, but may be quite frequently encountered in suitable habitats, e.g., five pairs breeding in 20 acres of the moist coastal thickets of Rocky Point, Plymouth. Little specific information is available about nestlings or fledglings in Massachusetts. Adults are observed carrying food in late June and early July, presumably for nestlings, and an adult and immature were observed together on July 17 (BOEM).

From mid-August until early September, adults replace all plumage in a complete postnuptial molt. Juveniles disperse and wander during this same period, and they also molt all feathers except the inner five or six primary and six outer secondary wing feathers. Fall migration is from mid-September to early October. A few stragglers have been recorded in November, but none have been reported to overwinter in Massachusetts. Normal wintering areas are in dense vegetation from the southern United States to Cuba and Nicaragua.

Trevor L. Lloyd-Evans

Yellow-throated Vireo
Vireo flavifrons

Egg dates: May 11 to June 17.

Number of broods: one.

Although it is the most colorful vireo found in New England, the Yellow-throated Vireo is unknown to most people. Generally uncommon, it summers over a large range in the eastern United States and Canada west to the prairies. In Massachusetts, this vireo was more abundant historically but steadily declined, especially in the eastern counties, in the period 1910 to 1960. While some authors have speculatively tied this decline to the increasing use of pesticides, the actual reasons remain obscure. The dramatic change of the Massachusetts landscape during this period has probably played a role as well. Since 1960, there has been some evidence of population recovery because these vireos have successfully exploited the forest openings created by Beavers.

During the Atlas period, Yellow-throated Vireos were closely associated with Beaver flowages, nesting primarily west of the Connecticut River and extending eastward very locally and sparsely through Worcester County and northern Middlesex County into central Essex County. Nesting is extremely rare in southeastern Massachusetts, with one isolated confirmation from Swansea. Since the Atlas period, there have been additional confirmations from Middlesex and Plymouth counties.

In Massachusetts, Yellow-throated Vireos are summer birds, arriving in the first week of May and departing by mid-September. During most of this period, except for a brief hiatus during molting, males vocalize from the forest canopy, delivering their slurred, two- to three-note phrases, slowly and repetitiously, with rising and falling inflections: *eeyay, ayo, oweeah.* Many of the call notes are similar to those of the Blue-headed Vireo. There is a low *hew* call, and excited birds give a chattering *chi-chi-cha-cha-chu-chu* series.

Yellow-throated Vireos are highly arboreal and are almost exclusively insectivorous, spending most of their time gleaning insects in the canopy of tall trees. They inhabit open deciduous and mixed forest and riparian woodlands, preferring oak and maple, and are often closely tied to wetlands, especially Beaver flowages. They are sometimes encountered, less often than formerly, about the shade trees of villages and farms. Mean territory size is reportedly 10 acres.

The nest, constructed by the female in about a week, is a typical vireo cup suspended from a fork of twigs on a horizontal branch of a hardwood 3 to 60 feet and typically over 20 feet above the ground. Nests in Massachusetts have been located in oak, maple, apple, and other deciduous trees. One nest was built at 20 feet in a Tulip-tree between some forked twigs on a horizontal

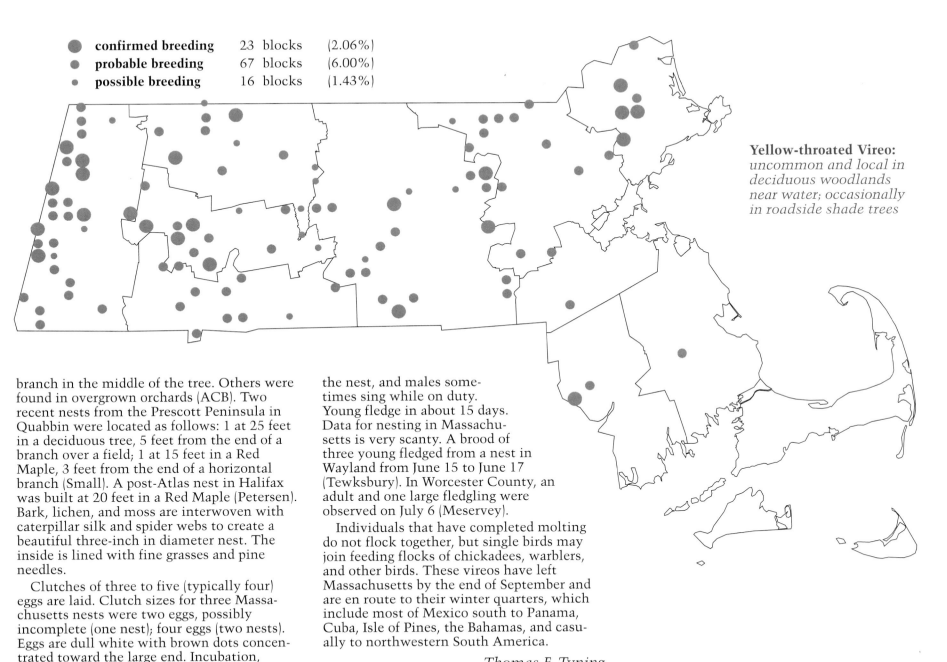

Yellow-throated Vireo: *uncommon and local in deciduous woodlands near water; occasionally in roadside shade trees*

- confirmed breeding — 23 blocks (2.06%)
- probable breeding — 67 blocks (6.00%)
- possible breeding — 16 blocks (1.43%)

branch in the middle of the tree. Others were found in overgrown orchards (ACB). Two recent nests from the Prescott Peninsula in Quabbin were located as follows: 1 at 25 feet in a deciduous tree, 5 feet from the end of a branch over a field; 1 at 15 feet in a Red Maple, 3 feet from the end of a horizontal branch (Small). A post-Atlas nest in Halifax was built at 20 feet in a Red Maple (Petersen). Bark, lichen, and moss are interwoven with caterpillar silk and spider webs to create a beautiful three-inch in diameter nest. The inside is lined with fine grasses and pine needles.

Clutches of three to five (typically four) eggs are laid. Clutch sizes for three Massachusetts nests were two eggs, possibly incomplete (one nest); four eggs (two nests). Eggs are dull white with brown dots concentrated toward the large end. Incubation, shared by both sexes, lasts about 14 days. Incubating birds reportedly sit very tightly on the nest, and males sometimes sing while on duty. Young fledge in about 15 days. Data for nesting in Massachusetts is very scanty. A brood of three young fledged from a nest in Wayland from June 15 to June 17 (Tewksbury). In Worcester County, an adult and one large fledgling were observed on July 6 (Meservey).

Individuals that have completed molting do not flock together, but single birds may join feeding flocks of chickadees, warblers, and other birds. These vireos have left Massachusetts by the end of September and are en route to their winter quarters, which include most of Mexico south to Panama, Cuba, Isle of Pines, the Bahamas, and casually to northwestern South America.

Thomas F. Tyning

Blue-headed Vireo
Vireo solitarius

Egg dates: May 13 to July 29.

Number of broods: one, sometimes two.

The Blue-headed Vireo, once known as the Solitary Vireo, is a fairly common summer resident of the high ground country of northern Worcester County westward throughout all of Franklin and Berkshire counties. In Hampshire and Hampden counties, it nests mostly on the Berkshire escarpment, more sparingly on the eastern side of the Connecticut River valley, and onward through southern Worcester County. Nesting is extremely rare in eastern Massachusetts, except for northernmost Middlesex County, where it probably nests regularly. The species was unreported in the Atlas period for Suffolk, Barnstable, Nantucket, and Dukes counties and occurred in only one block each in Norfolk and Plymouth counties. Blue-headed Vireos nested more commonly in southeastern Massachusetts prior to 1938, when a hurricane destroyed many mature stands of White Pines.

In the leafless April woods, sometimes when there is still snow on the ground, the song of the first solitary can be heard. Spring arrival occurs in mid-April in company with the Ruby-crowned Kinglet and Pine, Palm, and Yellow-rumped warblers, with widespread arrival by the month's end. Migration is protracted, and migrants appear throughout the state until mid-May.

The Blue-headed Vireo, described by many writers as a woodland recluse, dwells throughout the summer in cool, shady retreats in the solitude of the forests. It nests in coniferous or mixed woods with mature stands of White Pine or Eastern Hemlock invariably present. It is sometimes found in beech-hemlock woods. Interspersion of small openings with sapling hardwoods, blueberry, laurel, and other dense undergrowth is essential. Selective timber cuts sometimes create such openings, thereby benefiting the species.

Males seem to establish ill-defined territories of unknown size but that are out of hearing range of other pairs. Within these areas, singing birds wander about with no apparent singing tree, all the time deliberately gleaning insects among the foliage. The basic song is a series of rich, clear, two- to six-note phrases such as *chu-wee-cheerio.* Characteristically, any or all of these notes may be slurred. During courtship, numerous varied phrases are sometimes strung together and delivered rapidly in what is described as a rich warble. There are other contact and alarm calls, one of the latter being a harsh chatter.

Courting males, yellowish flank feathers fluffed out, are said to bob and bow before the female. Copulation may take place anywhere in the territory, often in low understory. Both sexes are involved in nest building, though the male's role may be limited to fetching materials. Always near White Pine or Eastern Hemlock, the nest may be situated in understory sapling hardwoods or the low branches of a deciduous tree; sometimes it is actually built in one of the evergreens. Nests from eastern Massachusetts have been found in White Pine, Red Cedar, Eastern Hemlock, and saplings of oak, Gray Birch, American Beech, hickory, and walnut. Heights ranged from 3.5 to 20 feet (ACB). The pendant nests are usually placed in the fork of a horizontal branch. The nest cup is fashioned from shreds of tissue paper, bark and leaves, bits of lichen, and moss, and is lined with fine grasses, hairs, and small pine needles.

Clutches of three to five (usually four) eggs are laid. Clutch sizes for 11 state nests were

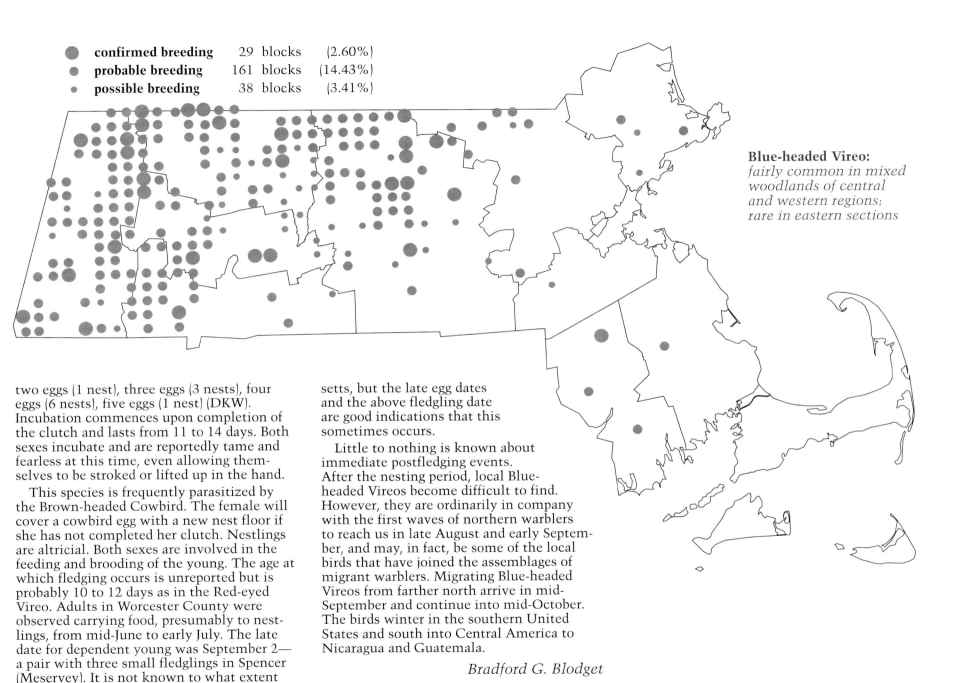

- confirmed breeding 29 blocks (2.60%)
- probable breeding 161 blocks (14.43%)
- possible breeding 38 blocks (3.41%)

Blue-headed Vireo: *fairly common in mixed woodlands of central and western regions; rare in eastern sections*

two eggs (1 nest), three eggs (3 nests), four eggs (6 nests), five eggs (1 nest) (DKW). Incubation commences upon completion of the clutch and lasts from 11 to 14 days. Both sexes incubate and are reportedly tame and fearless at this time, even allowing themselves to be stroked or lifted up in the hand.

This species is frequently parasitized by the Brown-headed Cowbird. The female will cover a cowbird egg with a new nest floor if she has not completed her clutch. Nestlings are altricial. Both sexes are involved in the feeding and brooding of the young. The age at which fledging occurs is unreported but is probably 10 to 12 days as in the Red-eyed Vireo. Adults in Worcester County were observed carrying food, presumably to nestlings, from mid-June to early July. The late date for dependent young was September 2— a pair with three small fledglings in Spencer (Meservey). It is not known to what extent this species is double brooded in Massachusetts, but the late egg dates and the above fledgling date are good indications that this sometimes occurs.

Little to nothing is known about immediate postfledging events. After the nesting period, local Blue-headed Vireos become difficult to find. However, they are ordinarily in company with the first waves of northern warblers to reach us in late August and early September, and may, in fact, be some of the local birds that have joined the assemblages of migrant warblers. Migrating Blue-headed Vireos from farther north arrive in mid-September and continue into mid-October. The birds winter in the southern United States and south into Central America to Nicaragua and Guatemala.

Bradford G. Blodget

Warbling Vireo
Vireo gilvus

Egg dates: May 20 to late July.

Number of broods: one; may re-lay if first attempt fails.

The Warbling Vireo is a fairly common summer resident except on the southeastern coastal plain, Cape, and Islands, where it is essentially absent. During the nineteenth century and the first half of the present century, this species was commonly found in parks and villages, and along tree-lined lanes of country towns. Although still located in similar areas of some of the outlying suburbs and more rural areas, the Warbling Vireo is now more apt to be encountered along the semi-open borders of river meadows, ponds, and streams. This riparian habitat may be the one that this species originally occupied before European settlement and the subsequent three centuries of development.

The Warbling Vireo generally arrives from its wintering grounds in early to mid-May. Residents seek open areas with mature deciduous trees, particularly along watercourses, on which to establish their territories. The males begin singing as soon as they arrive and will continue their performance for several months.

The warbling song of this species is both melodious and rolling. It is often compared to the song of the Purple Finch but is less forceful and has an evenly measured tempo. The male Warbling Vireo is a persistent singer whose song is given not only at the beginning of the nesting cycle but throughout the nesting season as he shares the incubation chores. Although there may be a brief hiatus of song while the fledglings develop, singing is often begun anew in August and continues into September. A coarse *zree-zree* call note is often given. The second common call note, used by both adults and young, is a *snip-snip* likened to the sound produced by garden shears.

Courtship displays are seldom noted, but the male may spread its wings and tail and hop around the female while producing a very soft version of the song. The Warbling Vireo builds a small, deeply cupped nest with fine strips of bark, grasses, and plant down. Spider's silk is often used to wrap and bind this structure, which may then be lined with

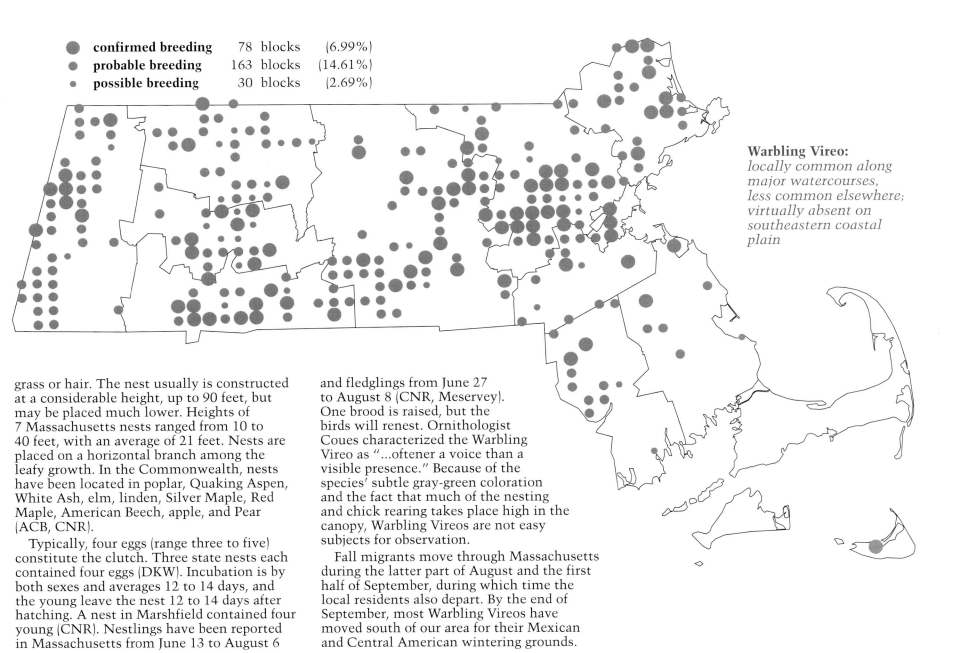

Warbling Vireo: *locally common along major watercourses, less common elsewhere; virtually absent on southeastern coastal plain*

- confirmed breeding 78 blocks (6.99%)
- probable breeding 163 blocks (14.61%)
- possible breeding 30 blocks (2.69%)

grass or hair. The nest usually is constructed at a considerable height, up to 90 feet, but may be placed much lower. Heights of 7 Massachusetts nests ranged from 10 to 40 feet, with an average of 21 feet. Nests are placed on a horizontal branch among the leafy growth. In the Commonwealth, nests have been located in poplar, Quaking Aspen, White Ash, elm, linden, Silver Maple, Red Maple, American Beech, apple, and Pear (ACB, CNR).

Typically, four eggs (range three to five) constitute the clutch. Three state nests each contained four eggs (DKW). Incubation is by both sexes and averages 12 to 14 days, and the young leave the nest 12 to 14 days after hatching. A nest in Marshfield contained four young (CNR). Nestlings have been reported in Massachusetts from June 13 to August 6 and fledglings from June 27 to August 8 (CNR, Meservey). One brood is raised, but the birds will renest. Ornithologist Coues characterized the Warbling Vireo as "...oftener a voice than a visible presence." Because of the species' subtle gray-green coloration and the fact that much of the nesting and chick rearing takes place high in the canopy, Warbling Vireos are not easy subjects for observation.

Fall migrants move through Massachusetts during the latter part of August and the first half of September, during which time the local residents also depart. By the end of September, most Warbling Vireos have moved south of our area for their Mexican and Central American wintering grounds.

Richard K. Walton

Red-eyed Vireo
Vireo olivaceus

Egg dates: mid-May to late August.

Number of broods: one; possibly sometimes two.

The Red-eyed Vireo, the most common and widely distributed vireo in Massachusetts, is primarily a bird of deciduous woodlands. It can be found nesting in mixed or deciduous forest growth, tree-bordered streets, and orchards, providing that these sites lie on the edge of extensive woodland. A break in the forest canopy created by the presence of a brook or bog increases the suitability of the locale for this species.

A few Red-eyed Vireos appear in late April and early May, but general arrival should not be expected until late May. The large number of transients seen is a reflection of this species' abundance both as a migrant and summer resident in all sections of the state. Males appear first on territory. Courtship, which includes soft singing and wing quivering, begins later with the arrival of the females. A typical, but poorly understood, behavior observed throughout the nesting season is characterized by a bird's drawing its feathers in tightly and swaying stiffly from side to side.

The male Red-eyed Vireo sings tirelessly from dawn until dusk, from spring until late summer, somewhat more sporadically through early autumn, and is one of a handful of birds that sings persistently through the heat of midday. The vocal effort of this vireo consists of short, abruptly rendered phrases, generally ending with a rising inflection and followed by a brief pause. The repetitive notes may be sung more than 40 times per minute and have earned the name Preacher Bird for the species. The field notes of the then young Edward Howe Forbush contain the following observation: "The longer pauses in song occurred only when a larva was captured and swallowed."

Eggs may be deposited and incubated in an unfinished nest while the male continues to bring construction material. The usual clutch consists of three or four ovoid, dull, white eggs with fine small spots of reddish to dark brown or blackish color on the large end. Clutch sizes for 11 state nests were three

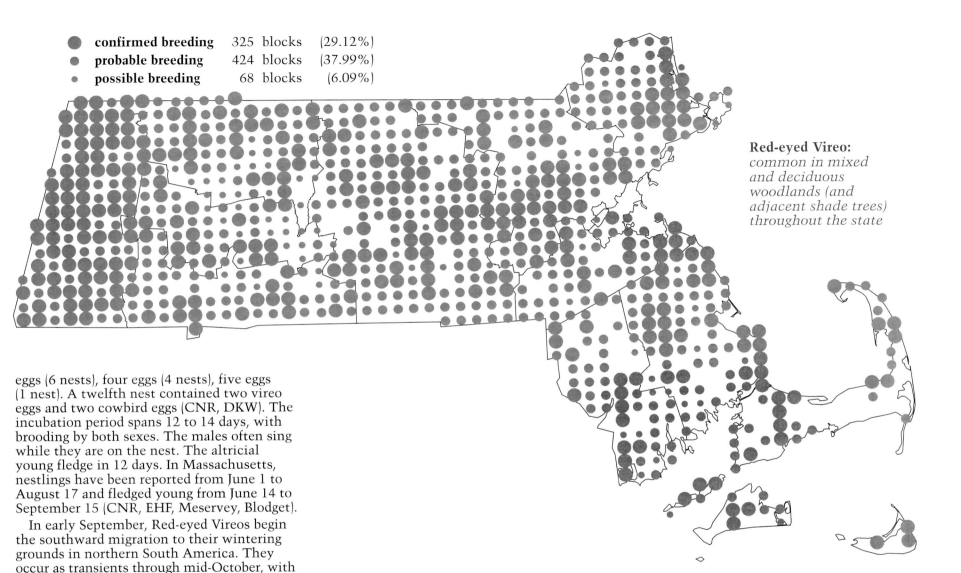

- ● **confirmed breeding** 325 blocks (29.12%)
- ● **probable breeding** 424 blocks (37.99%)
- · **possible breeding** 68 blocks (6.09%)

Red-eyed Vireo: *common in mixed and deciduous woodlands (and adjacent shade trees) throughout the state*

eggs (6 nests), four eggs (4 nests), five eggs (1 nest). A twelfth nest contained two vireo eggs and two cowbird eggs (CNR, DKW). The incubation period spans 12 to 14 days, with brooding by both sexes. The males often sing while they are on the nest. The altricial young fledge in 12 days. In Massachusetts, nestlings have been reported from June 1 to August 17 and fledged young from June 14 to September 15 (CNR, EHF, Meservey, Blodget).

In early September, Red-eyed Vireos begin the southward migration to their wintering grounds in northern South America. They occur as transients through mid-October, with occasional birds lingering into November.

Joseph F. Kenneally, Jr.

Blue Jay
Cyanocitta cristata

Egg dates: April 28 to June 28.

Number of broods: one; may re-lay if first attempt fails.

The Blue Jay is a common and conspicuous bird in Massachusetts. Traditionally a species of the forests, it has adapted well to civilization and is found presently throughout the state from the high mixed forests of the Berkshire Hills to low wooded thickets along the coast, suburban yards, and even densely populated urban areas.

Although Blue Jays can be found throughout the year in Massachusetts, migrations do occur as well as local movements within the population. Some individuals may remain in Massachusetts year-round. Spring migration in eastern sections begins in early March and ends in May. It is delayed a few weeks in the high western elevations and may continue into early June. Peak migration counts may total several hundred jays.

The Blue Jay has an impressive vocabulary. It is an outstanding mimic and ventriloquist as well. One of the commonest calls is described as a loud, harsh *jay* or *jeer*. Another familiar call is a creaking *wheedle-wheedle* or *whee-oodle*. The jay is also credited with a soft, delicate warble, which it delivers from some secluded spot. Among the calls imitated are those of several hawk species. Jays are normally quite loud and boisterous for most of the year, and they are quick to band together to announce the presence of an avian predator or other source of danger with their piercing cries. During the nesting season and the molt period that follows, they tend to be less conspicuous.

There is no elaborate courtship display. Instead, several males follow a female about, bobbing up and down before her and hopping among the branches of a tree. Little vocalizing is done during these activities. Courtship feeding of the female by the male has been observed. Some banding records indicate that mated pairs will remain together for several years.

Nests are built in a variety of trees and shrubs. Twenty-five Massachusetts nests were situated as follows: White Pine (4 nests), spruce (3 nests), yew (1 nest), unidentified conifers (5 nests), oak (3 nests), American Beech (2 nests), Black Cherry (2 nests), Multiflora Rose (2 nests), magnolia (1 nest), pyracantha (1 nest), and American Holly (1 nest). Heights ranged from 3 to 30 feet and averaged 11.5 feet (CNR). The nest is situated either in a crotch or among the outer branches. It is a bulky structure about 8 inches in diameter and 4 inches in height. The outer framework consists of dead twigs interlaced with bark, string, paper, or forest litter; and the inner cup, approximately 4 inches in diameter, is composed of fine rootlets.

Egg laying generally takes place from April into May, but if the eggs or nestlings are lost the birds will renest into June. Eggs are deposited one per day until a clutch of two to seven (usually four or five) eggs is completed. The clutch sizes in 22 Massachusetts nests were as follows: two eggs (1 nest), three eggs (3 nests), four eggs (12 nests), five eggs (5 nests), and six eggs (1 nest) (CNR). The eggs are variable, with olive or buff ground color and scattered, irregular, brownish spots.

Incubation lasts 16 to 18 days. Upon hatching, the young are blind and naked, but they develop rapidly and leave the nest at 17 to 21 days of age. The female performs most of the incubation and brooding. The male brings food to her initially and later helps feed the young. Jays are devoted parents and will defend their brood fiercely from

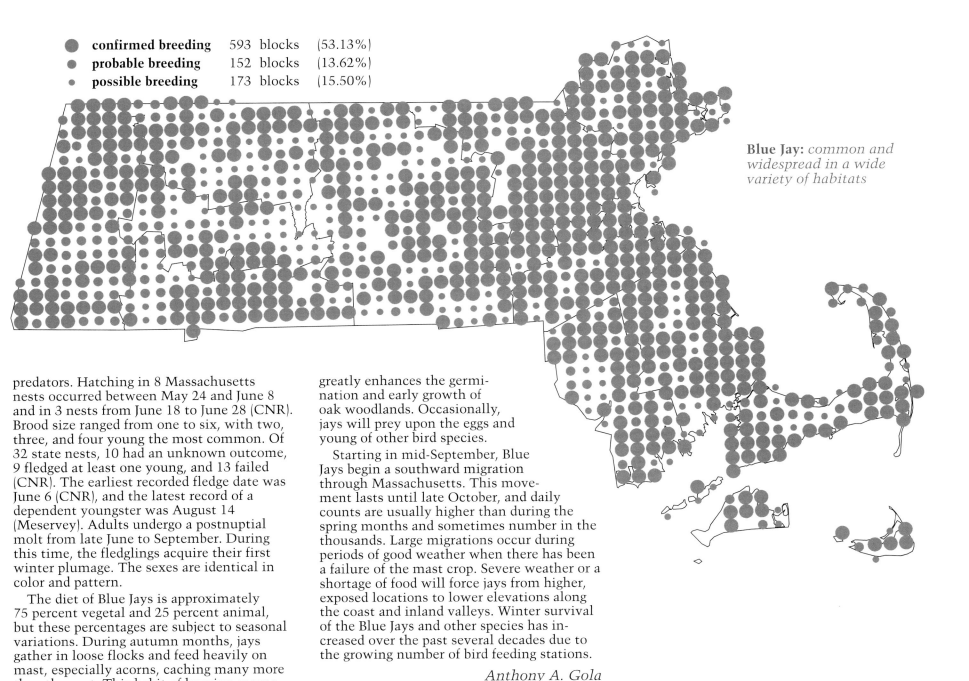

- ● **confirmed breeding** 593 blocks (53.13%)
- ● **probable breeding** 152 blocks (13.62%)
- · **possible breeding** 173 blocks (15.50%)

Blue Jay: *common and widespread in a wide variety of habitats*

predators. Hatching in 8 Massachusetts nests occurred between May 24 and June 8 and in 3 nests from June 18 to June 28 (CNR). Brood size ranged from one to six, with two, three, and four young the most common. Of 32 state nests, 10 had an unknown outcome, 9 fledged at least one young, and 13 failed (CNR). The earliest recorded fledge date was June 6 (CNR), and the latest record of a dependent youngster was August 14 (Meservey). Adults undergo a postnuptial molt from late June to September. During this time, the fledglings acquire their first winter plumage. The sexes are identical in color and pattern.

The diet of Blue Jays is approximately 75 percent vegetal and 25 percent animal, but these percentages are subject to seasonal variations. During autumn months, jays gather in loose flocks and feed heavily on mast, especially acorns, caching many more than they eat. This habit of burying acorns greatly enhances the germination and early growth of oak woodlands. Occasionally, jays will prey upon the eggs and young of other bird species.

Starting in mid-September, Blue Jays begin a southward migration through Massachusetts. This movement lasts until late October, and daily counts are usually higher than during the spring months and sometimes number in the thousands. Large migrations occur during periods of good weather when there has been a failure of the mast crop. Severe weather or a shortage of food will force jays from higher, exposed locations to lower elevations along the coast and inland valleys. Winter survival of the Blue Jays and other species has increased over the past several decades due to the growing number of bird feeding stations.

Anthony A. Gola

American Crow
Corvus brachyrhynchos

Egg dates: March to June 13.

Number of broods: one; may re-lay if first attempt fails.

The American Crow is one of the most familiar birds in the Commonwealth. It is found in forested areas, in fields and pastures, and along coastal beaches. During the Atlas period, nesting was confirmed in all sections of the state. The high number of confirmations in densely populated areas in eastern Massachusetts is a good indication of the crow's adaptability. Formerly, the species was held in general disdain and universally persecuted, but today, except during a brief open hunting season, crows are basically regarded with indifference or ignored, despite their occasional forays into cornfields and gardens.

In addition to the familiar *caw* or *cah* calls, given singly or in a series, the crow has a number of vocalizations used in different contexts and has the ability to imitate other birds and animals, including the human voice. The American Crow is a common year-round resident and shows a distinct preference for agricultural lands and farms, where food is readily available. Communal winter roosts begin to break up in late February and early March. Courtship behavior includes strutting, posturing, and vigorous, tumbling flights. There are many reports of two- and three-bird groups engaging in bill touching and mutual head and neck preening. Crows may breed in pairs or in cooperative groups, with offspring from previous years remaining to help their parents rear new broods. In a Massachusetts study, breeding group size ranged from 2 to 10 members, with an average of 4.4 birds. Such groups actively defend all-purpose territories averaging 104 acres throughout the year (Chamberlain-Auger).

Nests are constructed of sticks and twigs, generally broken from tree branches rather than selected from the ground litter, and are lined with dead grass. They are usually located in conifers, forested areas, or isolated woodlots, and are well concealed in the crotch of a high, horizontal branch. In Worcester County, nests are generally reported in White Pine (Blodget, TC). Forty-nine nests from Barnstable and Middlesex counties were located as follows: Pitch Pine (42 nests), White Pine (2 nests), spruce (2 nests), Red Cedar (2 nests), willow (1 nest) (Chamberlain-Auger). These nests averaged 16 inches in diameter, and the mean height was 32 feet. New nests are constructed each year, and in some instances a second is built if the first fails (Chamberlain-Auger). During all phases of the breeding cycle, crows are extremely wary and tend to make surreptitious trips to and from their nest.

Eggs are laid as early as March and as late as June (Chamberlain-Auger) but generally are produced in April or early May. The clutch size may vary from three to seven eggs, but the average is four to six. The clutch sizes for 7 Worcester County nests were four eggs (4 nests), five eggs (3 nests) (DKW). Incubation, mostly by the female, averaged 22.3 days for 12 state nests. Both parents feed the young. Nestlings have been reported in the state from April to July. The mean number of days until fledging for 16 Massachusetts nests was 30.1 days

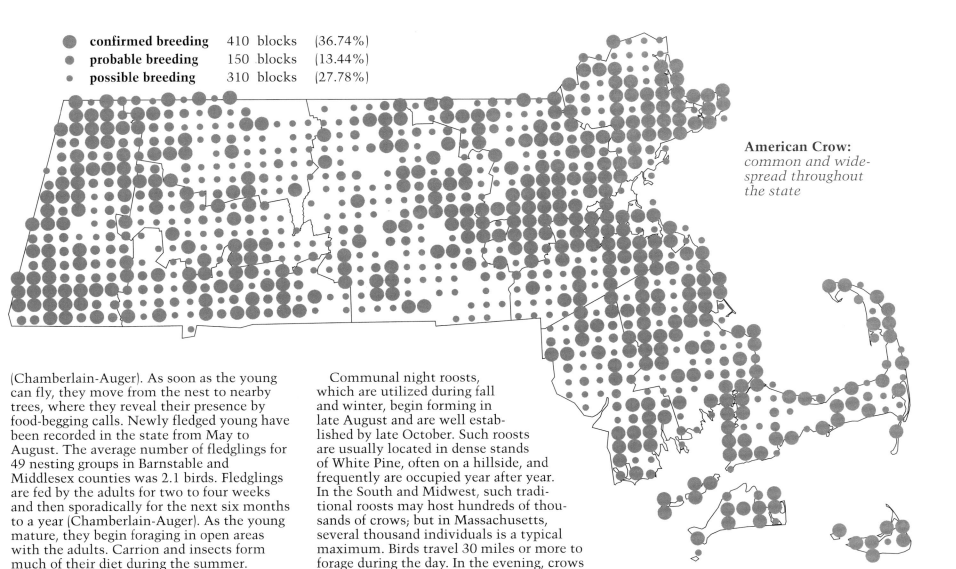

- confirmed breeding — 410 blocks (36.74%)
- probable breeding — 150 blocks (13.44%)
- possible breeding — 310 blocks (27.78%)

American Crow: *common and widespread throughout the state*

(Chamberlain-Auger). As soon as the young can fly, they move from the nest to nearby trees, where they reveal their presence by food-begging calls. Newly fledged young have been recorded in the state from May to August. The average number of fledglings for 49 nesting groups in Barnstable and Middlesex counties was 2.1 birds. Fledglings are fed by the adults for two to four weeks and then sporadically for the next six months to a year (Chamberlain-Auger). As the young mature, they begin foraging in open areas with the adults. Carrion and insects form much of their diet during the summer.

In fall, sizeable flocks of crows are a familiar sight in agricultural areas, where they feed on waste corn and chance animal tidbits. Modern harvesting practices leave substantial waste. The state crow population is apparently less migratory now than it was half a century ago, although birds from farther north may pass through or winter here.

Communal night roosts, which are utilized during fall and winter, begin forming in late August and are well established by late October. Such roosts are usually located in dense stands of White Pine, often on a hillside, and frequently are occupied year after year. In the South and Midwest, such traditional roosts may host hundreds of thousands of crows; but in Massachusetts, several thousand individuals is a typical maximum. Birds travel 30 miles or more to forage during the day. In the evening, crows may gather in a staging area located a mile or more from the roost before coordinated movement to the roost is made. When approaching the roost, crows are usually silent; in contrast, when departing in the morning, they are typically noisy and leave for feeding areas in a direct line or "as the crow flies."

In spite of an adverse reputation and intense persecution in the past, the crow flourishes today, adapting to the radical environmental alterations of humankind. The American Crow is, above all else, a survivor.

Richard A. Forster

Fish Crow
Corvus ossifragus

Egg dates: not available.

Number of broods: one.

The Fish Crow is an uncommon and local bird in Massachusetts, where it is near the northernmost part of its range. Stragglers have been observed here for over 100 years, but eventually the species became a permanent resident. It was only as recently as June 1973 that the first breeding record for Massachusetts was confirmed when two nests were found in Stony Brook Reservation in the Hyde Park section of Boston. During the Atlas period, Fish Crows were found nesting in several locations in the Boston metropolitan area, in southern Essex County, and in Plymouth County; and recently breeding has been suspected in Hampden County. Present breeding locations are in approximately the same areas as the historical sightings.

Throughout most of its range, this species favors coastal and brackish habitats, including wooded shorelines, marshes, beaches, and the vicinity of tidal rivers. In Massachusetts, nests have been found at inland locations, and some individuals are known to forage in large mall parking lots as far inland as Framingham (Blodget). The hoarse *caa* calls of the Fish Crow can best be described as sounding like those of a "crow with a Boston accent." In early spring, these calls can be a reliable identification clue, but caution must be exercised because female American Crows give similar sounds

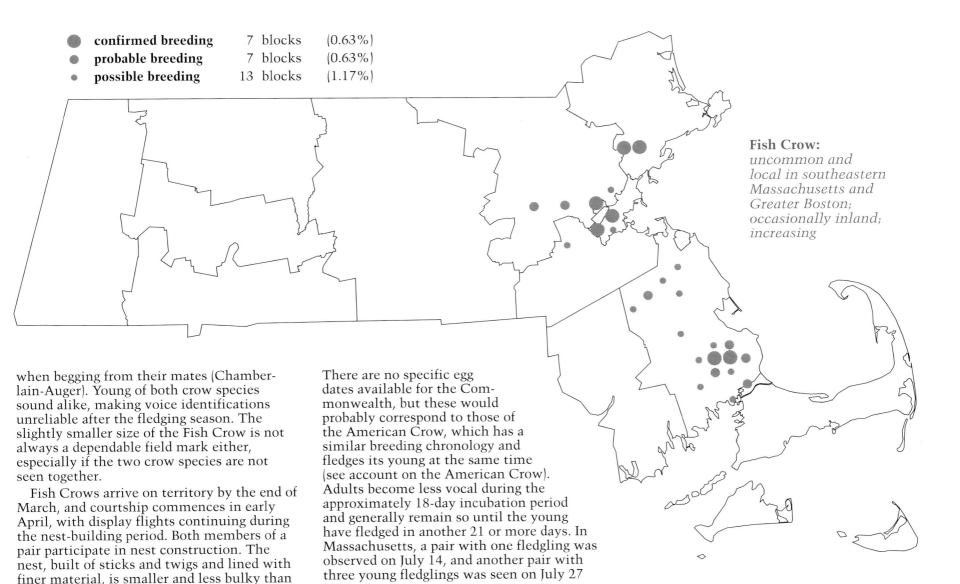

- ● **confirmed breeding**   7 blocks   (0.63%)
- ● **probable breeding**   7 blocks   (0.63%)
- • **possible breeding**   13 blocks   (1.17%)

Fish Crow: *uncommon and local in southeastern Massachusetts and Greater Boston; occasionally inland; increasing*

when begging from their mates (Chamberlain-Auger). Young of both crow species sound alike, making voice identifications unreliable after the fledging season. The slightly smaller size of the Fish Crow is not always a dependable field mark either, especially if the two crow species are not seen together.

Fish Crows arrive on territory by the end of March, and courtship commences in early April, with display flights continuing during the nest-building period. Both members of a pair participate in nest construction. The nest, built of sticks and twigs and lined with finer material, is smaller and less bulky than that of the American Crow. In Massachusetts, nests have been located in White Pine and Pitch Pine (Forster), and nest building has been observed from April 10 to April 19 (BOEM). Solitary nesting has been the rule in Massachusetts, but loose colonies are often reported in more southern parts of the range.

There are no specific egg dates available for the Commonwealth, but these would probably correspond to those of the American Crow, which has a similar breeding chronology and fledges its young at the same time (see account on the American Crow). Adults become less vocal during the approximately 18-day incubation period and generally remain so until the young have fledged in another 21 or more days. In Massachusetts, a pair with one fledgling was observed on July 14, and another pair with three young fledglings was seen on July 27 (BOEM).

During the summer the preponderance of records of the Fish Crow comes from Plymouth County; however, in winter almost all reports are confined to a narrow belt along the Sudbury River and particularly the Sudbury landfill. They seldom consort with the American Crows feeding in the surrounding, extensive agricultural fields. It is not known to what extent, if any, the state population is migratory.

Robert H. Stymeist

Horned Lark
Eremophila alpestris

Egg dates: early March to July 16.

Number of broods: one or two.

The Horned Lark enjoys a widespread, circumpolar range, breeding in North America from Alaska across Arctic Canada south to Mexico. One subspecies, *praticola*, the so-called "Prairie" Horned Lark, occurs in Massachusetts as a fairly common but locally distributed breeding bird. This race, best characterized by its pale plumage and whitish eye stripe, spread eastward into New England and markedly increased during the first half of the twentieth century. In Massachusetts, Horned Larks are most reliably found in the vicinity of coastal beaches, especially on the Cape and Islands, although suitable inland sites, particularly in the Connecticut River valley, include abandoned agricultural fields and pastures. The continued waning of agriculture and the loss of open land to forest regeneration and development have made sites such as golf courses and airports increasingly important to inland populations.

True harbingers of spring, local nesters return to the breeding grounds as early as late February but go largely undetected because they tend to mingle with the wintering or migrant "Northern" Horned Larks. This species' exacting habitat requirements are for the most part barren and unproductive environments. They will tolerate only sparse vegetation, readily deserting a site if weeds, grass, or other cover grow too dense. Coastal residents prefer dry, sandy, or gravelly upper portions of beaches, especially where there is a preponderance of wrack or debris that serves as singing posts and shelters for nests. Flats among or behind barrier dunes where only sparse growths of Dunegrass or heather exist also seem to be favored nesting and foraging sites. Most of the Horned Lark's diet is made up of various weed seeds, although a great variety of insects are also consumed during the warmer months.

During its nuptial flight, the male Horned Lark silently ascends to several hundred feet in altitude, where he begins his twittering *pit-wit, wee-pit, pit-wee, wee-pit* song while circling the nest site. This courtship flight is then abruptly halted with a dramatic plunge to earth on folded wings. A clear, high-pitched *tsee-titi* is the most frequent call note heard.

Horned Larks have a very protracted breeding season, nesting very early, with eggs laid in our area from the beginning of March to mid-July. From two to seven (usually four) pale greenish gray eggs, marked with fine brownish buff speckling, are deposited in a shallow scrape lined with fine grasses, plant down, and hair. The nest is usually placed in the shelter of a grass tuft, stone, or piece of driftwood. Three recent nests from eastern Massachusetts were located on beaches, 1 near a grass tuft, 1 near a pile of beach debris, and 1 near Beach Pea (CNR). A nest in Wellfleet contained one egg and three nest-

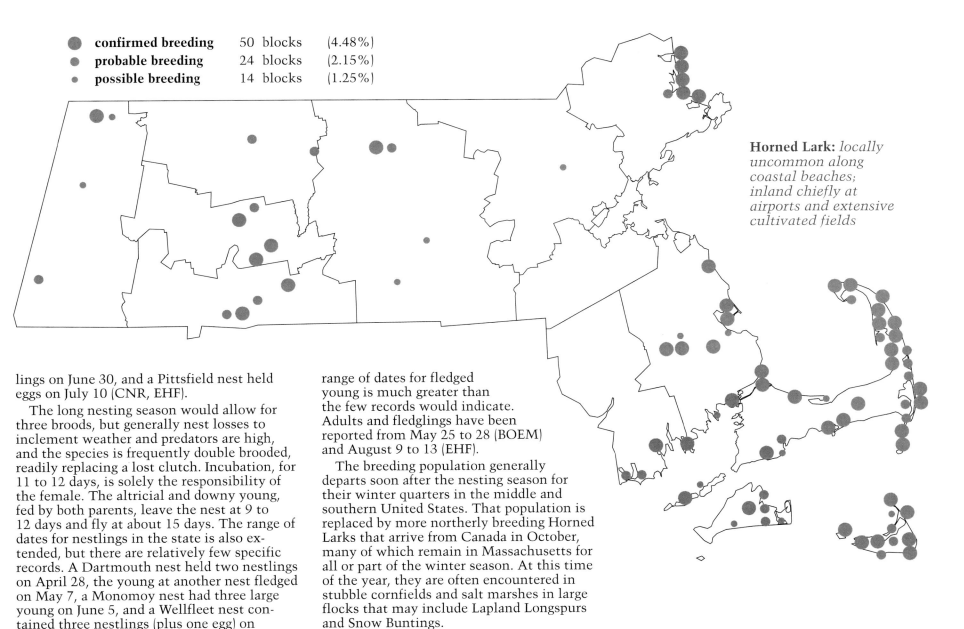

- confirmed breeding — 50 blocks (4.48%)
- probable breeding — 24 blocks (2.15%)
- possible breeding — 14 blocks (1.25%)

Horned Lark: *locally uncommon along coastal beaches; inland chiefly at airports and extensive cultivated fields*

lings on June 30, and a Pittsfield nest held eggs on July 10 (CNR, EHF).

The long nesting season would allow for three broods, but generally nest losses to inclement weather and predators are high, and the species is frequently double brooded, readily replacing a lost clutch. Incubation, for 11 to 12 days, is solely the responsibility of the female. The altricial and downy young, fed by both parents, leave the nest at 9 to 12 days and fly at about 15 days. The range of dates for nestlings in the state is also extended, but there are relatively few specific records. A Dartmouth nest held two nestlings on April 28, the young at another nest fledged on May 7, a Monomoy nest had three large young on June 5, and a Wellfleet nest contained three nestlings (plus one egg) on June 30 (BOEM, CNR).

The well-camouflaged young, unable to fly at first, will often crouch motionless to avoid detection by potential predators. Again, the range of dates for fledged young is much greater than the few records would indicate. Adults and fledglings have been reported from May 25 to 28 (BOEM) and August 9 to 13 (EHF).

The breeding population generally departs soon after the nesting season for their winter quarters in the middle and southern United States. That population is replaced by more northerly breeding Horned Larks that arrive from Canada in October, many of which remain in Massachusetts for all or part of the winter season. At this time of the year, they are often encountered in stubble cornfields and salt marshes in large flocks that may include Lapland Longspurs and Snow Buntings.

Richard S. Heil

Purple Martin
Progne subis

Egg dates: May 30 to July 10.

Number of broods: one; may re-lay if first attempt fails.

This largest of the swallows is an uncommon to locally common breeder; it is uncommon as a migrant in Massachusetts. Purple Martins are insectivorous, communal birds that are most common in Massachusetts in the vicinity of salt marshes, coastal farmland, and golf courses. They nest exclusively in multichambered nest boxes or hanging gourds, but their occupancy is dependent upon a strict maintenance regime that keeps the boxes in good repair and prevents House Sparrows and European Starlings from usurping the site. Successful boxes share several characteristics—open space in the immediate area, water close by, wires or dead trees to perch on, and a human steward to oversee the colony.

Though some colonies are well established, the species has had a history of fluctuating populations in Massachusetts. They were always absent from the heavy forests that characterized the precolonial period. Historically, small numbers of martins probably inhabited the larger river valleys and coastal marshes. They then became locally common, nesting throughout much of southern New England in the mid- to late 1800s; but by the turn of the century their numbers declined drastically. A major factor affecting Purple Martin numbers and distribution is weather; a wet, cool, nesting season or a storm soon after the arrival of the breeding birds can wreak havoc on the population. In 1903, a cold, wet spell in April decimated the local population, and numbers have never again reached the pre-1903 levels. Severe competition for nest sites from starlings and House Sparrows has also taken its toll, and loss of habitat has probably further contributed to the decline.

During the Atlas period, the Purple Martin was unrecorded as a breeder from all of inland Massachusetts. The last Worcester County nesters, a little-known colony in Southbridge, persisted through the 1960s but disappeared in the early 1970s, when the nest house was removed (Meservey). The locations of successful martin colonies are along portions of the North Shore, South Shore, and coastal plain of southeastern Massachusetts, including scattered sites on Cape Cod and Martha's Vineyard. To the north, the Parker River National Wildlife Refuge and vicinity are strongholds. To the south, small colonies are located in Wareham, Lakeville, Halifax, Hanson, Marshfield, and Middleboro. These latter colonies are on private land and are seemingly more likely to have populations that fluctuate than the colonies on the refuge. Away from these areas, martin houses placed in seemingly ideal locations and appearing to satisfy all breeding requirements are usually neglected. The small numbers (about 300 pairs) and dependency on humans make the future of the Purple Martin in Massachusetts uncertain.

Purple Martins begin to arrive in April and are well established at colonies by late in the month. The migration route from their South American wintering grounds is through Central America and north through the southern United States. Although martins

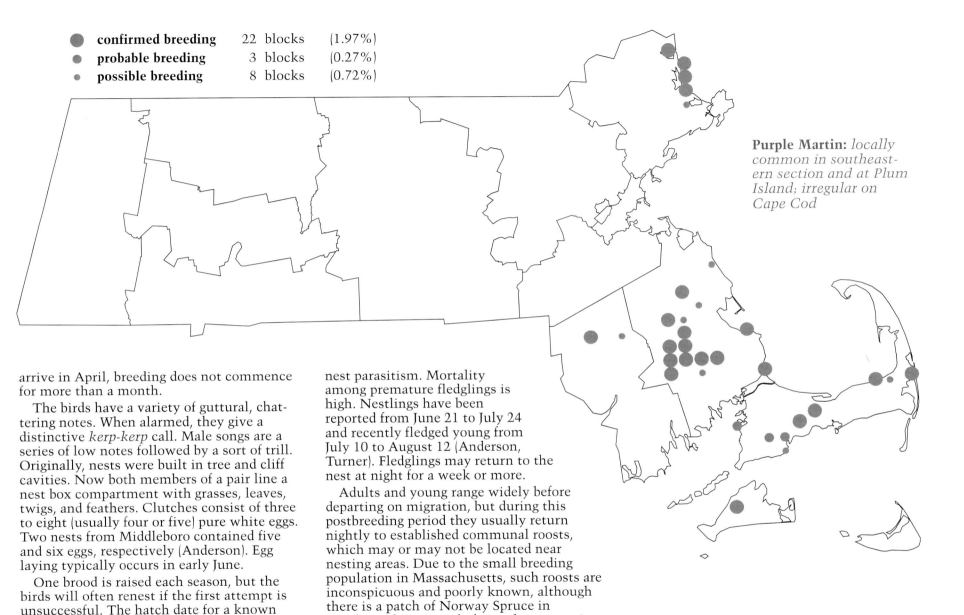

Purple Martin: *locally common in southeastern section and at Plum Island; irregular on Cape Cod*

arrive in April, breeding does not commence for more than a month.

The birds have a variety of guttural, chattering notes. When alarmed, they give a distinctive *kerp-kerp* call. Male songs are a series of low notes followed by a sort of trill. Originally, nests were built in tree and cliff cavities. Now both members of a pair line a nest box compartment with grasses, leaves, twigs, and feathers. Clutches consist of three to eight (usually four or five) pure white eggs. Two nests from Middleboro contained five and six eggs, respectively (Anderson). Egg laying typically occurs in early June.

One brood is raised each season, but the birds will often renest if the first attempt is unsuccessful. The hatch date for a known second nest attempt was July 10 (Anderson). Incubation by the female lasts about two weeks, sometimes slightly longer. The young can fly in about four weeks but sometimes fledge prematurely due to extreme heat and nest parasitism. Mortality among premature fledglings is high. Nestlings have been reported from June 21 to July 24 and recently fledged young from July 10 to August 12 (Anderson, Turner). Fledglings may return to the nest at night for a week or more.

Adults and young range widely before departing on migration, but during this postbreeding period they usually return nightly to established communal roosts, which may or may not be located near nesting areas. Due to the small breeding population in Massachusetts, such roosts are inconspicuous and poorly known, although there is a patch of Norway Spruce in Newbury that is regularly used as a premigration roost by martins, presumably from the Parker River National Wildlife Refuge colony.

Martins depart early from Massachusetts, generally during August, and September sightings are unusual. The wintering grounds are located in the Amazon basin of Brazil.

David E. Clapp

Tree Swallow
Tachycineta bicolor

Egg dates: April 19 to June 25.

Number of broods: one; may re-lay if first attempt fails.

The Tree Swallow is one of the more widely distributed nesting birds in Massachusetts. It is the most abundant species of swallow in the state, and during fall migration huge flocks gather along the coast, far outnumbering any other passerine migrant.

Spring migrants arrive in mid- to late March in the more sheltered swamps and river valleys, and during April they gradually spread out and reach the western and northern highlands by the end of the month. During cool, rainy periods in spring, the birds will be found almost exclusively flying low over ponds and lakes, hunting for insects close to the surface of the water. When the weather turns balmy, the birds move out over the countryside to flooded meadows, marshlands, swamps, farms, and open areas. Occasionally, spring snowstorms will bring considerable mortality to the earlier arrivals.

For nesting sites, the birds prefer wooded habitat near water, especially where dead trees are abundant, as in flooded swamps. Originally, they nested in natural cavities and old woodpecker holes in dead trees near ponds, streams, rivers, etc., but with the advent of civilization the birds readily adapted to nesting boxes erected for them on poles and trees. Tree Swallows are not colonial to any extent, although they will form loose groups when many dead trees or nesting boxes are located close together in favorable habitats such as swamps or salt marshes. The birds may be quite pugnacious in their claiming and defense of a nest site, battling individuals of both their own or other species. The song is a series of liquid twittering sounds uttered repeatedly on the wing or from a perch, and the common call note is a rapid *silip*, which becomes louder and harsher when the birds are agitated.

Nest height can vary from 3 feet to 60 feet or more and in Massachusetts ranged from 3 to 12 feet for 31 nest boxes. One nest was in a deciduous tree cavity at 9 feet (CNR). The actual nest is constructed from dry grasses and is lined with white feathers. The birds will go considerable distances to obtain the feathers and will sometimes play with them in the air, dropping them and then catching them in flight. Nest building activity reaches its peak during the last week of April and early May in Massachusetts, with egg laying soon afterward. Normally, four to six eggs are laid, but there have been instances when as many as ten eggs have been found when 2 females have laid in the same nest. Clutch sizes for 18 Massachusetts nests were as follows: one egg (1 nest), three eggs (2 nests), four eggs (4 nests), five eggs (6 nests), six eggs (4 nests), and seven eggs (1 nest) (CNR). Both sexes engage in nest building and feeding the young, but the female alone incubates for 13 to 16 days. The young remain in the nest 16 to 24 days, depending on brood size and food supply. Brood sizes for 21 state nests were: one young (1 nest), two young (1 nest), three young (6 nests), four young (6 nests), five young (6 nests), six young (1 nest) (CNR).

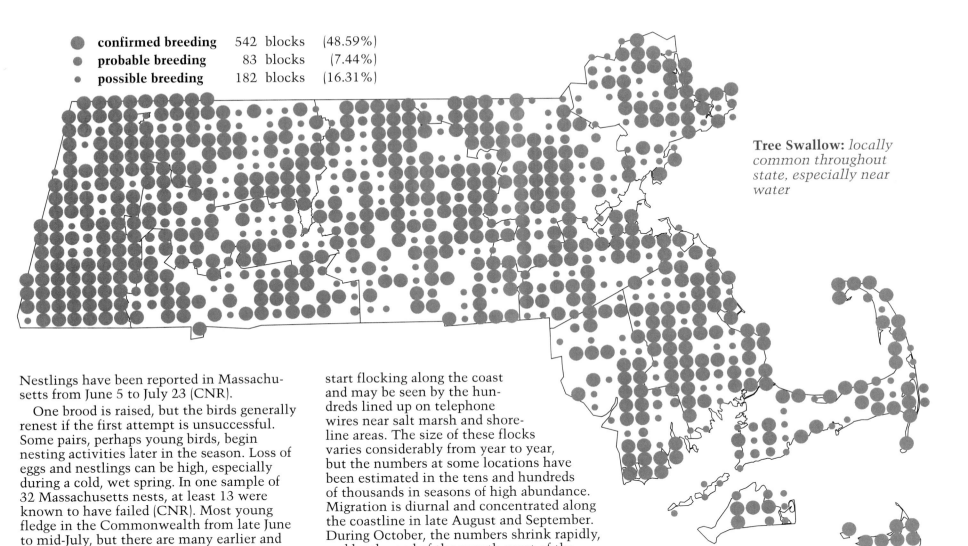

- confirmed breeding 542 blocks (48.59%)
- probable breeding 83 blocks (7.44%)
- possible breeding 182 blocks (16.31%)

Tree Swallow: *locally common throughout state, especially near water*

Nestlings have been reported in Massachusetts from June 5 to July 23 (CNR).

One brood is raised, but the birds generally renest if the first attempt is unsuccessful. Some pairs, perhaps young birds, begin nesting activities later in the season. Loss of eggs and nestlings can be high, especially during a cold, wet spring. In one sample of 32 Massachusetts nests, at least 13 were known to have failed (CNR). Most young fledge in the Commonwealth from late June to mid-July, but there are many earlier and later records (CNR, EHF). Fledglings may perch on a dead branch or wire to be fed or may receive food from their parents in flight. The young are distinguished from adults by their brownish juvenal plumage, a brownish wash across the lower breast, and stubbier wings and tail. They molt during the fall, and the first winter plumage is very similar to the adult winter plumage.

During July and August, Tree Swallows start flocking along the coast and may be seen by the hundreds lined up on telephone wires near salt marsh and shoreline areas. The size of these flocks varies considerably from year to year, but the numbers at some locations have been estimated in the tens and hundreds of thousands in seasons of high abundance. Migration is diurnal and concentrated along the coastline in late August and September. During October, the numbers shrink rapidly, and by the end of the month most of the birds have departed except for a few stragglers on the Cape and Islands. During spells of Indian summer in November and early December, reverse migration may bring a few birds back to coastal points. During extremely warm winters, small flocks of these birds may actually survive into January and February by feeding largely on bayberries and whatever insect life they can find along beaches and salt marshes. The regular winter range extends from the southern United States south to Central America.

David L. Emerson

Northern Rough-winged Swallow
Stelgidopteryx serripennis

Egg dates: May 7 to June 12.

Number of broods: one; may re-lay if first attempt fails.

The Northern Rough-winged Swallow was unrecorded anywhere in New England prior to 1850. The species appears to have undergone a significant range expansion through most of the historical period, apparently first entering Massachusetts via the Hudson and Hoosic river valleys in the late 1880s. The first Massachusetts nesting record was in 1890. By the 1940s, this swallow was well established throughout most of the state. During the Atlas period, it was found to have a sparse, yet generalized distribution in Massachusetts with no evident pattern. Since 1987, nesting has occurred sporadically on Nantucket (Perkins, Andrews).

One of the most interesting things about Northern Rough-winged Swallows is their adaptability to a wide variety of nesting situations, many of them artificial and frequently in close proximity to humans and often near water. They are also unique among swallow species in possessing a ragged "saw edge" of tiny hooklets along the outer web of the ninth primary. They are rather quiet, even phlegmatic, in disposition compared with other swallows. They typically spend many hours quietly perched on small branches or wires and are most active and noisiest in the hour or so after dawn and before dusk. When a nest is approached, both sexes often wheel about close to the intruder. Their flight is noticeably powerful and less erratic than that of other swallows, and their call is a rasping *brrt* or *brzzt*.

Northern Rough-winged Swallows return to Massachusetts in mid-April; however, they do not start nest building until May. Males sometimes arrive at a site first, although sometimes pairs appear to arrive already mated. They nest solitarily, though loose colonies of 3 to 20 pairs may establish themselves where conditions are favorable. One of the largest breeding groups in the state is at the West Hill Dam in Uxbridge, where at least 15 pairs have nested in some seasons (TC). There is little evidence of territorial behavior in this swallow. Nest building has been observed from May 10 to May 26 (Meservey). Females do virtually all of the actual nest construction, although the male generally accompanies his mate back to the nest hole, usually stopping short of entering but settling on a nearby perch.

Some females fashion a well-defined mud bowl, others a very disorganized, indifferently built saucer. Pine needles, bits of leaves, grasses, and rootlets are commonly added.

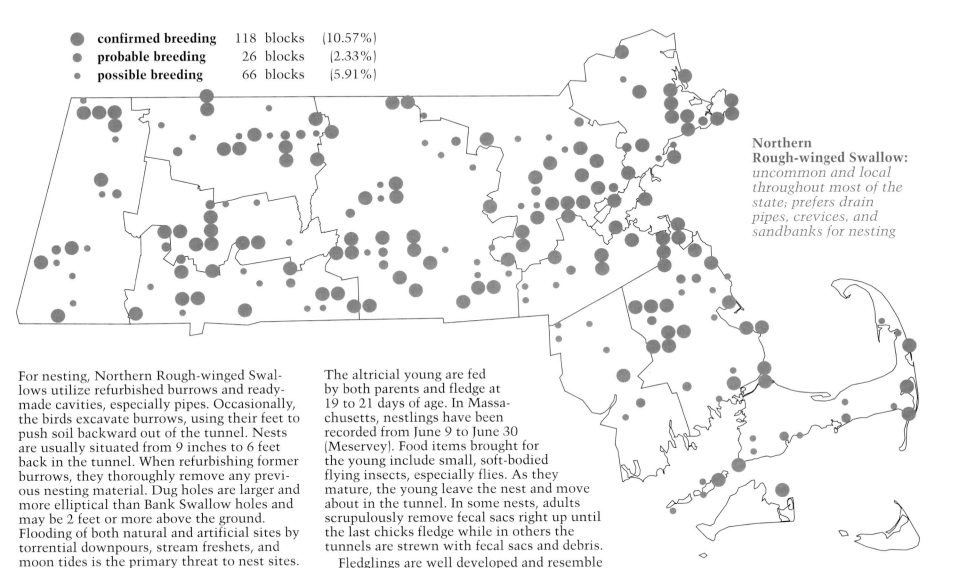

- confirmed breeding — 118 blocks (10.57%)
- probable breeding — 26 blocks (2.33%)
- possible breeding — 66 blocks (5.91%)

Northern Rough-winged Swallow: *uncommon and local throughout most of the state; prefers drain pipes, crevices, and sandbanks for nesting*

For nesting, Northern Rough-winged Swallows utilize refurbished burrows and ready-made cavities, especially pipes. Occasionally, the birds excavate burrows, using their feet to push soil backward out of the tunnel. Nests are usually situated from 9 inches to 6 feet back in the tunnel. When refurbishing former burrows, they thoroughly remove any previous nesting material. Dug holes are larger and more elliptical than Bank Swallow holes and may be 2 feet or more above the ground. Flooding of both natural and artificial sites by torrential downpours, stream freshets, and moon tides is the primary threat to nest sites.

Clutches of four to eight (usually six or seven) glossy white, unmarked eggs are produced. During egg laying, the female spends the night on the nest. The incubation period is 16 days, and only the female performs this task. The male remains nearby on an accustomed perch. If nesting is interrupted, second nesting attempts are common.

The altricial young are fed by both parents and fledge at 19 to 21 days of age. In Massachusetts, nestlings have been recorded from June 9 to June 30 (Meservey). Food items brought for the young include small, soft-bodied flying insects, especially flies. As they mature, the young leave the nest and move about in the tunnel. In some nests, adults scrupulously remove fecal sacs right up until the last chicks fledge while in others the tunnels are strewn with fecal sacs and debris.

Fledglings are well developed and resemble adults. After fledging, neither the young nor the adults return to the burrow, and they depart almost immediately for the south, seldom being reported after mid-August. Families with fledged young have been reported in the state from June 22 to July 25, and brood size for 9 of these family groups ranged from 3 to 9 (Meservey). They winter from southern Florida, southern Louisiana, and northern Mexico south to Panama.

Bradford G. Blodget

Bank Swallow
Riparia riparia

Egg dates: late April to mid-July.

Number of broods: one; may re-lay if first attempt fails.

Found over much of the world, the Bank Swallow (known as the Sand Martin in Eurasia) is one of the most wide ranging of the North American swallows. In Massachusetts, it is a locally common species but is limited in its distribution by the availability of suitable nesting sites. The smallest of the six swallow species that nest in the state, it is the only one that rarely uses artificial nesting sites.

In spring, Bank Swallows first arrive on the coast in mid-April, and by May they appear in numbers in the interior. Large bodies of water such as lakes, ponds, and rivers are favored feeding areas, especially during cold or inclement weather, but Bank Swallows also forage on the wing for small flying insects over fields, marshes, and other open habitats. They are often found far away from nesting colonies, which explains the high percentage of "possible" breeding records on the Atlas map.

Nesting colonies are restricted to steep, sandy, or gravelly banks. Bluffs along rivers or the shores of lakes, bays, and the ocean are the most typical natural sites. Today, Bank swallows also take advantage of steep slopes of gravel pits, banks along highway and railroad rights-of-way, and occasionally even large piles of sawdust at lumber mills. They do not usually use drainage pipes in walls or other similar artificial sites as Northern Rough-winged Swallows often do. In banks of loose sand, a single row of burrows may be dug at the top of the bank, where the humus and roots help strengthen the burrow. In banks of stratified materials, there may be several distinct rows of burrows at different levels of the bank in strata of preferred texture; and, in banks that are uniformly composed of more clay materials, burrows may be dug in irregular rows throughout. Nests may be less than a foot apart. Occasionally, a single pair will nest in a bank, but most often nesting is in colonies that number in the tens to hundreds of pairs.

Bank Swallows are generally rather silent, except when disturbed or around their nesting colony. Their buzzy call notes have been described as *speedz-sweet, speedz-sweet* or *speed-zeet, speed-zeet*, and their song as a twitter. Agitated birds give a shrill *te-a-rr* alarm call.

Both sexes participate in digging the burrow, building the nest, incubating the eggs, and feeding the young. The burrow is usually 3 or 4 feet long but may vary from 16 inches to 9 feet in length. A small cavity at the end contains a nest of grass, feathers, and similar materials. The first egg is laid before the nest cup has been built, but, by the time the eggs hatch, the nest is complete and ready for the nestlings. The clutch may vary from four to eight eggs but four or five are typical. Clutch sizes for 2 Massachusetts nests were six eggs each (DKW). The eggs are pure white and are incubated for 14 to 16 days before they hatch. Eggs have been found in Massachusetts nests from at least May 21 to July 3, but the egg-laying period probably ranges from late April to mid-July. At a colony in Spencer, adults were observed in two different years on May 8 carrying grass and twigs into their nest burrows, indicating that they had begun to lay eggs. Adults at some other burrows appeared to be incubating already at this time. The presence of

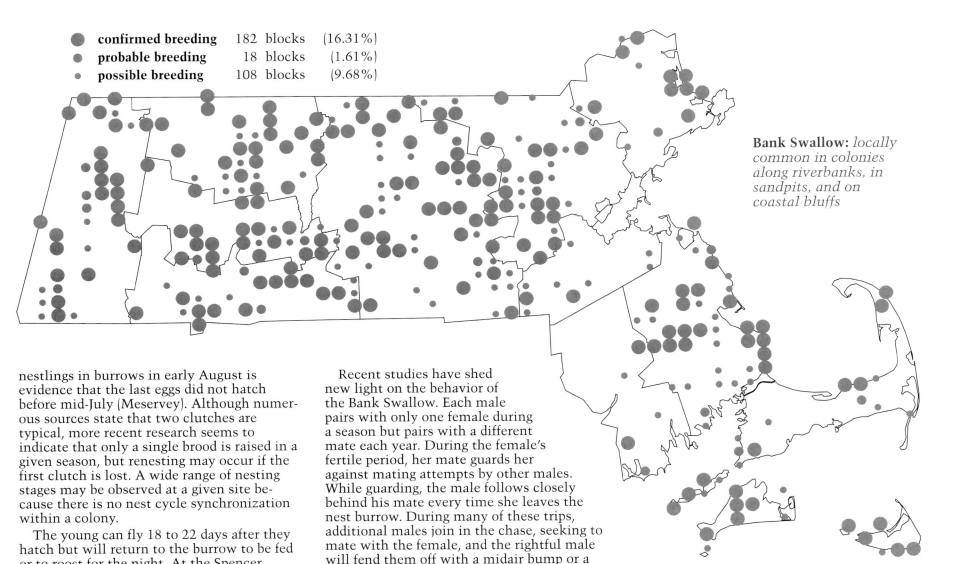

Bank Swallow: *locally common in colonies along riverbanks, in sandpits, and on coastal bluffs*

● confirmed breeding	182 blocks	(16.31%)
● probable breeding	18 blocks	(1.61%)
· possible breeding	108 blocks	(9.68%)

nestlings in burrows in early August is evidence that the last eggs did not hatch before mid-July (Meservey). Although numerous sources state that two clutches are typical, more recent research seems to indicate that only a single brood is raised in a given season, but renesting may occur if the first clutch is lost. A wide range of nesting stages may be observed at a given site because there is no nest cycle synchronization within a colony.

The young can fly 18 to 22 days after they hatch but will return to the burrow to be fed or to roost for the night. At the Spencer colony, adults were observed feeding nestlings from May 26 to August 5, with the largest numbers from June 17 to July 21. Newly fledged young were observed from June 16 to August 7 (Meservey). Both young and adults molt into a similar winter plumage after they arrive on their wintering grounds.

Recent studies have shed new light on the behavior of the Bank Swallow. Each male pairs with only one female during a season but pairs with a different mate each year. During the female's fertile period, her mate guards her against mating attempts by other males. While guarding, the male follows closely behind his mate every time she leaves the nest burrow. During many of these trips, additional males join in the chase, seeking to mate with the female, and the rightful male will fend them off with a midair bump or a face-to-face fight. If the competition is too stiff, he may force the female to turn back to the burrow.

At the end of the breeding season, Bank Swallows begin to flock together, often mixing with Tree, Barn, and Cliff swallows. Overnight roosts form just after sunset and are often in trees along riverbanks. Fall migration begins in mid-August, and the last transients are usually seen departing Massachusetts by early September. Bank Swallows winter in Brazil, northern Argentina, and central Chile.

Thomas W. French

Cliff Swallow
Hirundo pyrrhonota

Egg dates: May 22 to August 2.

Number of broods: one or two.

No historical data on the Cliff Swallow is available, but it was probably represented in Massachusetts originally in small numbers about steep cliff faces. In the early nineteenth century, the species expanded its range throughout New England as it shifted from nesting at cliff sites to human-built structures. A slow decline, probably linked to the introduction of the House Sparrow, began after 1880. In this century, the decline has continued with the disappearance of buildings that have overhanging eaves preferred by the birds; the gradual loss of open agricultural lands; and, in some areas, general suburbanization with its accompanying habitat loss, wetland filling, and pesticide use. In addition to these human-associated factors, the Cliff Swallow seems to be naturally vulnerable to prolonged periods of cold, rainy weather, when many are likely to starve to death.

Perhaps, due in part to these factors, the Cliff Swallow, like the Purple Martin, has a puzzling, uneven distribution within its range and may be absent from ostensibly suitable areas while abundant at others. During the Atlas period, nesting was confirmed in nine of the state's fourteen counties, with an overall sparse nesting distribution from Berkshire County generally eastward, sparingly into western and northern Worcester County, northern Middlesex County, and coastal Essex County. Except for isolated confirmations in two Plymouth County blocks, nesting was notably unrecorded in southeastern Massachusetts, owing, it is thought, to the sandy soil that is unsuitable for nest construction.

Although the Cliff Swallow is famed for its allegedly punctual arrival each March 10 at the mission of San Juan Capistrano in California, such regularity is seldom noted for this species in New England. Early arrivals are sometimes observed in late April, but general arrival at established colonies is usually in the first half of May. Late arrivals in June may be birds that have failed elsewhere and relocated, or birds whose migration was more protracted or began at a later date. Colonies, even large ones, can often exist undetected as birds remain close to these sites. Birds away from the colony are not particularly vocal but about their nests utter a constant social churring. The song is a series of unmusical creaking notes. When an intruder approaches the nests, the birds wheel about and protest with a loud, high *keer* alarm call.

The Cliff Swallow is highly colonial. Compared to western assemblages that sometimes contain thousands of birds, colonies in Massachusetts are small, ranging in size from 2 to 200 pairs, with few exceeding 100 pairs. The uniquely shaped nest is fastened to a vertical surface under wide eaves, such as on the sides of chicken coops, barns, or churches. The habit of building under bridges, common in the West, does not seem to have caught on widely in New England. More rarely, especially when there are wide doors, nests may be built inside a barn in close association with Barn Swallows.

Nest construction is usually underway in the third week of May. At several colonies in southern Worcester County, nests were under construction from May 14 to mid-July, with a peak from late May to mid-June. Some of the later nests were for second broods, but many represented renesting attempts. In cases of severe interference from humans and House Sparrows, the birds may relocate and build repeatedly. Both sexes participate in nest building, carrying mouthfuls of clay or mud to the nest site. Nest construction is accomplished in 5 to 14 days, with the speed influenced by weather conditions, mud supply, and amount of disturbance. House Sparrows, a menace at some colonies, attempt to appropriate nests for their own use, frequently breaking them in the process and resulting in disbandment of the colony. Nests resemble a gourd made of mud or clay pellets and are sometimes lined with fine grass or feathers. A necklike portion may protrude 5 to 6 inches or be absent altogether. Occa-

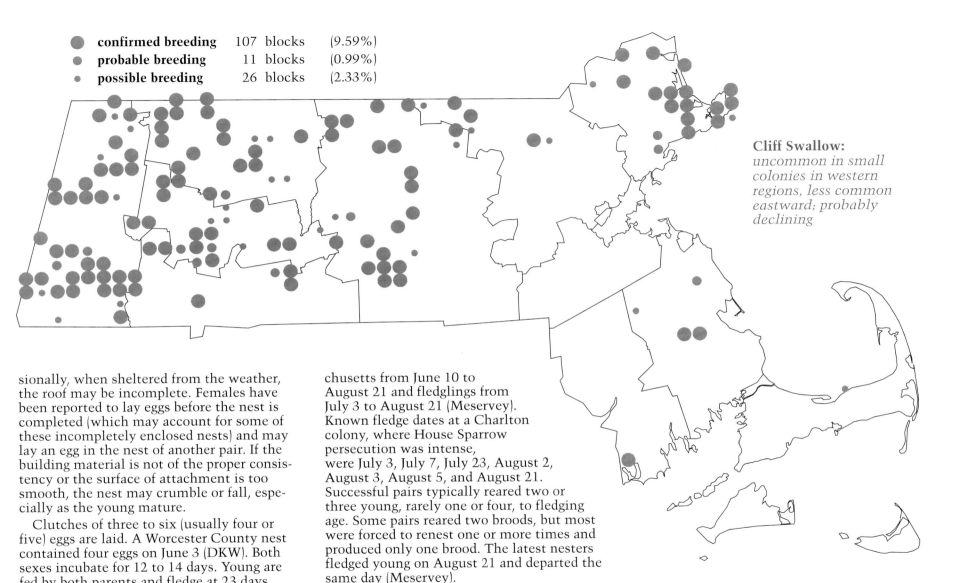

Cliff Swallow: *uncommon in small colonies in western regions, less common eastward; probably declining*

- ● confirmed breeding — 107 blocks (9.59%)
- ● probable breeding — 11 blocks (0.99%)
- • possible breeding — 26 blocks (2.33%)

sionally, when sheltered from the weather, the roof may be incomplete. Females have been reported to lay eggs before the nest is completed (which may account for some of these incompletely enclosed nests) and may lay an egg in the nest of another pair. If the building material is not of the proper consistency or the surface of attachment is too smooth, the nest may crumble or fall, especially as the young mature.

Clutches of three to six (usually four or five) eggs are laid. A Worcester County nest contained four eggs on June 3 (DKW). Both sexes incubate for 12 to 14 days. Young are fed by both parents and fledge at 23 days posthatching. They are fed for an undetermined period of time after fledging but presumably not more than a few weeks. Adults and young typically return to the nest to roost for a week or more. Once they have outgrown the nest, they will sit nearby or on it. Nestlings have been recorded in Massachusetts from June 10 to August 21 and fledglings from July 3 to August 21 (Meservey). Known fledge dates at a Charlton colony, where House Sparrow persecution was intense, were July 3, July 7, July 23, August 2, August 3, August 5, and August 21. Successful pairs typically reared two or three young, rarely one or four, to fledging age. Some pairs reared two broods, but most were forced to renest one or more times and produced only one brood. The latest nesters fledged young on August 21 and departed the same day (Meservey).

By the third week of August, southward migration is underway, and birds are seldom recorded after the first week of September. Cliff Swallows are diurnal migrants, moving south in small loose flocks but congregating in large numbers with other swallows at favored overnight roosting sites. Their migration path is remarkably long and eventually takes them to southeastern Brazil and central Argentina where they will spend the winter.

Hamilton Coolidge

Barn Swallow
Hirundo rustica

Egg dates: May 17 to July 24.

Number of broods: one or two.

The Barn Swallow is a widespread and common breeding bird throughout Massachusetts. To many people it is as reliable a harbinger of spring as the robin. Historically, Barn Swallows nested under cliffs and rock ledges, in caves (as at Nahant), and under branches of large trees; but the birds switched to using human-built structures with the advent of European civilization. This habit quickly became nearly universal, and today Barn Swallows utilize garages, carports, abandoned buildings, barns, sheds, porches, bridges, wharves, and boathouses. Open fields, bodies of water, or marshland provide favored feeding areas over which the birds catch insects. In many parts of Massachusetts, the Barn Swallow population has started to decline as urban and suburban sprawl engulfs farmlands, thus eliminating prime feeding and nesting areas.

Adult swallows return year after year to the same nesting site. Most young swallows return to the general area where they were born, but only occasional individuals return to their natal colony. Although there are records in Massachusetts of pairs nesting singly, most Barn Swallows gather in colonies of several to 45 pairs. Bent states that 7 or 9 nests per barn are typical, and he mentions a barn in Ipswich that held 55 nests.

The first birds to return in the spring are invariably males, which arrive from mid- to late April, depending on the advent of warm weather. April 18 is the most frequent date of arrival at one colony on upper Cape Cod. These early males may retreat if the weather turns cool but quickly return. The bulk of the nesting population has arrived by mid-May, but some continue to trickle in to the end of the month.

Barn Swallows keep up an almost constant chorus of chattering and twittering at their colonies from April through August. Songs are a combination of a series of jumbled, rapid, twittering *szee-szah* notes with harsh, rattling burps or trills interspersed or at the end. Calls to young and other members of the colony consist of soft *yit-yit* or *kyit-kyit* sounds. Colony defense consists of birds diving at intruders while uttering shrill *ee-tee* or *keet* alarm calls. Some well-established colonies will not tolerate any other type of bird, even Cliff Swallows, near the nests, but at other times Barn and Cliff swallows will share a site.

Courting males sing and pursue the paler colored females. Pairs may also perch close together and preen one another. Barn Swallows prefer to repair and reuse old nests rather than build new ones. When new construction is necessary, both sexes will work on fashioning the nest from layers of mud and grass with a lining of clean, white feathers. The birds generally choose beams, ledges, platforms, light fixtures, and even nails to provide some support for the nest. Overhead protection with 1 to 3 inches above the top of the nests is preferred. In places where nests are built with a lot of overhead space, additions in subsequent years will result in unusually high nests being built. Weather conditions and availability of building materials influence the rate of nest construction, which may take 12 days or more. As much as two weeks may pass between completion of the nest and the laying of eggs. The following breakdown for locations of 89 state nests will give some idea of the diversity of sites: shed (6 nests), barn (39 nests), garage (4 nests), cabin (1 nest), factory (4 nests), warehouse (10 nests), unspecified building (8 nests), porch/light fixture (5 nests), bridge (12 nests) (CNR, Meservey). Heights for 32 of these nests ranged from 3 to 25 feet, with an average of 10 feet (CNR).

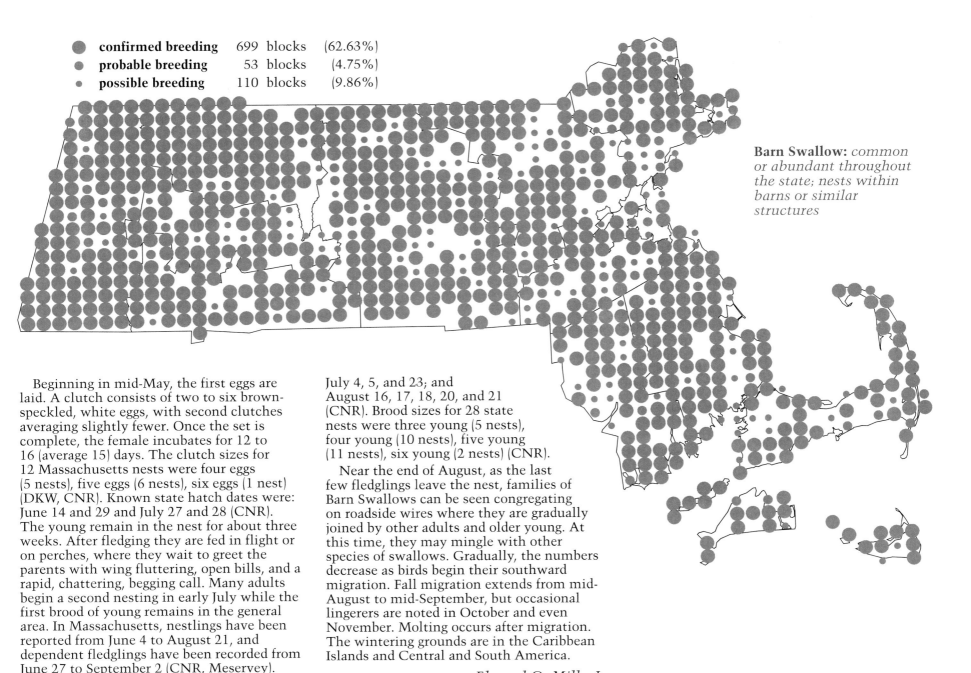

- confirmed breeding 699 blocks (62.63%)
- probable breeding 53 blocks (4.75%)
- possible breeding 110 blocks (9.86%)

Barn Swallow: *common or abundant throughout the state; nests within barns or similar structures*

Beginning in mid-May, the first eggs are laid. A clutch consists of two to six brown-speckled, white eggs, with second clutches averaging slightly fewer. Once the set is complete, the female incubates for 12 to 16 (average 15) days. The clutch sizes for 12 Massachusetts nests were four eggs (5 nests), five eggs (6 nests), six eggs (1 nest) (DKW, CNR). Known state hatch dates were: June 14 and 29 and July 27 and 28 (CNR). The young remain in the nest for about three weeks. After fledging they are fed in flight or on perches, where they wait to greet the parents with wing fluttering, open bills, and a rapid, chattering, begging call. Many adults begin a second nesting in early July while the first brood of young remains in the general area. In Massachusetts, nestlings have been reported from June 4 to August 21, and dependent fledglings have been recorded from June 27 to September 2 (CNR, Meservey). Known state fledging dates were June 27; July 4, 5, and 23; and August 16, 17, 18, 20, and 21 (CNR). Brood sizes for 28 state nests were three young (5 nests), four young (10 nests), five young (11 nests), six young (2 nests) (CNR).

Near the end of August, as the last few fledglings leave the nest, families of Barn Swallows can be seen congregating on roadside wires where they are gradually joined by other adults and older young. At this time, they may mingle with other species of swallows. Gradually, the numbers decrease as birds begin their southward migration. Fall migration extends from mid-August to mid-September, but occasional lingerers are noted in October and even November. Molting occurs after migration. The wintering grounds are in the Caribbean Islands and Central and South America.

Elwood O. Mills, Jr.

Black-capped Chickadee
Parus atricapillus

Egg dates: April 29 to July 12.

Number of broods: one; may re-lay if first attempt fails.

The Black-capped Chickadee—the Massachusetts state bird—is a permanent resident of the Commonwealth, though on occasion not all individuals remain throughout the year. It was confirmed as a breeder in all sections of the state, inhabiting woodlands, orchards, shade trees, yards, and city parks. In short, the adaptable chickadee will utilize any area with sufficient vegetation for cover and feeding. During the nesting season, it must find woodlands with suitable nest sites, but a birdhouse of the right dimensions will also accommodate it.

Adult Black-capped Chickadees are basically sedentary, but young birds sometimes wander, and, in some years, there may be definite migrations during which some of our breeders depart and the birds from more northerly regions pass through. The numbers involved in these movements vary greatly from year to year. Curiously, after such irruptions, few are noted returning during the following spring, and it is not known with certainty what percentage of these birds return north.

The most familiar vocalizations of this species are *chick-a-dee-dee-dee* calls of both short and long duration. Another well-known call is a sweet, whistled *fee-bee*. Though this sound is usually associated with the bonding of pairs, it may be heard at other times as well. Chickadees also produce several short, lisping notes and a rapid chatter when a predator is spotted. Foraging individuals often give a long series of *sizzle-ee, sizzle-oo* notes.

Spring courting is marked by a burst of agitated activity beginning in late March or April and extending into May. During this season, males actively pursue females. Once a pair bond has been formed, both sexes take part in carving out a cavity in a rotten stump, especially Gray Birch or White Pine. In Bristol County, 75 percent of the nests located were in dead branches or stubs of Gray Birch from 4 to 8 feet above the ground (ACB). Forbush also mentions nests in holes in fence posts, the tops of broken stumps, nest boxes, and cavities in elms as high as 50 feet above the ground. In another sample of 26 Massachusetts nests, 22 were in nest boxes in suburban and rural situations, 2 were in dead Gray Birches, 1 in a dead White Pine, and 1 in a dead snag. The heights ranged from 1 to 10 feet and averaged 5.6 feet, although in nest boxes nest height was determined by the positioning of the boxes (CNR).

The inner material of a stump must be soft and spongy in order to be chipped away and removed with the tiny beak of a chickadee. A week to 10 days may be required to complete the excavating task. When the cavity is ready, the female constructs the actual nest inside, gathering the fibers of Cinnamon Fern, moss, cottony vegetation, hair, insect cocoons, animal fur, and feathers. This work usually takes another 3 or 4 days to complete. The finished structure measures 2 or 3 inches in diameter and 1 inch deep.

Chickadees are generally single brooded, and late nesting records probably represent renesting efforts after an initial failure.

The clutch size averages six to eight eggs, although as many as thirteen eggs have been discovered, perhaps the result of 2 females laying in the same nest. For 19 Massachusetts nests, the clutch sizes ranged from two to nine eggs (CNR, DKW). The eggs are dull white, spotted with tiny reddish brown dots. When laying is completed, the female incubates for about 12 days, with the male feeding her during this time.

The nestlings are able to beg and open their beaks when they are 3 days old. At this point, they are covered with a sparse brownish gray down. Both parents feed the young, gathering small caterpillars, other insects, and insect eggs. The young usually fledge in 16 days. Nestling dates for Massachusetts ranged from May 18 to July 24, and fledging dates ranged from June 5 to July 24 (CNR). Family groups with dependent young were observed from early June to August 10 (Meservey).

- confirmed breeding 571 blocks (51.16%)
- probable breeding 224 blocks (20.07%)
- possible breeding 112 blocks (10.04%)

Black-capped Chickadee: *common and widespread throughout the state*

By the time they fledge, young chickadees resemble the adults, though their plumage is shaggier and appears cleaner. At the end of summer, they undergo a prebasic molt of the contour plumage and wing coverts, at which time they are nearly indistinguishable from adults. Adults undergo a complete molt during July and August, producing a richly colored plumage that is worn all winter. Subsequent wear and fading produces the paler plumage of spring.

The immatures remain in the company of their parents until late summer, moving only a mile or two from the nesting site during the first few months. By fall, the young begin to wander, and the winter flocks no longer consist of family groups. During the post-breeding period, chickadees are joined by a variety of other species, including titmice, woodpeckers, nuthatches, kinglets, brown creepers, and warblers. These mixed-species flocks are very characteristic of Massachusetts woodlands.

During the nonbreeding season, chickadee flocks consist of several pairs plus a variable number of floaters. Each flock maintains a hierarchy, with one pair ranked first, another second, and so on down the line through the floaters. Only the top two pairs get the chance to occupy the group's range for breeding. If a dominant male or female dies, the highest ranked floater of the right sex takes its place and pairs with the remaining partner. Floaters never pair with floaters; instead, they move from one flock to another in an attempt to raise their social standing. The lowest ranked floater moves up the social scale only if death or replacement removes all higher ranked individuals and a suitable opening occurs (Smith 1991).

Helen C. Bates

Tufted Titmouse
Parus bicolor

Egg dates: May 14 to July 15.

Number of broods: one; may re-lay if first attempt fails.

Like a number of other primarily southern species, the Tufted Titmouse has undergone a substantial northward range expansion in the last 50 years. Not even listed as an official state bird by Griscom and Snyder (1955), the species made an abrupt appearance in the state in 1959. In that year, the Springfield Christmas Count found 4 titmice followed by 38 in 1960 and 67 in 1961. Entry into the state apparently occurred mainly via the Connecticut River valley. From 1963 onward, the Massachusetts Audubon Society documented the spectacular range expansion into Massachusetts by a statewide census conducted each February. By 1970, the titmouse population was well established in most of its present state range.

During the Atlas period, the Tufted Titmouse was confirmed as a nester over most of the state but was encountered much less frequently in the central uplands of northern Worcester County and on the Berkshire escarpment westward through the Berkshires to the New York State line. It was also less frequent on Cape Cod and absent entirely from Martha's Vineyard and Nantucket.

The titmouse's loud melodic whistles of four to eight phrases, often rendered *peter-peter-peter*, *peto-peto-peto*, or *here-here*, may be heard in any month of the year. Both sexes sing. They also have a varied repertoire of notes, including a harsh *day-day-day* and some resembling the wheezy calls of the Black-capped Chickadee. Pairs may form in

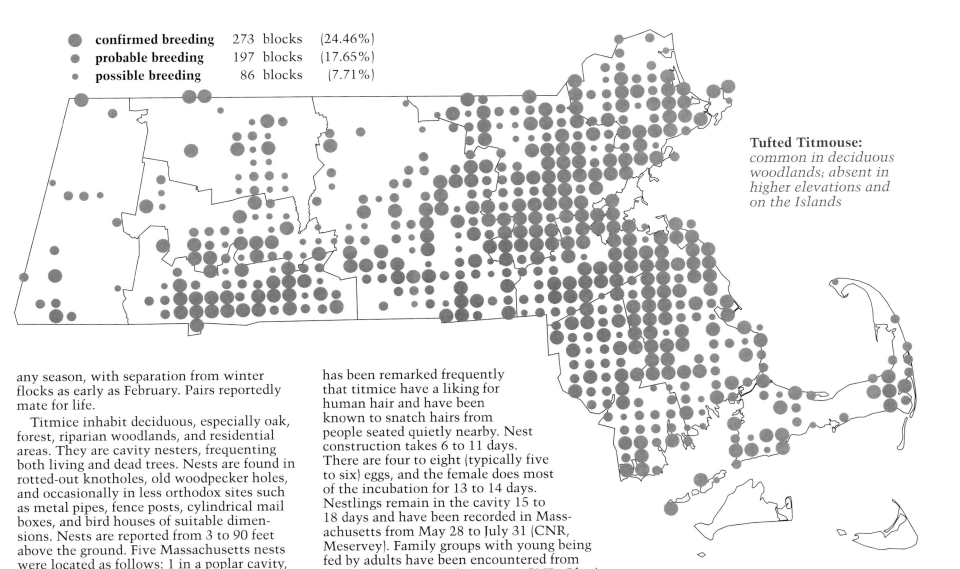

Tufted Titmouse: *common in deciduous woodlands; absent in higher elevations and on the Islands*

- confirmed breeding — 273 blocks (24.46%)
- probable breeding — 197 blocks (17.65%)
- possible breeding — 86 blocks (7.71%)

any season, with separation from winter flocks as early as February. Pairs reportedly mate for life.

Titmice inhabit deciduous, especially oak, forest, riparian woodlands, and residential areas. They are cavity nesters, frequenting both living and dead trees. Nests are found in rotted-out knotholes, old woodpecker holes, and occasionally in less orthodox sites such as metal pipes, fence posts, cylindrical mail boxes, and bird houses of suitable dimensions. Nests are reported from 3 to 90 feet above the ground. Five Massachusetts nests were located as follows: 1 in a poplar cavity, 1 in a maple cavity, 1 in a Gray Birch stump, 1 in a dead tree, and 1 in a screech-owl box. Heights ranged from 3 to 40 feet (CNR).

Nest building begins in late April, when birds may be seen carrying nesting material. Old leaves, strips of bark, bits of moss, and grass are all used to fill the cavity, with the nest bowl itself typically lined with hairs. It has been remarked frequently that titmice have a liking for human hair and have been known to snatch hairs from people seated quietly nearby. Nest construction takes 6 to 11 days. There are four to eight (typically five to six) eggs, and the female does most of the incubation for 13 to 14 days. Nestlings remain in the cavity 15 to 18 days and have been recorded in Massachusetts from May 28 to July 31 (CNR, Meservey). Family groups with young being fed by adults have been encountered from June 24 to August 14 (Meservey, CNR, Ober).

Immatures may accompany the adults for several months. Family groups roam through the woods, sometimes joining large multi-species flocks. Although the titmouse is generally nonmigratory, movements are known to occur, and rather large, loose flocks sometimes develop in the fall and wander extensively. Banded individuals over seven years of age have been reported.

Rudolph H. Stone

Red-breasted Nuthatch
Sitta canadensis

Egg dates: April 20 to June 20.
Number of broods: one.

Until the midtwentieth century, this species was considered a denizen of northern boreal forests and a regular breeder only in Berkshire County and only a casual to rare and local nester east to Plymouth and Essex counties. Historically, most observers east of Worcester County recorded it primarily as a migrant. The Red-breasted Nuthatch is also well known for its periodic and sometimes large winter irruptions. Interestingly, the Atlas project revealed it to breed fairly evenly throughout much of the state, in limited numbers even to the tip of Cape Cod at Provincetown, as well as at Martha's Vineyard and Nantucket. Following an irruption winter, it appears that some individuals may augment the local population by remaining in Massachusetts to breed.

Spring migration is generally lighter than that in fall, perhaps partly due to winter mortality. Movement begins in late March and continues into mid-May, by which time territories have usually been established. Late migrants are presumably birds that breed well to the north. The courtship of the Red-breasted Nuthatch is seldom noted, but the few references are consistent. The song consists of a rapid repetition of the short tin trumpet *wa-wa-wa* calls, and a series may be repeated as many as 16 times per minute. Sometimes, the wings and tail are held slightly open. Frequently, this activity is performed in a spruce or other conifer and is difficult to observe. Members of a pair or group communicate with various *hit, it,* or *yna* notes, given singly or in a series.

The nest cavity is excavated in soft, rotten snags or stumps from 5 to 70 feet above the ground. The birds generally construct their own chambers but may use old woodpecker holes or even nest boxes. Seven nests in Massachusetts were located as follows: 1 in a bird box at 7 feet (ACB), 1 in a dead poplar stub at 12 feet (Bagg & Elliot 1937), 1 in a dead Paper Birch, 1 in a dead Gray Birch (Brewster 1906), 2 in rotting White Pine stumps at 12 feet and 15 feet, respectively (Blodget), and 1 in a rotten branch of a live White Pine at 25 feet (Meservey). Pitch from nearby conifers typically is smeared around the hole, presumably as a protective measure, and the birds have been observed to fly straight into the hole without landing

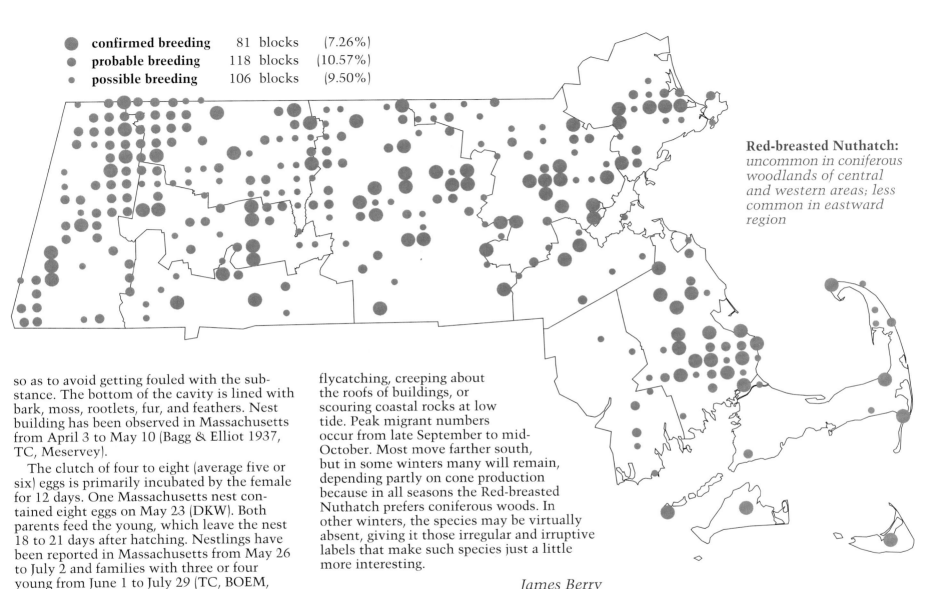

●	**confirmed breeding**	81 blocks	(7.26%)
●	**probable breeding**	118 blocks	(10.57%)
•	**possible breeding**	106 blocks	(9.50%)

Red-breasted Nuthatch: *uncommon in coniferous woodlands of central and western areas; less common in eastward region*

so as to avoid getting fouled with the substance. The bottom of the cavity is lined with bark, moss, rootlets, fur, and feathers. Nest building has been observed in Massachusetts from April 3 to May 10 (Bagg & Elliot 1937, TC, Meservey).

The clutch of four to eight (average five or six) eggs is primarily incubated by the female for 12 days. One Massachusetts nest contained eight eggs on May 23 (DKW). Both parents feed the young, which leave the nest 18 to 21 days after hatching. Nestlings have been reported in Massachusetts from May 26 to July 2 and families with three or four young from June 1 to July 29 (TC, BOEM, Brewster 1906, Meservey).

Birds often appear well away from breeding areas in late July, after the molt period. By August, when these nuthatches begin to move in earnest and numbers are swelled with the young of the year, interesting things can happen, and individuals can be seen flycatching, creeping about the roofs of buildings, or scouring coastal rocks at low tide. Peak migrant numbers occur from late September to mid-October. Most move farther south, but in some winters many will remain, depending partly on cone production because in all seasons the Red-breasted Nuthatch prefers coniferous woods. In other winters, the species may be virtually absent, giving it those irregular and irruptive labels that make such species just a little more interesting.

James Berry

White-breasted Nuthatch
Sitta carolinensis

Egg dates: April 3 to May 25.

Number of broods: one; may re-lay if first attempt fails.

One of our commoner woodland birds, the White-breasted Nuthatch is notable for its jerky, zigzag movements as it hitches itself up and down tree trunks in search of insects, regularly descending a tree headfirst. It is a species of mixed woodland, village trees, and orchards and is basically a permanent resident, breeding throughout the state, more sparingly on Cape Cod but absent from Nantucket.

White-breasted Nuthatches have a rather nasal *ank-ank-ank* call note. The number of notes in a sequence is an indication of the individual's state of excitement, with a double note and finally a rapid series of notes given during increasing agitation. Such excitement is often due to territorial conflict and may be accompanied by fanning of the tail and ruffling of the back feathers. As the pair, or the family, forages through the territory, the birds often keep contact with a soft *ip* or *it* note. In addition, the male has a simple song of six to eight *wee-wee-wee* or *wer-wer-wer-wer* notes.

Nuthatch pairs maintain a relationship throughout the year, inhabiting a territory of 25 to 50 acres. Pair bonds last for several seasons, perhaps for life. Courtship may begin during the winter or in early spring. The male repeatedly sings the *wer-wer-wer* song, bows to the female, and brings her items of food.

The nest is usually between 10 and 50 feet from the ground in a natural tree cavity, old woodpecker hole, or birdhouse. Five Massachusetts nests were discovered in nest boxes in suburban woods and 1 in an elm cavity 18 feet from the ground (CNR). The birds may use bluebird boxes. The cavity opening must be at least 1.5 inches in diameter, but they prefer one that is somewhat larger. Both sexes gather bark, grasses, rootlets, and fur for the nest. One Massachusetts nest was constructed of loose bark and a few small stones and was lined with soft feathers and fur (CNR). Near the opening the birds engage in a peculiar behavior called "bill-sweeping" (Stokes 1983), often with an insect held in the beak. The purpose of this is not clear.

Five to ten (average eight) eggs are laid. Each is short, oval in shape, and white with the larger end capped with light brown to lavender spots. Clutch sizes for 6 Massachu-

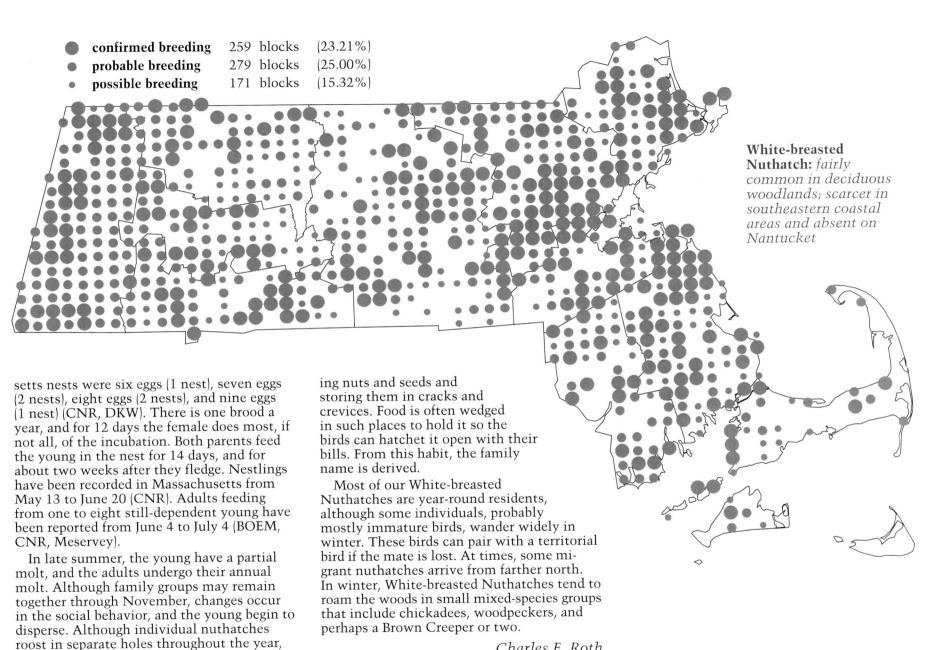

- confirmed breeding 259 blocks (23.21%)
- probable breeding 279 blocks (25.00%)
- possible breeding 171 blocks (15.32%)

White-breasted Nuthatch: *fairly common in deciduous woodlands; scarcer in southeastern coastal areas and absent on Nantucket*

setts nests were six eggs (1 nest), seven eggs (2 nests), eight eggs (2 nests), and nine eggs (1 nest) (CNR, DKW). There is one brood a year, and for 12 days the female does most, if not all, of the incubation. Both parents feed the young in the nest for 14 days, and for about two weeks after they fledge. Nestlings have been recorded in Massachusetts from May 13 to June 20 (CNR). Adults feeding from one to eight still-dependent young have been reported from June 4 to July 4 (BOEM, CNR, Meservey).

In late summer, the young have a partial molt, and the adults undergo their annual molt. Although family groups may remain together through November, changes occur in the social behavior, and the young begin to disperse. Although individual nuthatches roost in separate holes throughout the year, in autumn the members of a pair become increasingly independent in their foraging as well. They spend more of their time gathering nuts and seeds and storing them in cracks and crevices. Food is often wedged in such places to hold it so the birds can hatchet it open with their bills. From this habit, the family name is derived.

Most of our White-breasted Nuthatches are year-round residents, although some individuals, probably mostly immature birds, wander widely in winter. These birds can pair with a territorial bird if the mate is lost. At times, some migrant nuthatches arrive from farther north. In winter, White-breasted Nuthatches tend to roam the woods in small mixed-species groups that include chickadees, woodpeckers, and perhaps a Brown Creeper or two.

Charles E. Roth

Brown Creeper
Certhia americana

Egg dates: April 24 to July 5.

Number of broods: one; may re-lay if first attempt fails.

The sole representative of its family in North America, the Brown Creeper is a small, inconspicuous bird, whose nesting activities are easily overlooked. Were it not for the song—a short, wiry, musical jumble of notes—its presence might go completely undetected. Fortunately, the male's song period lasts from March through early July. Outside of the nesting season, it is equally obscure and gives a rather weak, sibilant *seeet*, similar to the call of the Golden-crowned Kinglet but longer, monosyllabic, and unaccented.

Prior to 1955, the Brown Creeper was largely restricted as a nester to the hill country of western and north-central Massachusetts; it occurs sporadically and is sometimes common locally eastward to the coast. Since 1960, the Brown Creeper seems to have increased markedly and expanded its regular nesting range. Nesting is now widespread and common throughout most of the state except in the northeast and southeast, where it remains only locally common, and in Nantucket County, where, lacking adequate forest, it is absent entirely as a nester.

The general maturing of forests throughout the state has benefited the Brown Creeper, which favors mature, mixed, swampy forest, including Atlantic White Cedar swamps, especially when there is an interspersion of dead trees with loose hanging bark. Following the return of the Beaver in the 1950s, a virtual population boom had occurred by the mid-1960s as the species opportunistically exploited the growing system of Beaver flowages and swamps and their attendant dead timber. Flourishing Brown Creeper populations exist in forests interlaced with Beaver impoundments, and these high populations have quite likely fueled some of the eastward range expansion evident in the last 20 years.

Creepers defend nesting territories of unknown size. Courting males sing and perform rapid, twisting flights. Nest building is at its peak in late April and early May but has been observed here as early as April 12 (BOEM). Nests are crescent shaped and built of mosses, twigs, dead leaves, and shredded bark, and lined with feathers and hair. They are almost invariably built 5 to 15 feet above the ground in standing or fallen dead trees beneath strips of bark that have become partially detached from the trunk and are curved inward. These nests, like the birds themselves, are extremely cryptic. Rotted-out cavities or old woodpecker holes are also reportedly used, although much less frequently. The female does most of the actual nest construction, although the male does fetch materials. While creepers are usually associated with mature, swampy forest, they will adapt to other sites, as evidenced by a nest in South Orleans 10 feet high in a Pitch Pine (CNR) and a nest in Charlton that was located 5 feet high behind the bark of a dead tree situated inside an old foundation in the woods and close to a driveway and several other buildings (Meservey).

Clutches of four to eight (average five to six) eggs are laid. Incubation by the female commences upon completion of the clutch

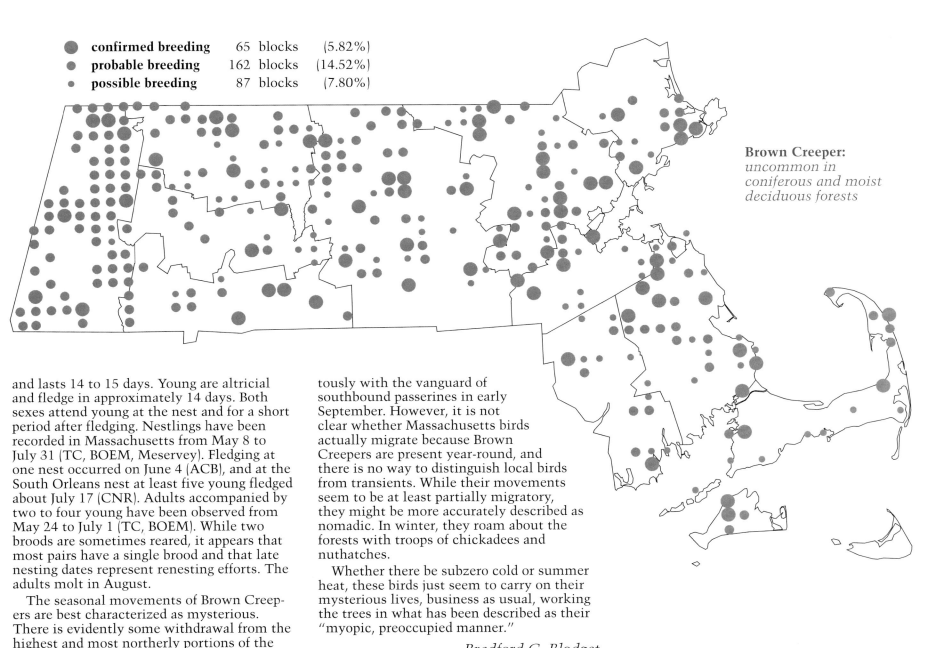

- **confirmed breeding**   65 blocks   (5.82%)
- **probable breeding**   162 blocks   (14.52%)
- **possible breeding**   87 blocks   (7.80%)

Brown Creeper: *uncommon in coniferous and moist deciduous forests*

and lasts 14 to 15 days. Young are altricial and fledge in approximately 14 days. Both sexes attend young at the nest and for a short period after fledging. Nestlings have been recorded in Massachusetts from May 8 to July 31 (TC, BOEM, Meservey). Fledging at one nest occurred on June 4 (ACB), and at the South Orleans nest at least five young fledged about July 17 (CNR). Adults accompanied by two to four young have been observed from May 24 to July 1 (TC, BOEM). While two broods are sometimes reared, it appears that most pairs have a single brood and that late nesting dates represent renesting efforts. The adults molt in August.

The seasonal movements of Brown Creepers are best characterized as mysterious. There is evidently some withdrawal from the highest and most northerly portions of the range in autumn, and birds show up ubiquitously with the vanguard of southbound passerines in early September. However, it is not clear whether Massachusetts birds actually migrate because Brown Creepers are present year-round, and there is no way to distinguish local birds from transients. While their movements seem to be at least partially migratory, they might be more accurately described as nomadic. In winter, they roam about the forests with troops of chickadees and nuthatches.

Whether there be subzero cold or summer heat, these birds just seem to carry on their mysterious lives, business as usual, working the trees in what has been described as their "myopic, preoccupied manner."

Bradford G. Blodget

Carolina Wren
Thryothorus ludovicianus

Egg dates: April 17 to June 20.

Number of broods: one or two.

Massachusetts represents the northern limit for the breeding range of this species, and in this state it is found nesting regularly only in the lower Connecticut River valley and the southeastern coastal plain from Norfolk County to Cape Cod. Curiously, it is absent on Nantucket but regular on Martha's Vineyard. Naushon Island is the original site for the first recorded nesting in Massachusetts in 1901. The Carolina Wren has been slowly increasing its numbers and expanding its range over the last forty years and has been recorded breeding as far north as Essex County, but the expansion has not been as dramatic as for many other southern species—e.g., Northern Cardinal and Northern Mockingbird. Breeding was confirmed in Worcester County after the Atlas period.

The preferred habitats for this species are wet woodland with dense underbrush and uprooted trees, brushy thickets, grapevine tangles, farmland brush piles, and bushy swamp and stream edges. Nest locations are quite varied because the species shows considerable adaptation in its site selection. Nests are normally located fairly close to the ground under upturned roots of trees, in thick brush, or in holes in trees or stumps. The species is also well known for nesting in less traditional sites such as fence post holes, mailboxes, farm buildings, empty tin cans, wooden boxes, and old clothing left lying around inside barns and garages. Some birds have shown extreme tolerance for human activity close to the nest. Forbush reported Massachusetts nests in baskets inside buildings and in a beam recess of a sawmill. More recently, a pair in Worcester was observed building a nest on the rear wheels of a parked truck, and in East Longmeadow a pair nested in a paint bucket on a shelf 5 feet up in a garage (CNR).

The Carolina Wren is one of the few birds that will sing during all 12 months of the year, and pairs remain together during the winter. The song is a loud, ringing series of phrases of two to four notes repeated several times, usually with a strong accent on the first or second note. A variety of alarm, call, and scolding notes are produced, and occasionally the bird will mimic other species.

The nests are built with grass, leaves, rootlets, bark fibers, twigs, and moss, and are

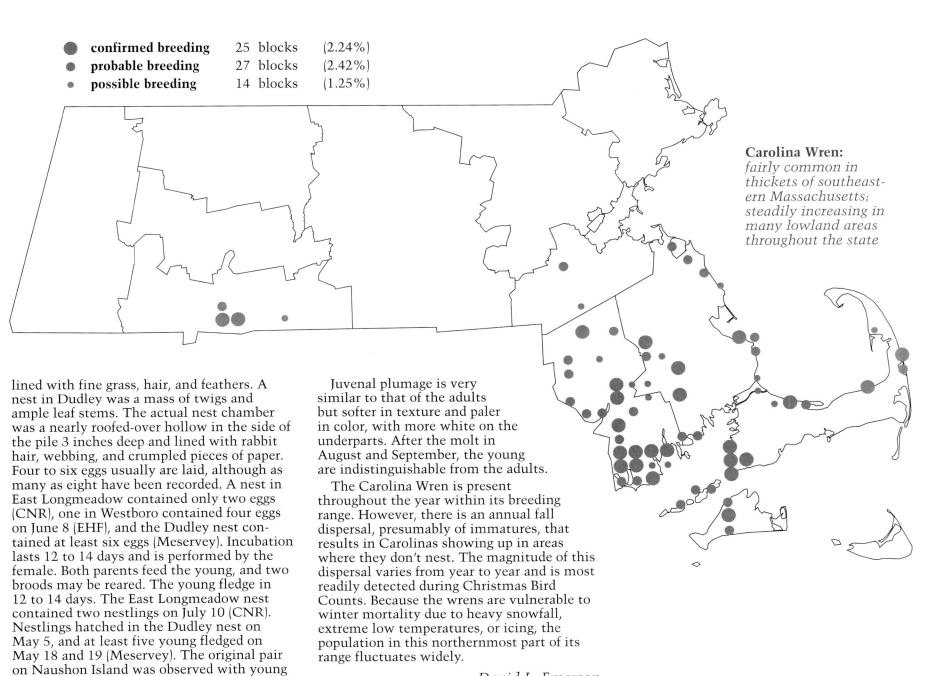

- confirmed breeding — 25 blocks (2.24%)
- probable breeding — 27 blocks (2.42%)
- possible breeding — 14 blocks (1.25%)

Carolina Wren: *fairly common in thickets of southeastern Massachusetts; steadily increasing in many lowland areas throughout the state*

lined with fine grass, hair, and feathers. A nest in Dudley was a mass of twigs and ample leaf stems. The actual nest chamber was a nearly roofed-over hollow in the side of the pile 3 inches deep and lined with rabbit hair, webbing, and crumpled pieces of paper. Four to six eggs usually are laid, although as many as eight have been recorded. A nest in East Longmeadow contained only two eggs (CNR), one in Westboro contained four eggs on June 8 (EHF), and the Dudley nest contained at least six eggs (Meservey). Incubation lasts 12 to 14 days and is performed by the female. Both parents feed the young, and two broods may be reared. The young fledge in 12 to 14 days. The East Longmeadow nest contained two nestlings on July 10 (CNR). Nestlings hatched in the Dudley nest on May 5, and at least five young fledged on May 18 and 19 (Meservey). The original pair on Naushon Island was observed with young on July 7 (EHF).

Juvenal plumage is very similar to that of the adults but softer in texture and paler in color, with more white on the underparts. After the molt in August and September, the young are indistinguishable from the adults.

The Carolina Wren is present throughout the year within its breeding range. However, there is an annual fall dispersal, presumably of immatures, that results in Carolinas showing up in areas where they don't nest. The magnitude of this dispersal varies from year to year and is most readily detected during Christmas Bird Counts. Because the wrens are vulnerable to winter mortality due to heavy snowfall, extreme low temperatures, or icing, the population in this northernmost part of its range fluctuates widely.

David L. Emerson

House Wren
Troglodytes aedon

Egg dates: May 9 to July 25.

Number of broods: one or two.

The cocked tail characteristic of this species is an apt symbol of its cheeky attitude, loud and aggressive behavior, and polygamous mating habits. Indeed, the Chippewa called it "big noise for its size." Yet the House Wren is less aggressive than the larger (and louder) Carolina Wren and the House Sparrow, which, at the turn of the century, was generally blamed for a marked decrease in the wren population due to competition for available nest sites. After 1917, the House Wren increased from being rare and local to locally common, and by the 1930s it had made a comeback. The species now breeds fairly evenly throughout Massachusetts except on Cape Cod, where its numbers are sparser, and Nantucket, where no evidence of nesting was found during the Atlas period.

House Wrens inhabit suburbs, parks, gardens, farms, orchards, wood edges, and open forests, frequenting low woody vegetation, vine tangles, and brushy areas. They appear in Massachusetts in late April and early May. The average arrival date for territorial males at a site in Ipswich from 1974 to 1985 was May 3 (range April 26 to May 10). Females generally arrive about 10 days later than the males.

A complete account of the courting and nesting behavior of this adaptable species could fill an entire book. Unlike most other birds (except some of the other wrens), male House Wrens, upon establishing their .5- to 3.5-acre territories, immediately set to work building stick nests in every suitable cavity. These may include birdhouses, natural cavities, woodpecker holes, or less orthodox sites including boxes, baskets, clothespin bags, pockets of clothing hanging on a line, old hats, blankets, pumps, radiators, flowerpots, shoes, cans, pipes, skulls, nests of other birds including kingfishers and Ospreys, and even wasp nests. One pair reportedly kept trying to build in the axle of a car in daily use, and another pair filled a two-gallon bucket with sticks. This last behavior, filling the cavity entirely, is typical, and while sticks form the usual base, pieces of metal are also sometimes used. The nest is then built on top of the pile. When a female arrives, the loud, bubbling song of the male becomes higher and squeakier; then, with quivering wings and tail cocked, he tries to lead the prospective mate to his various nests. If she likes one, she may proceed to add lining material; at other times, she will remove all of the sticks and rebuild the entire structure.

Habitats for 48 Massachusetts nests were as follows: wooded area (11 nests), field or field edge (10 nests), suburban (16 nests), old orchard (4 nests), hedgerow (3 nests), marsh edge (1 nest), farm (3 nests) (CNR). Of 51 state nests, 49 were in nest boxes, 1 was under the eaves of a shed, and 1 was in a natural cavity in a Gray Birch stump. The boxes were located from 3.5 to 10 feet in height, with an average of 5.8 feet. One box was placed at 25 feet. The nest under the eaves and the nest in the natural cavity were each at 8 feet in height (CNR).

A typical clutch of six to eight (range three to twelve) eggs is incubated by the female alone for about 13 days, starting after the last egg is laid. Clutch sizes for 36 Massachusetts nests were three eggs (1 nest), four eggs (6 nests), five eggs (6 nests), six eggs (13 nests), seven eggs (9 nests), eight eggs (1 nest) (CNR). At 3 nests in North

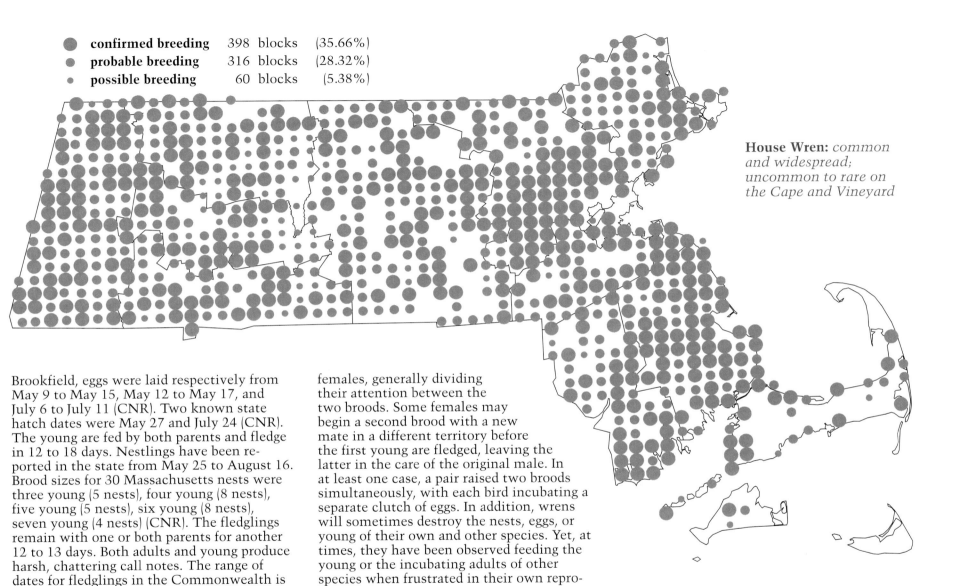

- **confirmed breeding**   398 blocks   (35.66%)
- **probable breeding**   316 blocks   (28.32%)
- **possible breeding**   60 blocks   (5.38%)

House Wren: *common and widespread; uncommon to rare on the Cape and Vineyard*

Brookfield, eggs were laid respectively from May 9 to May 15, May 12 to May 17, and July 6 to July 11 (CNR). Two known state hatch dates were May 27 and July 24 (CNR). The young are fed by both parents and fledge in 12 to 18 days. Nestlings have been reported in the state from May 25 to August 16. Brood sizes for 30 Massachusetts nests were three young (5 nests), four young (8 nests), five young (5 nests), six young (8 nests), seven young (4 nests) (CNR). The fledglings remain with one or both parents for another 12 to 13 days. Both adults and young produce harsh, chattering call notes. The range of dates for fledglings in the Commonwealth is June 7 to August 17. Known state fledge dates are June 12, June 15, June 25, July 2, July 6, August 2 (including a brood of seven from Taunton), and August 16 (CNR). Of 47 state nests, 18 resulted in fledged young, 13 failed, and 16 had an unknown outcome (CNR).

A small percentage of males mate with two females, generally dividing their attention between the two broods. Some females may begin a second brood with a new mate in a different territory before the first young are fledged, leaving the latter in the care of the original male. In at least one case, a pair raised two broods simultaneously, with each bird incubating a separate clutch of eggs. In addition, wrens will sometimes destroy the nests, eggs, or young of their own and other species. Yet, at times, they have been observed feeding the young or the incubating adults of other species when frustrated in their own reproductive attempts.

Following nesting, House Wrens move to deeper woods until departing in September or October and are thus less often encountered at this time. The adults have a complete molt and the young a partial one after the breeding season. By mid-October, most have departed New England. A few stragglers are noted into winter, but it is unlikely that many overwinter successfully. House Wrens winter in the southern United States and Mexico.

James Berry

Winter Wren
Troglodytes troglodytes

Egg dates: not available.

Number of broods: one; possibly two.

The Winter Wren is a jewel in the heart of the forest. Anyone who has seen this tiny, inquisitive bird and heard its ecstatic, sparkling song accompanied by a murmuring brook would be quick to agree. It is a regular migrant throughout Massachusetts and is best known as a breeder in Berkshire County and northern Worcester County. It is decidedly uncommon in eastern portions of the state and breeds only sparingly in Atlantic White Cedar swamps to the southeast. The Winter Wren is one of the least-known nesting birds in Massachusetts.

As winter retreats, early arrivals may be seen by the first week in April, but most migrants and breeding birds appear during the third week and continue through the second week in May. Its summer haunts are cool, moist, coniferous or mixed forests, where it frequents the edges of swamps, bogs, streams, and mountain brooks. In other seasons, it may be found in similar situations in deciduous forests. True to its scientific name, which means "caveman," it spends its time bobbing about and investigating the roots of fallen trees, decaying logs, brush piles, tangles of low undergrowth, and mossy stumps and banks. It is there that it finds the varied insects upon which it feeds and also a suitable cavity in which to build its bulky nest.

The song of the Winter Wren almost defies description, although many have been written. It is a long, ebullient series of rising and falling notes and tinkling trills sung from near the ground or the top of a tree or dead stub. All day long and sometimes into the evening, its haunting song may be heard. The whisper song is a soft rendition of the full melody, and the alarm or call notes are an abrupt *tick, chirr,* or *kip-kip.*

Upturned roots of fallen trees seem to provide cavities most favored for nest building, although other nooks and crannies may be chosen. The well-concealed nest is made of mosses, small sticks, grasses, and rootlets. It is quite large considering the size of its maker, is shaped to fit the cavity, has a small entrance on the side, and is lined with soft feathers and animal hairs. Often, as with some other species of wrens, dummy nests are built, probably by the male.

Four to seven white eggs, sparsely to heavily dotted with brownish specks, are probably laid from May through July. Three Vermont egg dates range from May 24 to June 14 (Laughlin & Kibbe 1985). The incubation period is from 14 to 16 days, and the young leave the nest about 19 days after hatching. Early and late breeding dates indicate that two broods may be reared in a

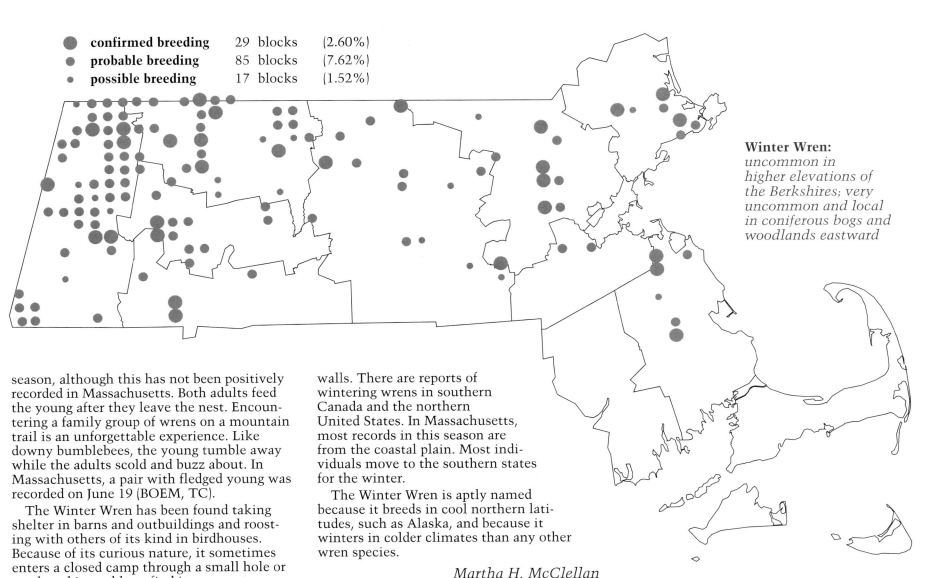

- **confirmed breeding**   29 blocks   (2.60%)
- **probable breeding**   85 blocks   (7.62%)
- **possible breeding**   17 blocks   (1.52%)

Winter Wren: *uncommon in higher elevations of the Berkshires; very uncommon and local in coniferous bogs and woodlands eastward*

season, although this has not been positively recorded in Massachusetts. Both adults feed the young after they leave the nest. Encountering a family group of wrens on a mountain trail is an unforgettable experience. Like downy bumblebees, the young tumble away while the adults scold and buzz about. In Massachusetts, a pair with fledged young was recorded on June 19 (BOEM, TC).

The Winter Wren has been found taking shelter in barns and outbuildings and roosting with others of its kind in birdhouses. Because of its curious nature, it sometimes enters a closed camp through a small hole or crack and is unable to find its way out.

By September, a few will have started to migrate, but most will move south with the first cold weather in October through the second week in November. During the fall migration, this wren may be found in more open or even unexpected habitats, such as gardens and yards, and it may explore stone walls. There are reports of wintering wrens in southern Canada and the northern United States. In Massachusetts, most records in this season are from the coastal plain. Most individuals move to the southern states for the winter.

The Winter Wren is aptly named because it breeds in cool northern latitudes, such as Alaska, and because it winters in colder climates than any other wren species.

Martha H. McClellan

Sedge Wren
Cistothorus platensis

Egg dates: May 25 to late July.

Number of broods: one or two.

Formerly an uncommon and local nester, the Sedge Wren is now one of our rarest breeding species and is not recorded every year. The erratic habits of this retiring bird defy logic. Perfectly suitable habitat will support one or more pairs one year, and then no pairs will be present the following season even though conditions appear unchanged. Most breeding records in the past quarter-century have been concentrated in the vicinity of the Connecticut River valley. At one time, the species was encountered regularly in portions of eastern Massachusetts, even in the vicinity of Boston, but much of the habitat has since been drained and developed or otherwise altered.

Migrants are likely to appear during the third week of May. However, prospecting males may appear at any time in June or early July. The preferred habitat is wet meadows, but territories will sometimes be established in cultivated fields of hay or Alfalfa. Contrary to its former name, Short-billed Marsh Wren, this species seldom resides in actual marshes, where cattails and bulrushes predominate, but may on occasion occupy the drier edges of grass or sedge meadows.

The male selects a low bush or prominent grass stalk as a perch from which to repeatedly deliver his song, which consists of several sharp introductory staccato notes followed by a descending chatter. Territorial males sing throughout the day, most intensely in the morning and evening, and also frequently at night. Between song bouts, the wren creeps mouselike through the meadows in search of food and is nearly impossible to detect.

Perhaps more than any other species, the Sedge Wren is renowned for its construction of dummy nests. Such structures, built by the male, even in the absence of a female, are globular in shape and loosely made of dried grasses. The active nest, built by the female, is a better constructed ball of dry and green grasses with an opening on the side and a lining of fine grasses, catkins, and feathers. It is well concealed and usually placed no more than 1 or 2 feet above the ground or water. Four eastern Massachusetts nests were located as follows: 2 were in freshwater marshes on the side of tall grass tussocks, 2 feet and 1 foot above the water, respectively; 1 was in a damp meadow in the base of a grass tuft a few inches above the ground; and 1 was in short green marsh grass 4 inches above the mud (ACB).

A clutch of four to eight (average six or seven) white eggs is incubated by the female for 12 to 14 days. The young remain in the nest for another 12 to 14 days, during which time they are fed by both parents, although

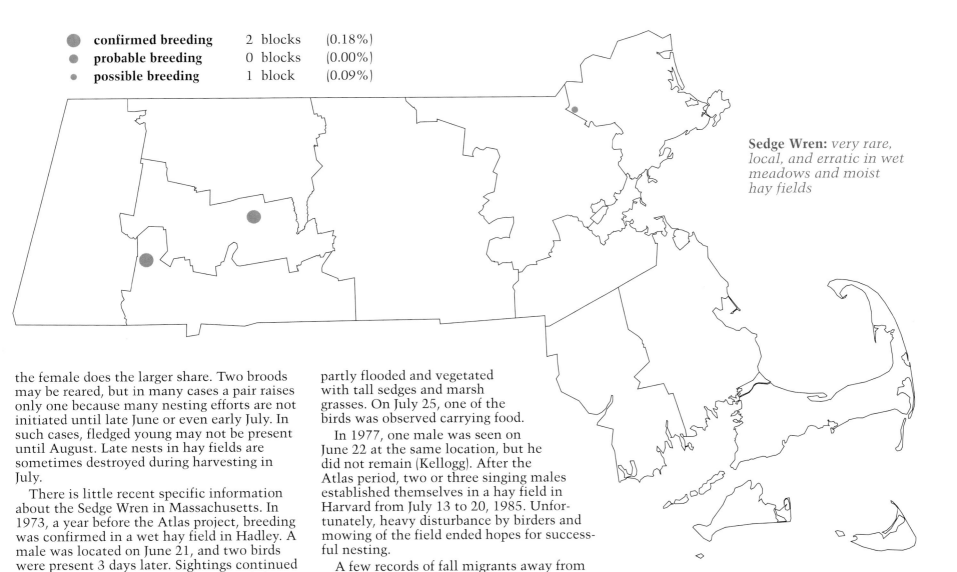

- confirmed breeding   2 blocks   (0.18%)
- probable breeding   0 blocks   (0.00%)
- possible breeding   1 block   (0.09%)

Sedge Wren: *very rare, local, and erratic in wet meadows and moist hay fields*

the female does the larger share. Two broods may be reared, but in many cases a pair raises only one because many nesting efforts are not initiated until late June or even early July. In such cases, fledged young may not be present until August. Late nests in hay fields are sometimes destroyed during harvesting in July.

There is little recent specific information about the Sedge Wren in Massachusetts. In 1973, a year before the Atlas project, breeding was confirmed in a wet hay field in Hadley. A male was located on June 21, and two birds were present 3 days later. Sightings continued until September 23, and at least one adult was seen carrying food (Gagnon). The following year, during the Atlas period, no birds returned to this field, but nesting was confirmed at a nearby location. The second Atlas confirmation occurred in 1976 in Blandford. Two pairs of wrens were recorded from July 5 to August 15 at an abandoned Beaver pond, partly flooded and vegetated with tall sedges and marsh grasses. On July 25, one of the birds was observed carrying food.

In 1977, one male was seen on June 22 at the same location, but he did not remain (Kellogg). After the Atlas period, two or three singing males established themselves in a hay field in Harvard from July 13 to 20, 1985. Unfortunately, heavy disturbance by birders and mowing of the field ended hopes for successful nesting.

A few records of fall migrants away from nesting sites indicate that Sedge Wrens are on the move from mid-September until early October. They may even sing occasionally during fall migration. The bulk of the population winters in coastal areas in the southern United States. There are two recent state winter records: one from Cape Cod and the other from Nantucket.

The Sedge Wren is listed as an endangered species in Massachusetts.

Richard A. Forster

Marsh Wren
Cistothorus palustris

Egg dates: May 27 to August 1.

Number of broods: one or two.

The Marsh Wren is a common but local summer resident of various wetland habitats. Inland populations are associated with extensive cattail marshes along major river valleys. On the coast, this species is found in the marsh grasses bordering tidal creeks and in the sedge or rush growth of brackish marshes. One of the best descriptions of the habitat affinities of this species was given by Forbush, who called the Marsh Wren a bird of the "oozy slough."

During the Atlas period, breeding was confirmed in 9 of the state's 14 counties, with the preponderance of records in eastern Massachusetts. In recent years in Worcester County, Marsh Wrens have declined in the Quaboag River marshes, a former stronghold, and are rare along the Still River in Bolton.

Bent described the vegetation for four different habitats in which Marsh Wrens nest in Massachusetts. In large freshwater marshes, the dominant plants are Broad-leaved and Narrow-leaved cattails, bulrush, sedge, Wild Rice, and Buttonbush. Along tidal riverbanks, the birds inhabit tall reeds and marsh grass, and the wrens sometimes utilize zones of pure Wild Rice along some inland rivers. In another type of marsh, particularly in Essex County, the characteristic flora is a mixture of bulrush, horsetail, Sweet Flag, bluejoint, and Reed Canary-grass.

The first males typically arrive on their breeding grounds in early May, but the bulk of the population does not appear until late in the month. The birds soon begin to establish nesting territories through various visual and vocal displays. Males may also at this time begin to build the dummy nests for which the species is noted. Some males are polygamous and must defend a large territory to accommodate the females, which will not tolerate one another in their own smaller territories. Marsh Wrens assemble in loose colonies and where sufficient habitat exists can occur in sizable numbers.

The reedy song of the Marsh Wren matches its chosen habitat. Several introductory rasping *chip*s are followed quickly by an extended and rapid rattle. Although this performance is often given from dense cover, the male also performs fairly spectacular, though brief, vocal flights above the nesting area. These aerial displays typify the exuberant energy of this species. Several call notes are given, including a low *tchuck-tchuck* and a scolding chatter. Courting males puff up their feathers, cock the tail, and quiver the wings.

The female Marsh Wren builds the active or brood nest, usually among cattail flags (typically the new green ones), sedges, or rushes from 1 to 3 feet above the water. The somewhat elongated woven structure, between 6 and 8 inches high, with a diameter of approximately 3 inches, is hitched to two or more plant stems. Various materials, including cattails and marsh grasses, are used for construction, and cattail down or feathers form the lining. A side hole provides access to the nest. A female in Bolton was recorded constructing a nest one foot above the water

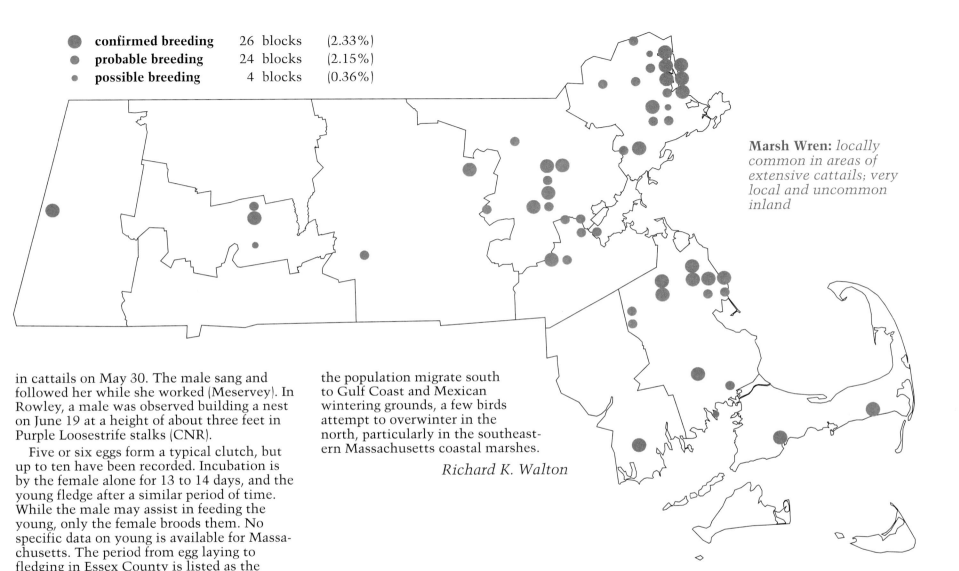

- ● **confirmed breeding**   26 blocks   (2.33%)
- ● **probable breeding**   24 blocks   (2.15%)
- · **possible breeding**   4 blocks   (0.36%)

Marsh Wren: *locally common in areas of extensive cattails; very local and uncommon inland*

in cattails on May 30. The male sang and followed her while she worked (Meservey). In Rowley, a male was observed building a nest on June 19 at a height of about three feet in Purple Loosestrife stalks (CNR).

Five or six eggs form a typical clutch, but up to ten have been recorded. Incubation is by the female alone for 13 to 14 days, and the young fledge after a similar period of time. While the male may assist in feeding the young, only the female broods them. No specific data on young is available for Massachusetts. The period from egg laying to fledging in Essex County is listed as the second week of June to the second week of August (ECOC). The extended range of egg dates would allow for two broods. However, flooding along rivers during June can cause heavy losses of both eggs and young.

Marsh Wrens often remain in and near their nesting habitats throughout September and into October. While most members of the population migrate south to Gulf Coast and Mexican wintering grounds, a few birds attempt to overwinter in the north, particularly in the southeastern Massachusetts coastal marshes.

Richard K. Walton

Golden-crowned Kinglet
Regulus satrapa

Egg dates: May to June 29.

Number of broods: one; may re-lay if first attempt fails.

The diminutive Golden-crowned Kinglet is found in Massachusetts throughout the year, but as a breeding bird it is largely limited to the spruce-clad hill country of the Berkshires. When found apart from our coniferous uplands, it is inevitably in plantings of tall spruces, often the Norway Spruces that have been popular for landscaping since the mid-nineteenth century and were planted in watersheds surrounding reservoirs built in the 1930s. Many of these spruces now exceed a hundred years of age and reach 75 feet or more in height, and have been found in many northeastern states to be attracting kinglets far from the usual breeding grounds. Nesting kinglets are inconspicuous birds, but careful searches of mature spruce plantings in eastern Massachusetts might increase the number of known breeding sites.

Golden-crowned Kinglets are best known as transients, more common in fall than spring. They are seldom abundant and are generally found in conifers, often in White Pines. Both as migrants and as less common wintering birds, they are difficult to find high in evergreens, and their high-pitched, lispy notes are hard to hear as the active wing-flickering kinglets flutter about the tips of branches overhead. They are often members of mixed foraging flocks of chickadees, titmice, nuthatches, and creepers. The kinglets remaining in Massachusetts by winter's end are augmented in April and early May by transients. Any Golden-crowned Kinglet discovered by late May or June in conjunction with ornamental spruces or watershed spruce plantations should be studied for indications of nesting. Some kinglets are believed to be permanent residents of their breeding sites, with the migrants being from more northern nesting areas.

The lisping, often trisyllabic call notes of the Golden-crowned Kinglet, difficult as they are to hear, are better known than the high squeaky song, which is a far less impressive effort than the loud, rollicking, musical song of the Ruby-crowned Kinglet. Unlike the Ruby-crowned Kinglet, which regularly sings during migration, the Golden-crowned Kinglet is seldom heard except on the breeding grounds.

Nesting begins in mid- to late May and extends through early July. The small, deep, cup nest is suspended from a branch, usually near the tip and often at a great height. Three nests in Winchendon were located in evergreen woods as follows: one 60 feet high and 2 feet from the top of a spruce, one 50 feet high and 20 feet from the top of a spruce, and one at a height of 30 feet. These nests were constructed of moss, lichen, bark, rootlets, and feathers, and each contained deep hollows for the eggs (Brewster 1888). An interesting record for a nest in eastern Massachusetts is that of one in Stoughton discovered in a Red Cedar. The tree was 20 feet in height, and the nest was located 18 feet up on a small branch near the trunk (Blake 1916).

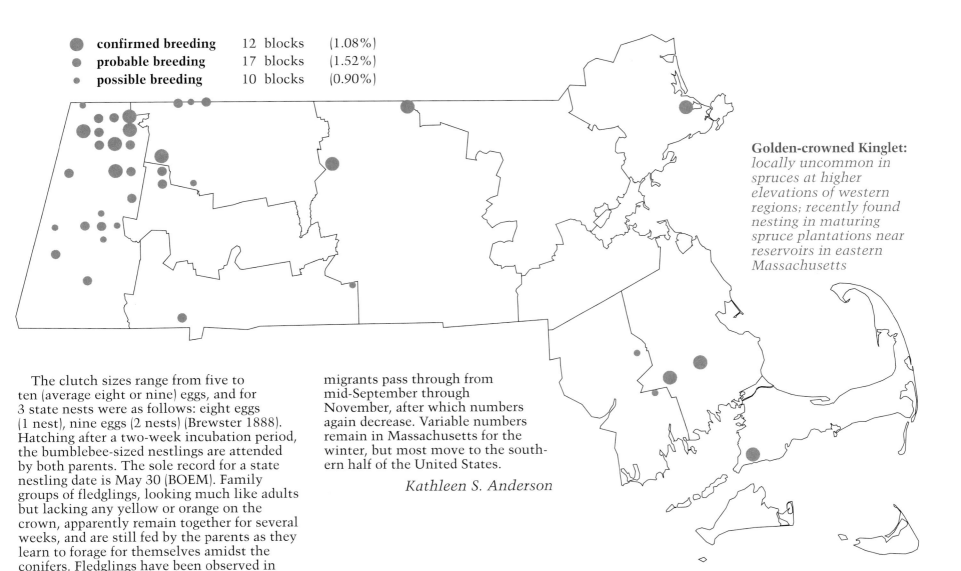

- confirmed breeding — 12 blocks (1.08%)
- probable breeding — 17 blocks (1.52%)
- possible breeding — 10 blocks (0.90%)

Golden-crowned Kinglet: *locally uncommon in spruces at higher elevations of western regions; recently found nesting in maturing spruce plantations near reservoirs in eastern Massachusetts*

The clutch sizes range from five to ten (average eight or nine) eggs, and for 3 state nests were as follows: eight eggs (1 nest), nine eggs (2 nests) (Brewster 1888). Hatching after a two-week incubation period, the bumblebee-sized nestlings are attended by both parents. The sole record for a state nestling date is May 30 (BOEM). Family groups of fledglings, looking much like adults but lacking any yellow or orange on the crown, apparently remain together for several weeks, and are still fed by the parents as they learn to forage for themselves amidst the conifers. Fledglings have been observed in Massachusetts on July 4 and July 9 (BOEM, Blodget).

After completion of the August molt, the adults and young are indistinguishable. By mid-September, they begin appearing in areas where they have not nested, and in this season they may be found in thickets and in deciduous woods, as well as in conifers. Fall migrants pass through from mid-September through November, after which numbers again decrease. Variable numbers remain in Massachusetts for the winter, but most move to the southern half of the United States.

Kathleen S. Anderson

Ruby-crowned Kinglet
Regulus calendula

Egg dates: not available.

Number of broods: one.

The little but energetic Ruby-crowned Kinglet is a rare breeding species in Massachusetts. It is much better known as a transient, especially in spring when its distinctive song is a prominent feature of songbird migration in late April. The song is remarkable both for its volume and for its lengthy duration and variety of notes. The nineteenth-century ornithologist Eliot Coues, commenting on the voice of the Ruby-crowned Kinglet, said, "If the strength of the human voice were in the same proportion to the size of the larynx, we could converse with ease at a distance of a mile or more" (EHF). Once an observer has heard the song and identifies the songster, the sound is not easily forgotten.

The Ruby-crowned Kinglet is at home in the boreal forest that stretches from Alaska east across Canada to Newfoundland. In New England, it breeds south regularly to central Maine and northern New Hampshire and Vermont. There are historical records of breeding from an area known as Hill 51 in Savoy, Massachusetts. Adults with recently fledged young were reported at this location in July three times from 1915 to 1932. Thus, it should come as no surprise that the only confirmed breeding during the Atlas period came from this same general area and was a report of an adult feeding a fledgling. Whether the Ruby-crowned Kinglet is an erratic, opportunistic breeder in this region or whether a small breeding population is permanently established remains to be determined. A singing male present during June 1977 at Hog Island in Essex was apparently unsuccessful in finding a mate to share his adventure.

During courtship the male sings loudly and exposes his bright crown feathers. The three-part song has been described as *eee-tee-tee-tee-too-too-tu-tu-ti-ta-tidaweet-tidaweet-tidaweet*. Both sexes have a *did-it* contact note and a *peu-peu* alarm call. The pensile nest is a globular affair constructed mainly of mosses. The site is a conifer, most often a spruce, where the nest is placed well out toward the tip of a branch and often at a considerable height of 30 feet or more.

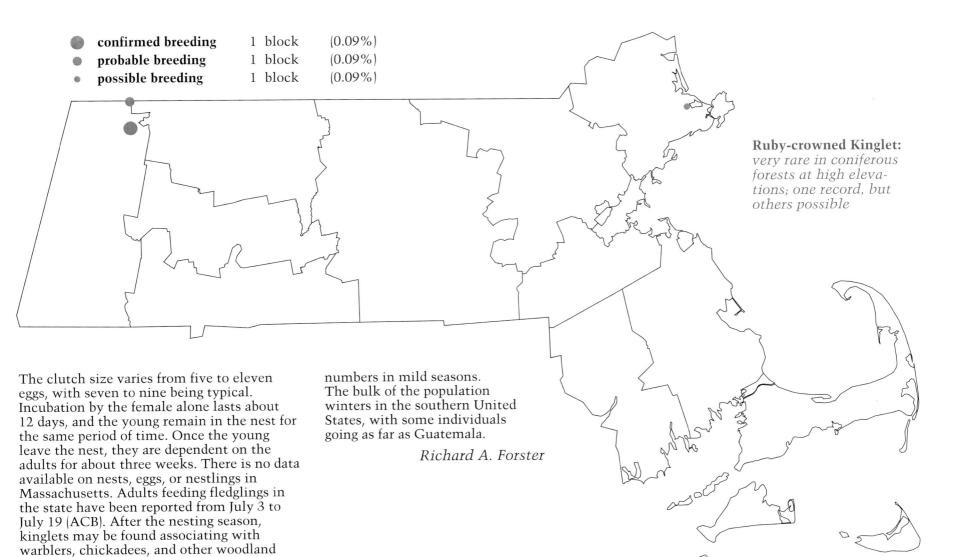

- confirmed breeding 1 block (0.09%)
- probable breeding 1 block (0.09%)
- possible breeding 1 block (0.09%)

Ruby-crowned Kinglet: *very rare in coniferous forests at high elevations; one record, but others possible*

The clutch size varies from five to eleven eggs, with seven to nine being typical. Incubation by the female alone lasts about 12 days, and the young remain in the nest for the same period of time. Once the young leave the nest, they are dependent on the adults for about three weeks. There is no data available on nests, eggs, or nestlings in Massachusetts. Adults feeding fledglings in the state have been reported from July 3 to July 19 (ACB). After the nesting season, kinglets may be found associating with warblers, chickadees, and other woodland species.

The first southbound migrants appear in the middle of September. Peak numbers are encountered during the first half of October. Scattered individuals or small groups are found throughout fall into early winter, especially in milder southeastern coastal regions where they can be found in small numbers in mild seasons. The bulk of the population winters in the southern United States, with some individuals going as far as Guatemala.

Richard A. Forster

Blue-gray Gnatcatcher
Polioptila caerulea

Egg dates: May 10 to June 18.

Number of broods: one; may renest if first attempt fails.

The Blue-gray Gnatcatcher is yet another member of the southern avifauna that has recently spread into Massachusetts and the northeastern United States. Prior to 1960, it was regarded only as a transient, with the bulk of the observations from late August to early September. Since 1960, the gnatcatcher has experienced a steady increase, both as a spring and fall migrant and also as an uncommon and somewhat local summer resident in every geographical region of the state. The gnatcatcher has a distinct affinity for water margins—wooded edges along ponds, streams, and the edges of Red Maple swamps. Nesting records are concentrated along the major river valleys and at Quabbin, where the increase in Beaver numbers and their associated wetlands provide ideal habitat. According to Atlas figures, it was estimated that in 1976 there were 20 to 30 pairs of gnatcatchers nesting in Quabbin.

The first resident gnatcatchers arrive in mid-April, and from then until the end of the month the remaining residents filter in, although migrants may be noted until mid-May. Observers familiar with the high, thin *speee* call note are likely to be the first to encounter this sprite because it is constantly vocal while flitting about in the treetops. The song is a lengthy series of thin, squeaky, wheezy notes often easily overlooked but occasionally sung with remarkable intensity. Shortly after the arrival of the female, as much as a week later than the male, nest construction begins. The nest is an architectural wonder, a compact, cuplike structure affixed firmly to a horizontal branch at heights of up to 70 feet but generally under 25 feet. Massachusetts nests have been found in Red Maple at heights ranging from 10 to 20 feet, and one was discovered at 16 feet in an oak in a dry area (Meservey, Kellogg). The exterior of the nest is adorned with lichens held in place by spider silk, while the nest consists mostly of plant down and similar components, including the buffy wood from the stalk of Cinnamon Fern. Once the nest is completed, it may be as long as two weeks before the eggs are laid, when the trees have leafed out. The delay between nest completion and egg laying has been attributed to the fact that essential nesting material is available only during this short time span. If the nest site is later deemed unsuitable or a renesting attempt occurs, the material from the original nest is

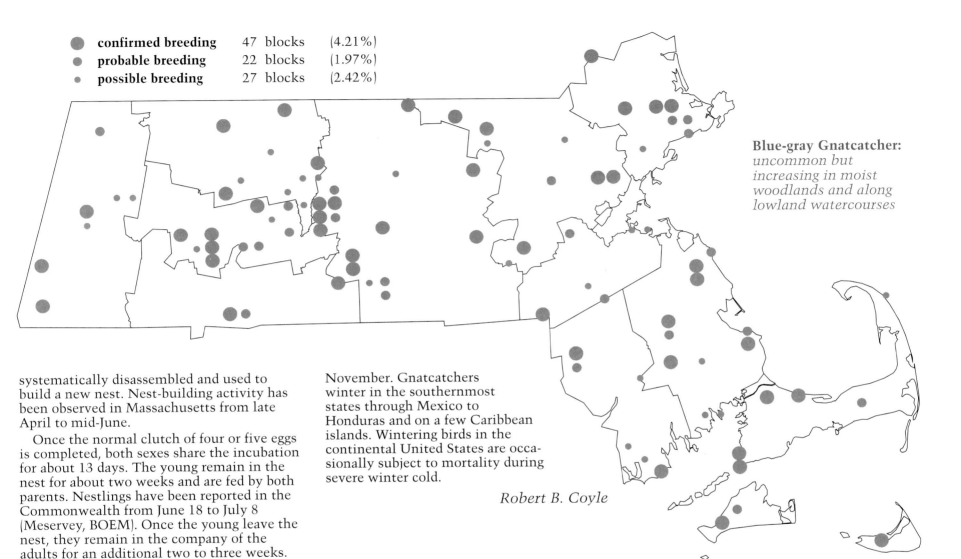

Blue-gray Gnatcatcher: *uncommon but increasing in moist woodlands and along lowland watercourses*

● **confirmed breeding**	47 blocks	(4.21%)
● **probable breeding**	22 blocks	(1.97%)
· **possible breeding**	27 blocks	(2.42%)

systematically disassembled and used to build a new nest. Nest-building activity has been observed in Massachusetts from late April to mid-June.

Once the normal clutch of four or five eggs is completed, both sexes share the incubation for about 13 days. The young remain in the nest for about two weeks and are fed by both parents. Nestlings have been reported in the Commonwealth from June 18 to July 8 (Meservey, BOEM). Once the young leave the nest, they remain in the company of the adults for an additional two to three weeks. Pairs or single adults have been observed in the state feeding fledglings from July 2 to August 1 (Meservey). By late July onward, gnatcatchers begin to roam the woodlands with groups of chickadees and warblers prior to migration.

Most residents depart from mid-August to mid-September, although scattered individuals are occasionally reported in October and November. Gnatcatchers winter in the southernmost states through Mexico to Honduras and on a few Caribbean islands. Wintering birds in the continental United States are occasionally subject to mortality during severe winter cold.

Robert B. Coyle

Eastern Bluebird
Sialia sialis

Egg dates: April 1 to August 7.

Number of broods: two; rarely three.

The Eastern Bluebird became a common species during the historical period when settlers cleared the forest and created fields and orchards. During the twentieth century, bluebird populations in the Northeast declined by about 90 percent due to a combination of factors. Changing land uses and habitat destruction severely reduced the number of nest sites, and there was severe competition for the remaining cavities from native bird species as well as the alien House Sparrow and European Starling. Adverse weather, pesticides, and a decline of the winter food supply further added to the bird's problems. However, since the 1970s, the bluebird has made a promising comeback as a result of more effective conservation and the establishment of bluebird trails with specially designed nest boxes.

During the Atlas period, the bluebird was most common in rural areas of central and western Massachusetts and in southeastern portions of the state, where frequent forest fires help maintain suitable habitat requirements. Limited breeding also occurred in Essex County, the outer Cape, and Martha's Vineyard, but not on Nantucket.

The first spring migrants usually appear in the middle of March, but arrival may continue well into April, depending upon weather conditions. By mid-April, most of the residents are well established on territory. The preferred nesting habitats—open areas with scattered trees where the ground is not covered with tall undergrowth—include farmland, orchards, open woodlands, swamps, pastures, golf courses, large lawns, country cemeteries, and military reservations.

The bluebird's song is a rich, musical, warbling *cheuerly-cheuerly*, often repeated, and is basically an extension of the common *chur-lee* call note. Vocalization is most intense shortly after males arrive and begin to establish territories a few days in advance of the females. Calls are given frequently in flight. The alarm note is a chatter. Courtship is a thing of gentle beauty, with the male hovering before the female or perching near her and singing. It may be brief or last as much as a week before pairing occurs. The male locates a suitable nesting cavity, but the female makes the final choice of a site. Bluebirds are very territorial because a pair must maintain an area with an adequate food supply to rear the young.

Eastern Bluebirds are cavity nesters, utilizing natural tree cavities, old woodpecker holes in trees and fence posts, and suitable bird boxes. Heights of 44 Massachusetts nests in bird houses ranged from 3 to 10 feet, and heights of 2 natural deciduous cavities were 10 feet and 20 feet, respectively (CNR). The highest recorded state nest was at 35 feet in an old woodpecker hole in a power line pole (Meservey). In 4 or 5 days, the female builds a nest of fine dried grasses or pine needles. Bluebirds generally lay four or five short, oval, pale blue (occasionally white) eggs. Each successive clutch in a season tends to average one egg less than the previous one. Clutch sizes for 35 state nests were two eggs, possibly incomplete clutches (2 nests), three eggs (2 nests), four eggs (12 nests), five eggs (17 nests), six eggs (2 nests) (CNR). The female incubates for approximately 14 days, and the young remain in the nest for about 18 days before fledging.

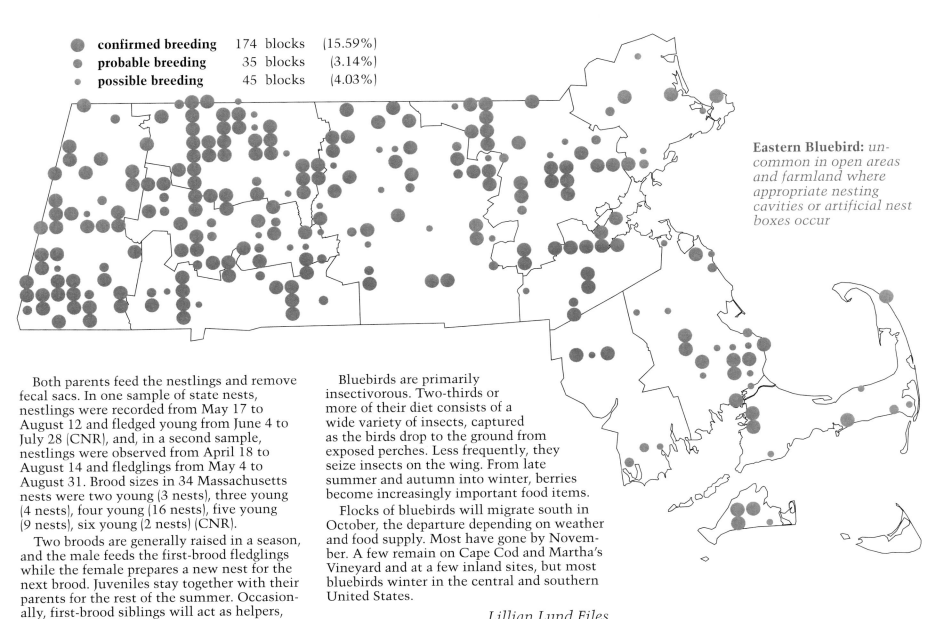

Eastern Bluebird: *uncommon in open areas and farmland where appropriate nesting cavities or artificial nest boxes occur*

●	**confirmed breeding**	174 blocks	(15.59%)
•	**probable breeding**	35 blocks	(3.14%)
·	**possible breeding**	45 blocks	(4.03%)

Both parents feed the nestlings and remove fecal sacs. In one sample of state nests, nestlings were recorded from May 17 to August 12 and fledged young from June 4 to July 28 (CNR), and, in a second sample, nestlings were observed from April 18 to August 14 and fledglings from May 4 to August 31. Brood sizes in 34 Massachusetts nests were two young (3 nests), three young (4 nests), four young (16 nests), five young (9 nests), six young (2 nests) (CNR).

Two broods are generally raised in a season, and the male feeds the first-brood fledglings while the female prepares a new nest for the next brood. Juveniles stay together with their parents for the rest of the summer. Occasionally, first-brood siblings will act as helpers, feeding the second-brood nestlings. Most late broods are out of the nests by the end of August. Third broods are rare but do occur in Massachusetts, in which case young fledglings may be seen in September.

Bluebirds are primarily insectivorous. Two-thirds or more of their diet consists of a wide variety of insects, captured as the birds drop to the ground from exposed perches. Less frequently, they seize insects on the wing. From late summer and autumn into winter, berries become increasingly important food items.

Flocks of bluebirds will migrate south in October, the departure depending on weather and food supply. Most have gone by November. A few remain on Cape Cod and Martha's Vineyard and at a few inland sites, but most bluebirds winter in the central and southern United States.

Lillian Lund Files

Veery
Catharus fuscescens

Egg dates: May 20 to June 30.
Number of broods: one or two.

The Veery is a common forest bird. It ranges throughout most of Massachusetts but is more abundant in the western part of the state. Though the Veery is a "probable" nester on the Elizabeth Islands, it has not been recorded breeding on Martha's Vineyard or Nantucket, is found only on the western part of Cape Cod, and is rare in southern Plymouth County. The species inhabits moist deciduous and coniferous woods, wooded swamps, thickets, and dry oak-pine woods.

Veeries arrive in Massachusetts during the first two weeks of May. During migration, this normally shy and solitary thrush is described as being quite tame and sometimes migrates in nocturnal flights made up of over a thousand individuals. Males may sing occasionally during migration but are generally silent for several days after reaching their nesting grounds. They sing throughout the day, but most intensively at dawn and dusk, and many observers know the Veery largely by the haunting song, four or five phrases descending in a spiral. A good rendition is *da-vee-ur-vee-ur-veer-veer*. The song period ends about the middle of July. The common call note is a sharp *wheew*. When the birds are agitated, they give a loud *whuck* and a screaming cry.

The nest is built in 6 to 10 days on the ground or in low shrubs or a brush pile. Of 5 Massachusetts nests, 3 were on the ground in open woodlands, 1 was on the ground at the edge of a path between dense woods and a gravel pit, and 1 was in a low thicket (Blodget, CNR). Rarely, nests are placed in

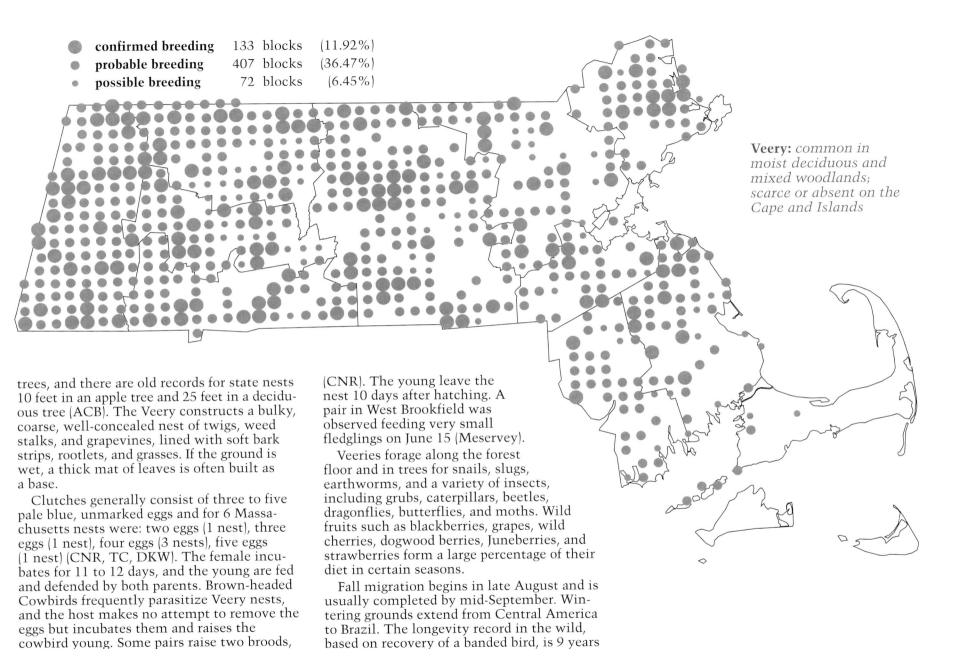

Veery: *common in moist deciduous and mixed woodlands; scarce or absent on the Cape and Islands*

●	confirmed breeding	133 blocks	(11.92%)
●	probable breeding	407 blocks	(36.47%)
·	possible breeding	72 blocks	(6.45%)

trees, and there are old records for state nests 10 feet in an apple tree and 25 feet in a deciduous tree (ACB). The Veery constructs a bulky, coarse, well-concealed nest of twigs, weed stalks, and grapevines, lined with soft bark strips, rootlets, and grasses. If the ground is wet, a thick mat of leaves is often built as a base.

Clutches generally consist of three to five pale blue, unmarked eggs and for 6 Massachusetts nests were: two eggs (1 nest), three eggs (1 nest), four eggs (3 nests), five eggs (1 nest) (CNR, TC, DKW). The female incubates for 11 to 12 days, and the young are fed and defended by both parents. Brown-headed Cowbirds frequently parasitize Veery nests, and the host makes no attempt to remove the eggs but incubates them and raises the cowbird young. Some pairs raise two broods, but it is believed that many late nesting efforts are renesting attempts. One Massachusetts nest contained five young on June 9 (CNR). The young leave the nest 10 days after hatching. A pair in West Brookfield was observed feeding very small fledglings on June 15 (Meservey).

Veeries forage along the forest floor and in trees for snails, slugs, earthworms, and a variety of insects, including grubs, caterpillars, beetles, dragonflies, butterflies, and moths. Wild fruits such as blackberries, grapes, wild cherries, dogwood berries, Juneberries, and strawberries form a large percentage of their diet in certain seasons.

Fall migration begins in late August and is usually completed by mid-September. Wintering grounds extend from Central America to Brazil. The longevity record in the wild, based on recovery of a banded bird, is 9 years 11 months.

Elissa M. Landre

Swainson's Thrush
Catharus ustulatus

Egg dates: early June to mid-July.

Number of broods: one.

The Swainson's Thrush is a bird of the Canadian Zone, including the scattered upland coniferous woods that dot the western limits of our state. There, it is a fairly common summer resident, its calls being among the most characteristic sounds of the habitat. Its distribution since historical times remains basically unchanged. Since the Atlas period, nesting birds have been discovered in western Hampshire County, but there have been no recent summer records from northern Worcester County.

In spring, the Swainson's Thrush is among the later migrants to make an appearance in the Northeast. Seldom does it arrive before the second week of May, with peak numbers of transients occurring later in the month. In this season, the species is often seen with its migrant companions, the wood-warblers, feeding in the branches of the blossoming spring woodland as well as on the ground. Migration in the spring occurs on a broad front, evidence of which is to be heard on quiet evenings when the notes of the Swainson's Thrush may be easily distinguished overhead. Listening for the diagnostic flight call of this species, observers have recorded hundreds or even thousands in a single evening.

The five eastern North American forest thrushes are renowned for their beautiful songs. Compared with the Hermit Thrush, with which it often shares breeding habitat, the Swainson's Thrush is not nearly the virtuoso songster. Yet, though its song may lack the celebrated ethereal quality of the Hermit Thrush's, its musical strain remains one of our woodlands' finest. The Swainson's Thrush sings quite commonly during spring migration but is more vocal on its breeding grounds, where its song spirals upward from the dense, evergreen woods. The common call note is a short *whit,* and the flight call is a sharp *queep.*

In Massachusetts, Swainson's Thrushes commonly choose an area of dense, young spruce woods for a breeding site. There, often near a stream or boggy area, a territory is established. At lower elevations, such as the Huntington-Chester area, the breeding territories center on stands of Eastern Hemlock surrounded by primarily deciduous forest, including such species as Yellow Birch, Black Cherry, and American Beech (Blodget, Meservey). Nests are invariably placed upon horizontal branches, usually 3 to 6 feet, rarely up to 20 feet, from the ground and are so well made as to have commanded the praises of many early naturalists. The nest has been described as more elaborately and neatly constructed than that of any of our other thrushes. Conspicuous among the materials are the *Hypnum* mosses, which by their dark fibrous masses give a very distinctive character to these nests and distinguish

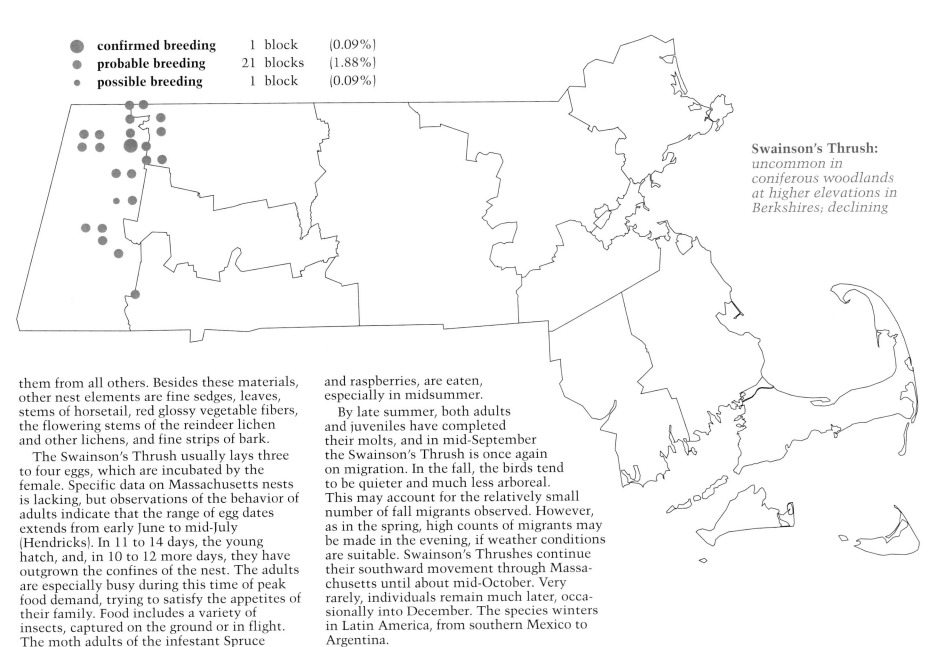

Swainson's Thrush: *uncommon in coniferous woodlands at higher elevations in Berkshires; declining*

them from all others. Besides these materials, other nest elements are fine sedges, leaves, stems of horsetail, red glossy vegetable fibers, the flowering stems of the reindeer lichen and other lichens, and fine strips of bark.

The Swainson's Thrush usually lays three to four eggs, which are incubated by the female. Specific data on Massachusetts nests is lacking, but observations of the behavior of adults indicate that the range of egg dates extends from early June to mid-July (Hendricks). In 11 to 14 days, the young hatch, and, in 10 to 12 more days, they have outgrown the confines of the nest. The adults are especially busy during this time of peak food demand, trying to satisfy the appetites of their family. Food includes a variety of insects, captured on the ground or in flight. The moth adults of the infestant Spruce Budworm may be taken. A good percentage of wild fruits, such as soft-skinned blackberries and raspberries, are eaten, especially in midsummer.

By late summer, both adults and juveniles have completed their molts, and in mid-September the Swainson's Thrush is once again on migration. In the fall, the birds tend to be quieter and much less arboreal. This may account for the relatively small number of fall migrants observed. However, as in the spring, high counts of migrants may be made in the evening, if weather conditions are suitable. Swainson's Thrushes continue their southward movement through Massachusetts until about mid-October. Very rarely, individuals remain much later, occasionally into December. The species winters in Latin America, from southern Mexico to Argentina.

Brian E. Cassie

Hermit Thrush
Catharus guttatus

Egg dates: May 3 to August 9.

Number of broods: two.

Considered by many to be the finest songster in North America, the Hermit Thrush has a wide, though sporadic, breeding distribution across the state. Viewed as rare at the turn of the century, the species has since expanded its breeding range and become common in many locales. During the Atlas period, it was common in the hills of western Massachusetts, scarce to absent in central and northeastern sections, and fairly common but local in portions of the southeastern coastal plain. Nesting was "confirmed" on Martha's Vineyard.

Migrating Hermit Thrushes first appear in Massachusetts during the last half of April, during which time many will continue farther north. By early May, breeding birds can be heard singing in a variety of secluded woodland habitats, from the damp mixed forests of the western hills to the dry pine barrens along the coastal plain. The common characteristics of these nesting areas are a dense understory of young vegetation and an abundance of evergreens, although woodlands with a good mix of deciduous trees will sometimes suffice.

The song, given primarily at dawn and dusk, is a lovely, ethereal series of flutelike phrases, ascending and descending randomly. The early naturalists were so impressed by the musical refrains of the Hermit Thrush that they called it the "American Nightingale." The male's song period may extend from April to late August. A distinctive, soft, low *chuck* note is given throughout the year and is often the first indication of this reclusive species' presence. Another common note of both sexes is a *tway* sound.

The nest is a deep cup of dried grasses, bark fiber, and small twigs, often lined with evergreen needles, and is usually on the ground, although some are built in a small tree up to about 8 feet above the ground. Four recent Massachusetts nests were located as follows: 2 on the ground concealed by low vegetation in mixed forest, 1 on the ground beneath a Mountain Laurel, and 1 on the ground in a fairly open cutover mixed forest under a clump of small White Pine seedlings (CNR, Blodget, Meservey). The normal clutch is three to four eggs, but as many as six have been recorded. Clutch sizes for 9 state nests were as follows: three eggs (4 nests), four eggs (5 nests) (Nice 1933, DKW, CNR, Meservey).

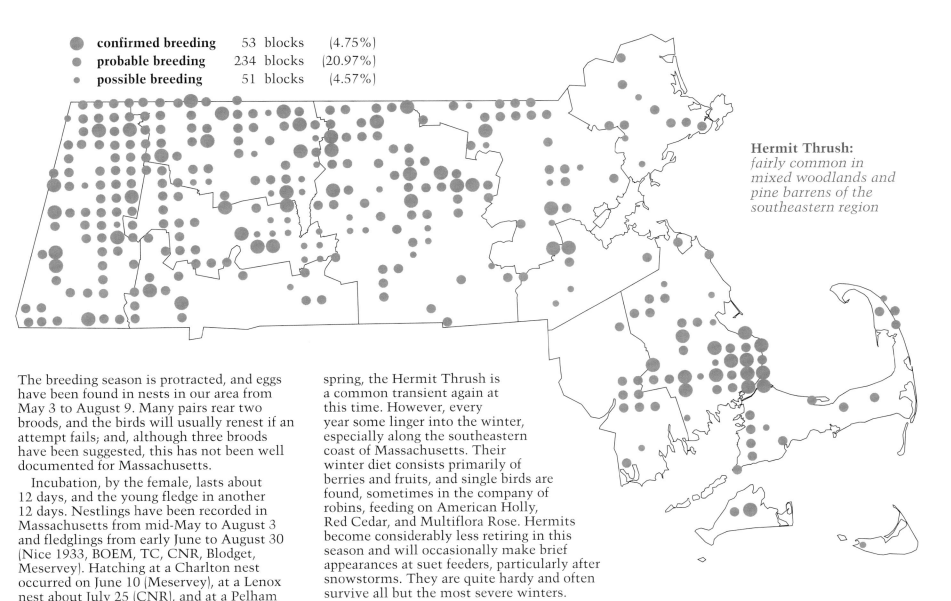

- confirmed breeding — 53 blocks (4.75%)
- probable breeding — 234 blocks (20.97%)
- possible breeding — 51 blocks (4.57%)

Hermit Thrush: *fairly common in mixed woodlands and pine barrens of the southeastern region*

The breeding season is protracted, and eggs have been found in nests in our area from May 3 to August 9. Many pairs rear two broods, and the birds will usually renest if an attempt fails; and, although three broods have been suggested, this has not been well documented for Massachusetts.

Incubation, by the female, lasts about 12 days, and the young fledge in another 12 days. Nestlings have been recorded in Massachusetts from mid-May to August 3 and fledglings from early June to August 30 (Nice 1933, BOEM, TC, CNR, Blodget, Meservey). Hatching at a Charlton nest occurred on June 10 (Meservey), at a Lenox nest about July 25 (CNR), and at a Pelham nest on August 3 (Nice 1933). Both sexes care for the young, bringing them a variety of insects, worms, and other small invertebrates.

The southbound migration occurs in October and early November and takes most birds to the southern United States. As in the spring, the Hermit Thrush is a common transient again at this time. However, every year some linger into the winter, especially along the southeastern coast of Massachusetts. Their winter diet consists primarily of berries and fruits, and single birds are found, sometimes in the company of robins, feeding on American Holly, Red Cedar, and Multiflora Rose. Hermits become considerably less retiring in this season and will occasionally make brief appearances at suet feeders, particularly after snowstorms. They are quite hardy and often survive all but the most severe winters.

Blair Nikula

Wood Thrush
Hylocichla mustelina

Egg dates: May 14 to June 26.

Number of broods: one, rarely two; may renest if first attempt fails.

The Wood Thrush has become, in recent decades, a common breeding species throughout most of Massachusetts, where it now nests regularly in suitable habitat. The species is notably scarce only on Cape Cod, the Islands, and coastal areas in general where appropriate woodlands for breeding are lacking. Unfortunately, at the same time that the Wood Thrush has been extending its breeding range northward in New England, the species is doubtlessly experiencing loss of winter habitat due to deforestation in Central America.

The Wood Thrush is still a fairly common spring migrant. Peak movement is during the second and third weeks of May, although some individuals arrive in late April. Males begin singing almost immediately upon arrival. Like that of other thrush species, the song is flutelike, loud, and somewhat ventriloquial. The song, a series of phrases consisting of four notes followed by a short, high, wispy trill, is most frequently given at dawn and dusk. The common call is a series of sharp *pit-pit-pit* notes. Courtship includes rapid pursuit flights of the female by the male.

The Wood Thrush's preferred nesting habitat is mature, moist deciduous forest with an understory of shrubs and young trees, although the species can also be found in dry woodland. Apparently, the Wood Thrush can tolerate some human activity and has adapted to nesting in small forest patches and even in shrubby vegetation in suburban areas. Habitats for 12 Massachusetts nests were mixed woodland (8 nests), deciduous woods (1 nest), suburban yard (3 nests) (CNR). The cup-shaped nest, built by the female, is placed in a notch or forked branch of a shrub or sapling, usually 5 to 10 feet above the ground. Occasionally, the nest is built higher up in a mature tree.

In Massachusetts, 20 nests were located as follows: deciduous trees, including apple and oak (7 nests); conifers, including Red Cedar, Red Pine, White Pine, and Jack Pine (7 nests); shrubs and bushes, including laurel, viburnum, honeysuckle, and lilac (6 nests) (CNR). Heights of 17 of these nests ranged from 3 to 20 feet, with an average of 6.9 feet (CNR). Grasses and weed stalks are used to construct the outer layer; a middle layer of mud, wet leaves, and grass is then added. Finally, the cup is lined with moss or fine rootlets. Also, bits and pieces of paper are frequently incorporated into the structure.

Three or four (sometimes five) eggs are laid, and clutch sizes for 17 Massachusetts nests were three eggs (8 nests), four eggs (8 nests), three eggs plus one cowbird egg (1 nest) (CNR, DKW). Known state laying dates were May 20 to May 23 and May 23 to May 26, and known state hatch dates were June 4, June 5, June 17, and June 18 (CNR). The female incubates the eggs for 12 to 14 days, but the nestlings are tended by both parents. Food brought to the young includes insects and other invertebrates as well as berries later on. If the nest or young are threatened, the adults give a loud distress call, which is a *tswack* note repeated four or five times. Nestlings have been recorded in Massachusetts from June 3 to June 24. Brood sizes for 10 state nests were three young (3 nests), four young (6 nests), one youngster in

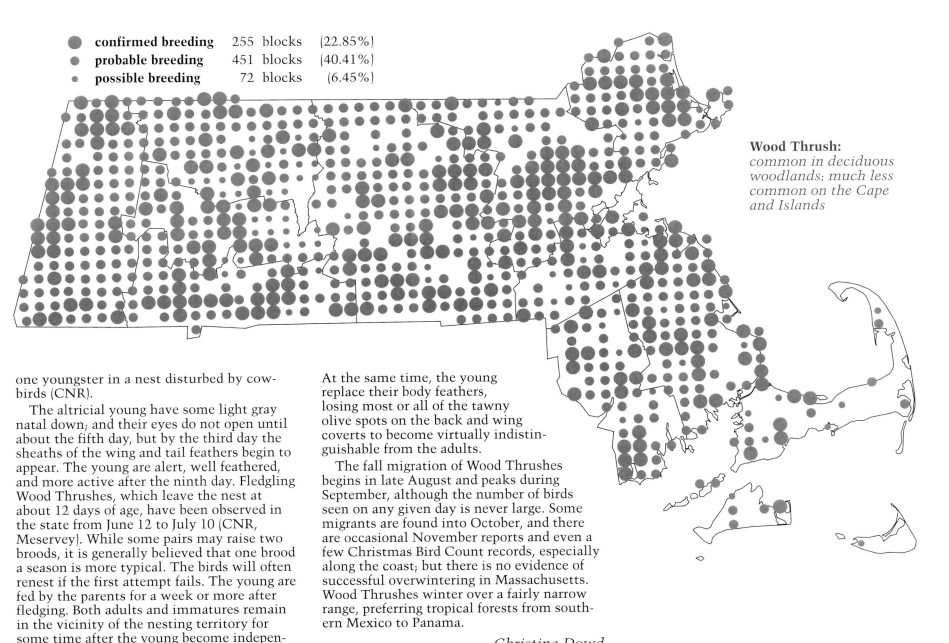

- **confirmed breeding** 255 blocks (22.85%)
- **probable breeding** 451 blocks (40.41%)
- **possible breeding** 72 blocks (6.45%)

Wood Thrush: *common in deciduous woodlands; much less common on the Cape and Islands*

one youngster in a nest disturbed by cowbirds (CNR).

The altricial young have some light gray natal down; and their eyes do not open until about the fifth day, but by the third day the sheaths of the wing and tail feathers begin to appear. The young are alert, well feathered, and more active after the ninth day. Fledgling Wood Thrushes, which leave the nest at about 12 days of age, have been observed in the state from June 12 to July 10 (CNR, Meservey). While some pairs may raise two broods, it is generally believed that one brood a season is more typical. The birds will often renest if the first attempt fails. The young are fed by the parents for a week or more after fledging. Both adults and immatures remain in the vicinity of the nesting territory for some time after the young become independent. During this period, adults undergo a complete postnuptial molt, replacing wing and tail feathers as well as the body plumage. At the same time, the young replace their body feathers, losing most or all of the tawny olive spots on the back and wing coverts to become virtually indistinguishable from the adults.

The fall migration of Wood Thrushes begins in late August and peaks during September, although the number of birds seen on any given day is never large. Some migrants are found into October, and there are occasional November reports and even a few Christmas Bird Count records, especially along the coast; but there is no evidence of successful overwintering in Massachusetts. Wood Thrushes winter over a fairly narrow range, preferring tropical forests from southern Mexico to Panama.

Christina Dowd

American Robin
Turdus migratorius

Egg dates: April 12 to August 10.
Number of broods: two; rarely three.

American Robins are among the most ubiquitous nesting species in Massachusetts, occupying open lands of all description and forested areas, as well, usually where there is scant understory and interspersions of small clearings. Originally birds of the forest, robins responded positively to the clearing of land and settlement and are thought to be many times more common today than they were in the colonial period. They are especially abundant in suburban areas. However, in heavily suburbanized regions characterized by intensive human activities and large numbers of domestic predators, reproductive success may be quite poor, and such populations may be maintained only by immigration.

True northbound migrants appear in early March, and passage occurs throughout April, with peaks in late March and early April. Local males waste no time in establishing their territories. The defended territory, over which the male stands guard throughout the nesting season, includes the immediate vicinity of the nest and usually an additional ill-defined area of lawns and gardens. The territorial song of the robin—a loud and continuous rich caroling, rising and falling in pitch and usually described as *cheerily-cheerily-cheerily-cheerrio*—is first heard on warm mornings in late March and early April.

Dawn and dusk choruses reach a peak in late April but continue, gradually diminishing in intensity, through mid-July. The familiar call notes of the robin, often rendered *kwee-kwee-kuk-kuk-kuk* or sometimes *puck-kuk-kuk-kuk*, are uttered with emphatic jerks of the tail and at varying speeds and intensity of delivery, depending on the bird's emotional state. The flight call, a thin *see-lip*, may be heard at any time of the year.

Robin nests are commonly built in trees and shrubs, reportedly as high as 70 feet but usually 10 to 25 feet. A preference for coniferous sites is evident early in the season. Robins occasionally nest on porches, sheltered windowsills, and the eaves of buildings, and artificial nesting shelves are sometimes accepted. Rarely, nests may be found in the crevices of natural cliffs or boulders, in stone walls, and even on the ground.

In Massachusetts, 118 nests were located as follows: 65 on conifer branches at an average height of 11.3 feet, 24 on deciduous branches at an average height of 11.7 feet, 13 in shrubs at an average height of 6.5 feet, 4 in vines at an average height of 10.5 feet, and 12 on buildings at an average height of 15 feet (CNR). The nest is a deep bowl of coarse twigs and grasses, with the inside walls and bottom reinforced substantially with mud and lined on the inside with fine grasses and other pliable fibrous material. Construction is done almost entirely by the female and typically takes 4 to 6 days, occasionally longer if heavy rains wash out the mud work. Usually, a new nest is built for each brood; however, occasionally a nest may be refurbished for a second brood.

Up to 5 days may elapse before the first egg is deposited in a completed nest. Once egg laying begins, eggs are laid on successive days, and incubation starts on the evening following the deposition of the second egg. The female, which usually spends 80 percent of her time on the eggs, accomplishes incubation. Clutches of three or four eggs are the rule; about 2 percent of clutches contain five eggs. The incubation period is 12 to 14 days. In Massachusetts, eggs have been found in nests from the second week of April until a late hatch date of August 10 (Nice 1933).

Upon hatching, the young are immediately tended with great interest by both parents. The young are capable of begging for food

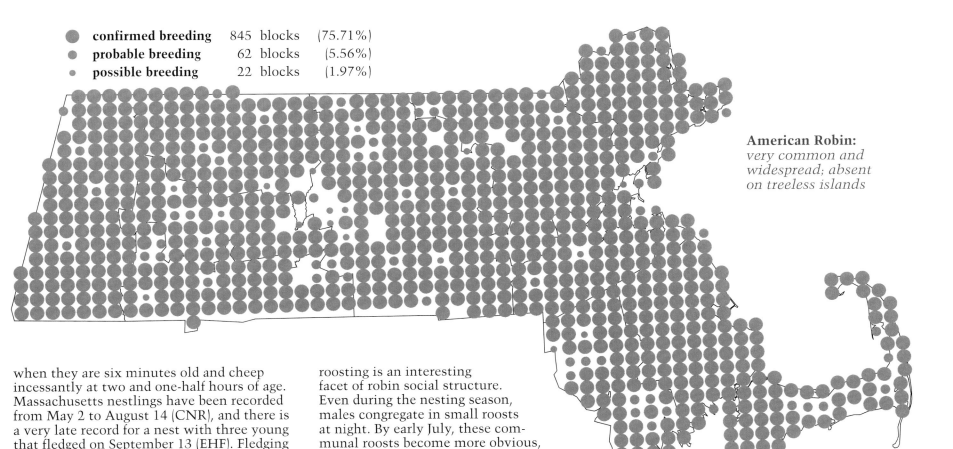

American Robin: *very common and widespread; absent on treeless islands*

- confirmed breeding — 845 blocks (75.71%)
- probable breeding — 62 blocks (5.56%)
- possible breeding — 22 blocks (1.97%)

when they are six minutes old and cheep incessantly at two and one-half hours of age. Massachusetts nestlings have been recorded from May 2 to August 14 (CNR), and there is a very late record for a nest with three young that fledged on September 13 (EHF). Fledging occurs at 14 to 16 days, at which point the male largely takes charge of the young, freeing the female to prepare a new nest and lay a second clutch of eggs. Of 125 nests in the Commonwealth, 41 fledged young, 31 failed, and 53 had an unknown outcome (CNR). Because there is a high loss of both eggs and young, renesting attempts are frequent, and most pairs are able to raise only two broods in a season.

Dispersal from the parents' territory is rather slow, with many juveniles remaining within a half-mile of their birthplace two months after fledging. Most juveniles eventually join feeding flocks of adult robins in areas of abundant fruit. Communal night roosting is an interesting facet of robin social structure. Even during the nesting season, males congregate in small roosts at night. By early July, these communal roosts become more obvious, and the number of robins using them swells throughout the summer as more and more non-nesting birds join the ranks. These roosts, in which juveniles often heavily outnumber adults, are used nightly until mid-September, after which time the roosts break down and the habits of the birds change markedly. Small flocks of robins then begin radiating across the countryside searching for wild fruit. Southward migration is most pronounced between the third week of September and the third week of October, with the last migrating robins usually gone by the end of November. In some years, however, large numbers may linger into early winter.

Massachusetts robins winter from the Carolinas southward throughout the Florida peninsula and westward through the Gulf states to eastern Texas. Most return to nest within 25 miles of their birthplace.

Bradford G. Blodget

Gray Catbird
Dumetella carolinensis

Egg dates: May 3 to August 15.
Number of broods: one or two.

The Gray Catbird is a common and well-known summer resident throughout the state—a reflection of the ubiquity of its preferred habitat of suburbs, overgrown fields, and woodland edges. It is especially abundant along the coastal plain, such as on Plum Island and Cape Cod, where the profusion of berry-producing shrubs provides both food and nesting cover.

The first migrants show up in early May, but the bulk of the residents do not arrive until mid- to late May. Adult male catbirds arrive first on their breeding ground and immediately begin to establish territories through vigorous singing from exposed perches. A catbird's territory ranges from about one to three acres, and the bird usually concentrates its activities inside these boundaries. Disputes with neighboring catbirds over territorial boundaries are settled with loud song and chases.

The song of the catbird is given only by the male and consists of a long series of different phrases, some of which imitate the songs and calls of other birds. The best-known call of the Gray Catbird is the *meow* call, which is given by both sexes and sounds like a cat. It is most often given during aggressive encounters and in response to predators. Other calls include a grating *kak-kak-kak* and a soft *chuck*.

Females arrive on the breeding grounds one to two weeks after the males, and initial courtship consists of the male's repeatedly chasing the female for short distances within the territory. During these chases, which may last up to one-half hour, both the male and female may do a display in which they fluff out all of their contour feathers so that they appear to be about the size of a softball.

Soon after the establishment of the pair bond, nest building begins. At first, both the male and female may carry twigs about, placing them in several trial locations. But, within a few days, the nest site is selected and nest building begins in earnest by the female. It takes 5 to 8 days to complete the nest. The nest is usually located in dense vegetation, 2 to 10 feet high, and is a bulky collection of twigs, grape or Red Cedar bark, and fern rootlets as lining. Locations for 39 state nests were as follows: shrubs and low vegetation (32 nests), conifer (4 nests), hedge (3 nests) (CNR). Heights of these nests ranged from 2 to 9 feet, with an average of 5.2 feet (CNR).

Egg laying begins as soon as the nest is completed. Three to four turquoise blue eggs is the usual clutch, and incubation begins with the laying of the last egg. The female does all of the incubation. When she is not at

- **confirmed breeding** 643 blocks (57.62%)
- **probable breeding** 241 blocks (21.59%)
- **possible breeding** 29 blocks (2.60%)

Gray Catbird: *very common in shrubby areas and thickets throughout the state*

the nest, the male replaces her to protect the nest, but he does not incubate. This coordination of activities around the nest is accomplished with short, quiet fragments of song, a call that sounds like *kwut,* a wing-raised display, and the male presenting food to the female. Clutch sizes for 32 state nests were two eggs (2 nests), three eggs (11 nests), four eggs (18 nests), five eggs (1 nest) (CNR).

The young hatch in about 13 days. For the first few days, the female broods the nestlings and the male supplies her with food, which she in turn feeds to the young. In a few days, both sexes bring food to the nestlings. After about two weeks in the nest, the young birds fledge but are still fed by the parents for another two weeks. A second brood may occur, in which case the female builds a new nest in a different part of the territory. Brood sizes for 20 state nests were two young (5 nests), three young (8 nests), four young (7 nests) (CNR). Nestlings have been reported in Massachusetts from June 9 to August 13 and fledglings from June 19 to August 12 (CNR). Outcomes for 39 state nests were unknown (11 nests), failed (15 nests), young fledged (13 nests) (CNR).

After the last brood has become independent, catbirds tend to move to lowland areas of dense cover and abundant food. At this time, they undergo their yearly molt. Fall migration begins in late August and continues well into October. Almost all catbirds spend the winter well to the south of Massachusetts (southern United States to Panama and West Indies), but there are always a few lingerers found in the early winter months, although few of them survive.

Donald W. Stokes and Lillian Q. Stokes

Northern Mockingbird
Mimus polyglottos

Egg dates: May 11 to August 22.
Number of broods: two.

The Northern Mockingbird is a permanent resident in Massachusetts, but this was not always the case. It was considered a rare visitor in the state during the mid- to late 1800s. During the 1920s, Forbush reported that it had become less rare and that breeding was rare but regular—a status that remained unchanged for the next quarter-century. A gradual increase was noted during the 1950s and 1960s, which greatly accelerated during the 1970s. This growth and range expansion owed much to the widespread planting of Multiflora Rose since the 1930s and the growth of suburbia along the Atlantic coast.

As a breeder, the mockingbird is widely distributed in lowland portions of the state and is less common or absent at higher elevations in central and western Massachusetts. It favors nesting sites near yards, especially those with lawns, shrubs, vines, and conifers, and also inhabits brushy forest edges or clearings and farm hedgerows.

The mockingbird is a talented songster and a clever mimic. Frank M. Chapman—a noted twentieth-century ornithologist—considered its reputation for mimicry overrated; but Forbush described it as a "most gifted singer" whose song "equals and even excels the whole feathered choir." Male mockingbirds often choose a prominent singing perch and during courtship may also hop, bound, flutter, and flash their wings. Songs are sometimes given in flight. In addition to its own varied notes, mockingbirds mimic other birds, insects, animals, human whistles, and even vehicle and machine noises. Various authors and growing numbers of unhappy sleepless auditors state that mockingbirds give their finest and most persistent vocal performances on moonlit nights during the nesting season. In Massachusetts, males sing occasionally at any time of year but are generally silent when they are molting. The common call note is a loud *smack* or *chuck*.

Male mockingbirds vigorously claim and defend breeding territories, sometimes fighting and other times using a ritualized dance involving hopping, stepping stiffly, raising the tail, and flashing the wings. Polygamy has also been documented for this species. Nesting habitats for 28 state nests were as follows: suburban (18 nests), old field (9 nests), orchard (1 nest) (CNR). Both sexes collect nesting material, including small twigs, grasses, rootlets, and dried leaves, but the female apparently does most of the actual construction. The bulky nest takes about two days to build and is generally placed in a

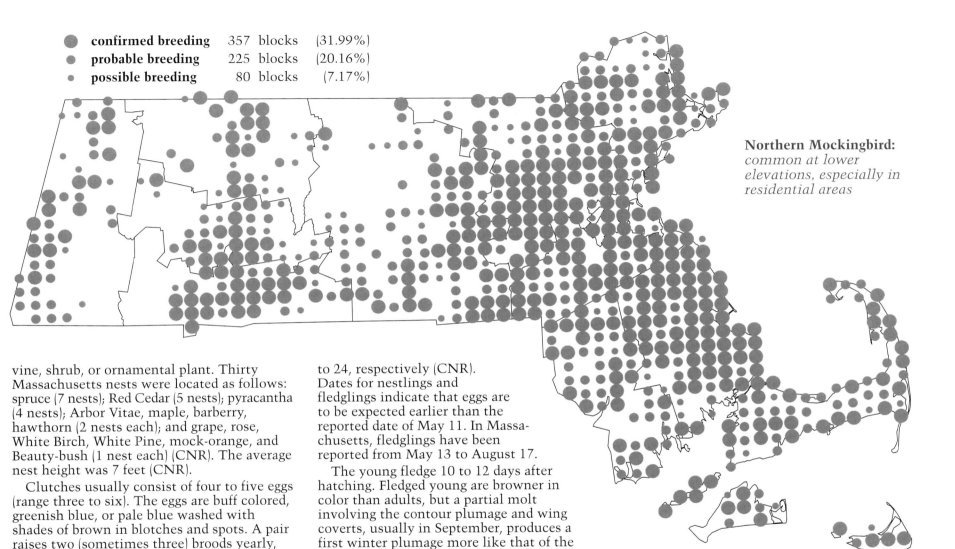

Northern Mockingbird: *common at lower elevations, especially in residential areas*

- confirmed breeding — 357 blocks (31.99%)
- probable breeding — 225 blocks (20.16%)
- possible breeding — 80 blocks (7.17%)

vine, shrub, or ornamental plant. Thirty Massachusetts nests were located as follows: spruce (7 nests); Red Cedar (5 nests); pyracantha (4 nests); Arbor Vitae, maple, barberry, hawthorn (2 nests each); and grape, rose, White Birch, White Pine, mock-orange, and Beauty-bush (1 nest each) (CNR). The average nest height was 7 feet (CNR).

Clutches usually consist of four to five eggs (range three to six). The eggs are buff colored, greenish blue, or pale blue washed with shades of brown in blotches and spots. A pair raises two (sometimes three) broods yearly, and they will renest after a failure. The female incubates for 12 days or slightly longer, and, after the young have hatched, both parents participate in their feeding and care, dive-bombing any bird, animal, or human that approaches them. Nestlings have been observed in the state from May 11 to September 1. Known state hatch dates are May 21 and June 4, and 2 three-egg clutches were laid from May 18 to 20 and May 22 to 24, respectively (CNR). Dates for nestlings and fledglings indicate that eggs are to be expected earlier than the reported date of May 11. In Massachusetts, fledglings have been reported from May 13 to August 17.

The young fledge 10 to 12 days after hatching. Fledged young are browner in color than adults, but a partial molt involving the contour plumage and wing coverts, usually in September, produces a first winter plumage more like that of the adult birds. Adult mockingbirds undergo a complete postnuptial molt. Following nesting and molt, some adults remain to set up a winter territory in the area where they bred while others move varying distances. Most immatures engage in a migratory dispersal that is as likely to take them north as south; this dispersal results in a wintering population that is concentrated along the coast. When a mockingbird establishes a winter territory, it makes its presence known by monitoring its territory from a conspicuous perch throughout the day. If the territory is small, the occupant may attempt to drive other birds away from it, but individual mockingbirds are variable in this behavior.

Helen C. Bates

Brown Thrasher
Toxostoma rufum

Egg dates: May 9 to June 24.

Number of broods: one, possibly two; may re-lay if first attempt fails.

The well-known and distinctive Brown Thrasher is widespread in Massachusetts but is more common in the eastern part of the state and in the southern Connecticut River valley than in the central and western sections. As a rule, the thrasher is shy and retiring, a bird of the countryside rather than the shrubbery about the lawn or garden. Brown Thrashers breed in areas of dry secondary growth, such as is found along high-tension-line clearings and overgrown pastures; in dry forests bordering fields; and in coastal thickets. They are absent in deep forests and in the higher mountains. Thrashers are most abundant in the oak-pine habitat of southeastern portions of the state.

The first spring migrants begin arriving in our area during the latter part of April and early May. Males may not sing for several days after settling into an area but become conspicuous once they begin to vocalize from the top of a shrub or tall deciduous tree that has yet to leaf out. They assume a characteristic singing pose with the tail directed downward. At planting time, a thrasher sitting on some treetop near a busy farmer sings, so the country people say, *Drop it, drop it, cover it, cover it, I'll pull it up, I'll pull it up,* and so it has been called the "Planting Bird." Singing occurs from late April to early July, but the frequency diminishes as nesting progresses. The song is bold and emphatic, and the phrases are abrupt and often distinctly double phrased. Although it may mimic the songs and calls of other birds, the thrasher is much less versatile in this regard than the familiar mockingbird. The common calls are several whistled notes and a loud *smack.*

During courtship, the male follows the female about, singing softly. The bulky nest is usually built in a bush or low tree, often on the ground, and is composed largely of twigs and roots and lined with fine rootlets. Bent reported that, in southeastern Massachusetts, half of the nests that he found were on the ground under bushes, trees, or thickets and half were in bushes, small trees, and brush heaps. The highest was located at 4 feet in an Arbor Vitae. In the vicinity of Boston, 10 of 23 nests were built on the ground (ACB). Nine recent state nests were recorded as follows: crabapple (1 nest), Multiflora Rose (2 nests), Red Cedar (1 nest), Bog Rosemary (1 nest), Pasture-juniper (1 nest), honeysuckle (2 nests), White Pine sapling (1 nest). Heights ranged from 2 to 6 feet, with an average of 3.6 feet (CNR).

Incubation of the three to five eggs is shared by both adults for 11 to 14 days. Clutch sizes for 7 Massachusetts nests were three eggs (2 nests), four eggs (4 nests), five eggs (1 nest) (CNR, DKW). The young remain

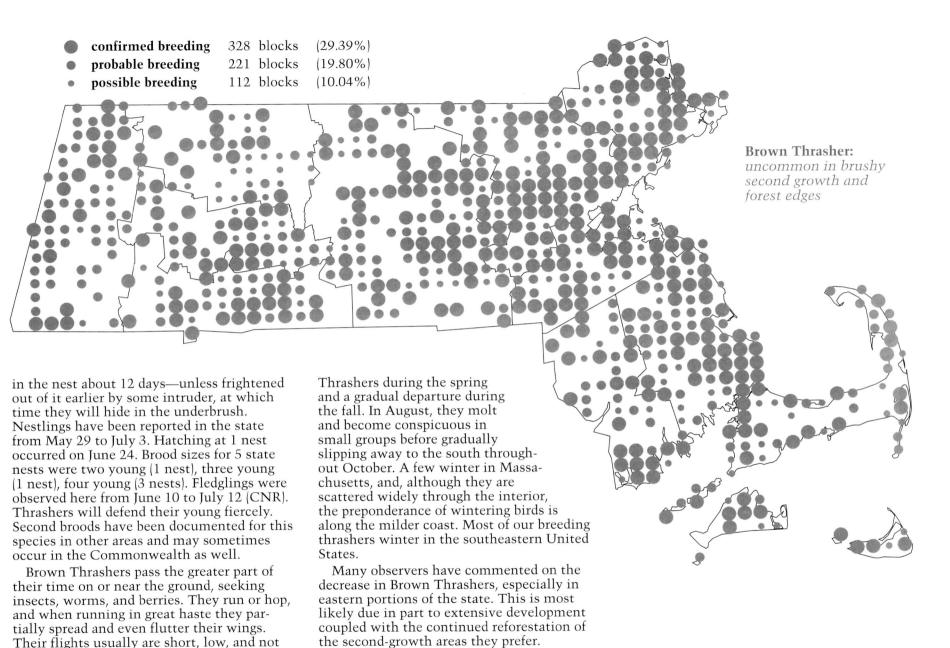

- **confirmed breeding** 328 blocks (29.39%)
- **probable breeding** 221 blocks (19.80%)
- **possible breeding** 112 blocks (10.04%)

Brown Thrasher: *uncommon in brushy second growth and forest edges*

in the nest about 12 days—unless frightened out of it earlier by some intruder, at which time they will hide in the underbrush. Nestlings have been reported in the state from May 29 to July 3. Hatching at 1 nest occurred on June 24. Brood sizes for 5 state nests were two young (1 nest), three young (1 nest), four young (3 nests). Fledglings were observed here from June 10 to July 12 (CNR). Thrashers will defend their young fiercely. Second broods have been documented for this species in other areas and may sometimes occur in the Commonwealth as well.

Brown Thrashers pass the greater part of their time on or near the ground, seeking insects, worms, and berries. They run or hop, and when running in great haste they partially spread and even flutter their wings. Their flights usually are short, low, and not very rapid. Thrashers like to dust themselves, and they are fond of bathing. There is a gradual accumulation of resident Brown Thrashers during the spring and a gradual departure during the fall. In August, they molt and become conspicuous in small groups before gradually slipping away to the south throughout October. A few winter in Massachusetts, and, although they are scattered widely through the interior, the preponderance of wintering birds is along the milder coast. Most of our breeding thrashers winter in the southeastern United States.

Many observers have commented on the decrease in Brown Thrashers, especially in eastern portions of the state. This is most likely due in part to extensive development coupled with the continued reforestation of the second-growth areas they prefer.

Ruth P. Emery

European Starling
Sturnus vulgaris

Egg dates: early April to July.

Number of broods: one or two.

Perhaps no other species of bird in Massachusetts is more widespread or familiar than the European Starling. It is not universally admired and may, in fact, be held in public contempt more than any other bird. The starling was reportedly first introduced in Cincinnati, Ohio, in 1872. From then until 1900, at least ten additional attempts at introduction were tried, including two in Massachusetts: Worcester (1884) and Springfield (1897). The first apparently successful introductions were in New York City's Central Park in 1890 and 1891.

Charles W. Townsend reported the first Essex County starling in 1908, and the species was first seen in the Concord region by Brewster in 1913. Beginning about 1920, there was a dramatic increase that continued unabated, and today the starling is widespread and abundant throughout most of inhabited North America in a wide variety of habitats. Currently, there are only a few locations in Massachusetts where it might conceivably not breed or be seen, such as dense woodlands.

The breeding chronology of resident birds is more subtle and prolonged than for migrant species from the south. One sign of approaching spring is a change in the bill color from dusky to yellow. The white-spotted winter plumage is transformed through wear into a shiny black with iridescent highlights on the head and body. In late February or early March, the winter flocks and small groups begin to disperse, and mated pairs start to explore nesting sites. At this time, the males become vocal, the song being little more than a varied series of squeaks, whistles, and rattles. However, the starling is greatly underappreciated as a mimic. Birdsongs that are frequently imitated tend to be the ones with simple, distinct phrases, such as those of the Eastern Meadowlark, Eastern Wood-Pewee, Killdeer, Eastern Phoebe, and Red-tailed Hawk. Also diagnostic in the song is the so-called "wolf whistle." The alarm calls, rapid chatters and rasping notes, are a sure indication of the presence of a predator or intruder. Young birds utter harsh, grating sounds.

Nest building may commence in early April, with most nests underway by late in the month. The nest site is almost any suitably sized cavity, either constructed by humans or natural. Old woodpecker holes and the rotted eaves of ill-kept houses offer excellent nest sites. There is some competition with native species for cavities, and the starling has been considered a major factor in the decline of the Eastern Bluebird. Twelve state nests were located as follows: holes or nooks in buildings (5 nests); nest box (2 nests); cavity in tree, including elm and Sugar Maple (5 nests). Heights ranged from 8 to 33 feet, with an average of 19 feet (CNR). The nest cavity is rather casually filled with grasses and occasional twigs and is often lined with feathers.

A clutch of four to six white, pale blue, or greenish eggs is typical. Clutch sizes for 2 state nests were four eggs each (CNR). Both sexes share in the approximately 12-day incubation period and in feeding the young. At some nests, the females were observed to assume the greater role in caring for the nestlings. Once the young have hatched, their constant begging for food reveals the location of the nest. Brood sizes for 4 state nests were two young (1 nest), three young (2 nests), four young (1 nest) (CNR). The young remain in the nest for about three weeks and can fly well once they depart. Sometimes the young fledge prematurely, perhaps due in part to the heavy parasite infestation typical of the nests. There is some variation in the breeding cycle, but in general in Massachusetts the young of the first brood begin to fledge in mid-May, with large numbers appearing by the end of the month into early June. Second-brood young fledge during July and sometimes during early August (EHF, CNR). Once fledged, the young are a distinctive uniform gray-brown. Both immatures and adults congregate on lawns, short grass fields, and especially mown hay fields, where they often associate with blackbirds.

As summer progresses, the feeding flocks increase in size as second broods augment the numbers. At this time, flocks can be seen in the early evening flying to communal roosts, which they share with robins and blackbirds. These summer roosts are usually established

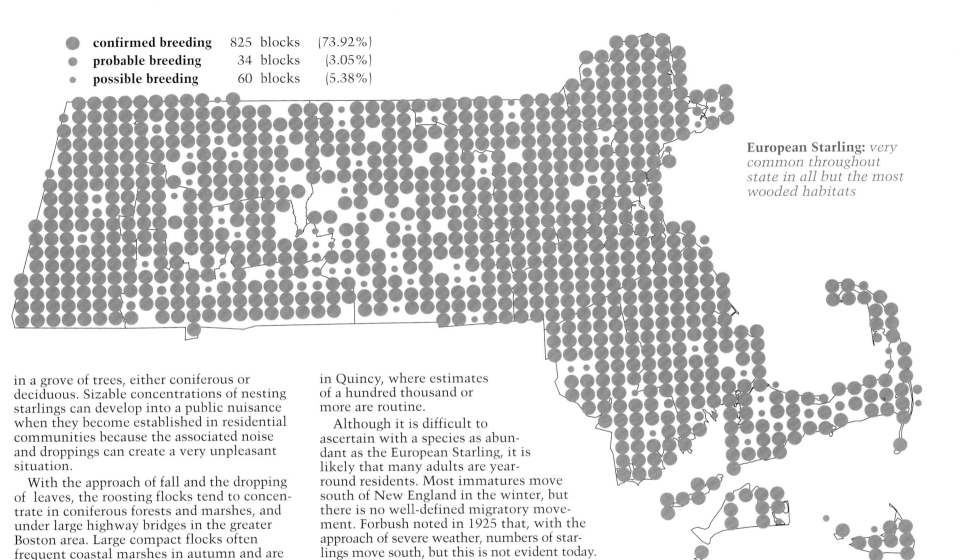

- confirmed breeding 825 blocks (73.92%)
- probable breeding 34 blocks (3.05%)
- possible breeding 60 blocks (5.38%)

European Starling: *very common throughout state in all but the most wooded habitats*

in a grove of trees, either coniferous or deciduous. Sizable concentrations of nesting starlings can develop into a public nuisance when they become established in residential communities because the associated noise and droppings can create a very unpleasant situation.

With the approach of fall and the dropping of leaves, the roosting flocks tend to concentrate in coniferous forests and marshes, and under large highway bridges in the greater Boston area. Large compact flocks often frequent coastal marshes in autumn and are sometimes seen twisting and turning in unison in a manner similar to shorebirds. During the winter, the roosting flocks gather in larger numbers in fewer locations. Starlings will roost in downtown Boston behind signs and on ledges of older buildings. Abandoned buildings or warehouses will often host thousands of starlings in a roost. Perhaps the best-known roost is the Fore River Bridge in Quincy, where estimates of a hundred thousand or more are routine.

Although it is difficult to ascertain with a species as abundant as the European Starling, it is likely that many adults are year-round residents. Most immatures move south of New England in the winter, but there is no well-defined migratory movement. Forbush noted in 1925 that, with the approach of severe weather, numbers of starlings move south, but this is not evident today.

The ecological amplitude of starlings must be admired, and it is a major contributing factor in their success in this country. They can be seen hawking through the air for insects like a swallow or flycatcher, feeding on berries in fall and winter like robins and waxwings, or gleaning agricultural fields like blackbirds. Further evidence of their adaptability is the way they crowd just ahead of a bulldozer at a dump, or they ignore trucks and cars whizzing past them as they feed on highway median strips. Lastly, it should be noted that starlings are directly beneficial to suburban homeowners because they frequently feed on grubs found in lawns.

Richard A. Forster

Cedar Waxwing
Bombycilla cedrorum

Egg dates: May 30 to August 13.

Number of broods: one; possibly sometimes two.

The occurrence of Cedar Waxwings throughout the year in Massachusetts is notoriously erratic. They breed throughout the state but are not evenly distributed. Nesting is widespread in the highlands of the western sections and, to a lesser extent, in the central uplands. The distribution in the coastal plain is more spotty, local, and often irregular from year to year. During the Atlas period, portions of Cape Cod and Bristol County had the fewest reported nest confirmations.

Wandering flocks of this species may be present in the state at any time, but it is generally believed that the individuals that breed here arrive in late May or June and that those seen earlier are northbound migrants. However, waxwings do not adhere strictly to any regular migration schedule and often do not nest immediately following their arrival. More confusion results from the fact that waxwings tend to form small feeding flocks even while they are breeding.

Cedar Waxwings occur in open scrub, old pastures, second-growth woods, orchards, gardens, and along the edges of waterways, but they avoid the open expanses of large fields and marshes as well as forest interiors. Though they are well-marked, distinctive birds, waxwings can be difficult to see in the foliage, at which time their thin, wheezy calls are often the best indication of their presence. These calls are high, lisping hisses represented as *see, see* or *ssse, sssee*. A quiet song of mellow full notes has been described but it is rarely heard.

Cedar Waxwings are late nesters, often not beginning to nest until midsummer. Courtship behavior includes wing quivering and food begging by the female, as well as bill rubbing and an exchange of berries between the sexes. This exchanging of berries from one bird to another is also seen at other times of the year and often involves several members of a flock.

Nests are typically located in orchards, in shade trees, at forest edges, or in suburban plantings. They tend to be on horizontal branches of either coniferous or deciduous trees, and often range from 4 to 40 feet off the ground. Cedar Waxwing nests are built by

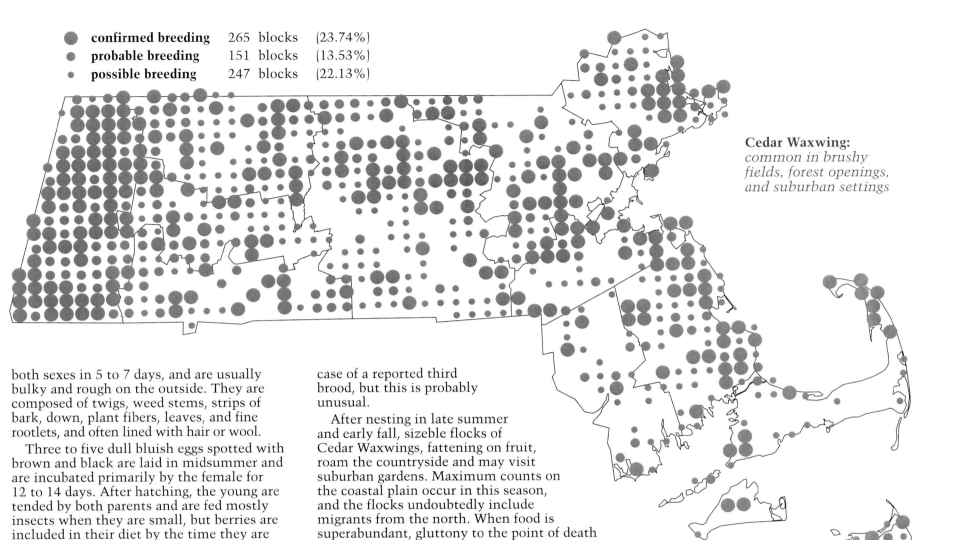

Cedar Waxwing: *common in brushy fields, forest openings, and suburban settings*

- confirmed breeding — 265 blocks (23.74%)
- probable breeding — 151 blocks (13.53%)
- possible breeding — 247 blocks (22.13%)

both sexes in 5 to 7 days, and are usually bulky and rough on the outside. They are composed of twigs, weed stems, strips of bark, down, plant fibers, leaves, and fine rootlets, and often lined with hair or wool.

Three to five dull bluish eggs spotted with brown and black are laid in midsummer and are incubated primarily by the female for 12 to 14 days. After hatching, the young are tended by both parents and are fed mostly insects when they are small, but berries are included in their diet by the time they are several days old. The young ordinarily fledge in 14 to 16 days, with the latest date for a nestling in Massachusetts being September 19 (EHF).

There are conflicting reports in the literature about the number of broods reared in a season, and this has not been well studied in Massachusetts, but the extended nesting period would certainly allow for pairs to rear two broods. Forbush actually mentions one case of a reported third brood, but this is probably unusual.

After nesting in late summer and early fall, sizeble flocks of Cedar Waxwings, fattening on fruit, roam the countryside and may visit suburban gardens. Maximum counts on the coastal plain occur in this season, and the flocks undoubtedly include migrants from the north. When food is superabundant, gluttony to the point of death has been described in this species; and stories of waxwings becoming intoxicated from eating fermented fruit are often quoted. Improbable as this seems, it does actually occur. After the fall molt, the young lose their breast streaks and resemble the adults.

During the winter, variable numbers of waxwings may show up anywhere in the state. They concentrate in sites where fruit remains on the trees, often traveling long distances in search of favorable areas. In this season, a careful search of the flocks may reveal one or more individuals of their rarer, wintering relative, the Bohemian Waxwing.

Norman P. Hill

Blue-winged Warbler
Vermivora pinus

Egg dates: May 20 to June 8.

Number of broods: one.

The Blue-winged Warbler is a relatively recent addition to the avifauna of Massachusetts. It was first recorded in 1857 in Dedham, followed by 1878 in West Roxbury, and was strictly accidental until after the turn of the century. Nesting was first confirmed in 1909 in Sudbury. By 1913, hybrids were noted in Lexington, evidence that interbreeding with the Golden-winged Warbler had begun. The next nest was discovered in Brockton in 1923.

By 1930, the species was considered a rare summer resident in the eastern part of the state and was accidental or absent from the western portions. The invading population became established in a band extending from southern Worcester County east to the coast and reaching north to Essex County and south to Cape Cod. During the 1930s, the Connecticut River valley became another population center. Blue-winged Warbler numbers climbed slowly until about 1955 and then rose markedly, until there was a veritable population explosion during the 1970s. The Berkshires and other western areas were the last regions to be colonized.

This range expansion received a great impetus resulting from the large patches of suitable habitat available after the widespread abandonment of farmland. Massachusetts traditionally was the breeding ground of the Golden-winged Warbler, a generally uncommon species with a few pockets of local abundance. Blue-winged and Golden-winged warblers are close relatives with very similar behaviors and requirements for nesting, and they interbreed freely where their ranges overlap. Crossbreeding and the subsequent backcrossing of hybrids produces a range of genetically mixed individuals that may look like either the parent species or may be either of two distinctive forms, the "Brewster's" and "Lawrence's" warblers. For reasons that are not yet completely understood, the Blue-winged Warbler seems to have replaced the Golden-winged Warbler in approximately 50 years after initial contact (see Golden-winged Warbler account).

During the Atlas period, the Blue-winged Warbler was a common breeder throughout the south-central and southeastern sections of the state and in portions of Essex County. It was less common in the Berkshires and sparsely distributed on upper Cape Cod and Martha's Vineyard. Although the species is now regularly reported from areas of northern New England, Massachusetts still basically marks the limit of the northeastern breeding range, and the population drops off very rapidly toward the northern borders of the state.

Favored breeding sites are old, brushy fields with scattered clumps of saplings and trees, wood clearings, border areas with low undergrowth, power-line cuts, and the edges of wooded swamps. Breeders arrive from late April through the first two weeks of May, when males carve out territories with much singing, chasing, and fighting. Unmated males sing tirelessly from elevated, and often bare, song perches and generally actively pursue females when they appear. These same song perches are also used later by mated males patrolling their territories. Males sing two types of songs that are used in different contexts. The more familiar pattern is the *bee-buzz* or *swee-zee* song, which occurs in many variations. The second song is longer and also variable and can be represented as *wee-chi-chi-chi-chur-chee-chur*. Both sexes have several *tzip* or *buzz* call notes and a thin *seet* note, and the young utter a long series of buzz notes as a begging call.

Nests, built mainly if not entirely by the female, are placed on or very close to the ground, often on a foundation of dry leaves, and are usually among or attached to the upright stems of weeds or grass clumps. They are composed of grass and pieces of leaves and are lined with finer material. A 1909 Sudbury nest was in mixed woods between the exposed roots of a decayed stump and partially concealed by ferns (ACB). Six recent nests from Worcester County were all located on the ground as follows: 1 in grass beneath saplings at the edge of a narrow opening next to dense brush, 2 in clumps of goldenrod, 1 in a grass clump in a field 5 or 6 feet out from the surrounding trees, 1 in a grass clump in a clearing with many tree seedlings, and 1 in ferns in a small overgrown orchard near woods. One of these nests opened to the side

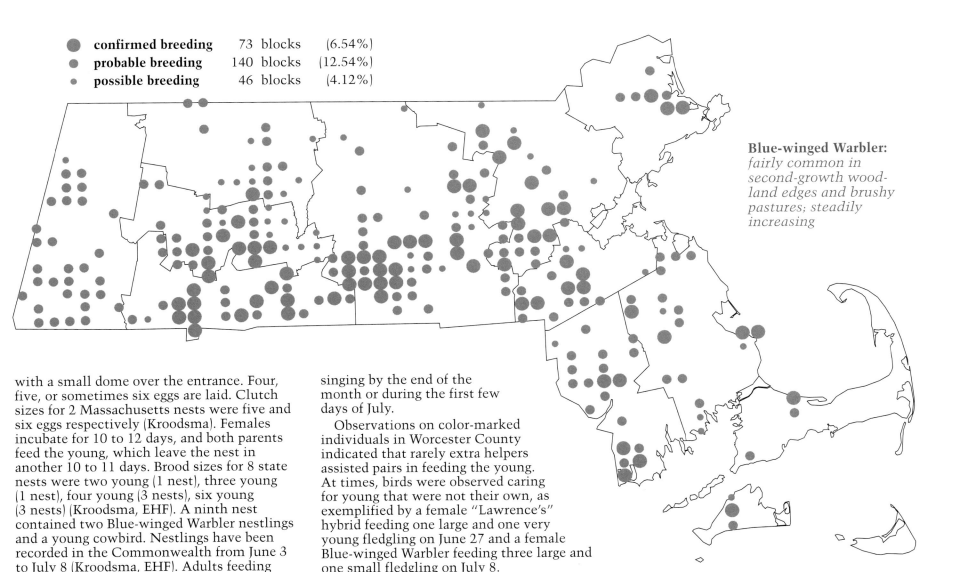

Blue-winged Warbler: *fairly common in second-growth woodland edges and brushy pastures; steadily increasing*

- confirmed breeding — 73 blocks (6.54%)
- probable breeding — 140 blocks (12.54%)
- possible breeding — 46 blocks (4.12%)

with a small dome over the entrance. Four, five, or sometimes six eggs are laid. Clutch sizes for 2 Massachusetts nests were five and six eggs respectively (Kroodsma). Females incubate for 10 to 12 days, and both parents feed the young, which leave the nest in another 10 to 11 days. Brood sizes for 8 state nests were two young (1 nest), three young (1 nest), four young (3 nests), six young (3 nests) (Kroodsma, EHF). A ninth nest contained two Blue-winged Warbler nestlings and a young cowbird. Nestlings have been recorded in the Commonwealth from June 3 to July 8 (Kroodsma, EHF). Adults feeding fledglings have been observed in the state from June 11 to August 16, with most records from June 27 to July 18.

One brood is raised each season, and a paucity of late nesting dates seems to indicate that second attempts are rare. Females are occasionally seen carrying nest materials in mid-June, but even unmated males cease singing by the end of the month or during the first few days of July.

Observations on color-marked individuals in Worcester County indicated that rarely extra helpers assisted pairs in feeding the young. At times, birds were observed caring for young that were not their own, as exemplified by a female "Lawrence's" hybrid feeding one large and one very young fledgling on June 27 and a female Blue-winged Warbler feeding three large and one small fledgling on July 8.

Adults undergo a complete molt in late July and early August, during which time the young attain their yellowish winter plumage. Throughout August, Blue-winged Warblers forage in mixed-species flocks, which they accompany into wooded areas as well as border zones. Most depart on their southward flight at the end of August and during the first week of September. Only a few stragglers are encountered after that time. The wintering grounds include Mexico, Central America, and northern South America.

W. Roger Meservey

Golden-winged Warbler
Vermivora chrysoptera

Egg dates: May 20 to June 24.

Number of broods: one.

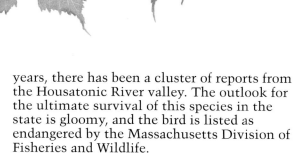

The Golden-winged Warbler was formerly an uncommon migrant and uncommon to locally common breeding resident throughout much of Massachusetts, with the exception of the southeastern portion of the state and the Islands, where it is absent. The first state nest was found in 1869, and after 1874 the population apparently increased as more habitat became available. Reports indicate that the species was more prevalent in the eastern sections of the state at that time.

A population decline of the Golden-winged Warbler throughout much of its range has been well documented, and the major reasons for that decline are twofold. First, for much of the nineteenth century, the Massachusetts landscape was approximately 85 percent cleared of forests for agricultural and logging purposes. Subsequent changing land use patterns allowed abandoned pasture to grow into second-growth woodlands, ideal habitat for Golden-winged Warblers. Over time, these woodlands matured into hardwood (primarily oak) forests that were unsuitable for Golden-winged Warblers. Second, since the late 1950s, the population of the closely related Blue-winged Warbler has increased dramatically and spread into what little Golden-winged Warbler habitat remained (see Blue-winged Warbler account). In addition to the well-known hybridization of the two species, the Blue-winged Warbler is behaviorally more aggressive and eventually replaces the Golden-winged Warbler.

The situation in Worcester County illustrates this pattern of replacement. Prior to the late 1930s, the Golden-winged Warbler was a regular, but uncommon, nester throughout the county. The first Blue-winged Warbler was reported in 1938, and by the early 1960s the species had established itself as a common breeder. "Brewster's" hybrids have been reported from 1943 onward, and the rarer "Lawrence's" hybrid (which indicates an advanced state of contact between the two parent species) has been reported from 1968 to the present. By the late 1970s, reports of successful nesting of the Golden-winged Warbler practically ceased, and most reports were of unmated males, with a high proportion giving Blue-winged Warbler type songs and aberrant songs. During the Atlas period, the Golden-winged Warbler was rare and declining. Since the close of the Atlas period, there have been a few records of migrants and lone territorial males.

In eastern Massachusetts, the decline in Golden-winged Warblers since the late 1950s has been equally serious. As an example, there were approximately 25 pairs (based on singing males) in Wellesley in 1954, where none could be found by 1980. The last strongholds for them were in northern Essex County from North Andover to West Newbury, with another population pocket persisting in the Franklin-Medfield area, but both of these have since been replaced by Blue-winged Warblers and hybrids.

Although the Golden-winged Warbler was not confirmed breeding in Berkshire County during the Atlas period, the number of probable reports indicated that nesting was occurring. In western Massachusetts, in alarmingly low numbers, the Golden-winged Warbler is making its last stand. In recent years, there has been a cluster of reports from the Housatonic River valley. The outlook for the ultimate survival of this species in the state is gloomy, and the bird is listed as endangered by the Massachusetts Division of Fisheries and Wildlife.

The favored habitat of the Golden-winged Warbler is second-growth woodlands, especially on hillsides, where Gray Birch and associated shrubby vegetation predominate. Other suitable nesting areas are damp meadows overgrown with weeds and saplings, power-line cuts, and abandoned orchards. Male Golden-winged Warblers arrive on the breeding grounds during the first week of May. They immediately set up territories, which include the brushy areas nearby and their forest edge, and announce their presence by singing their distinctive buzzy song. The song is frequently given from an exposed dead branch of a tree at the edge of the woodland. Typical songs consist of three or four notes, the introductory note higher than the succeeding ones, which are all on the same pitch: *zeee zer-zer-zer* or *zee-ze-ze*. Songs of a second type are used in different

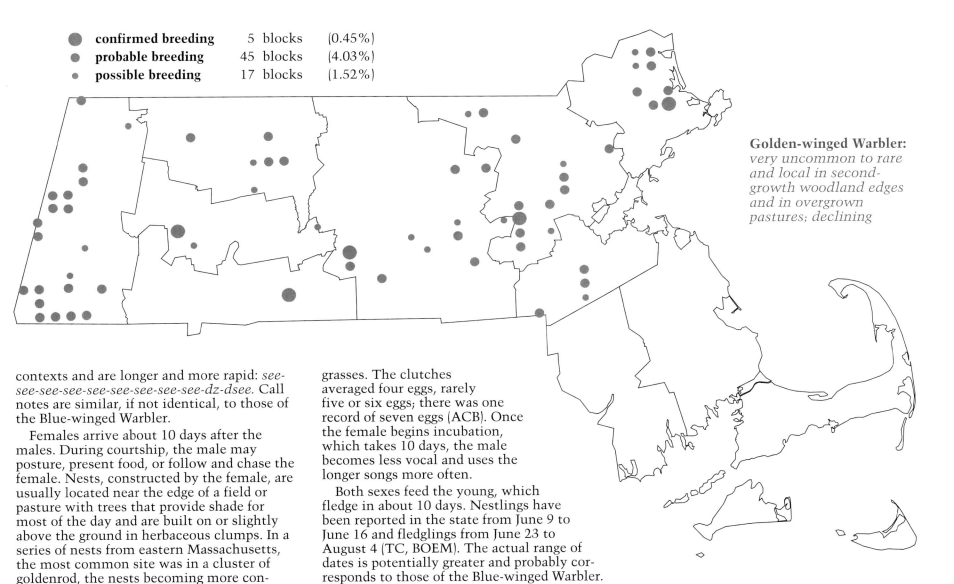

Golden-winged Warbler: *very uncommon to rare and local in second-growth woodland edges and in overgrown pastures; declining*

● confirmed breeding	5 blocks	(0.45%)
● probable breeding	45 blocks	(4.03%)
• possible breeding	17 blocks	(1.52%)

contexts and are longer and more rapid: *see-see-see-see-see-see-see-see-see-dz-dsee*. Call notes are similar, if not identical, to those of the Blue-winged Warbler.

Females arrive about 10 days after the males. During courtship, the male may posture, present food, or follow and chase the female. Nests, constructed by the female, are usually located near the edge of a field or pasture with trees that provide shade for most of the day and are built on or slightly above the ground in herbaceous clumps. In a series of nests from eastern Massachusetts, the most common site was in a cluster of goldenrod, the nests becoming more concealed as the weeds grew. Other nests were found at the bases of briers and small saplings, in tussocks of meadow grass, and in clumps of low weeds such as ironweed (ACB). The nest itself is a bulky structure, usually set on a base of dead leaves and constructed of dry leaves with a coarse lining of bark and grasses. The clutches averaged four eggs, rarely five or six eggs; there was one record of seven eggs (ACB). Once the female begins incubation, which takes 10 days, the male becomes less vocal and uses the longer songs more often.

Both sexes feed the young, which fledge in about 10 days. Nestlings have been reported in the state from June 9 to June 16 and fledglings from June 23 to August 4 (TC, BOEM). The actual range of dates is potentially greater and probably corresponds to those of the Blue-winged Warbler. One brood is raised each season, and there is no information on renesting. The adults undergo a complete molt after nesting, and the young acquire their first winter plumage about a month after fledging. Until they depart in late August and early September, Golden-winged Warblers join mixed-species flocks. The wintering grounds are from Central America to northern South America.

W. Roger Meservey and Richard A. Forster

Nashville Warbler
Vermivora ruficapilla

Egg dates: May 21 to July 21.

Number of broods: one.

A century and a half ago, two-thirds of the land in Massachusetts had been cleared for agriculture. Accordingly, the Nashville Warbler, essentially a bird of habitat that was formerly farmland, especially of fields grown up with saplings and shrubs, was comparatively rare and little known. The species was also a scarce migrant throughout the East. Alexander Wilson, who inappropriately named the species based on a specimen he secured in spring migration in Nashville, Tennessee, hardly knew the bird in life.

As increasing amounts of farmland were abandoned and allowed to regenerate to woodland, the Nashville Warbler began to increase in numbers, and by the turn of the century it had become quite common in many areas. Today, it is a widespread breeding bird, with its Massachusetts stronghold in the drier uplands of the western counties; it is now less common than formerly in eastern sections of the state.

The first Nashville Warblers arrive in Massachusetts in late April, and spring migration continues throughout May. During this time, they can be found in a wide range of habitats, including parks, orchards, and woodland edges. Like other warblers, the Nashville seems to be in perpetual motion, flitting about in the mid- and upper-level branches, now plucking an inchworm from the edge of an emerging leaf, now pausing to break into ebullient song, now dashing out of sight into the thick foliage. Almost completely insectivorous, the Nashville Warbler takes a variety of adult and immature insects, including beetles, flies, and grasshoppers. Caterpillars make up a significant part of the diet.

The song of the Nashville Warbler is a typical wood-warbler "nonwarbled song." It is regularly comprised of two parts, the first a series of high, double-note phrases and the second a series of lower-pitched notes delivered at about twice the speed of the first part. To human ears, the song approximates the following: *see it, see it, see it, see it, ti, ti, ti, ti, ti, ti, ti*. Both sexes give *chip* or *pink* call notes.

On its breeding ground, the male is in full song, advertising his presence from early May to early June. Uncharacteristic of warblers once nesting is underway, the singing drops off markedly and the birds become very inconspicuous. Some males exhibit another brief burst of singing once the young have fledged (Meservey). Although this species occupies diverse types of breeding habitats across its North American range, in Massachusetts it is generally found nesting only in two types. In the southeastern part of the state, open Scrub Oak woodlands seem to be preferred, while, in the central and western sections, Nashville Warblers most often choose pastures and fields overgrown with Red Cedar and dense stands of young Gray Birch and Quaking Aspen.

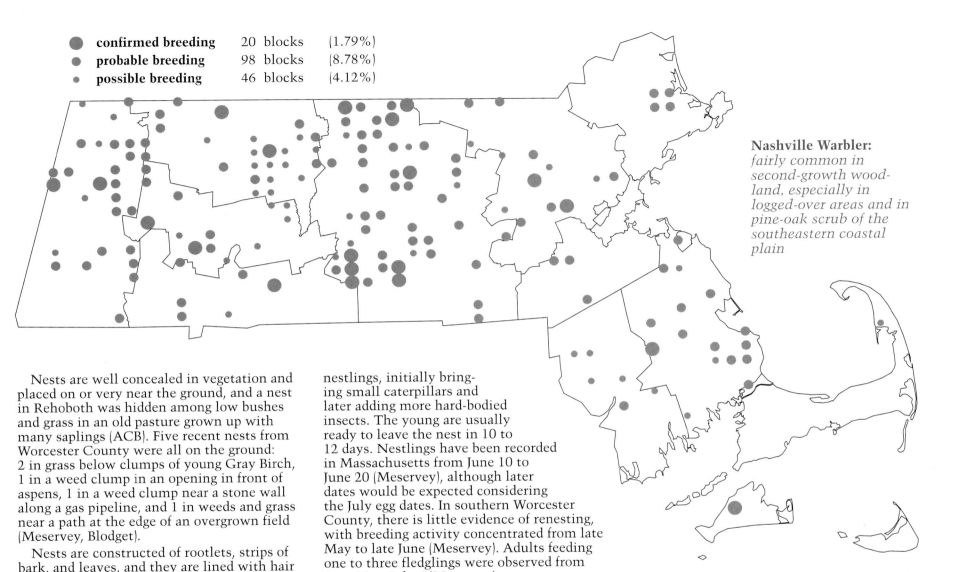

- confirmed breeding — 20 blocks (1.79%)
- probable breeding — 98 blocks (8.78%)
- possible breeding — 46 blocks (4.12%)

Nashville Warbler: *fairly common in second-growth woodland, especially in logged-over areas and in pine-oak scrub of the southeastern coastal plain*

Nests are well concealed in vegetation and placed on or very near the ground, and a nest in Rehoboth was hidden among low bushes and grass in an old pasture grown up with many saplings (ACB). Five recent nests from Worcester County were all on the ground: 2 in grass below clumps of young Gray Birch, 1 in a weed clump in an opening in front of aspens, 1 in a weed clump near a stone wall along a gas pipeline, and 1 in weeds and grass near a path at the edge of an overgrown field (Meservey, Blodget).

Nests are constructed of rootlets, strips of bark, and leaves, and they are lined with hair and fine grasses. The clutch of four or five creamy, lightly speckled eggs is incubated by the female for 11 to 12 days. A nest in Princeton contained three eggs on May 25 (Blodget), and 1 nest from the Millbury area held four eggs on June 8 (DKW). Three nests from Pelham were all discovered during June (Nice 1933). Both parents provide food for the nestlings, initially bringing small caterpillars and later adding more hard-bodied insects. The young are usually ready to leave the nest in 10 to 12 days. Nestlings have been recorded in Massachusetts from June 10 to June 20 (Meservey), although later dates would be expected considering the July egg dates. In southern Worcester County, there is little evidence of renesting, with breeding activity concentrated from late May to late June (Meservey). Adults feeding one to three fledglings were observed from June 28 to July 6 (Meservey).

With the advent of early fall and completion of the molt, Nashville Warblers are on the move. Some begin to depart in late August, and many transients pass throughout September and early October. As in spring, migrants can be observed in a variety of habitats. Occasionally, individuals have attempted to overwinter in thick coastal vegetation, but few, if any, survive through December. The wintering grounds extend from the southern United States south through Mexico to Guatemala.

Brian E. Cassie

Northern Parula
Parula americana

Egg dates: May 20 to July 7.

Number of broods: one; sometimes two.

This diminutive warbler is a regular spring and fall migrant throughout the state, but the number of local breeding pairs is decreasing. In the late 1800s and early 1900s, the Northern Parula was a common breeder in Bristol County and parts of Cape Cod and nested locally in other areas of eastern Massachusetts. Loss of habitat from development and the creation of cranberry bogs contributed to the decline. Another factor was the unexplained disappearance of old man's beard lichen, which seems to be a requisite for nesting in many sections of the state.

During the Atlas period, nesting was confirmed only along the Mashpee River and in the Harwich marshes on Cape Cod. A few other possible sites were in Plymouth County and on Martha's Vineyard and the Elizabeth Islands. Much of the area bordering the Mashpee River has suffered encroachment by homes, roads, and developments. An incomparably rich estuary and waterway has been reduced to a tattered remnant of its former self. What remains is still probably the best place in Massachusetts to look for this smallest of all eastern warblers during the nesting season.

The first spring migrants appear in late April and early May in a variety of habitats but seem to prefer wet areas near bogs or streams or along rivers and swamps. They often frequent one of the larger trees, maybe an oak putting out tassels, high up and difficult to spot, the males singing sporadically. Often, an individual will remain in the same tree for an hour or more, busily pursuing small insects and in no hurry to fly farther on. Birch, Pitch Pine, and even cedar may harbor the inchworms and grubs that parulas prefer as food. Northbound migrants continue to pass through during May, with the last stragglers continuing even into early June.

The thin, wheezy, and ascending song is more buzz than trill and usually ends, after a slight crescendo, quite abruptly, as if cut off with a snap. Variations on this vocalization, heard during the nesting season, retain the same unmusical buzzy quality. In common with most species of warblers, the Northern Parula has several call notes. A *chip* similar to that of the Yellow Warbler is given and also a high thin *sip* note.

Little has been recorded about territory establishment or courtship in this species. Gray-green, feathery lichen commonly known as old man's beard is the preferred nesting material of the parula. Where this lichen is found, this warbler may nest, especially if the area is moist. Both the lichen and the parula have been declining in many parts of the Northeast for much of the twentieth century. In Bristol County, many nests were formerly found in neglected orchards where apple trees were festooned

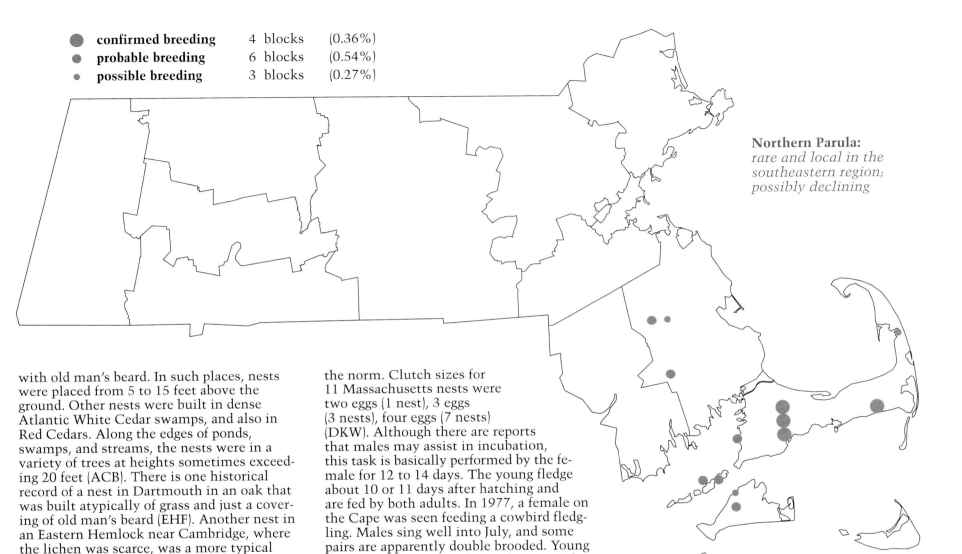

Northern Parula: *rare and local in the southeastern region; possibly declining*

- confirmed breeding — 4 blocks (0.36%)
- probable breeding — 6 blocks (0.54%)
- possible breeding — 3 blocks (0.27%)

with old man's beard. In such places, nests were placed from 5 to 15 feet above the ground. Other nests were built in dense Atlantic White Cedar swamps, and also in Red Cedars. Along the edges of ponds, swamps, and streams, the nests were in a variety of trees at heights sometimes exceeding 20 feet (ACB). There is one historical record of a nest in Dartmouth in an oak that was built atypically of grass and just a covering of old man's beard (EHF). Another nest in an Eastern Hemlock near Cambridge, where the lichen was scarce, was a more typical nest, the birds having managed to gather enough of the material for construction (ACB). On Cape Cod, nests have been found in maples, oaks, and pines.

The nest is usually a hanging basket of lichen with a side door. It may be unlined or contain small amounts of fine grass, hair, pine needles, or plant down. Four or five white eggs dotted with brownish spots are the norm. Clutch sizes for 11 Massachusetts nests were two eggs (1 nest), 3 eggs (3 nests), four eggs (7 nests) (DKW). Although there are reports that males may assist in incubation, this task is basically performed by the female for 12 to 14 days. The young fledge about 10 or 11 days after hatching and are fed by both adults. In 1977, a female on the Cape was seen feeding a cowbird fledgling. Males sing well into July, and some pairs are apparently double brooded. Young birds and females may be confusing to identify but share with the male a greenish yellow patch between the scapulars and extending a short way down the back—hence the old name of "Blue Yellow-backed Warbler."

Fall migrants may appear in late July, but most of the movement occurs from late August through September. Records for October are not uncommon, and there are some reports even through early November. The species winters from Florida south through the West Indies, Mexico, and Central America.

The Northern Parula is listed as a threatened species in Massachusetts.

Robert Pease

Yellow Warbler
Dendroica petechia

Egg dates: May 17 to July 3.

Number of broods: one; may re-lay if first attempt fails.

This brilliant little bird is one of the most widespread wood-warblers in North America, having a virtually continentwide breeding range. Its distribution, combined with its bright color, its loud, sweet, persistent singing, and its accommodation to a variety of habitats, makes the Yellow Warbler one of the most familiar of its tribe. The species was confirmed nesting statewide during the Atlas period, and was most abundant in eastern Massachusetts. Its chosen habitats vary but are most frequently wet meadows, streamsides, swamp or marsh borders, thickets, hedgerows, roadsides, or brushy bottomlands. It may also choose orchards, cutover lands, or even open forest.

After initiating a northward migration on the heels of winter, the Yellow Warbler does a remarkably good job of not crossing the latitudes faster than spring advances, rarely arriving in Massachusetts before the last week of April. General arrival is the first week of May, with the largest numbers, many probably transients, coming the second and third weeks.

Yellow Warbler songs are variable but as is the case with many other warbler species can be divided into two main types used in different contexts. The emphatic songs of the first type can be represented as *tseet-tseet-tseet-sitta-sitta-see* while the second type, which predominates as nesting progresses, can be phrased as *sweet-sweet-sweet-sweet-sweeter-sweeter*. Both sexes have a loud *chip* note and a high *zeet* call given in flight.

Yellow Warbler territories are variable in size and may be as small as two-fifths of an acre, but some pairs in southern Worcester County held territories of at least 2 acres (Meservey). In territories where trees or other landmarks are absent, the males engage in much chasing back and forth, but, when the boundaries are more readily defined, fewer disputes occur.

After a noisy courtship involving much chipping by both sexes and chasing of the female by the male, the former selects a nest site in a low bush or shrub, generally no higher than 3 to 8 feet. The nest may be placed higher in a sapling or tree if an understory growth is absent. Six nests from eastern Massachusetts ranged in height from 1 to 5 feet, with an average of 2.5 feet, and were placed in forsythia, Red Maple, honeysuckle, barberry, Speckled Alder, and spiraea (CNR). In Berkshire County, 1 nest was located at 8 feet in an unidentified bush; 2 others in Box Elder were at 9 and 10 feet (CNR). Five southern Worcester County nests averaged 4 feet in height; 3 of these were in spiraea and 2 in raspberry (Meservey). In the Amherst area, nest construction was noted for 6 nests from May 20 to June 12, with a median date of May 24. The heights of these and other nests in the region ranged from 1 to 8.5 feet and averaged 4 to 5 feet. They were placed generally in the outer portions of the bush or tree, often in an isolated plant. The most commonly used species was Multiflora Rose; other identified substrates included Elderberry, willow, raspberry, honeysuckle, and blueberry (Spector). In Essex County, spiraea and Speckled Alder were favored nest sites.

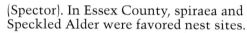

The female constructs the nest while the male sings as many as 3,240 songs in one day. Nests typically contain a large proportion of plant down and spider silk and are easily spotted because they are so white. Three to six eggs are laid, commonly four or five, and incubation is by the female for about 11 days. Clutch sizes for 28 state nests were two eggs, probably incomplete (1 nest); three eggs (1 nest); four eggs (16 nests); five eggs (10 nests) (DKW, CNR, Spector, Meservey). In an Amherst study, cowbird eggs were present in 4 of the 9 nests found in the egg stage (excluding the clutch of two). The mean clutch size for 6 nests with only warbler eggs (again excluding the two-egg clutch) was 4.5 (Spector). Known state hatch dates were June 5, June 6, June 7, and June 13 (Spector, Meservey).

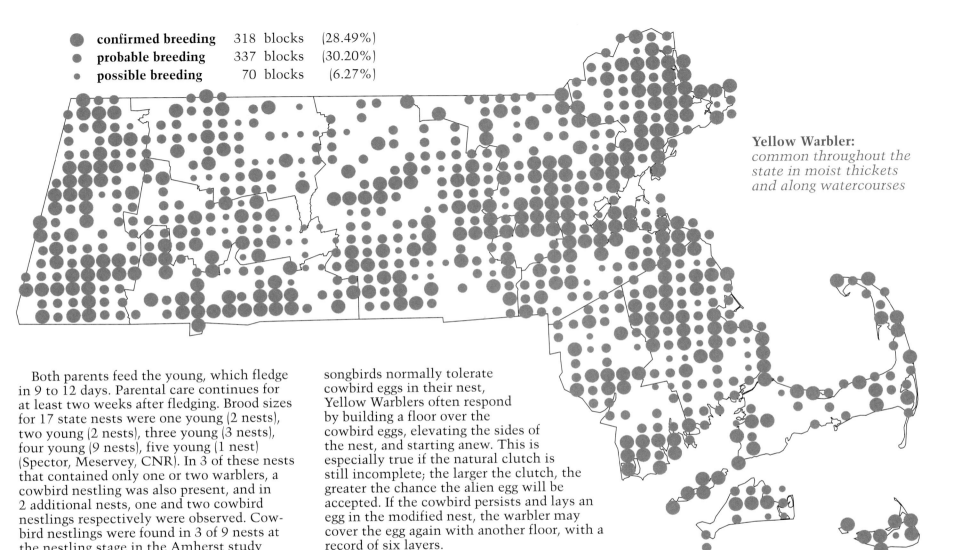

Yellow Warbler: *common throughout the state in moist thickets and along watercourses*

- confirmed breeding — 318 blocks (28.49%)
- probable breeding — 337 blocks (30.20%)
- possible breeding — 70 blocks (6.27%)

Both parents feed the young, which fledge in 9 to 12 days. Parental care continues for at least two weeks after fledging. Brood sizes for 17 state nests were one young (2 nests), two young (2 nests), three young (3 nests), four young (9 nests), five young (1 nest) (Spector, Meservey, CNR). In 3 of these nests that contained only one or two warblers, a cowbird nestling was also present, and in 2 additional nests, one and two cowbird nestlings respectively were observed. Cowbird nestlings were found in 3 of 9 nests at the nestling stage in the Amherst study (Spector). The mean brood size for 8 nests with only warbler nestlings was 3.375 (Spector). Nestlings were reported in Massachusetts from June 5 to July 4 and dependent fledglings from June 14 to July 26 (Spector, Meservey, CNR).

Perhaps the most interesting aspect of this species' nesting behavior is its defense against cowbird predation. While many small songbirds normally tolerate cowbird eggs in their nest, Yellow Warblers often respond by building a floor over the cowbird eggs, elevating the sides of the nest, and starting anew. This is especially true if the natural clutch is still incomplete; the larger the clutch, the greater the chance the alien egg will be accepted. If the cowbird persists and lays an egg in the modified nest, the warbler may cover the egg again with another floor, with a record of six layers.

The Yellow Warbler often begins its southbound migration as soon as the young can fend for themselves. Some adults molt on or near the breeding areas. Many are gone by the end of July, and most local birds depart by mid-August. Subsequent records, which continue throughout September, are likely of more northern birds. After this, only rare stragglers are noted. In its winter range, from Mexico and the Bahamas to northern South America, where it may arrive as early as August, the species prefers habitats similar to those used in its breeding range. It is also territorial on the wintering grounds.

James Berry

Chestnut-sided Warbler
Dendroica pensylvanica

Egg dates: May 22 to July 19.

Number of broods: one or two.

The Chestnut-sided Warbler is an excellent example of a species that has prospered because of human influence on the environment. Because it requires forest edge and regions of secondary growth for nesting, it was rare in the heavy forests that covered much of Massachusetts before European settlement. The birds benefited as large areas of forest were cut because of the increase in edge. With the decline of agriculture and subsequent abandonment of farmland beginning in the late nineteenth century, the species rapidly became a common breeder as widespread acreage began to revert to forest.

The heyday of the Chestnut-sided Warbler is past. As the Massachusetts forests matured during the twentieth century, the population of this warbler declined, particularly in eastern portions of the state. It is still common today through western and central Massachusetts and eastward to Essex County, but it seems a certainty that these numbers will be severely reduced in the future.

Breeding habitats are areas of secondary growth, shrubs, and a dense understory of herbaceous plants. The most attractive sites are old, abandoned fields and pastures, where patches of trees intermix with stands of saplings and clumps of briers and other low-growing plants. Other favorable locales are woodland clearings, brook and roadside borders, and power-line cuts. Territories generally include portions of the surrounding mature forest as well as brushy areas.

Most residents arrive during the first week of May, although some return later, and a few appear during the last days of April. Because the breeding grounds extend north to southern Canada, migrants continue to pass through during May. The males are highly territorial, and, because the species is so common, the claiming of an area usually involves extensive periods of singing interspersed with chases and fights. This time-consuming activity continues throughout the breeding season.

Males have two basic types of songs, both of which are variable. The first is characterized by a distinctive, accented end note and is represented as *chee, chee, chee, chee, swee, beat, you,* while the second is a more rambling song without the accented ending, i.e., *chew, chew, chew, chew, chee, chee, chitchee, witchee.* The two song types are apparently used in different contexts. Males also have definite favored song perches. These are typically elevated and exposed; and, if a dead snag or bare branch is available, it will almost certainly be used. Courtship usually includes chases, and the male often follows the female, hopping about in his characteristic manner with wings drooped and tail slightly cocked. Both sexes have the typical warbler *chip* and *seet* notes. In addition, the young give a series of *chip* notes as a begging call.

Nests, constructed by the females, are composed of coarse grass, plant fibers, bark strips, and spider webs and have a lining of fine grass, rootlets, or hair. These are placed in a low bush, sapling, brier, or other herbaceous plant from 1 to 6 feet above the ground, with 2 feet being the average height. In eastern Massachusetts, nests have been found from 14 inches to 3 feet high in hazelnut and also in huckleberry, blackberry, Hardhack, and saplings (ACB). Worcester County nests have been located in blackberry and other brambles (6 nests), low bushes in old fields or wood borders (4 nests), and spiraea (1 nest). The earliest of these nests was under construction on May 13 and the last on July 3.

Four eggs are usually produced, but there may be three or five. Clutch sizes for 9 state nests were three eggs (4 nests), four eggs (5 nests) (DKW). The female incubates for 10 to 12 days, and the young remain in the nest for 9 to 12 days, during which time they are cared for by both parents. By the end of June, most of the first-brood young have fledged. They continue to be fed for the next several weeks as they gradually attain independence. Two broods are commonly reared, and during early July the female builds a

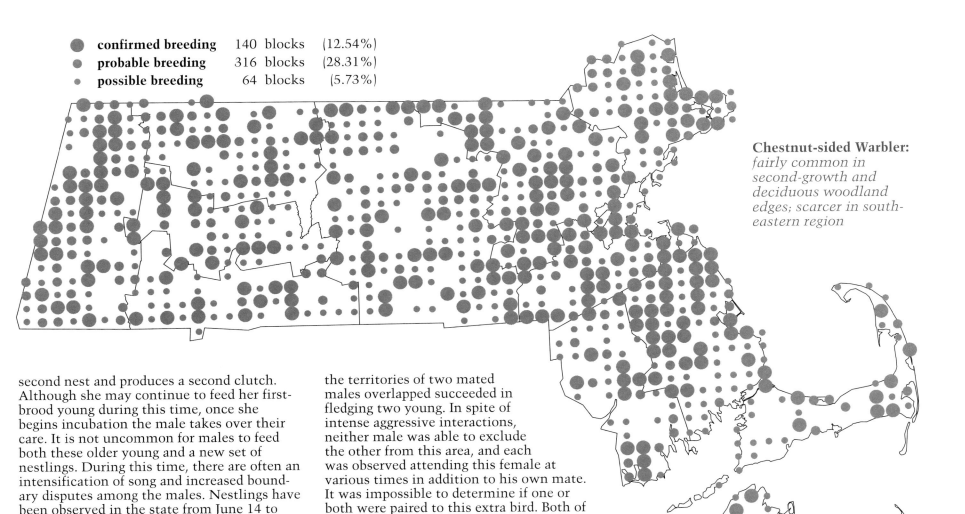

- confirmed breeding — 140 blocks (12.54%)
- probable breeding — 316 blocks (28.31%)
- possible breeding — 64 blocks (5.73%)

Chestnut-sided Warbler: *fairly common in second-growth and deciduous woodland edges; scarcer in southeastern region*

second nest and produces a second clutch. Although she may continue to feed her first-brood young during this time, once she begins incubation the male takes over their care. It is not uncommon for males to feed both these older young and a new set of nestlings. During this time, there are often an intensification of song and increased boundary disputes among the males. Nestlings have been observed in the state from June 14 to August 3. A Charlton nest contained four large young on July 2. Three known fledging dates are June 27, July 3, and July 26. Fledged dependent young were recorded here from June 18 to August 26.

Intensive studies of color-banded individuals in Worcester County revealed that many birds formed monogamous pairs, but there were also some interesting discoveries. On a number of occasions, adults were observed feeding fledglings that were not their own. In one case, a female nesting in a zone where the territories of two mated males overlapped succeeded in fledging two young. In spite of intense aggressive interactions, neither male was able to exclude the other from this area, and each was observed attending this female at various times in addition to his own mate. It was impossible to determine if one or both were paired to this extra bird. Both of the resident females responded aggressively toward her.

By mid-August, many young are attaining winter plumage and are dispersing, and the adults are in a complete molt, some still tending late fledglings. From early August onward, individuals join the mixed flocks of Black-capped Chickadees and warblers that occur in this season. Chestnut-sided Warblers are characteristic members of these flocks and accompany them into deep woods as well as more open situations.

Most of the resident birds leave at the end of August and the first week of September; migrants continue to be observed until the end of September. The wintering range is in Central America from Guatemala to Panama.

W. Roger Meservey

Magnolia Warbler
Dendroica magnolia

Egg dates: June 15 to July 30.

Number of broods: one or two.

The Magnolia Warbler's attractive plumage colors and patterns make it a favorite of most observers. Although the species is a common spring and fall migrant throughout the state, it is a much less common and often markedly local breeder. It favors regions of higher elevations and is invariably attracted to areas supporting coniferous trees such as spruce, Eastern Hemlock, Balsam Fir, or Red Cedar. The breeding range in Massachusetts centers in Berkshire County and the western portions of Franklin, Hampshire, and Hampden counties. To the east, the species is encountered only locally, with the exception of northwestern Worcester County, where it is well distributed but uncommon.

The current state breeding range of the Magnolia Warbler almost exactly matches that occupied in the 1920s, but Forbush also showed a few widely scattered breeding season records for Middlesex, Essex, and Plymouth counties. Historically, the species may have been more widely distributed in Massachusetts, hanging on for a time in a few isolated eastern pockets, but some combination of factors has since led to the disappearance of these outposts.

During the first week of May, breeding birds begin to appear, and by the middle of the month the species becomes common as later arrivals show up along with many northbound migrants. By late May, the migrant wave has passed, but a few stragglers are observed into the first week of June. Resident males claim and defend breeding territories, which are occasionally in deep coniferous forest but are more commonly in more open areas in one of two types of situations. The first is a forest clearing where young spruces and hemlocks can be found in addition to more mature trees; the second is an abandoned field or pasture grown up to Red Cedars, pines, and Pasture-junipers. In certain favorable locales, several territories may abut one another. It is difficult to understand why some areas support breeding populations of this species while nearby zones of apparently suitable habitat are not occupied.

As is the case with many warbler species, unmated male Magnolia Warblers often sing repeatedly for long periods of time, and once a female arrives it may be pursued or followed about the territory. Songs are variable but can be divided into two main types. The more familiar pattern can be represented as *wee-o, wee-o, wee-chy* or *wichy, wichy, wee-see,* and the alternate as *weetee, weet, wur,* with a low ending note. Calls include a loud *chip* note, a characteristic softer *tlep* note, a scolding *tit-tit-tit,* and a unique and surprisingly jaylike *de kay, kay, kay,* uttered when the birds are agitated. The young have a begging call consisting of a long series of two- or three-note phrases: *tsee-tsee, tsee, tsee, tu,* etc.

Nests are almost always situated in a conifer, from 1 to 35 feet high but more commonly not above 6 feet, and may be positioned either at the top of a small tree or out from the trunk on a horizontal branch. A nest in Pelham, believed to be a second attempt following failure, was constructed from June 17 to June 21 in a clump of juniper. It was 33 inches from the ground and well hidden from view (Nice 1926). While there are reports of males helping to gather materials for building the nest, the actual work of construction is done by the female. Pieces of dry grass, spruce, twigs, pine needles, and spiderwebs may be used for the outside of the structure, which is lined with hair, moss, or fine rootlets, especially black ones. The Pelham nest was typical, being composed of dried grass stems with a lining of pine needles and black horsehair (Nice 1926).

The most common clutch size is four, with the range being three to six eggs. A Worcester County nest contained four eggs on June 15 (DKW), and at a Pelham nest three eggs were laid on June 22, 23, and 24, respectively (Nice 1926). At a second Pelham nest, laying was estimated to have occurred between July 15 and 18 (Nice 1928). Incubation lasts from

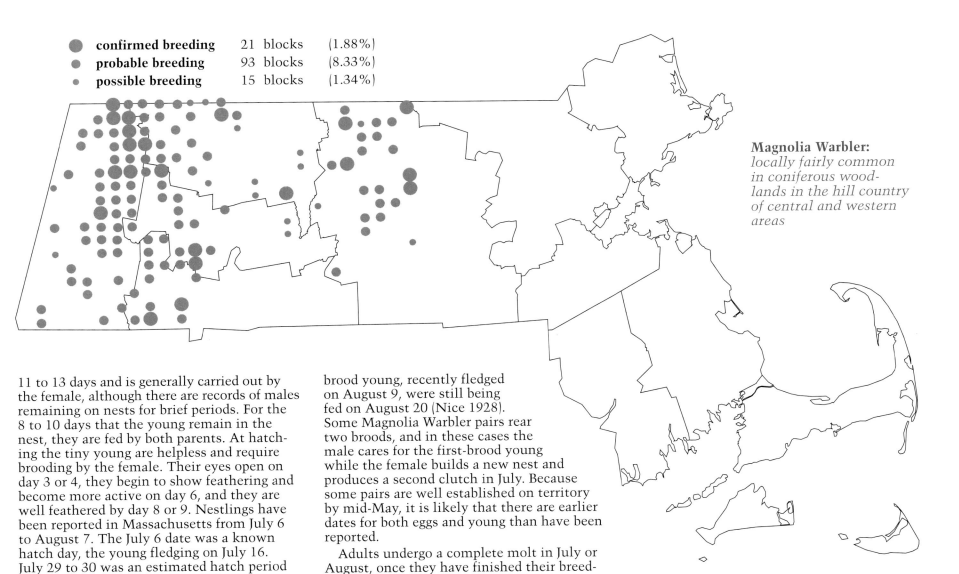

- confirmed breeding — 21 blocks (1.88%)
- probable breeding — 93 blocks (8.33%)
- possible breeding — 15 blocks (1.34%)

Magnolia Warbler: *locally fairly common in coniferous woodlands in the hill country of central and western areas*

11 to 13 days and is generally carried out by the female, although there are records of males remaining on nests for brief periods. For the 8 to 10 days that the young remain in the nest, they are fed by both parents. At hatching the tiny young are helpless and require brooding by the female. Their eyes open on day 3 or 4, they begin to show feathering and become more active on day 6, and they are well feathered by day 8 or 9. Nestlings have been reported in Massachusetts from July 6 to August 7. The July 6 date was a known hatch day, the young fledging on July 16. July 29 to 30 was an estimated hatch period for another state nest (Nice 1926, 1928).

Parental care continues for several weeks as the fledglings gradually become self-sufficient. At this time, the family groups may wander away from the breeding territory. In Massachusetts, adults have been observed feeding fledglings from July 14 to August 20 (Nice 1926, 1928). Two second-brood young, recently fledged on August 9, were still being fed on August 20 (Nice 1928). Some Magnolia Warbler pairs rear two broods, and in these cases the male cares for the first-brood young while the female builds a new nest and produces a second clutch in July. Because some pairs are well established on territory by mid-May, it is likely that there are earlier dates for both eggs and young than have been reported.

Adults undergo a complete molt in July or August, once they have finished their breeding duties. Both young and old birds join mixed flocks of Black-capped Chickadees, warblers, and other species and forage in these groups throughout August. By the first week of September, most of the resident birds have departed on their southward flight. Migrants from farther north continue to pass throughout the month, with small numbers being regularly encountered into the third week of October. After this date, only a few stragglers are observed. Winter quarters are in Mexico, Central America, and the West Indies.

W. Roger Meservey

Black-throated Blue Warbler
Dendroica caerulescens

Egg dates: May 23 to June 10.

Number of broods: one; occasionally two.

This is a warbler that is generally associated with the northern mixed forest, and in fact it is found breeding in or near deep woods. The preferred sites seem to be in openings such as forest clearings, stream or wood borders, or hillsides of secondary-growth maples and beeches. In more southern sections, the birds are especially attracted to wooded regions with a heavy undergrowth of Mountain Laurel, while in more northern areas an understory of American Yew seems to be most desirable.

The nesting distribution of the Black-throated Blue Warbler in Massachusetts is determined by the occurrence of these special types of forest. They are found from Berkshire County east to Worcester County but are most abundant in the more western regions. Some breeding occurs in northwestern Middlesex County, but the species is totally absent as a nester from all eastern parts of the state.

The return of the resident birds in the spring occurs mainly from the end of April through the first week of May, with stragglers continuing to arrive later in the month. Because much of the breeding range lies to the north of Massachusetts, many migrants continue to pass through the state during May, at which time they may be found in regions where they do not nest. Breeding males are very territorial, and, although they may chase and fight one another, much of their behavior at this time consists of long song bouts. Males have at least two main song types with many variations. Three representations are an ascending *zee-o, zee-o, zee-o, zee*; a more rapid *hur, hur, hur, hur, weeee*; and a husky *dzeurr, dzeurr, dzee*. Males with atypical songs are regularly encountered. The common call notes of both sexes include a sharp *chuck*, similar to that of the Dark-eyed Junco, and a high, thin *seet*. In addition, the males have a call consisting of a series of *bzz* notes, and the young have a three-syllable, trilling, begging call.

Nests, constructed by the female, are generally placed from 1 to 4 feet from the ground in a low tree or shrub, most often Mountain Laurel, Eastern Hemlock, spruce, or yew. They are composed of twigs, bark, and leaves, with a lining of rootlets and hair. Two nests in Pelham were situated as follows: 1, 9 inches up on two dead branches hidden in a mass of ferns; 1, 15 inches up on two dead branches in the center of a nearly dead alder. The first was covered with strips of birch bark and lined with pine needles, and the second was constructed almost entirely of pieces of bark of the Yellow Birch (Nice 1926, 1930). A nest in Rowe was located 3 feet up in a hemlock sapling growing in a thick clump (EHF).

The usual clutch is four eggs, but three and rarely five are produced on occasion. Five clutches from Worcester County each contained four eggs (DKW). Incubation, by the female, lasts about 12 days, and the young

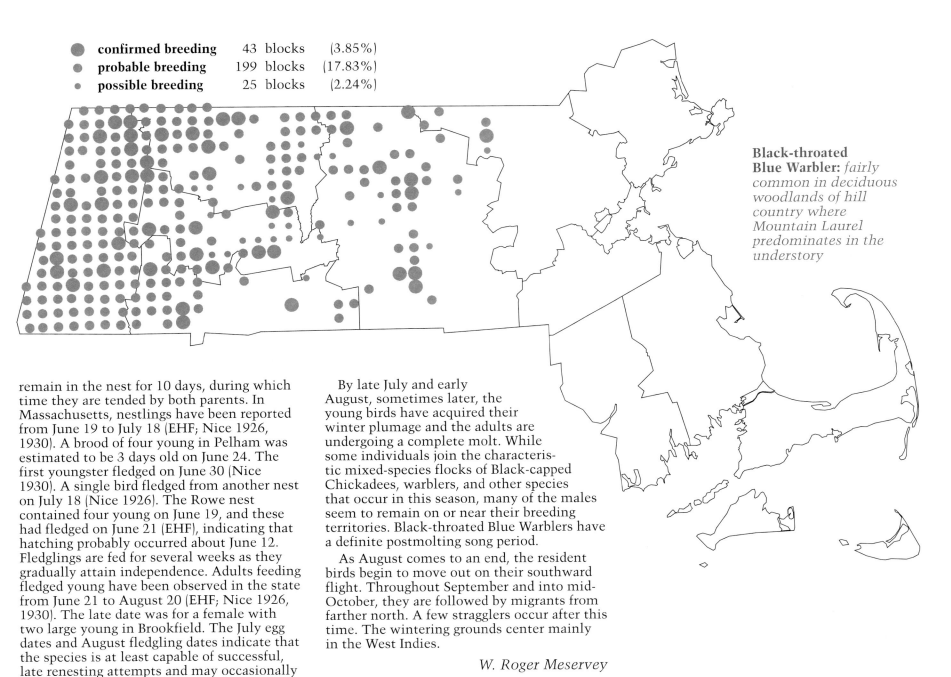

- **confirmed breeding** 43 blocks (3.85%)
- **probable breeding** 199 blocks (17.83%)
- **possible breeding** 25 blocks (2.24%)

Black-throated Blue Warbler: *fairly common in deciduous woodlands of hill country where Mountain Laurel predominates in the understory*

remain in the nest for 10 days, during which time they are tended by both parents. In Massachusetts, nestlings have been reported from June 19 to July 18 (EHF; Nice 1926, 1930). A brood of four young in Pelham was estimated to be 3 days old on June 24. The first youngster fledged on June 30 (Nice 1930). A single bird fledged from another nest on July 18 (Nice 1926). The Rowe nest contained four young on June 19, and these had fledged on June 21 (EHF), indicating that hatching probably occurred about June 12. Fledglings are fed for several weeks as they gradually attain independence. Adults feeding fledged young have been observed in the state from June 21 to August 20 (EHF; Nice 1926, 1930). The late date was for a female with two large young in Brookfield. The July egg dates and August fledgling dates indicate that the species is at least capable of successful, late renesting attempts and may occasionally rear two broods.

By late July and early August, sometimes later, the young birds have acquired their winter plumage and the adults are undergoing a complete molt. While some individuals join the characteristic mixed-species flocks of Black-capped Chickadees, warblers, and other species that occur in this season, many of the males seem to remain on or near their breeding territories. Black-throated Blue Warblers have a definite postmolting song period.

As August comes to an end, the resident birds begin to move out on their southward flight. Throughout September and into mid-October, they are followed by migrants from farther north. A few stragglers occur after this time. The wintering grounds center mainly in the West Indies.

W. Roger Meservey

Yellow-rumped Warbler
Dendroica coronata

Egg dates: May 20 to late July.

Number of broods: one or two.

The hardy Yellow-rumped Warbler is the only member of its tribe that can be found in significant numbers somewhere in the state in any season. It is the most abundant warbler during both spring and fall migrations, when many individuals pass through Massachusetts en route to and from the extensive northern breeding grounds. Forbush gave the status of the species during the breeding season as "uncommon to rare and local...chiefly on the high lands of the western counties east to eastern Worcester County" and "casual in the eastern counties south to Plymouth County." This is the same nesting distribution that was established for the species during the Atlas period. The change is that, within this range, the Yellow-rumped Warbler has apparently increased its numbers. It is now widely distributed throughout the western and northern portions of the state, and it may be common in favorable habitats. Yellow-rumped Warblers also appear to be established breeders, rather than merely casual, in northern Middlesex County and in Plymouth County.

Yellow-rumped Warblers will nest in several forest types, but the common factor is that some kind of evergreen needs to be present. Breeding territories may be located in coniferous woods of spruce, hemlock, or pine; in young conifers at the edge of more mature forest; or in mixed woods containing stands of conifers, especially mature White Pines.

The song of the Yellow-rumped Warbler is a variable, musical trill, sometimes broken into separate notes, often containing double elements and sometimes changing pitch near the end. The resident males sing from late April to early August. Both sexes give a *seet* note, a diagnostic *check* call, and a scolding *tchip*. Young birds have a characteristic food-begging call.

Spring migration extends from the second week of April to the end of May, with the biggest flights occurring in late April and early May. Territorial males court prospective mates in typical warbler fashion, chasing them or approaching closely with the plumage fluffed. Nest building and incubation are carried out primarily by the female. Nests are usually placed on a horizontal branch of an evergreen, although sometimes a hardwood tree will be selected. These cups of bark, twigs, and plant fibers may be located from 4 to 50 feet high, and they are unique among warbler nests. In addition to the usual lining of fine grasses and hair, they contain many feathers with the shafts woven into the nest structure and the tips forming a screen over the eggs.

Three Massachusetts nests were located as follows: one not far from the end of a long branch near the top of a tall White Pine in a pasture in Webster, one in Carver 25 feet up in a White Pine (EHF), and one 6 feet up on a branch next to the trunk of a small Red Cedar in Pelham. This last nest was a shallow cup of cedar twigs, bark, plant fibers, string, and pine needles lined with horsehair and many grouse feathers.

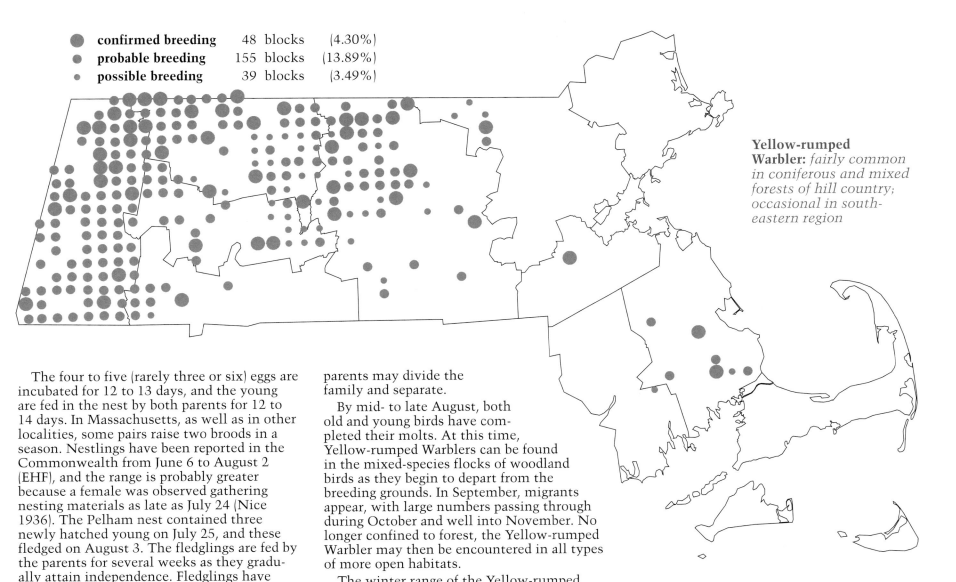

Yellow-rumped Warbler: *fairly common in coniferous and mixed forests of hill country; occasional in southeastern region*

- confirmed breeding 48 blocks (4.30%)
- probable breeding 155 blocks (13.89%)
- possible breeding 39 blocks (3.49%)

The four to five (rarely three or six) eggs are incubated for 12 to 13 days, and the young are fed in the nest by both parents for 12 to 14 days. In Massachusetts, as well as in other localities, some pairs raise two broods in a season. Nestlings have been reported in the Commonwealth from June 6 to August 2 (EHF), and the range is probably greater because a female was observed gathering nesting materials as late as July 24 (Nice 1936). The Pelham nest contained three newly hatched young on July 25, and these fledged on August 3. The fledglings are fed by the parents for several weeks as they gradually attain independence. Fledglings have been observed in the state from June 14 (a female and one very young fledgling) to August 21 (a male and two large fledglings). Pairs or single birds were feeding from one to three young (Nice 1926, Meservey). Once the second set of youngsters has left the nest, the family group may wander together or the parents may divide the family and separate.

By mid- to late August, both old and young birds have completed their molts. At this time, Yellow-rumped Warblers can be found in the mixed-species flocks of woodland birds as they begin to depart from the breeding grounds. In September, migrants appear, with large numbers passing through during October and well into November. No longer confined to forest, the Yellow-rumped Warbler may then be encountered in all types of more open habitats.

The winter range of the Yellow-rumped Warbler is extensive, reaching from Pennsylvania, New York, and coastal Maine south through the southern states, West Indies, and Mexico to Panama. During the winter in Massachusetts, the species is common locally along the coast and rare west to central Worcester County and the Connecticut River valley. The birds are able to withstand northern winters by switching from a primarily insectivorous diet to one of Bayberry and other small fruits.

W. Roger Meservey

Black-throated Green Warbler
Dendroica virens

Egg dates: May 21 to August 6.

Number of broods: one; occasionally two.

As with many forest species, on a statewide basis the Black-throated Green Warbler is probably no more uncommon now than it was in presettlement times, due to the extensive reforestation that has taken place since the 1880s. In proper habitat, it is not at all difficult to find. It is most prevalent in the northern hardwood-hemlock forest of Berkshire County, less common in the White Pine-hemlock-oak forests of Worcester County and northeastern sections, and locally distributed in the Red Maple-Atlantic White Cedar swamps of the southeastern portion of the state.

Black-throated Green Warblers are a conspicuous and relatively common migrant throughout Massachusetts. The first wave of spring arrivals usually occurs in the first week of May, although individuals rarely may appear as early as mid-April. Migration usually ceases by the end of May. Unlike many other wood-warblers, whose voices are less distinct and memorable, the Black-throated Green Warbler sings a song that is instantly recognizable by its vocal quality as well as by its pattern. When this warbler is heard in early May, it heightens the senses, for it is immediately evident that, if a wave of warblers has arrived, spring is truly here.

Males establish territories with much singing and excited chipping, sometimes chasing and fighting for possession of an area. The songs are of two types and are used in different contexts. One type is *zee-zee-zoo-zoo-zee*, the third and fourth notes lower in pitch. The alternate song seems to be uttered with a quicker tempo, *zee-zee-zee-zee-zee-zoo-zee*. Slight emphasis is placed on the fifth note. The dreamy, lispy quality is unmistakable, and, to quote Maine pioneer ornithologist Cordelia Stanwood, "His voice is suggestive of the drowsy summer days, the languor of the breeze dreamily swaying the pines, spruces, firs, and hemlocks." The common call notes are a *chet*, similar to that of the Black-and-White Warbler, and a lisping *tsip*.

Breeding pairs prefer to place their nests in White Pine or Eastern Hemlock, but there is considerable individual variation as to tree species, nest height and location, and materials used. A nest in Rehoboth was placed 8 inches above the ground in a Red Cedar (ACB), but they are generally located higher and may be 50 to 70 feet up. On Martha's Vineyard, a nest was discovered at 8 feet in a whorl of small branches in a Pitch Pine. The nest was the typical cup constructed of grass, seaweed, and bark lined with fine grasses, cow hair, horsehair, and feathers (ACB). In Pelham, several nests were located in Eastern Hemlock, including one 15 feet up and 6 feet from the trunk and one 30 feet up and 15 feet from the trunk (Nice 1932). A nest in West Brookfield was constructed about 60 feet high in a large White Pine (Meservey). In Winchendon, 1 nest was located in a spruce sapling, 1 in a spruce at 12 feet high and 10 feet from the trunk, and 2 against an Eastern Hemlock trunk at 12 feet and 15 feet, respectively (Brewster 1888).

Four to five eggs constitute a typical clutch, which is incubated mostly by the female for about 12 days. Clutch sizes for 7 state nests were as follows: three eggs (1 nest), four eggs (5 nests), five eggs (1 nest) (ACB, Brewster 1888, DKW). The Martha's Vineyard nest contained four fresh eggs on June 8 (ACB). The period from hatching to fledging is the easiest time to confirm breeding because adults are continuously foraging and returning to the nest to feed the young, with males usually, but not always, assisting in this care. Food consists of insects, including caterpillars, larvae, and beetles gleaned

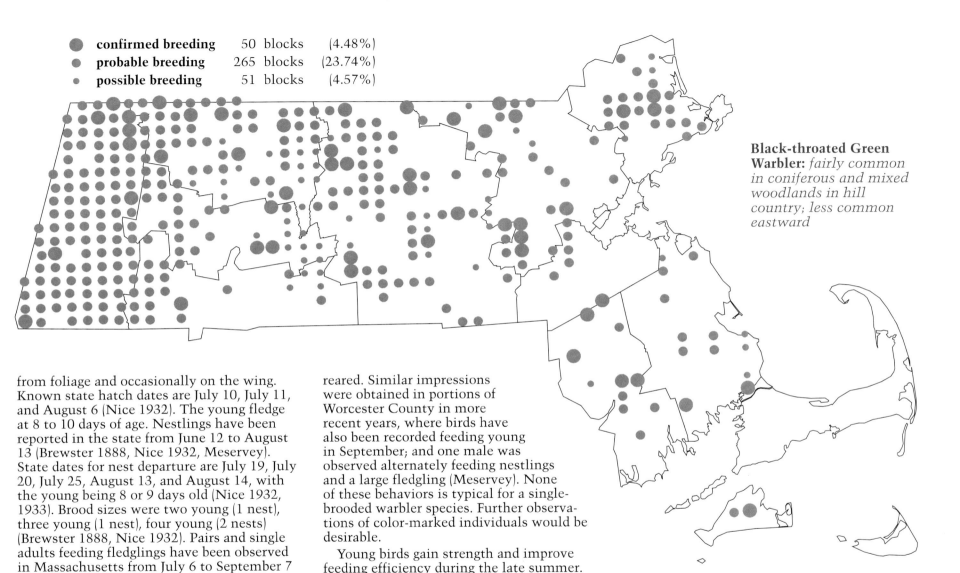

Black-throated Green Warbler: *fairly common in coniferous and mixed woodlands in hill country; less common eastward*

- confirmed breeding — 50 blocks (4.48%)
- probable breeding — 265 blocks (23.74%)
- possible breeding — 51 blocks (4.57%)

from foliage and occasionally on the wing. Known state hatch dates are July 10, July 11, and August 6 (Nice 1932). The young fledge at 8 to 10 days of age. Nestlings have been reported in the state from June 12 to August 13 (Brewster 1888, Nice 1932, Meservey). State dates for nest departure are July 19, July 20, July 25, August 13, and August 14, with the young being 8 or 9 days old (Nice 1932, 1933). Brood sizes were two young (1 nest), three young (1 nest), four young (2 nests) (Brewster 1888, Nice 1932). Pairs and single adults feeding fledglings have been observed in Massachusetts from July 6 to September 7 (Nice 1936, Meservey).

Although much literature reports that this species is not double brooded, there is evidence that some pairs do in fact raise two sets of young. Nice (1932) observed that the prolonged song period of the male (into August) and the late dates for fledglings were highly suggestive that two broods may be reared. Similar impressions were obtained in portions of Worcester County in more recent years, where birds have also been recorded feeding young in September; and one male was observed alternately feeding nestlings and a large fledgling (Meservey). None of these behaviors is typical for a single-brooded warbler species. Further observations of color-marked individuals would be desirable.

Young birds gain strength and improve feeding efficiency during the late summer. At this time a molt occurs, complete in the adults and partial (body feathers) in the young. As in other wood-warblers, considerable wandering occurs during this season, bringing individuals into areas not normally frequented by the species. Fall migration is underway by late August, peaks at the end of September, and generally ends by late October. Stragglers have been found as late as mid-December, but there is no record of overwintering. The wintering grounds extend from Texas and Florida through Mexico to Panama and the Greater Antilles.

Bruce A. Sorrie

Blackburnian Warbler
Dendroica fusca

Egg dates: June 6 to July.

Number of broods: one; sometimes two; may re-lay if first attempt fails.

Because the Blackburnian Warbler prefers either deep evergreen woods of spruce, fir, pine, and hemlock or mixed forest with large stands of hemlock and White Pine as a breeding site, it is primarily a bird of the more northern and western parts of the state. It is widely distributed in the higher elevations of Worcester County and is well established throughout Franklin County and in the western portions of Hampshire and Hampden counties. The species is found throughout Berkshire County, which represents its center of abundance in the state due to extensive areas of suitable habitat. In eastern Massachusetts, the Blackburnian Warbler is found breeding only in a few areas of Middlesex County and also in portions of Essex County; it is absent from the entire southeastern section of the state. Although there are literature reports that this species avoids White Pine, this is certainly not the case in many sections of the Commonwealth.

Breeding birds begin to appear in late April, with the majority arriving during the first week of May. However, many migrants continue to pass through the state during May en route to the extensive nesting grounds to the north. The highly territorial males fight and chase one another through the treetops, and the density of breeders is determined by the suitability of the habitat. In especially attractive areas, the birds are often found clustered in a series of adjacent territories, but in less desirable regions the birds are more widely spaced and territories may be isolated.

During courtship, the males may chase the females on rapid, twisting flights through the foliage. Their thin, wiry song is delivered from elevated, exposed perches. As is the case for many warbler species, male blackburnians have two main types of songs, both of which are highly variable. A good representation of each song type would be *sleecha, sleecha, sleecha, sleecha, slee* (upslurred ending) and *tche-see, tche-see, tche-see, tche*. Both sexes have a *chip* note and a high, thin *seet* note. Young birds give a begging call, *zee-zee, zee-zee, zee-zee*, very similar to that of the young of the Magnolia Warbler.

Blackburnian Warbler nests are almost always built in a conifer, from 6 to 80 feet above the ground, and are positioned well out from the trunk on a horizontal limb or in a small fork of branches near the treetop. Two females in Worcester County were observed carrying nesting materials on May 22 and June 9, respectively. In each case, the birds disappeared into the high branches of White Pines at least 40 feet above the ground. A nest found near Winchendon was located 30 feet up in a Black Spruce and 6 feet out from the trunk. It was well hidden by the tree's needles and old man's beard lichens. In Pelham, a nest was discovered in a somewhat unusual situation 18 feet high near the top of a Red Cedar in a fairly open area of young growth 150 yards from pine-hemlock woods where the male usually sang (Nice 1932). Nests are constructed of twigs and lichen and are lined with fine hair, dry grass, and rootlets. Four eggs make up the usual clutch, but five are sometimes produced. A Worcester County nest contained four eggs on June 26 (DKW).

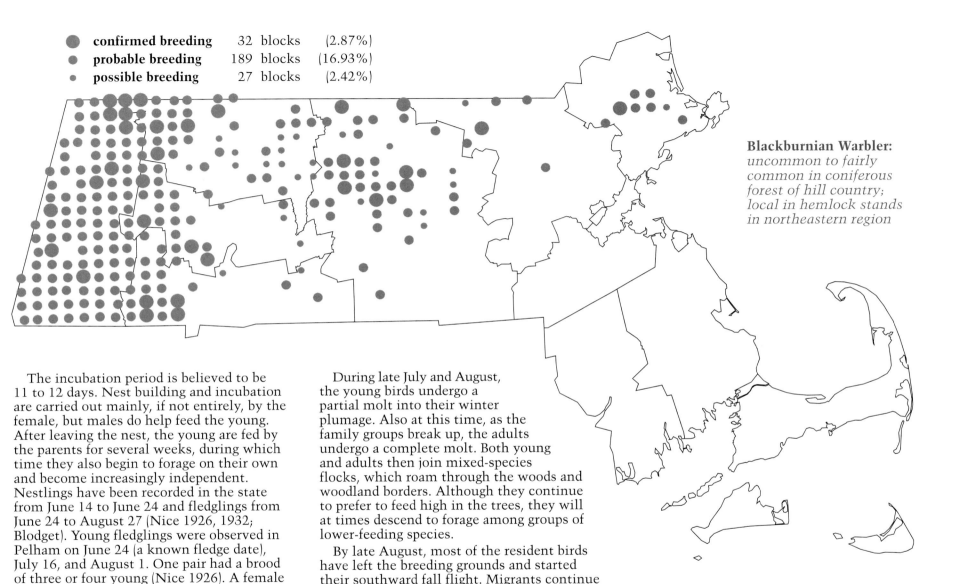

Blackburnian Warbler: *uncommon to fairly common in coniferous forest of hill country; local in hemlock stands in northeastern region*

- confirmed breeding — 32 blocks (2.87%)
- probable breeding — 189 blocks (16.93%)
- possible breeding — 27 blocks (2.42%)

The incubation period is believed to be 11 to 12 days. Nest building and incubation are carried out mainly, if not entirely, by the female, but males do help feed the young. After leaving the nest, the young are fed by the parents for several weeks, during which time they also begin to forage on their own and become increasingly independent. Nestlings have been recorded in the state from June 14 to June 24 and fledglings from June 24 to August 27 (Nice 1926, 1932; Blodget). Young fledglings were observed in Pelham on June 24 (a known fledge date), July 16, and August 1. One pair had a brood of three or four young (Nice 1926). A female in Boylston was feeding one large youngster on August 27. Blackburnian Warblers are known to renest if the first attempt fails, but, as is the case also with the Black-throated Blue Warbler, the late dates for dependent young suggest that two broods may be reared on occasion.

During late July and August, the young birds undergo a partial molt into their winter plumage. Also at this time, as the family groups break up, the adults undergo a complete molt. Both young and adults then join mixed-species flocks, which roam through the woods and woodland borders. Although they continue to prefer to feed high in the trees, they will at times descend to forage among groups of lower-feeding species.

By late August, most of the resident birds have left the breeding grounds and started their southward fall flight. Migrants continue during September, with a few still present in early October. After this, the species is encountered only as a straggler. During both spring and fall migration, Blackburnian Warblers occur throughout the state. They are less restricted to conifers during these periods and may be found in deciduous woods, shade trees, and even shrubby, secondary-growth vegetation. The wintering grounds are mainly in Colombia, Venezuela, and Peru.

W. Roger Meservey

Pine Warbler
Dendroica pinus

Egg dates: May 8 to late July.

Number of broods: one or two.

The Pine Warbler is aptly named because for much of the year it seldom wanders far from pine woodlands. In the Pitch Pine barrens of Cape Cod and the Islands, a traditional stronghold, Pine Warblers are common and widespread throughout the summer. Their numbers diminish as sandy lands become valuable real estate and pines fall to the bulldozer. More rarely, pines yield to the natural succession of oaks, and breeding birds look elsewhere for favored habitat. There are good populations in southern Bristol and Plymouth counties, but north of this the species becomes very spotty and local in distribution. In recent years, Pine Warblers have increased in some sections of interior Massachusetts. There, the preferred tree is the White Pine. These birds are rare to absent at the higher elevations and in areas of deciduous forest.

The ragged Pitch Pine forest, an anathema to foresters, suits the Pine Warbler just fine. Sparse growth makes for sunny glades and sheltered clearings, and wind damage adds the insects of decay to the multitudinous pests of pines. The understory of Scrub Oak and huckleberry increases the variety of insect and vegetal food. The interior White Pine forest, with taller trees and little understory, is an altogether different type of habitat. Most characteristically, Pine Warblers work along trunks and branches, leisurely probing for eggs and larvae in bark crevices, extracting an inchworm from the needles, plucking a spider from its filament, and earning their other name of "Pine-creeping Warbler."

The Pine Warbler is one of the earliest of migrant spring warblers. Because the birds winter in the southern third of the breeding range, the distance for spring migration is comparatively short. First arrivals in late March cheer the waking woods with a plangent trill. Massachusetts breeding birds are on territory by the third week of April, although migrants heading farther north continue into early May. Their passage takes them into orchards and deciduous thickets, but for nesting they seek conifers, usually the pines mentioned previously, but occasionally another type of pine and rarely Red Cedar or hemlock. They prefer not the deep, dark woods but rather the edges and openings.

The breeding male sings persistently. His is one of the least versatile of warbler songs, a changeless paean to the pines. While authorities disagree on the quality of the song, it is similar to that of the Chipping Sparrow but generally more musical, sweeter, slower, and softer. The Pine Warbler sings into summer, pausing during molt and singing again during warm autumn days. Call notes include a lisping *seet* and a soft *chip*. Rapid and insistent chipping is common during courtship and territory defense and is given also by begging young. Courting males may show some pugnacity.

Nests recorded in southeastern Massachusetts were all in Pitch Pines at heights of 10 to 25 feet and were saddled on horizontal limbs, generally hidden in clumps of needles but sometimes exposed. One nest near Mount Auburn Cemetery was located at 40 feet in a Red Cedar (ACB). Two recent Worcester County nests were, respectively, at 15 and 30 feet near the ends of White Pine branches (Meservey). The female

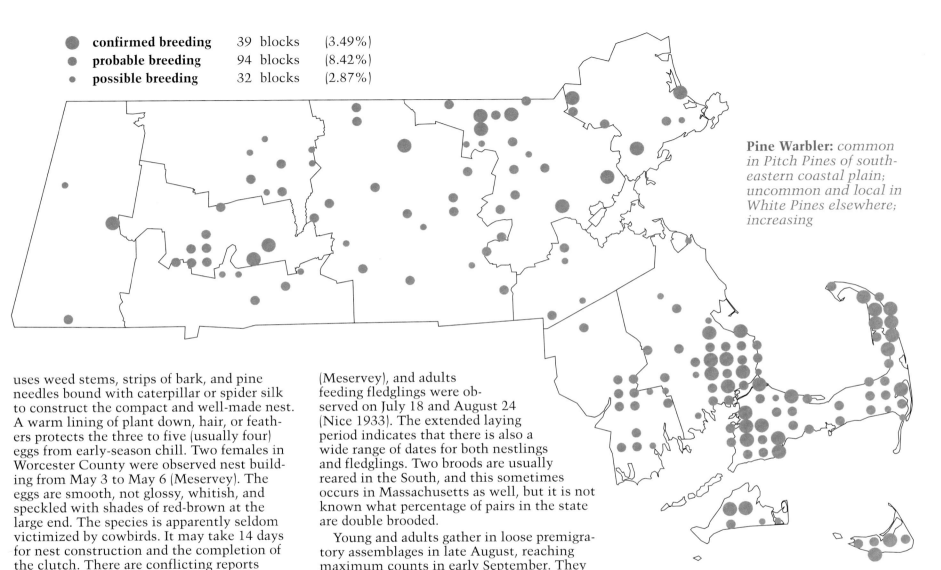

● confirmed breeding	39 blocks	(3.49%)
● probable breeding	94 blocks	(8.42%)
· possible breeding	32 blocks	(2.87%)

Pine Warbler: *common in Pitch Pines of southeastern coastal plain; uncommon and local in White Pines elsewhere; increasing*

uses weed stems, strips of bark, and pine needles bound with caterpillar or spider silk to construct the compact and well-made nest. A warm lining of plant down, hair, or feathers protects the three to five (usually four) eggs from early-season chill. Two females in Worcester County were observed nest building from May 3 to May 6 (Meservey). The eggs are smooth, not glossy, whitish, and speckled with shades of red-brown at the large end. The species is apparently seldom victimized by cowbirds. It may take 14 days for nest construction and the completion of the clutch. There are conflicting reports about the male's assisting in incubation, but this task is probably performed mainly by the female. Both sexes tend the young. The incubation and nestling periods are unrecorded, but each probably averages about 12 days.

Nesting data for Massachusetts is very scanty. A pair was feeding nestlings on June 6 (Meservey), and adults feeding fledglings were observed on July 18 and August 24 (Nice 1933). The extended laying period indicates that there is also a wide range of dates for both nestlings and fledglings. Two broods are usually reared in the South, and this sometimes occurs in Massachusetts as well, but it is not known what percentage of pairs in the state are double brooded.

Young and adults gather in loose premigratory assemblages in late August, reaching maximum counts in early September. They may flock with chickadees and other warblers but often associate with Chipping Sparrows, Palm Warblers, and Eastern Bluebirds feeding in thickets. Most have departed by mid-October. A few linger along the coast until early November, sometimes attempting to winter. There are a few records of inland wintering individuals. Survival is rare except during a mild season aided by feeding stations stocked with suet, seeds, and doughnuts. The regular winter grounds are in the Gulf states north to Virginia and Tennessee.

Priscilla Bailey

Prairie Warbler
Dendroica discolor

Egg dates: May 22 to June 21.

Number of broods: one; possibly two; will renest if first attempt fails.

A species that prefers disturbed areas, the Prairie Warbler has probably expanded its range since the 1600s. In Massachusetts, settler-related activities and subsequent land use patterns, which persisted well into the twentieth century, greatly increased its favored shrubby habitats. The present status is uncertain, but populations may be declining as the forests of Massachusetts mature and suburban sprawl engulfs breeding habitat. This tendency was first noted on Cape Cod and is clearly accelerating in the pine barrens of Plymouth County today. Future changes in the distribution and abundance of Prairie Warblers in the Commonwealth may well continue to reflect this balance of habitat creation and elimination by human activities both here and on the wintering grounds, where rapid human population growth has also resulted in widespread habitat destruction.

Typical Massachusetts habitats are scrub, open woodlands, burned or logged areas, overgrown fields, power-line corridors, thickets of young pines, and, above all, Pitch Pine-Scrub Oak barrens. The patchy distribution of confirmed breeding records on the map reflects the discontinuous nature of this range of temporary habitats. However, the species can be astonishingly abundant in such optimum habitat as the Myles Standish State Forest in Plymouth County. There, Manomet Center for Conservation Sciences staff has estimated a population of 5,000 to 10,000 pairs, with a maximum density of 112 pairs per 100 acres in areas burned ten to twenty years previously.

Breeding plumage is attained by a partial molt on the wintering grounds in February and March, and spring migrants arrive in Massachusetts from late April to the end of May. Although Prairie Warblers are regularly observed during spring migration, the numbers recorded are usually low (less than 10 per day). Their return is often first noticed only when residents reach their nesting areas.

The song, a good test of the upper limits of human hearing, consists of a series of high, musical, buzzy notes that steadily ascend the scale. There are slower and faster versions of the song, the former resembling that of the Field Sparrow, a species often found breeding in the same habitats. Songs are usually delivered from a vantage point, such as the upper branches of a Pitch Pine, with maximum intensity just after dawn and very little in the middle of the day. The common call note of both sexes is a weak *chip*.

The nest, built solely by the female, has an outer shell of grasses, plant materials, and fibers. The cup is lined with plant down, cobwebs, hair, or feathers. In Plymouth County nests are located between 1 and 10 feet up in low scrub and second growth, particularly Scrub Oak, in forked stems or on lateral branches (MCCS). Bent reports other state nests in barberry; hazelnut at 2 or 3 feet; Sweet Fern at 20 inches; saplings of oak, poplar, apple, cherry, and maple, usually not

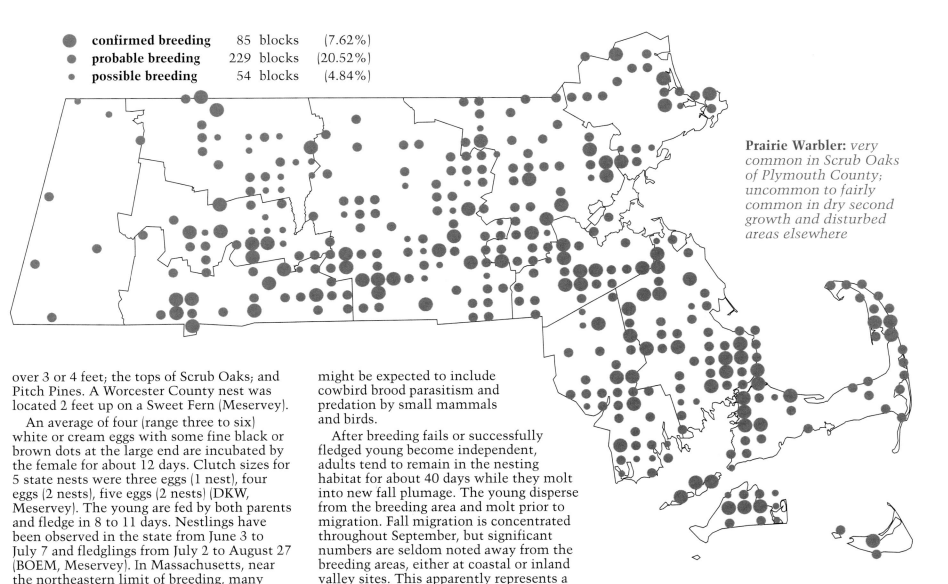

Prairie Warbler: *very common in Scrub Oaks of Plymouth County; uncommon to fairly common in dry second growth and disturbed areas elsewhere*

over 3 or 4 feet; the tops of Scrub Oaks; and Pitch Pines. A Worcester County nest was located 2 feet up on a Sweet Fern (Meservey).

An average of four (range three to six) white or cream eggs with some fine black or brown dots at the large end are incubated by the female for about 12 days. Clutch sizes for 5 state nests were three eggs (1 nest), four eggs (2 nests), five eggs (2 nests) (DKW, Meservey). The young are fed by both parents and fledge in 8 to 11 days. Nestlings have been observed in the state from June 3 to July 7 and fledglings from July 2 to August 27 (BOEM, Meservey). In Massachusetts, near the northeastern limit of breeding, many pairs are probably single brooded, but replacement nestings after failure are usual.

Prairie Warblers are known to raise two broods in some areas, and the late dates for dependent fledglings suggest that some birds are double brooded. Causes of nest failure might be expected to include cowbird brood parasitism and predation by small mammals and birds.

After breeding fails or successfully fledged young become independent, adults tend to remain in the nesting habitat for about 40 days while they molt into new fall plumage. The young disperse from the breeding area and molt prior to migration. Fall migration is concentrated throughout September, but significant numbers are seldom noted away from the breeding areas, either at coastal or inland valley sites. This apparently represents a decline from the "abundant transient" status, suggested by Griscom and Snyder (1955), prior to the mid-1950s. Early fall migrants can be observed in July, while late individuals are reported through October, with stragglers into December. The wintering grounds are in southern Florida and the West Indies, and on islands off the east coast of northern Central America.

Trevor L. Lloyd-Evans

Blackpoll Warbler
Dendroica striata

Egg dates: mid-June to early July.

Number of broods: one.

The Blackpoll Warbler is a rare and peripheral breeding species in Massachusetts, nesting in recent years only at a high elevation on Mount Greylock in northern Berkshire County. There is a historical breeding record from Savoy. This warbler occurs in the Commonwealth at the southern limit of its breeding range, which extends from the higher elevations of northern New England to the extreme northern edge of the boreal forests from Labrador to Alaska. In recent years, the number of local breeders has been declining, perhaps due in part to increasing human disturbance.

Blackpoll Warblers are common migrants throughout the state in late spring, and they are among the last wood-warblers to pass northward in late May and early June. Only a few early individuals appear before the middle of May. Their large numbers are often difficult to detect as they forage in the tops of the newly leafed-out trees.

The song of the blackpoll is one of the highest pitched songs of any North American passerine. It is an insectlike, high, thin *tsit-tsit-tsit-tsit-tsit* given on one pitch, becoming slightly louder and more emphatic in the middle and diminishing toward the end. The last notes are sometimes run rapidly together, with almost a sputtering. Another song type is a rapid *ti-ti-ti-ti-ti-ti-ti-ti*. The call notes are a high-pitched lisp, which resembles *zeet,* and several emphatic *chip* notes.

On Mount Greylock, breeding Blackpoll Warblers are found in patches of stunted Balsam Fir, where they glean insects from needles and twigs and sometimes dart out from branches to catch flying insects. Males establish territories during June. Nests are built in branches 1 to 10 feet above the ground and are constructed of small twigs, pieces of bark, dried grass, moss, and lichen and lined with feathers, plant fibers, and fine rootlets. The clutch size is four to five eggs, which are creamy or grayish white and speckled and blotched with varying shades of chestnut, gray, or purple. Observations of the birds indicate that nests with eggs may be found from mid- to late June, occasionally into early July (Hendricks). Females construct the nests and incubate the eggs while males will bring food to their mates, and later to the young. The incubation period is about 11 days, and the young fledge in another 11 to 12 days. A single brood is raised per year.

During late summer, Blackpolls roam the woodlands in association with chickadees and other warbler species. At this time, the adults undergo a complete molt, resulting in the drab greenish autumn plumage, which, for the male, is in stark contrast to the natty spring plumage. The first migrants begin to appear in Massachusetts in early September,

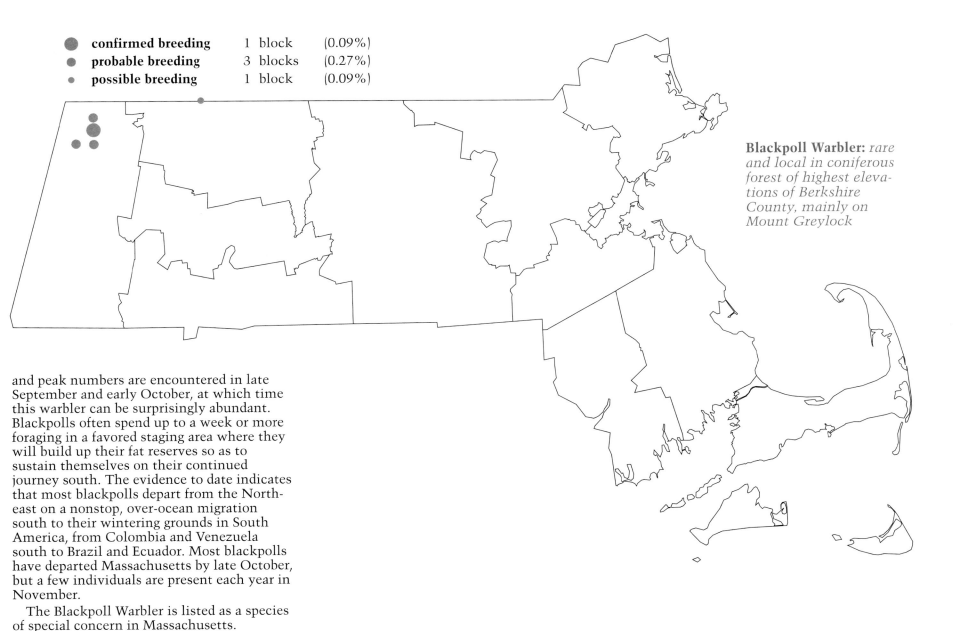

● **confirmed breeding**	1 block	(0.09%)
● **probable breeding**	3 blocks	(0.27%)
· **possible breeding**	1 block	(0.09%)

Blackpoll Warbler: *rare and local in coniferous forest of highest elevations of Berkshire County, mainly on Mount Greylock*

and peak numbers are encountered in late September and early October, at which time this warbler can be surprisingly abundant. Blackpolls often spend up to a week or more foraging in a favored staging area where they will build up their fat reserves so as to sustain themselves on their continued journey south. The evidence to date indicates that most blackpolls depart from the Northeast on a nonstop, over-ocean migration south to their wintering grounds in South America, from Colombia and Venezuela south to Brazil and Ecuador. Most blackpolls have departed Massachusetts by late October, but a few individuals are present each year in November.

The Blackpoll Warbler is listed as a species of special concern in Massachusetts.

Scott M. Melvin

Black-and-white Warbler
Mniotilta varia

Egg dates: May 18 to June 14.

Number of broods: one.

The Black-and-white Warbler is one of the most common warblers in broad-leaved woodlands throughout the state, occurring in both mature and young deciduous forest. It is less common in primarily coniferous woods. The species was confirmed as a breeder throughout Massachusetts, including Cape Cod and the Islands, wherever there was suitable habitat. It is a notably distinctive warbler, with its zebra-striped plumage and habit of creeping about nuthatchlike on tree trunks as it searches for insects.

The Black-and-white Warbler is among the first of the wood-warblers to return in the spring, arriving near the end of April with the Yellow-rumped Warbler and Palm Warbler. Its distinctive song is a familiar element of the dawn chorus of early spring warbler waves. Migration continues into early May, with only stragglers passing through by the end of the month. During migration, the birds may be seen anywhere that there are sufficient trees for foraging.

The song, which is among the easiest warbler songs to learn, is a high-pitched *squeeky, squeeky, squeeky* or *wee-see, wee-see, wee-see*. A longer and more complex version is given during the breeding season. Two common call notes of both sexes are a sharp *chink* and a weak *tsip*. The birds also string a series of calls together to produce a rattling chatter, which is commonly used as a begging call by the young.

Males generally arrive first and claim a territory, which they defend vigorously. During courtship, they chase prospective mates and perch near them with fluttering wings. Females are the principal nest builders, constructing a small, well-concealed structure of dry leaves and grasses on the ground at the base of a tree trunk or fallen log. Nearly all Massachusetts nests reported were in such locations, but there are a few rare state records for nests located several feet up on dead stumps or tree trunks (ACB). A recent nest in Charlton was built at the base of the trunk of one of a group of young White Pines near a field grown up to saplings (Meservey). Females in southern Worcester County were observed gathering nesting material from May 27 to June 9 (Meservey).

The female incubates the eggs, which are whitish with small brown splotches, for 10 to 12 days. Clutch sizes for 4 Massachusetts nests were four eggs (1 nest), five eggs (3 nests) (DKW). Like other ground-nesting birds, the female Black-and-white Warbler will engage in a broken-wing display if a potential predator approaches the nest. Males assist in feeding the young and may also attempt to aid in luring intruders away. The young fledge in 8 to 12 days. In Massachusetts, adults were observed feeding nestlings from June 6 to

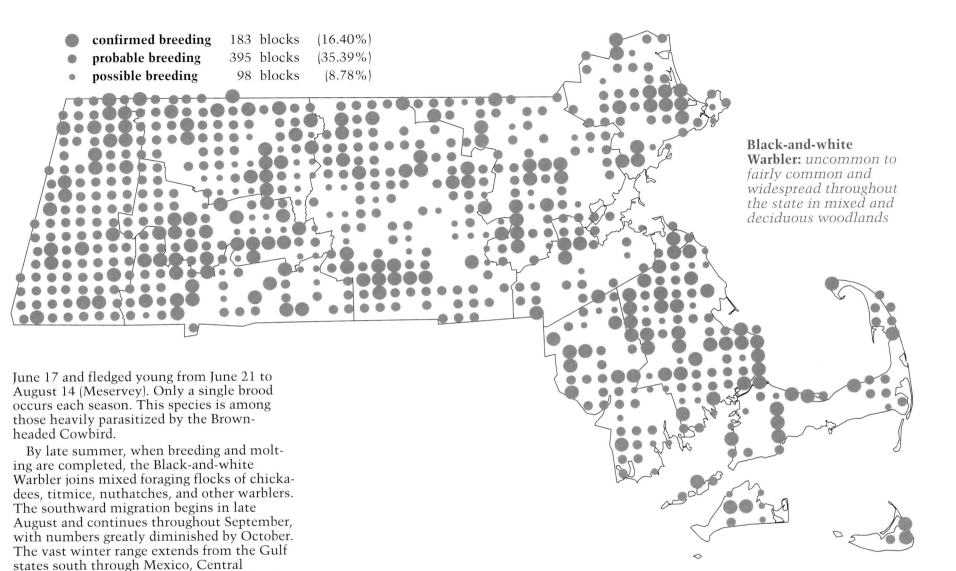

Black-and-white Warbler: *uncommon to fairly common and widespread throughout the state in mixed and deciduous woodlands*

June 17 and fledged young from June 21 to August 14 (Meservey). Only a single brood occurs each season. This species is among those heavily parasitized by the Brown-headed Cowbird.

By late summer, when breeding and molting are completed, the Black-and-white Warbler joins mixed foraging flocks of chickadees, titmice, nuthatches, and other warblers. The southward migration begins in late August and continues throughout September, with numbers greatly diminished by October. The vast winter range extends from the Gulf states south through Mexico, Central America, and some of the Caribbean Islands to South America.

John C. Kricher

American Redstart
Setophaga ruticilla

Egg dates: May 22 to July 20.

Number of broods: one or two.

Attractive in color patterns and lively in actions, the American Redstart is easily recognized and is certainly one of the better known warblers. It is a common breeder throughout the state. Although the species is most abundant from Worcester County west to Berkshire County, it also nests in limited numbers in coastal Essex, Plymouth, and Bristol counties and is found throughout Cape Cod, the Elizabeth Islands, Martha's Vineyard, and Nantucket. This extensive breeding range is made possible by the redstart's flexibility in the choice of nesting areas. These can be in heavily wooded regions of mixed or deciduous forest, woodland openings and forest edges, or in low-growth saplings and shrubs. Some individuals choose more open situations such as orchards and even stands of large shade trees. Clumps of saplings bordering fields or more mature forest seem to be especially favored sites for nesting.

Breeding birds begin to arrive in the first week of May, and are joined by others during the next several weeks. The redstart is also common as a migrant at this time throughout the state. Males are very territorial, and in favorable areas several may stake out claims in close proximity. They are persistent singers and will chase one another and also display by flying in wide arcs from perch to perch in a distinctive stiff-winged manner. During courtship, they may pursue the females or follow them about the territory. Constant patrolling may be required to keep out intruders, especially if the latter are unmated.

Redstart songs are extremely variable, but, as is the case with many warbler species, they can be divided into two main types, one with a loud ending note and one without. Two renditions are *zee, zee, zee, zee, zwee-oo* and *wee-see, wee-see, wee*. Both sexes have a loud *chip* note and a softer, thin *seet* note. The young have a chipping, food-begging call.

Eleven Massachusetts nests were located as follows: apple (1 nest), lilac (1 nest), Sugar Maple (1 nest), Red Maple (2 nests), Black Cherry (1 nest), unspecified deciduous trees or saplings (5 nests) (CNR, Meservey). Heights for 7 of these ranged from 8 to 20 feet and averaged 16 feet (CNR), but the range from other areas is 3 to 35 feet. Nests, constructed by the females, are neatly woven cups of grass, bark strips, plant fibers, and spiderweb lined with rootlets, fine grass, and hair. They are placed in a crotch of a tree, sapling, or bush. The normal clutch is four eggs, but two, three, and rarely five may be produced. Clutch sizes for 10 Massachusetts nests were four eggs (9 nests), five eggs (1 nest) (DKW, CNR). Two other nests were parasitized by cowbirds; one was abandoned on May 24 with a single cowbird egg, and one contained a large cowbird nestling on July 20 (CNR).

Incubation, also by the female, lasts from 12 to 14 days, and the young are fed in the nest by both parents for 8 to 10 days. When hatched, they have only sparse down. At 3 days the eyes begin to open and the wing quills show as tiny papillae, and by 8 days the birds are well feathered. Nestlings have been recorded in Massachusetts from June 10 to July 25 (Meservey, CNR). Brood sizes for 3 state nests were four young (1 nest), five young (2 nests) (EHF, CNR). Fledglings are fed for several weeks, but by the time they have attained winter plumage and appear fully grown, they are independent. At this time, aggressive interactions split up the family groups. Pairs or single adults feeding one to three fledglings have been reported in the Commonwealth from June 24 to August 6 (Meservey, CNR).

Although commonly considered to rear only one brood, in fact, some pairs do produce

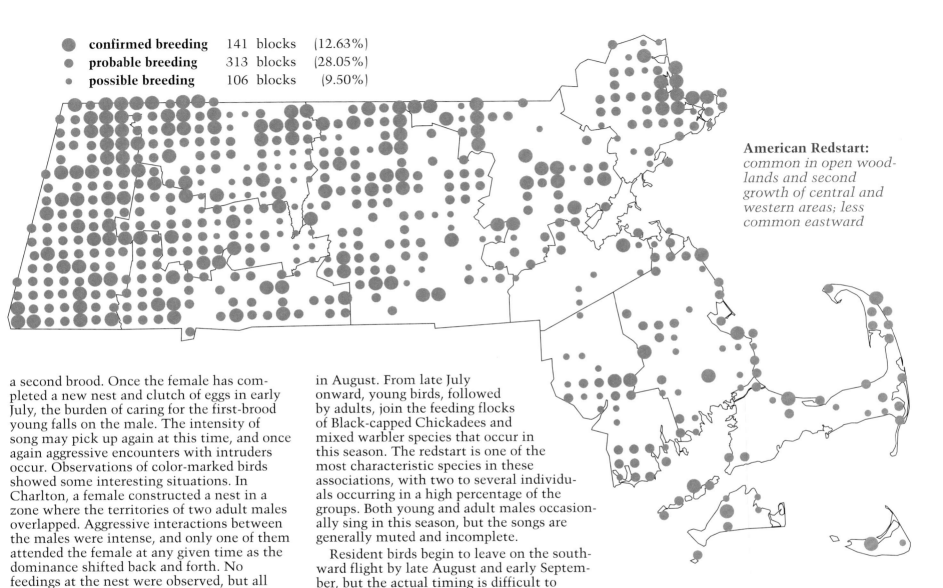

- **confirmed breeding** 141 blocks (12.63%)
- **probable breeding** 313 blocks (28.05%)
- **possible breeding** 106 blocks (9.50%)

American Redstart: *common in open woodlands and second growth of central and western areas; less common eastward*

a second brood. Once the female has completed a new nest and clutch of eggs in early July, the burden of caring for the first-brood young falls on the male. The intensity of song may pick up again at this time, and once again aggressive encounters with intruders occur. Observations of color-marked birds showed some interesting situations. In Charlton, a female constructed a nest in a zone where the territories of two adult males overlapped. Aggressive interactions between the males were intense, and only one of them attended the female at any given time as the dominance shifted back and forth. No feedings at the nest were observed, but all three adults shared in the care of the fledglings. The female then moved well into the territory of one male and raised a second brood there with him alone (Meservey).

Adults that rear one brood are usually in heavy molt by late July. Those that are successful in producing two broods will molt in August. From late July onward, young birds, followed by adults, join the feeding flocks of Black-capped Chickadees and mixed warbler species that occur in this season. The redstart is one of the most characteristic species in these associations, with two to several individuals occurring in a high percentage of the groups. Both young and adult males occasionally sing in this season, but the songs are generally muted and incomplete.

Resident birds begin to leave on the southward flight by late August and early September, but the actual timing is difficult to determine because of the arrival of transients from farther north. Redstarts are common migrants throughout the state during September. The species occurs regularly to mid-October, with occasional stragglers noted thereafter. The wintering grounds are extensive and include Mexico, the West Indies, Central America, and northern South America. A few birds even winter in southern Florida.

W. Roger Meservey

Worm-eating Warbler
Helmitheros vermivorus

Egg dates: June 7 to June 19.

Number of broods: one.

Forbush reported the Worm-eating Warbler to be casual or a straggler in Massachusetts and a rare summer resident, locally not uncommon, in Connecticut. Nesting was first suspected in the Commonwealth when a singing male was discovered at Bash Bish Falls, Mount Washington, on June 24, 1923. Subsequently, nesting was reported from three different stations in southern Berkshire County (including from 1939 on at South Egremont), at Mount Tom, Holyoke, and several locales in the lower Connecticut River valley (1952 onward). In Worcester County, the species was rare before the 1970s but has been recorded annually since 1975. Territorial males were recorded in Uxbridge (1978), Southbridge and Petersham (1979), and Upton (1981), with breeding finally confirmed in Petersham in 1982. During subsequent years, the birds have continued to nest in Petersham and also at a station in Uxbridge. Sightings have occurred from other parts of Quabbin Reservoir away from the Petersham site. In eastern Massachusetts, breeding was first confirmed in Weston (1962) and Dover (1975). In mid-June 1981, four birds found in the Westport area were strongly suggested to be nesting.

During the Atlas period, nesting Worm-eating Warblers were confirmed at two sites in Southwick, Easthampton, Weston, and Dover. What emerges from the available information is a picture of slow increase and very gradual eastward range expansion. Despite a pattern of isolated occurrences suggested by the Atlas data, partly an artifact of the species' low detectability, the true range in Massachusetts is probably broader. The species should be sought and expected from southern Berkshire County eastward across the lower Connecticut River valley, across Worcester County (north to Petersham and Royalston) to southern Middlesex County, central Norfolk County, and Bristol County.

The Worm-eating Warbler starts its northward migration in late March, reaches Texas and Florida in early April, and arrives at the northern fringe of its nesting range here in Massachusetts by mid-May, with individuals appearing from late April onward. As a terminal migrant, it is seldom recorded away from the nesting stations. Six records for spring transients in Worcester County range from May 2 to May 19 (TC).

Worm-eating Warblers occupy the brushy undergrowth of leaf-covered, rocky, wooded hillsides and ravines, usually near water. Advertising males deliver an unvarying and unmusical buzz, often rendered *chee-e-e-e-e-e*, which experienced observers readily distinguish from the similar sounding song of the Chipping Sparrow. Males sing while perched at 10 to 20 feet in low understory or while on the ground. In Massachusetts, some males seem to sing very infrequently, perhaps due to the absence of adjacent territorial birds. Both sexes have a loud *chip* call note and a softer *dzt*.

The nest, presumably built mainly by the female, is well concealed in the leaf litter on the forest floor, often at the base of a sapling or bush. Nest building was observed in Southwick on May 24, 1976. A nest in Weston (1962) was described as typical, lined with stems of hairy-cap moss and built on the ground beneath an overhanging Lowbush

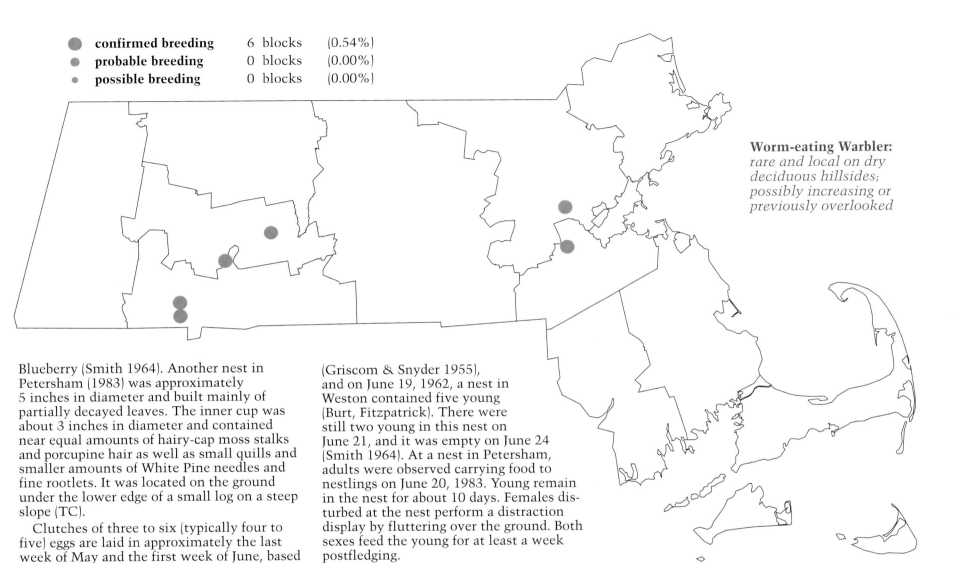

Worm-eating Warbler: *rare and local on dry deciduous hillsides; possibly increasing or previously overlooked*

●	confirmed breeding	6 blocks	(0.54%)
●	probable breeding	0 blocks	(0.00%)
●	possible breeding	0 blocks	(0.00%)

Blueberry (Smith 1964). Another nest in Petersham (1983) was approximately 5 inches in diameter and built mainly of partially decayed leaves. The inner cup was about 3 inches in diameter and contained near equal amounts of hairy-cap moss stalks and porcupine hair as well as small quills and smaller amounts of White Pine needles and fine rootlets. It was located on the ground under the lower edge of a small log on a steep slope (TC).

Clutches of three to six (typically four to five) eggs are laid in approximately the last week of May and the first week of June, based on observations of adults and young. Observed egg dates in Massachusetts have been between June 7 and June 19 (BOEM, Veit). Connecticut egg dates are May 25 to June 19 (EHF) and May 27 to June 29 (ACB). Eggs are white with brown speckles. The incubation period is about 13 days.

On June 19, 1949, a nest containing four young was discovered in South Egremont (Griscom & Snyder 1955), and on June 19, 1962, a nest in Weston contained five young (Burt, Fitzpatrick). There were still two young in this nest on June 21, and it was empty on June 24 (Smith 1964). At a nest in Petersham, adults were observed carrying food to nestlings on June 20, 1983. Young remain in the nest for about 10 days. Females disturbed at the nest perform a distraction display by fluttering over the ground. Both sexes feed the young for at least a week postfledging.

Adults feeding fledglings have been reported in Massachusetts from June 19 to July 11 (Smith 1964, TC). A molting adult feeding two large fledglings was observed on June 29 at the Petersham nest site. Fledglings calling to be fed give constant *chip* notes. These warblers become virtually undetectable after July, though there are scattered records through August to mid-September. The known winter range is from Venezuela to Panama and in the Bahamas and Greater Antilles east to the Virgin Islands.

Thomas H. Gagnon

Ovenbird
Seiurus aurocapillus

Egg dates: May 17 to July 16.

Number of broods: one.

The Ovenbird, a bird of mature deciduous and coniferous forests, is a common Massachusetts nester, a common spring migrant, and a regular though not abundant fall migrant. During the breeding season, it prefers open forests with little or no understory vegetation and ample leaf litter; during migration it can be found in woods, thickets, and scrubby habitats. Although undoubtedly scarce in sparsely wooded nineteenth-century Massachusetts, the Ovenbird is now a common nesting species throughout the forested regions of the state. It is a regular, though local resident of Cape Cod, Martha's Vineyard, and the predominantly scrubby coastal regions. The explosive suburban development that has taken place in recent years has had an adverse effect on eastern breeding populations of Ovenbirds by removing and fragmenting their nesting habitat, which has resulted in disturbance and increased predation by dogs, cats, and natural enemies. Similarly, clear-cutting practices in Central American forests, where the Ovenbird is found during the winter, may adversely affect nonbreeding populations.

In spring, Ovenbirds begin to appear in late April, though the peak of spring migration does not occur until the second half of May. Males can be heard singing during early May, and territories of residents are well established by the middle of the month. The male's territorial song, a loud *teacher, teacher, teacher*, increases in volume from the first to the last phrase and may be accented on either syllable *(tea-cher* or *tea-cher)*. In forests lacking a dense understory, the song resonates quite a distance, thus making if difficult for the human observer to pinpoint the source. This song is given throughout June and into August but becomes much more intermittent after mid-July. Males also perform an aerial song display (often crepuscular or nocturnal) that has been variously described as a courtship display, sheer jubilation, or a territorial display. It can be heard well into August. Both sexes have several call notes, including a *tchip* contact note and a *tick* or *tchuck* of alarm.

During courtship, the male may strut before the female on the ground or chase her vigorously in flight. When the male has defined the territory, the female selects the nest site and builds the nest on the forest floor. The nest is a dome-shaped, Dutch-ovenlike construction made of dried leaves, twigs, and grasses, cryptically concealed by the abundant leaf litter. Two Pelham nests were made of dead leaves and dry White Pine needles (Nice 1931). Of 11 Massachusetts nests, 10 were located on the ground in deciduous or mixed forests (Nice 1931, CNR, Blodget). The remaining nest was built atypically among green grasses in an open space and was soon preyed upon. (Nice 1931, DKW, CNR). The eggs are incubated for 11 to 14 days by the female only. If an intruder approaches the nest too closely, an adult will perform a distraction display, fluttering across the forest floor with outspread wings and uttering loud *ticks*. These *ticks* are also voiced from low branches or shrubs.

The altricial young are helpless upon hatching. At 4 days of age their eyes are opening and the feather papillae are showing. By the sixth day, the feathers have broken through their sheaths, and the young become more active. They are well feathered at 8 days and are ready to fledge then or during

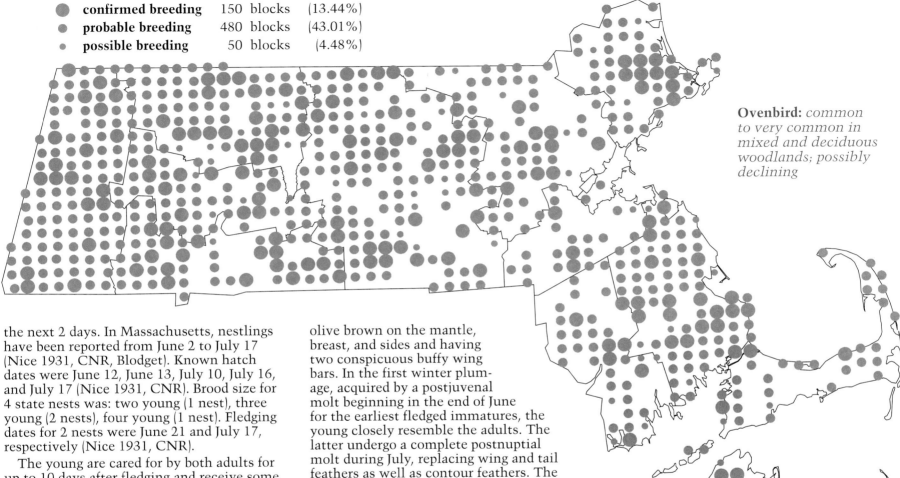

Ovenbird: *common to very common in mixed and deciduous woodlands; possibly declining*

● confirmed breeding	150 blocks	(13.44%)
● probable breeding	480 blocks	(43.01%)
· possible breeding	50 blocks	(4.48%)

the next 2 days. In Massachusetts, nestlings have been reported from June 2 to July 17 (Nice 1931, CNR, Blodget). Known hatch dates were June 12, June 13, July 10, July 16, and July 17 (Nice 1931, CNR). Brood size for 4 state nests was: two young (1 nest), three young (2 nests), four young (1 nest). Fledging dates for 2 nests were June 21 and July 17, respectively (Nice 1931, CNR).

The young are cared for by both adults for up to 10 days after fledging and receive some food until they are five weeks old. Food brought to the young consists of insects, spiders, slugs, centipedes, etc., picked up from the forest floor. Fledglings with adults have been reported in the state from June 21 to August 5 (Nice 1931, CNR, Meservey), but the actual range is probably greater. Data on second broods is lacking, but the range of dates for eggs and young suggests that the birds may renest after a failure.

In the juvenal plumage, the young differ from adults in being streaked or spotted with olive brown on the mantle, breast, and sides and having two conspicuous buffy wing bars. In the first winter plumage, acquired by a postjuvenal molt beginning in the end of June for the earliest fledged immatures, the young closely resemble the adults. The latter undergo a complete postnuptial molt during July, replacing wing and tail feathers as well as contour feathers. The young remain on the territory until after their molt. Adults may disperse before their molt is entirely complete; however, males can be heard singing territorial song well into August and may disperse later than their families.

The fall movement of Ovenbirds is less noticeable than that of spring. What little migration is observed begins in August and extends through September; stragglers are seen regularly during October, and occasional individuals survive into the early winter. The normal wintering grounds extend from the southern United States through Mexico and Central America to northern South America.

Christina Dowd

Northern Waterthrush
Seiurus noveboracensis

Egg dates: May 21 to June 15.

Number of broods: one.

The loud, declamatory but musical song of the Northern Waterthrush startles the ear when it is first heard in late April. It rings out in urban areas over the noise of traffic and is easily heard coming from dense shrubbery in suburban gardens or from thickets in the swamps and bogs of the more rural areas of Massachusetts. Uncommon and local breeders in the state, waterthrushes nest here in wooded swamps. The most numerous sites are in the Berkshires, with fair numbers in Franklin, Hampshire, and Worcester counties. Others are scattered to the east wherever swamp land or other stagnant-water habitat occurs. The species does not nest on Cape Cod or the Islands.

The Northern Waterthrushes that nest here arrive in late April and early May, but the species is a common migrant throughout the state, with the greatest passage occurring in mid-May. The male waterthrush sings readily in migration but is not always easy to see. As he walks along, teetering with tail bobbing among the tree roots on a muddy bank, he stays well within the shadowed protection of overhanging grasses or low branches. In a number of locales in Massachusetts, the breeding territories of the Northern Waterthrush overlap with those of the closely related Louisiana Waterthrush. The former species occupies the more sluggish portions of streams and adjacent swamps while the latter frequents the swift-flowing sections of brooks.

The name waterthrush is well deserved for this species resembles a thrush, and it feeds and nests on or near the ground in the immediate vicinity of water. In addition to swamps and sluggish streams, it frequents quiet or isolated ponds and lakes and even temporary pools of standing water such as rain-filled ditches. In such areas, these birds feed on insects—mosquitoes, flies, ants, etc.—and their larvae, slugs, and even tiny fish.

Several vocalizations of the Northern Waterthrush have been described. The familiar song, a loud, ringing, bubbling warble, emphatically delivered, may be heard until the middle of July. There is also a flight song that has been described as a longer, hurried, and jumbled version of the usual song with call notes interspersed. This is sung as the bird flies through or just above the woods. The common call or *chip* note is a sharp, thin, and metallic *clink*.

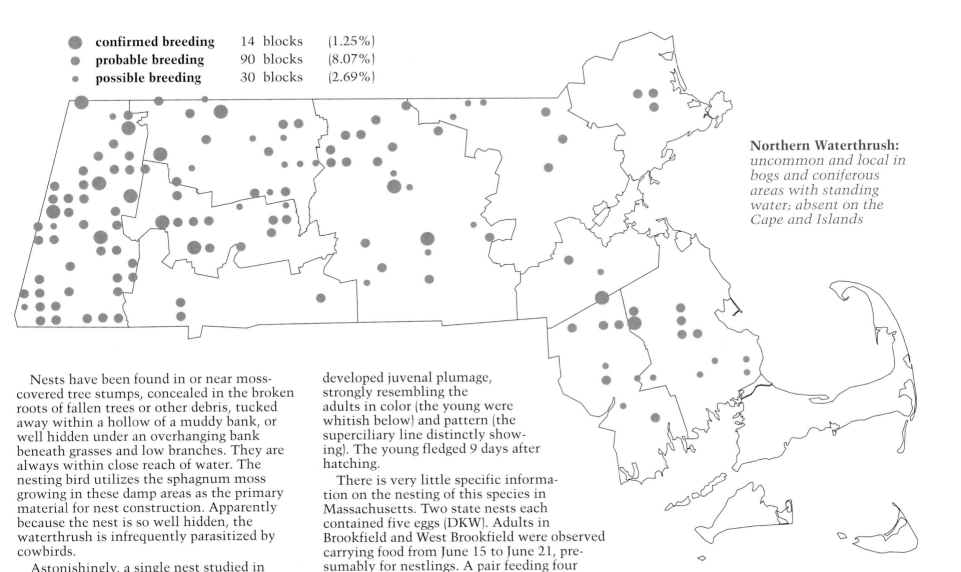

Northern Waterthrush: *uncommon and local in bogs and coniferous areas with standing water; absent on the Cape and Islands*

Nests have been found in or near moss-covered tree stumps, concealed in the broken roots of fallen trees or other debris, tucked away within a hollow of a muddy bank, or well hidden under an overhanging bank beneath grasses and low branches. They are always within close reach of water. The nesting bird utilizes the sphagnum moss growing in these damp areas as the primary material for nest construction. Apparently because the nest is so well hidden, the waterthrush is infrequently parasitized by cowbirds.

Astonishingly, a single nest studied in 1985 in New Jersey (Wander & Wander 1985) constitutes the only report in the literature on incubation and fledging age in this species. The nest contained a five-egg clutch, and the young hatched after 13 days of incubation (thought to have begun with the third egg laid). At 8 days of age, the five nestlings were banded and showed well-developed juvenal plumage, strongly resembling the adults in color (the young were whitish below) and pattern (the superciliary line distinctly showing). The young fledged 9 days after hatching.

There is very little specific information on the nesting of this species in Massachusetts. Two state nests each contained five eggs (DKW). Adults in Brookfield and West Brookfield were observed carrying food from June 15 to June 21, presumably for nestlings. A pair feeding four large fledglings was recorded on July 17 in Charlton (Meservey).

Waterthrushes begin their southward migration in July, with the peak occurring in late August to mid-September, but stragglers are found, especially along the coast, into early October and rarely to December. The winter range extends from Texas, Florida, and Bermuda through Central America and the West Indies south to Peru and Ecuador. Once in its winter quarters, a Northern Waterthrush quickly establishes a territory near water in which to lead a solitary existence.

Dorothy Rodwell Arvidson

Louisiana Waterthrush
Seiurus motacilla

Egg dates: not available.
Number of broods: one.

The Louisiana Waterthrush is a fairly common summer resident in the hill country of central and western Massachusetts but becomes decidedly rare east of Worcester County. It is absent entirely from Cape Cod and the Islands. General expansion northward and eastward has been documented in the historical period. It was first found nesting in Massachusetts on Mount Tom, Holyoke, in 1869. Forbush's (1929) map, though giving fewer nesting records, is remarkably similar to the Atlas map and documents that, even before 1930, eastward movement in the hills was well underway. The greater number of confirmed nestings in the Atlas period may, in part, be an artifact of more determined systematic coverage of the state than was the case in the years before 1930.

Louisiana Waterthrushes arrive in mid-April. Their wild, ringing, sweet song carries for a considerable distance through the still-leafless April woods. The song characteristically starts with three loud, clear, slurred, and downward-inflected notes and ends as a jumble of musical notes. Like its congener, the Ovenbird, this species also reportedly has a complex and little-known flight song, which is sometimes given at night. Singing continues through June and is rarely heard to early August. In territorial defense, waterthrushes sometimes pour out song repeatedly in flight.

The preferred habitat is along the Blackfly- and mosquito-infested reaches of both swift and sluggish brooks and streams running through deep mixed woodlands. Unique among wood-warblers, pairs occupy long linear territories that approximate the thread of the stream, sometimes for several miles of quiet sluggish headwater swamps as well as swiftly flowing reaches. Territories are often so long that singing birds cannot be heard from one end to the other.

Typically, within these linear territories, the birds spend most of their time foraging about low in the stream itself or on the stream's banks. Birds walk nonchalantly about, rather than hopping, and wag their tails continuously. The birds sing while walking about, during flight up and down the

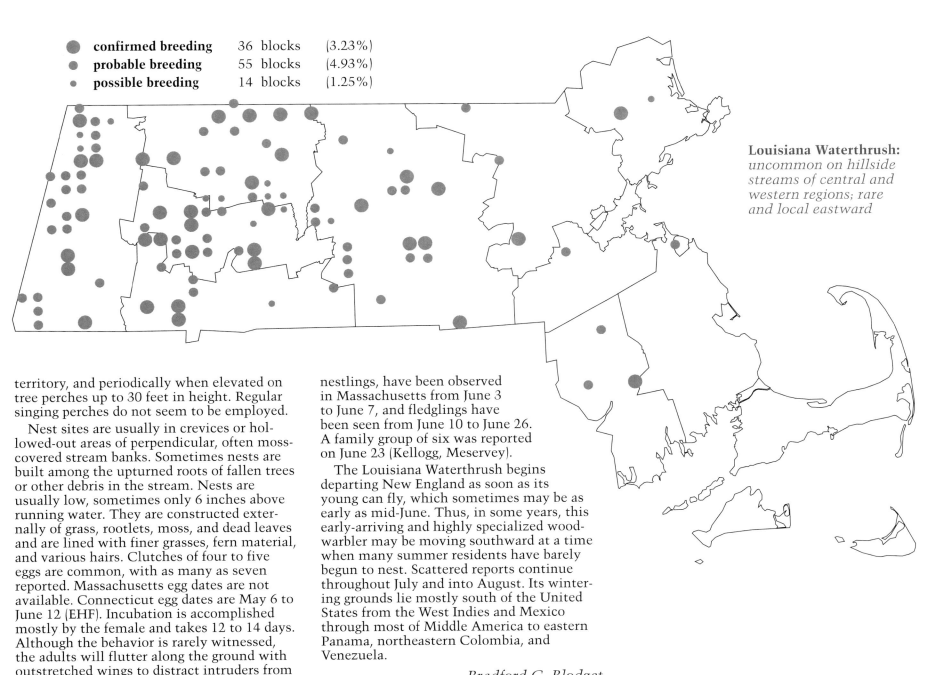

● confirmed breeding	36 blocks	(3.23%)
● probable breeding	55 blocks	(4.93%)
· possible breeding	14 blocks	(1.25%)

Louisiana Waterthrush: *uncommon on hillside streams of central and western regions; rare and local eastward*

territory, and periodically when elevated on tree perches up to 30 feet in height. Regular singing perches do not seem to be employed.

Nest sites are usually in crevices or hollowed-out areas of perpendicular, often moss-covered stream banks. Sometimes nests are built among the upturned roots of fallen trees or other debris in the stream. Nests are usually low, sometimes only 6 inches above running water. They are constructed externally of grass, rootlets, moss, and dead leaves and are lined with finer grasses, fern material, and various hairs. Clutches of four to five eggs are common, with as many as seven reported. Massachusetts egg dates are not available. Connecticut egg dates are May 6 to June 12 (EHF). Incubation is accomplished mostly by the female and takes 12 to 14 days. Although the behavior is rarely witnessed, the adults will flutter along the ground with outstretched wings to distract intruders from young. Adults carrying food, presumably to nestlings, have been observed in Massachusetts from June 3 to June 7, and fledglings have been seen from June 10 to June 26. A family group of six was reported on June 23 (Kellogg, Meservey).

The Louisiana Waterthrush begins departing New England as soon as its young can fly, which sometimes may be as early as mid-June. Thus, in some years, this early-arriving and highly specialized wood-warbler may be moving southward at a time when many summer residents have barely begun to nest. Scattered reports continue throughout July and into August. Its wintering grounds lie mostly south of the United States from the West Indies and Mexico through most of Middle America to eastern Panama, northeastern Colombia, and Venezuela.

Bradford G. Blodget

Mourning Warbler
Oporornis philadelphia

Egg dates: not available.

Number of broods: one.

This handsome bird is regarded as one of the rarest of the warblers occurring regularly in Massachusetts. The opinion is in part due to the bird's shyness, its fondness for skulking in dense briers and other shrubbery, and its arrival later in the spring than most birds, at a time when the foliage is often fully developed.

The Mourning Warbler received its scientific name because Alexander Wilson discovered the type specimen near Philadelphia. He never saw another, and Audubon encountered few. The male does wear a veil of crepe, but, as Edward Howe Forbush pointed out, this is the only thing about the bird that would suggest mourning, calling its song "a paean of joy."

Like other members of its genus, the Mourning Warbler has a loud, penetrating song, often described as *chirry, chirry, chory, chory*. Some of the less typical notes can be confused with the song of the Common Yellowthroat. The Mourning Warbler's alarm note is a distinctive, sharp *chip*.

Mourning Warblers were first discovered at 3,491 feet on Mount Greylock in the summer of 1883 by William Brewster, who termed them "abundant." They have been seen there ever since, although in diminishing numbers, probably as a result of the maturing of the sprout growth on the upper mountain. During the 1940s and 1950s, from two to eight singing males occurred along the upper portions of the roads leading to the summit. Today, possibly as a result of the enormous increase in auto traffic, Mourning Warblers are now more likely to be found lower down, especially where slash growth occurs. In recent summers, birds have been seen as low as 1,500 feet along Rockwell and Notch roads, and, in 1983, males were found in July as low as 1,200 feet in the Hopper, on the west side of Mount Greylock.

Historically, Mourning Warblers also occurred in North Adams, Savoy, and Florida. In recent summers, they have been located in Cheshire, Dalton, and Hancock. All locations are in northern Berkshire County at an elevation greater than 1,500 feet.

Spring migrants may appear as early as the second week of May, but the bulk of the birds do not arrive until the end of the month or the first week of June. Some of the latter are probably northbound transients. Males

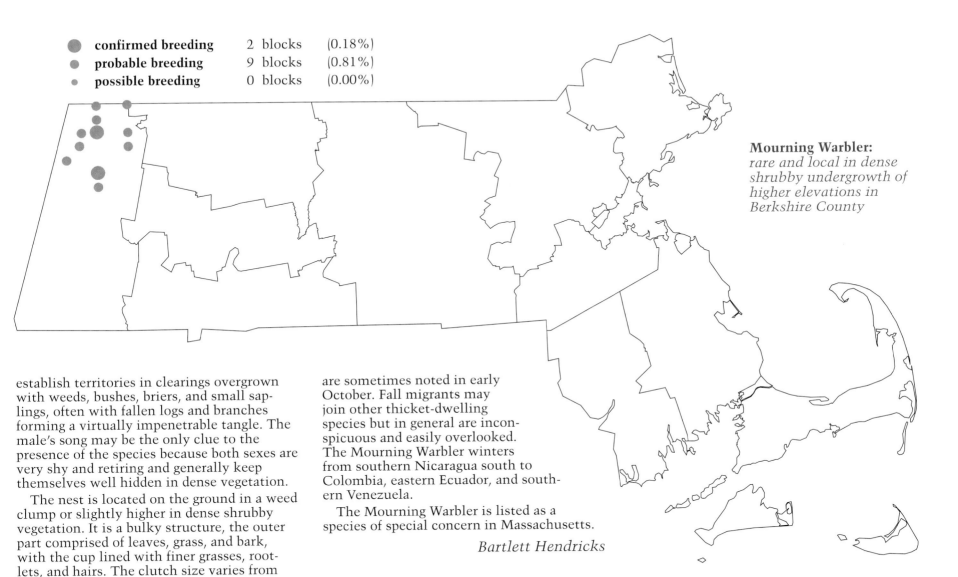

- confirmed breeding — 2 blocks (0.18%)
- probable breeding — 9 blocks (0.81%)
- possible breeding — 0 blocks (0.00%)

Mourning Warbler: *rare and local in dense shrubby undergrowth of higher elevations in Berkshire County*

establish territories in clearings overgrown with weeds, bushes, briers, and small saplings, often with fallen logs and branches forming a virtually impenetrable tangle. The male's song may be the only clue to the presence of the species because both sexes are very shy and retiring and generally keep themselves well hidden in dense vegetation.

The nest is located on the ground in a weed clump or slightly higher in dense shrubby vegetation. It is a bulky structure, the outer part comprised of leaves, grass, and bark, with the cup lined with finer grasses, rootlets, and hairs. The clutch size varies from three to five eggs, with four eggs being average. The eggs are incubated solely by the female for about 12 days. The young leave the nest about 10 days after hatching and are tended by both parent birds for several weeks.

Migration commences by mid-August and continues through September with the passage of more northerly breeders. Stragglers are sometimes noted in early October. Fall migrants may join other thicket-dwelling species but in general are inconspicuous and easily overlooked. The Mourning Warbler winters from southern Nicaragua south to Colombia, eastern Ecuador, and southern Venezuela.

The Mourning Warbler is listed as a species of special concern in Massachusetts.

Bartlett Hendricks

Common Yellowthroat
Geothlypis trichas

Egg dates: May 19 to June 27.

Number of broods: two.

The Common Yellowthroat may be the best-known warbler breeding in Massachusetts due to its inquisitive behavior, a loud and recognizable song, the male's distinctive plumage of a black mask and a bright yellow throat and chest, and its wide distribution throughout the Commonwealth. This warbler is found nesting in a wide variety of habitats, many characterized as both brushy and damp—swampy thickets, cattail marshes, overgrown meadows, and tangles near streams. The birds also utilize dry habitats such as roadside thickets, woodland clearings, and overgrown upland pastures. Seemingly, they occupy any small niche of suitable habitat, many of which are successional.

While most references stress the traditional relationship between yellowthroats and damp thickets, recent research has documented great concentrations of yellowthroats breeding in the Scrub Oaks of the sandy pinelands of southern Plymouth County, where the Prairie Warbler also occurs in high density. The attraction of this habitat appears to be the brushlike thickets of Scrub Oak and the openness of a canopy limited to scattered Pitch Pines. The highest breeding density recorded was 75 territories per one hundred acres unburned for 40 years. In contrast, a nearby coastal area that has the appearance of being good yellowthroat habitat has territories ranging from only 11 to 27 per 100 acres.

Yellowthroats generally begin to appear in early May. During spring flights, they may seem to be everywhere, although many of these birds are migrants that will move on. Banding has revealed that local breeding adults are the first to arrive and hold territories and transients bound for the North pass through a week or more later. Breeding yellowthroats are easy to find because the male sings his *witchety-witchety-witchety* or *wee-chee-te, wee-chee-te, wee-chee-te* song loudly and nearly continuously. Both sexes respond immediately to intruders with a scolding *check* or *quit* note and a rasping, chattering call, usually keeping close to the ground and cocking their tails in a wrenlike manner. The male has a less-known aerial flight song, a rapid jumble of short notes incorporating a few *witchety*s, most often given in the evening at a height of 10 to 15 feet. Both types of songs are heard regularly throughout July and sporadically to the end of August.

Despite the conspicuous behavior of the birds, the rather large and bulky nest is well hidden and difficult to find. It is built on or just above the ground in thick vegetation and sometimes over water. Details about nests in Massachusetts are scarce. Bent mentions two nests from Arlington built in Red Cedar. Six Worcester County nests were all on the ground and concealed in clumps of grass or weeds. All were in old fields and were 6 feet or less from a wood border or row of trees and saplings (Meservey). The cupped nest is constructed of leaves, grass, and bark strips and lined with fine grass and rootlets. The female incubates the clutch of three to five (rarely six) eggs for 12 days. Clutch sizes for 11 state nests were three eggs (1 nest), four eggs (10 nests) (DKW). Eggs have been recorded in the Commonwealth from mid-May to early

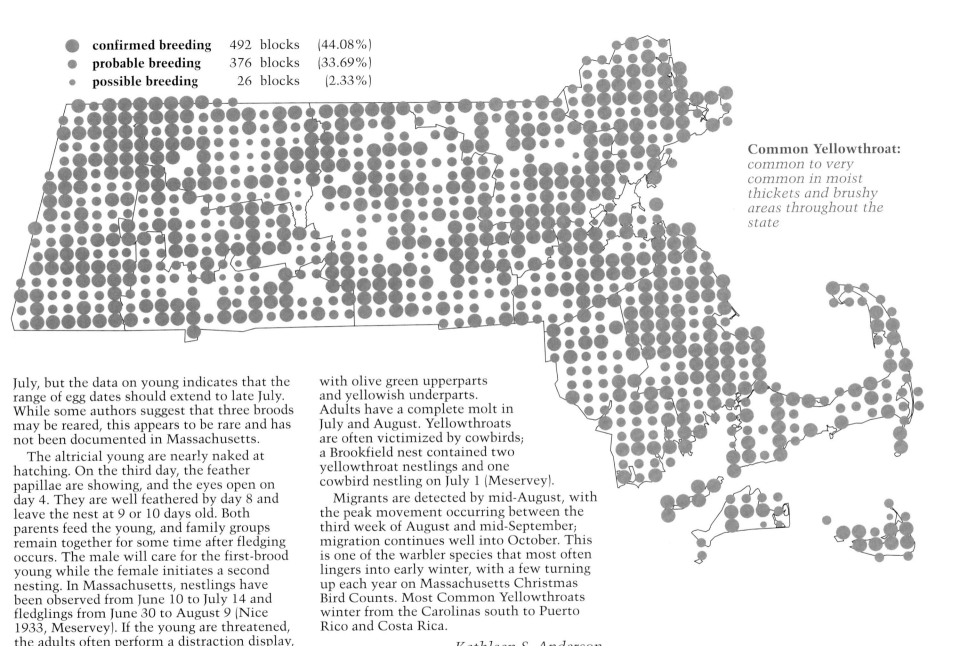

- **confirmed breeding** 492 blocks (44.08%)
- **probable breeding** 376 blocks (33.69%)
- **possible breeding** 26 blocks (2.33%)

Common Yellowthroat: *common to very common in moist thickets and brushy areas throughout the state*

July, but the data on young indicates that the range of egg dates should extend to late July. While some authors suggest that three broods may be reared, this appears to be rare and has not been documented in Massachusetts.

The altricial young are nearly naked at hatching. On the third day, the feather papillae are showing, and the eyes open on day 4. They are well feathered by day 8 and leave the nest at 9 or 10 days old. Both parents feed the young, and family groups remain together for some time after fledging occurs. The male will care for the first-brood young while the female initiates a second nesting. In Massachusetts, nestlings have been observed from June 10 to July 14 and fledglings from June 30 to August 9 (Nice 1933, Meservey). If the young are threatened, the adults often perform a distraction display, hopping about with quivering and spread wings. Following the postjuvenal molt, young birds closely resemble the adult females, with olive green upperparts and yellowish underparts. Adults have a complete molt in July and August. Yellowthroats are often victimized by cowbirds; a Brookfield nest contained two yellowthroat nestlings and one cowbird nestling on July 1 (Meservey).

Migrants are detected by mid-August, with the peak movement occurring between the third week of August and mid-September; migration continues well into October. This is one of the warbler species that most often lingers into early winter, with a few turning up each year on Massachusetts Christmas Bird Counts. Most Common Yellowthroats winter from the Carolinas south to Puerto Rico and Costa Rica.

Kathleen S. Anderson

Hooded Warbler
Wilsonia citrina

Egg dates: June 3.

Number of broods: one or two.

No springtime warbler is listed more frequently on the most-wanted list drawn up by Massachusetts birders than the elusive Hooded Warbler. Forbush noted only 20 Massachusetts records between 1900 and 1927, and Wallace Bailey wrote, "Any observer who discovers one in a whole year may consider himself fortunate" (Bailey 1955). In 1978, the Hooded Warbler was described as rare, a local breeder, and a summer vagrant (Blodget 1978).

The preferred habitat of this warbler includes moist thickets in woodlands of second-growth such as oak, beech, ash, gum, and Red Maple, which afford a fairly open understory over the shrubs and rather dense mixed vegetation below. In Massachusetts, Swamp Azalea, viburnum, Catbrier, and ferns figure prominently in the understory, and these plants are used advantageously to provide both support and cover for the nest.

The breeding of Hooded Warblers in southeastern Massachusetts was first documented in 1967 and 1968, when several nests were monitored in Westport. Unfortunately, human encroachment on the breeding territories during the ensuing years undermined the possibility of their continued nesting success. The Atlas failed to produce any confirmed nesting attempts, although singing males were occasionally heard and believed to be on territory.

Hooded Warblers generally arrive in southeastern Massachusetts in early May. Some authors have expressed the belief that males return in spring before females, but that conclusion may have been reached because the latter are inconspicuous and not usually sighted until the courtship singing of the male leads the observer to a pair's nesting territory. Females are definitely more secretive than males, and they tend to move furtively about in habitat where observation is not easy.

Both male and female Hooded Warblers give a call note best described as the ringing chink of a silver spoon striking a thin china cup. The male's note is stronger and louder than that of the female, whose note can be likened to the sweet little chip of a cardinal. The literature seems full of different descriptions of the male's song, although all authors agree that it is uncommonly beautiful and that there are two versions that differ primarily in length. Most also agree that the next-to-last note is noticeably emphasized, while the final note falls away in a soft slur. The series of rapidly whistled phrases is usually given 3 times in succession, but occasionally only 2. The song is somewhat comparable to the vocalization of a Chestnut-sided Warbler but is delivered with greater velocity, clarity, and assertiveness.

Nests are usually located low in the undergrowth, not less than 1.5 feet and seldom more than 3 feet above ground. The cup is quite deep, given such a small resident, and is structurally very strong, albeit made of fine materials. The exterior of the wall is generally formed with thin strips of the outer layers of grapevine bark. In Westport, where a tidal river reached to within 300 feet of one nest site, some of the lower lashings of a 1960s nest were of eelgrass and the inside was finished with closely packed, horizontally curved fine plant fibers in a tightly structured cup. The coarser exterior is bonded and made water-resistant with a generous application of tough, adhesive cobwebs.

Nest building usually takes 6 to 8 days and may, ironically, contribute to the misfortunes that frequently befall nesting success. Cowbirds are notorious brood parasites of Hooded Warbler nests, a practice that seems remarkable because the little nests are so well concealed. One must suppose that gravid cowbirds see the countless comings and goings of the nest builder, thus learning the whereabouts of the nest. Also, such extended activity may arouse the curiosity of domesticated cats and dogs along with snakes, Raccoons, and Opossums. If predatory destruction interrupts the first nesting, the female will quickly rebuild then produce a

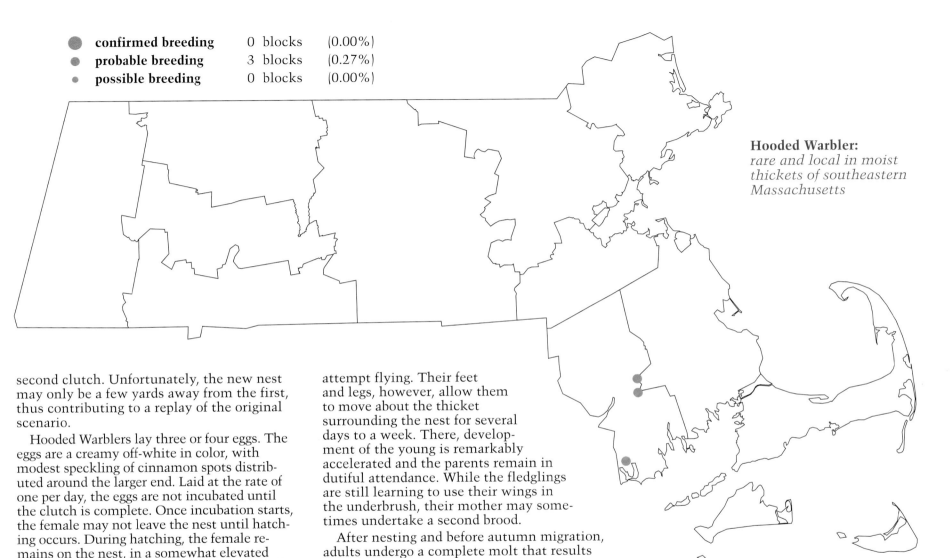

Hooded Warbler: *rare and local in moist thickets of southeastern Massachusetts*

- confirmed breeding — 0 blocks (0.00%)
- probable breeding — 3 blocks (0.27%)
- possible breeding — 0 blocks (0.00%)

second clutch. Unfortunately, the new nest may only be a few yards away from the first, thus contributing to a replay of the original scenario.

Hooded Warblers lay three or four eggs. The eggs are a creamy off-white in color, with modest speckling of cinnamon spots distributed around the larger end. Laid at the rate of one per day, the eggs are not incubated until the clutch is complete. Once incubation starts, the female may not leave the nest until hatching occurs. During hatching, the female remains on the nest, in a somewhat elevated position. She reaches down with her bill, pulls up the shell halves, and proceeds to break them down into bite-size pieces, which she swallows. Both parents feed the young a variety of small caterpillars, plus insects including crane flies, moths, mosquitoes, and gnats.

Fledging occurs when the young are 8 or 9 days old, although the wings of the juveniles are not developed enough for them to attempt flying. Their feet and legs, however, allow them to move about the thicket surrounding the nest for several days to a week. There, development of the young is remarkably accelerated and the parents remain in dutiful attendance. While the fledglings are still learning to use their wings in the underbrush, their mother may sometimes undertake a second brood.

After nesting and before autumn migration, adults undergo a complete molt that results in a winter plumage similar to the breeding plumage except that there are usually some yellowish tips to the black feathers of the crown and throat. Juveniles also undergo a complete molt but often retain some of their juvenile flight feathers until their first spring, as well as exhibiting considerable yellowish feathering in the black hood and throat feathers. Fall migrants bound for wintering grounds extending from southern Mexico to Panama are on the move by mid-August. The species is rare in Massachusetts after mid-September. Small numbers also winter in the Bahamas and Greater Antilles.

Gilbert Fernandez and Josephine Fernandez

Canada Warbler
Wilsonia canadensis

Egg dates: May 20 to June 26.

Number of broods: one.

The Canada Warbler is a fairly common summer resident in Massachusetts where there are open, damp, swampy areas within extensive tracts of mixed evergreen and deciduous woods. There it is partial to the lower story of the forest and the undergrowth. Most of the suitable breeding habitat is in the Berkshires and northern and western Worcester County, but there are breeding outposts around the Commonwealth east to Essex County and southeast to the Red Maple and Atlantic White Cedar swamps of Bristol and Plymouth counties. There is no evidence that its status has changed significantly since Forbush's day, despite statements by both Bailey (1955) and Griscom and Snyder (1955) that it had increased, especially in the eastern portion of its range.

A few Canada Warblers arrive in early May, but most appear during the last half of the month. Since they have a penchant for well-leafed-out, swampy undergrowth, they conceal themselves from all but the most persistent observers. The loud song is often the only clue to the birds' presence. Many migrants pass through to more northerly breeding grounds.

Nesting territories may be somewhat isolated and centered on a damp swale in a forest, but, in favorable areas of dense low vegetation along sluggish streams and swamps, several males may hold adjacent territories. The song is described as a short, explosive jumble of notes, frequently introduced by a single *chip*. When the birds are agitated, they give a sharp *chick* or *chink* call. Courtship is seldom observed, but males are known to pursue females in usual warbler fashion.

The nest typically is placed on or near the ground, often in a streamside bank, in a hummock, on a mossy stump or log, or in the exposed roots of a fallen tree. One of the few state nests recorded was in Bridgewater in "some mixed, moist woods, mostly white pines with a few oaks and other deciduous trees, near a swampy place…in plain sight at the foot of a clump of brakes *(Pteridium*

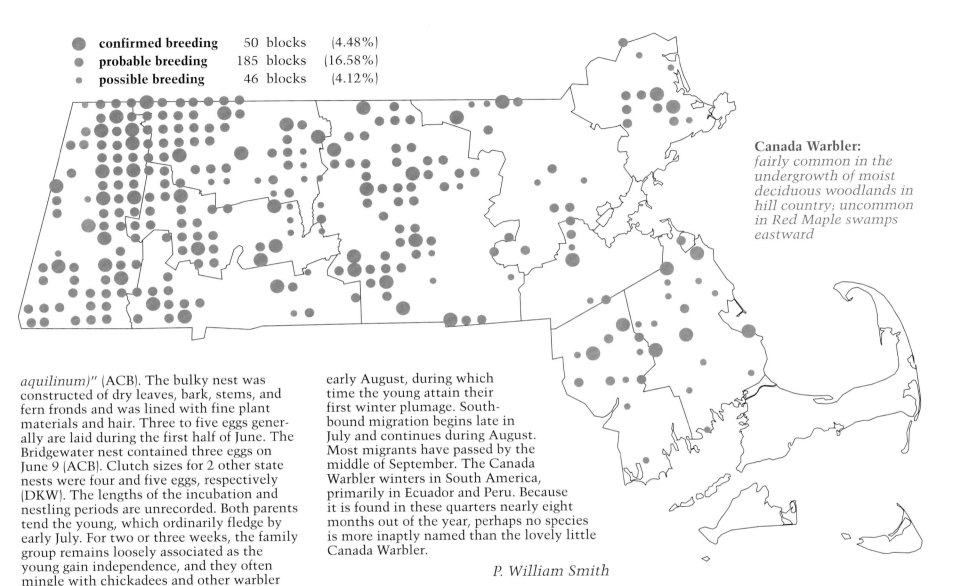

Canada Warbler: *fairly common in the undergrowth of moist deciduous woodlands in hill country; uncommon in Red Maple swamps eastward*

● confirmed breeding	50 blocks	(4.48%)
● probable breeding	185 blocks	(16.58%)
· possible breeding	46 blocks	(4.12%)

aquilinum)" (ACB). The bulky nest was constructed of dry leaves, bark, stems, and fern fronds and was lined with fine plant materials and hair. Three to five eggs generally are laid during the first half of June. The Bridgewater nest contained three eggs on June 9 (ACB). Clutch sizes for 2 other state nests were four and five eggs, respectively (DKW). The lengths of the incubation and nestling periods are unrecorded. Both parents tend the young, which ordinarily fledge by early July. For two or three weeks, the family group remains loosely associated as the young gain independence, and they often mingle with chickadees and other warbler species. These generally retiring birds become very bold when their young are threatened, approaching and performing elaborate distraction displays. In Massachusetts, adults have been observed feeding fledglings from June 24 to July 29 (Meservey, BOEM).

Adults have a complete molt in July or early August, during which time the young attain their first winter plumage. Southbound migration begins late in July and continues during August. Most migrants have passed by the middle of September. The Canada Warbler winters in South America, primarily in Ecuador and Peru. Because it is found in these quarters nearly eight months out of the year, perhaps no species is more inaptly named than the lovely little Canada Warbler.

P. William Smith

Yellow-breasted Chat
Icteria virens

Egg dates: May 18 to June 18.

Number of broods: one.

A basically southern species, the Yellow-breasted Chat is at the northern limit of its range in Massachusetts. Never numerous, and limited in its distribution by habitat, the chat historically has been found regularly only in the eastern lowlands and western river valleys. Its numbers have varied dramatically over the years. Forbush's map shows summer records from all counties except Nantucket, with a concentration from southern Essex County through the Boston area to Norfolk County. At that time, there were even some records from the Worcester County uplands. During the Atlas period, the chat was absent as a breeder from most of the state. The southeastern coast seems to be the only area where it continues to nest with any regularity. Post-Atlas records from western Massachusetts include pairs from Northampton, Agawam, and Lenox.

A few males arrive in May with waves of other migrants in various locales, but usually disappear after a day or two of fruitless singing. Residents are on territory by late May or early June. The chat is a secretive, thicket-loving bird, and prefers a dry, mixed transition habitat with tall grasses and weeds alongside briers and grape tangles as well as scattered small trees and bushes. Such habitat occurs here only in transition as pastures and fields revert to forests, and that process on a wide scale has almost ended in Massachusetts.

In its favored tangles, the chat lurks and feeds, sending out his extraordinary song of loud, clear hoots, deep caws, and nasal mews in measured and unending succession, usually two or three similar notes at a time in an unpredictable sequence. They make short, furtive flights and confound the listener when the calls start again from a completely different place. Once the male has attracted a mate and nesting begins, his song quickens and becomes more continuous and fervent, and he rises from the thickets to the top of a bush or nearby tree. If there are other males present, the singer is especially apt to perform his supreme display, mounting into the air to hover with dangling legs as he pours forth his song. Much of the vocalizing continues during the night.

Up to six eggs are laid in a well-hidden, bulky nest of leaves and bark strips lined with small grasses and placed about 3 feet above the ground. Incubation lasts 11 to

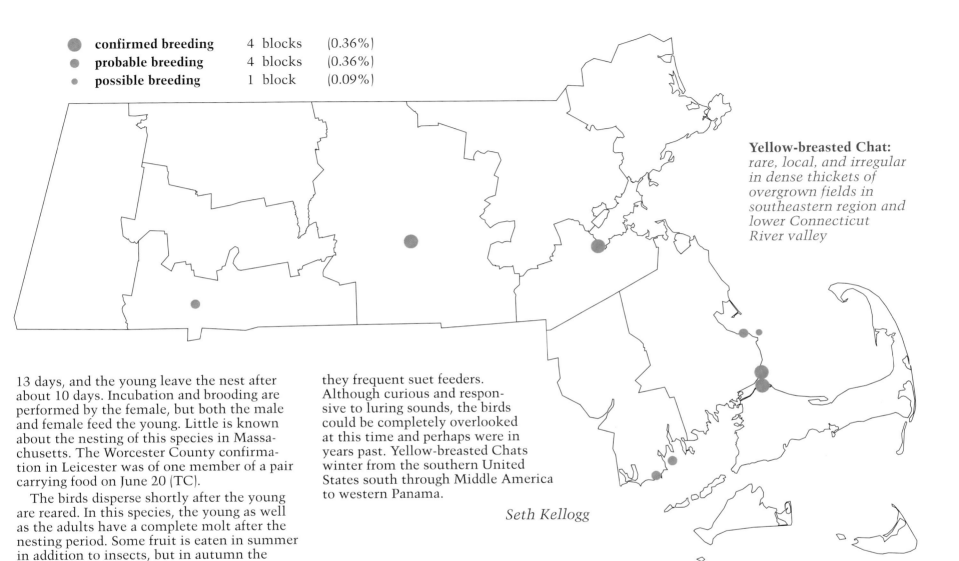

- confirmed breeding — 4 blocks (0.36%)
- probable breeding — 4 blocks (0.36%)
- possible breeding — 1 block (0.09%)

Yellow-breasted Chat: *rare, local, and irregular in dense thickets of overgrown fields in southeastern region and lower Connecticut River valley*

13 days, and the young leave the nest after about 10 days. Incubation and brooding are performed by the female, but both the male and female feed the young. Little is known about the nesting of this species in Massachusetts. The Worcester County confirmation in Leicester was of one member of a pair carrying food on June 20 (TC).

The birds disperse shortly after the young are reared. In this species, the young as well as the adults have a complete molt after the nesting period. Some fruit is eaten in summer in addition to insects, but in autumn the birds become dependent on the former food source. Fall migrants are curiously abundant for a species at the northern edge of its range and begin to appear along the coast in mid- to late August, with numbers peaking in mid-September and continuing through October. Some individuals routinely linger into winter. Most do not survive the coldest weather, even on those rare occasions when they frequent suet feeders. Although curious and responsive to luring sounds, the birds could be completely overlooked at this time and perhaps were in years past. Yellow-breasted Chats winter from the southern United States south through Middle America to western Panama.

Seth Kellogg

Scarlet Tanager
Piranga olivacea

Egg dates: May 24 to June 25.

Number of broods: one.

Irrespective of the number of occasions that one watches this species, be it foraging for food high in an oak during the breeding season or in respite from migration in a scrubby coastal thicket, the splendid brilliance of the male Scarlet Tanager is always an awesome sight. Belonging to a large and colorful subfamily of Neotropical birds, the Scarlet Tanager is one of the few species that nest north of Mexico. In Massachusetts, it enjoys a widespread and fairly common breeding status with the exception of much of Cape Cod and the Islands. There, its habitat requirements for mature deciduous or mixed deciduous-coniferous woodlands are either scarce or lacking entirely. Throughout much of the state, mature White Oak-White Pine forests appear to be especially preferred.

Most Scarlet Tanagers arrive in our area in mid-May and reveal their presence perhaps more by songs and calls than by movements, which are characteristically sluggish and purposeful. Paradoxically, the striking coloration of the male seems at times to be well adapted to camouflage, as he disappears suddenly and without effort in the variously deceiving sunlight and shadows at play in the forest canopy.

The arrival of the female in spring evokes the courtship and display response from the male, which generally has established its territory a week or so earlier. In display, the male hops on branches with the wings outstretched, thus exposing his brilliant scarlet back in striking contrast to the jet black of the wings. The song of the Scarlet Tanager, likened to the phrases *querit, queer, queery, querit, queer,* is a rather hoarse robinlike caroling, which the male sings tirelessly in proclamation of territory, often even at midday, from May until late July or early August. Also heard frequently is a low, distinctive *chip-burr* call that is given by both sexes throughout the summer.

The nest is a shallow, loosely woven cup of small twigs, grasses, and rootlets lined with finer and softer material, typically being placed in the fork of a branch 15 to 45 feet high and well out toward the periphery of a tree. In Massachusetts, nests have been found in apple, Red Cedar (15 feet), American Beech

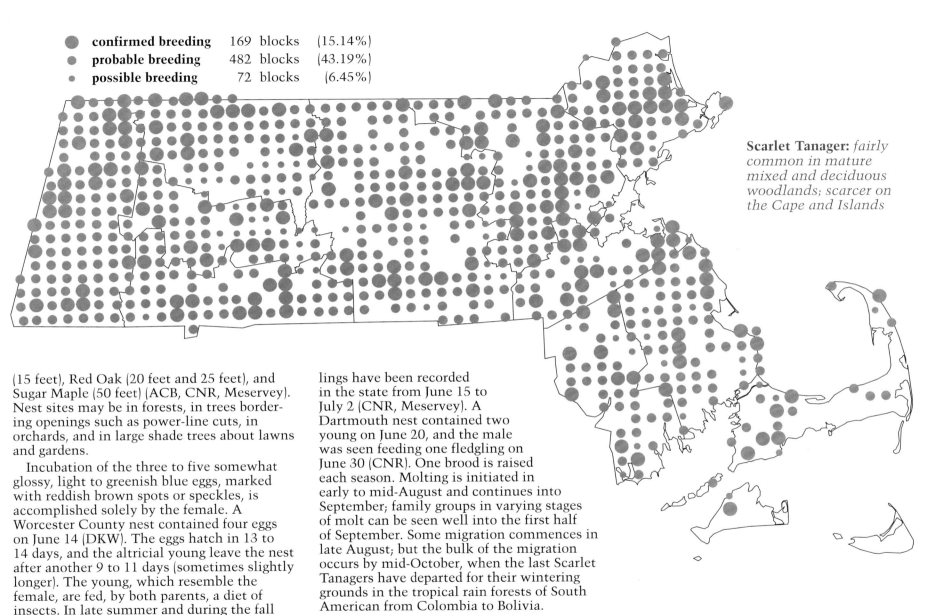

● confirmed breeding	169 blocks	(15.14%)
● probable breeding	482 blocks	(43.19%)
· possible breeding	72 blocks	(6.45%)

Scarlet Tanager: *fairly common in mature mixed and deciduous woodlands; scarcer on the Cape and Islands*

(15 feet), Red Oak (20 feet and 25 feet), and Sugar Maple (50 feet) (ACB, CNR, Meservey). Nest sites may be in forests, in trees bordering openings such as power-line cuts, in orchards, and in large shade trees about lawns and gardens.

Incubation of the three to five somewhat glossy, light to greenish blue eggs, marked with reddish brown spots or speckles, is accomplished solely by the female. A Worcester County nest contained four eggs on June 14 (DKW). The eggs hatch in 13 to 14 days, and the altricial young leave the nest after another 9 to 11 days (sometimes slightly longer). The young, which resemble the female, are fed, by both parents, a diet of insects. In late summer and during the fall migration, various fruits and berries are also readily ingested. Nest parasitism by cowbirds has been documented.

There is little specific information available about nesting in Massachusetts. Nestlings have been recorded in the state from June 15 to July 2 (CNR, Meservey). A Dartmouth nest contained two young on June 20, and the male was seen feeding one fledgling on June 30 (CNR). One brood is raised each season. Molting is initiated in early to mid-August and continues into September; family groups in varying stages of molt can be seen well into the first half of September. Some migration commences in late August; but the bulk of the migration occurs by mid-October, when the last Scarlet Tanagers have departed for their wintering grounds in the tropical rain forests of South American from Colombia to Bolivia.

Richard S. Heil

Eastern Towhee
Pipilo erythrophthalmus

Egg dates: May 11 to June 23.

Number of broods: one or two.

The bustling and energetic, yet often secretive, Eastern Towhee is one of the most common and widely distributed breeding birds in the state, but it is most abundant in the brushy areas of the southeastern region, particularly Cape Cod and the Islands. Although it occupies a variety of habitats, ranging from open forest and woodland edges to overgrown fields and power-line cuts, it is most at home in dense thickets and undergrowth. This ground feeder is often first noticed because of its characteristic noisy raking of the ground litter while foraging. The species was described in the early literature as "an exemplary woodland citizen" because of its diet of "harmful insects." Towhees also feed on other invertebrates, seeds, and small fruits.

Towhees are seldom seen in any numbers during the spring migration. Adult males return first in late April and early May and make their presence known by singing on or near the territory held the year before. Females and yearling males have settled in by mid-May. Towhees sing with vigor during spring and early summer and have quite a varied repertoire. The song is described as *drink-your, teeeee* or *sweet-bird, cheeeee*, sometimes shortened to *your-teee*. The first note is on a high pitch, and the song ends in a trill. Females also sing on occasion. The call note, quite important in the poor visibility conditions of their habitat, is a *che-'wink* or *tow-'whee*. Towhees are fairly quiet after the breeding season, and often the call note or the feeding kick noises are the best clues to their presence in late summer or fall.

Courtship activity includes singing sometimes very muted songs, and noticeable short-flight chasing of the female by the male. During such pursuits, the male frequently flashes the white outer tail feathers by fanning his tail. The nest is located on the ground or at low heights in vines and bushes and is made of grasses or woven leaves, sometimes lined with hairs. Males may present nesting material to their mates but do not perform any of the actual nest construction. Habitats for 16 Massachusetts nests were woods (8 nests), old field (3 nests), brush (2 nests), hedgerow (1 nest), old orchard (1 nest), and dune thicket (1 nest) (CNR). Locations for 22 state nests were the following: ground (21 nests), ranging from sites in sand and pine needles, clumps of Poison Ivy, False Solomon's Seal, Lowbush Blueberry, and grass to dense brush. One nest was 2 feet up in a bush (CNR, Meservey).

Clutch sizes for 27 state nests were one egg (1 nest, probably an incomplete clutch), two eggs (2 nests, probably also incomplete), three

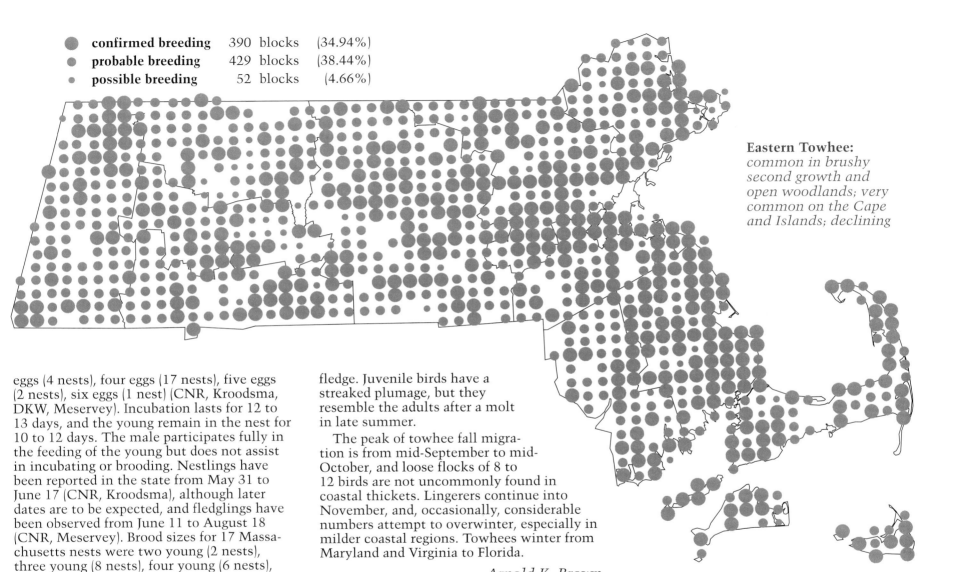

- **confirmed breeding** 390 blocks (34.94%)
- **probable breeding** 429 blocks (38.44%)
- **possible breeding** 52 blocks (4.66%)

Eastern Towhee: *common in brushy second growth and open woodlands; very common on the Cape and Islands; declining*

eggs (4 nests), four eggs (17 nests), five eggs (2 nests), six eggs (1 nest) (CNR, Kroodsma, DKW, Meservey). Incubation lasts for 12 to 13 days, and the young remain in the nest for 10 to 12 days. The male participates fully in the feeding of the young but does not assist in incubating or brooding. Nestlings have been reported in the state from May 31 to June 17 (CNR, Kroodsma), although later dates are to be expected, and fledglings have been observed from June 11 to August 18 (CNR, Meservey). Brood sizes for 17 Massachusetts nests were two young (2 nests), three young (8 nests), four young (6 nests), five young (1 nest) (CNR, Kroodsma, Meservey). One known state hatch date was June 11 (CNR). Outcomes for 15 sample state nests were young fledged (3 nests), unknown (3 nests), failure (9 nests) (CNR). Renesting usually occurs in the same territory, with the first egg of the second clutch deposited 10 to 20 days after the young of the first brood fledge. Juvenile birds have a streaked plumage, but they resemble the adults after a molt in late summer.

The peak of towhee fall migration is from mid-September to mid-October, and loose flocks of 8 to 12 birds are not uncommonly found in coastal thickets. Lingerers continue into November, and, occasionally, considerable numbers attempt to overwinter, especially in milder coastal regions. Towhees winter from Maryland and Virginia to Florida.

Arnold K. Brown

Chipping Sparrow
Spizella passerina

Egg dates: May 5 to June 26.

Number of broods: one or two.

The Chipping Sparrow is a common summer resident throughout most of the state. Because of this species' affinity for open and semiopen habitats, it is assumed that more optimum conditions existed during the period of maximum clearing, i.e., during the midnineteenth century, and that there has been some decline in the relative abundance of the Chipping Sparrow over the last century. However, the gradual creation of more and more suburban areas during the last three decades with their parklike aspect of grassy lawn and ornamental conifers has created favorable habitat. The chippy also sometimes frequents open woods or forest borders. It is less common at the higher altitudes of Worcester and Berkshire counties and is absent from heavily forested areas.

Chipping Sparrows depart from their southern coastal-plain wintering grounds in early spring, and, although occasional birds turn up here in early April, territorial residents as well as transients generally are noted between mid-April and mid-May. Breeding birds are found commonly around suburban residences, golf courses, and rural farmland.

The male chippy is an energetic and persistent, if somewhat limited, songster. As its name implies, the song of this species is an uninterrupted string of *chip*s. These are usually given at a rate fast enough to create a trill. Although the pitch and speed may vary, each phrase is normally steady, unvarying, and monotonous. A high *tsip* alarm note may be given about the nest, and a twittering conversational call is heard frequently.

Males advertise their presence by singing from elevated perches. During courtship, they intersperse bouts of singing with quick flights to the ground near a female. Favored nest sites include shrubs and trees, particularly evergreens, about dwellings, and orchards. Locations of 22 Massachusetts nests were cemetery (3 nests), yard (7 nests), hedge (3 nests), woods (6 nests), field (3 nests) (CNR). Twenty-six state nests were situated as follows: unspecified conifer (6 nests); Arbor Vitae (2 nests); pines, including Red, Scotch, and White (7 nests); Red Cedar (5 nests); Norway Spruce (3 nests); barberry (1 nest); apple (1 nest); unspecified deciduous tree (1 nest) (CNR). Heights ranged from 1 to 19 feet and averaged 7 feet (CNR).

Females build a tightly constructed nest consisting of dried grasses and rootlets. The fact that the entire structure may be lined with horsehair indicates that this species has close ties to the environs of people. The clutch size is normally four eggs. For 19 state nests clutch sizes were three eggs (4 nests), four eggs (14 nests), five eggs (1 nest) (CNR, DKW, ACB). Incubation is by the female alone for a period of 11 to 14 days, and during this time the male feeds her at the nest. Both

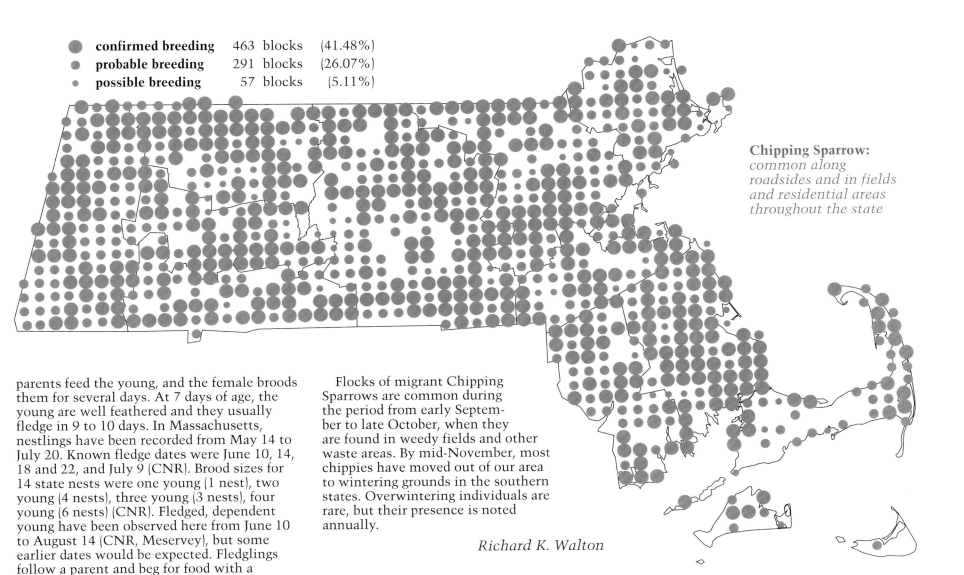

- confirmed breeding 463 blocks (41.48%)
- probable breeding 291 blocks (26.07%)
- possible breeding 57 blocks (5.11%)

Chipping Sparrow: *common along roadsides and in fields and residential areas throughout the state*

parents feed the young, and the female broods them for several days. At 7 days of age, the young are well feathered and they usually fledge in 9 to 10 days. In Massachusetts, nestlings have been recorded from May 14 to July 20. Known fledge dates were June 10, 14, 18 and 22, and July 9 (CNR). Brood sizes for 14 state nests were one young (1 nest), two young (4 nests), three young (3 nests), four young (6 nests) (CNR). Fledged, dependent young have been observed here from June 10 to August 14 (CNR, Meservey), but some earlier dates would be expected. Fledglings follow a parent and beg for food with a constant *zip-ip-zip-ip-zip-ip* series of notes. Two broods per season are usual. During the latter part of the summer, family groups are often observed together. Adults and young complete their molts at the end of the nesting season.

Flocks of migrant Chipping Sparrows are common during the period from early September to late October, when they are found in weedy fields and other waste areas. By mid-November, most chippies have moved out of our area to wintering grounds in the southern states. Overwintering individuals are rare, but their presence is noted annually.

Richard K. Walton

Field Sparrow
Spizella pusilla

Egg dates: May 8 to August 13.

Number of broods: one or two; possibly three; may re-lay if first attempt fails.

The Field Sparrow is a fairly common breeding species throughout much of the state. However, because its nesting ecology is closely associated with old field habitat, the Field Sparrow has a population that is on the decline in Massachusetts. During the latter half of the nineteenth century, midwestern and prairie farmlands offered favorable alternatives to the rocky New England soils, and many Massachusetts agricultural ventures were abandoned. The earlier clearing and subsequent second growth provided prime habitat for the Field Sparrow.

As these open, brushy areas have gradually given way to both reforestation and suburban development, optimal nesting areas for this species have been greatly reduced. A few human-made clearings, such as those associated with power lines through wooded areas, provide suitable nesting sites for the Field Sparrow. However, suburban development has not offered the same advantages to this species as it has to its congener, the Chipping Sparrow.

In the spring, Field Sparrow migrants and returning residents move into our area during the month of April. Territorial males are often heard and seen around abandoned pastures and other unworked rural farmlands. Fields containing some briers and shrub growth make ideal nesting habitat. In wooded regions, open areas that have been cleared or burned also provide nesting territory. The males chase one another and may fight before boundary lines are settled. As is the case with the robin, there may be neutral zones where several pairs will feed without conflict. When the female arrives about two weeks later, the male chases her somewhat aggressively at first, but this soon ceases and he then follows her closely as she moves about the territory.

The Field Sparrow's song begins with several slurred whistles, quickens to a series of rapid notes, and fades off at the end. This has been variously represented as *t-e-w, t-e-w, t-e-w, tew, tew tew, tew* or *seeea, seeea, wee, wee, we, we, we*. The rhythm is much like that of a Ping-Pong ball bouncing on a table. Once heard, the evocative song is unforgettable. There are a variety of other notes. A *zee-zee* or *tsee-tsee* series and a longer *zeeeeeeee* are used as warnings. Males use a *zip-zip-zip* call to attract females, and a loud *chip* is given by begging fledglings or alarmed adults. A common contact note is a soft *tsip*.

The earliest nests of the season normally will be constructed on the ground in a clump of grass. As the season progresses, the female is more likely to build a nest in shrubs or briers. Six nests in Massachusetts were located as follows: 1 on the ground in grass in a dry field, 1 on the ground in grass in a damp meadow, 1 close to the ground in goldenrod stalks, 1 near the ground in a shrub, 1 at 1.7 feet above the ground in Multiflora Rose, and 1 at 1.3 feet in Meadowsweet (CNR, Meservey). The meadow nest was completely covered over with both dry and green grass stalks and had a side entrance that was also roofed over (Meservey). The cup-shaped nest of the Field Sparrow is loosely constructed of grasses and twigs and is often lined with hair or other fine materials. Three to four eggs are usual. Clutch sizes for 9 state nests were three eggs (2 nests), four eggs (6 nests), five eggs (1 nest) (CNR, DKW, Meservey).

Incubation by the female lasts 11 to 14 days, sometimes longer. The nestlings are

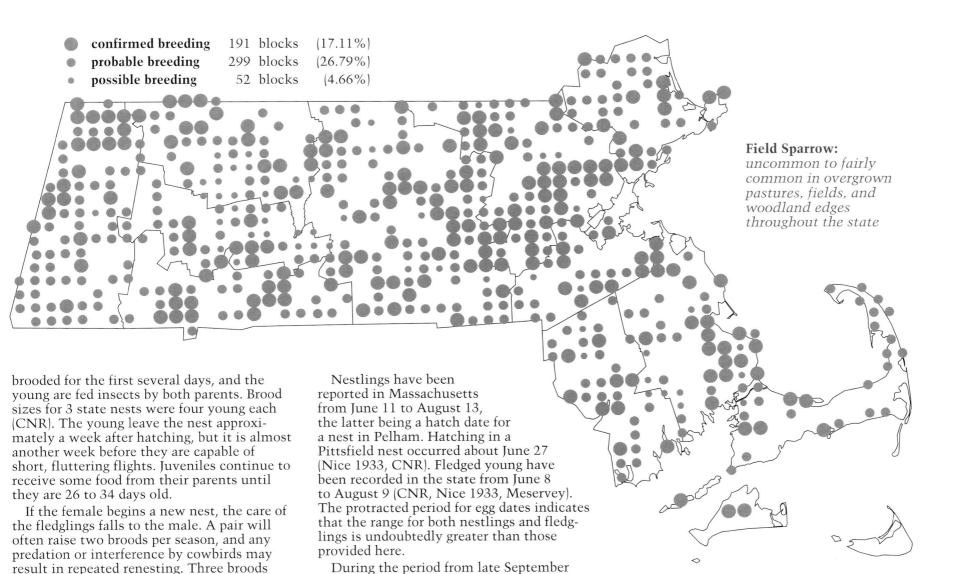

Field Sparrow: *uncommon to fairly common in overgrown pastures, fields, and woodland edges throughout the state*

- confirmed breeding — 191 blocks (17.11%)
- probable breeding — 299 blocks (26.79%)
- possible breeding — 52 blocks (4.66%)

brooded for the first several days, and the young are fed insects by both parents. Brood sizes for 3 state nests were four young each (CNR). The young leave the nest approximately a week after hatching, but it is almost another week before they are capable of short, fluttering flights. Juveniles continue to receive some food from their parents until they are 26 to 34 days old.

If the female begins a new nest, the care of the fledglings falls to the male. A pair will often raise two broods per season, and any predation or interference by cowbirds may result in repeated renesting. Three broods have been documented in some areas but not for Massachusetts. The young birds often remain together on or close by their home territory for the remainder of the season. Adults and young complete their molts by the end of the summer.

Nestlings have been reported in Massachusetts from June 11 to August 13, the latter being a hatch date for a nest in Pelham. Hatching in a Pittsfield nest occurred about June 27 (Nice 1933, CNR). Fledged young have been recorded in the state from June 8 to August 9 (CNR, Nice 1933, Meservey). The protracted period for egg dates indicates that the range for both nestlings and fledglings is undoubtedly greater than those provided here.

During the period from late September through late October, small flocks of migrant Field Sparrows can be found in weedy fields, along roadsides, and in waste places. By mid-November, most of the population has moved out of our area to wintering grounds in the southern United States. Some Field Sparrows winter in Massachusetts each year, and small flocks of overwintering birds are occasionally encountered in weedy habitats, especially in southeastern coastal areas.

Richard K. Walton

Vesper Sparrow
Pooecetes gramineus

Egg dates: April 11 to August 11.
Number of broods: one or two.

The Vesper Sparrow is an uncommon and local summer resident. In Massachusetts, the decline of this species, which Forbush (writing at the beginning of the last century) characterized as a common summer resident, is associated with the demise of agriculture. In the latter half of the nineteenth century, during the so-called "period of abandonment," many Massachusetts farms were let go. Typically, the first farms to be abandoned were the poor-soil farms in upland situations. It was these same upland areas of pasture and short grass fields that provided optimal habitat for the Vesper Sparrow.

Another factor that may have been involved in this sparrow's decline relates to its territorial requirements. Because the Vesper Sparrow needs a larger territory than many of the sparrow species, apparently suitable habitat may be unused due to its relatively small size or fragmentation. During the Atlas period, the species was found in widely scattered locales across the state, with most confirmations and probable nesting near the coast and in the Connecticut River valley.

Vesper Sparrows return to Massachusetts in late March or early April. Resident birds establish territories in pasturelands, on rural airports and military reservations, in agricultural fields, in cleared areas of pine barrens, and on coastal moors. These habitats are normally well drained, with sandy or rocky soils. Early in the season, the males patrol these territories, making frequent use of prominent perches to advertise their presence vocally. The Vesper Sparrow does not confine its singing to evening song. In fact, its vocal efforts may be heard throughout the day, particularly before the first eggs are laid. The song begins with two low, slurred notes, which are followed by two high notes, and concludes with a descending and rapid jumble of trills. A *chirp* call note is common.

During courtship, the male moves about the female on the ground with his wings held open and his tail spread. The nest, consisting of dried grasses and rootlets, is built in a slight depression in the ground. Typically, four or five eggs are laid, which are incubated by both parents (mostly the female). Clutch sizes for 6 Massachusetts nests were three eggs (1 nest), four eggs (4 nests), five eggs (1 nest) (DKW, ACB, BOEM). The incubation period has been reported as 11 to 13 days.

Disturbance of a nest after the young have hatched usually results in an injury-feigning distraction display. The female does most of the brooding of the small nestlings and also shields them from the sun. A nest in eastern Massachusetts contained four young (BOEM). Fledging occurs from 7 to 12 days of age. Vesper Sparrows often raise two broods per season, with the male taking charge of the

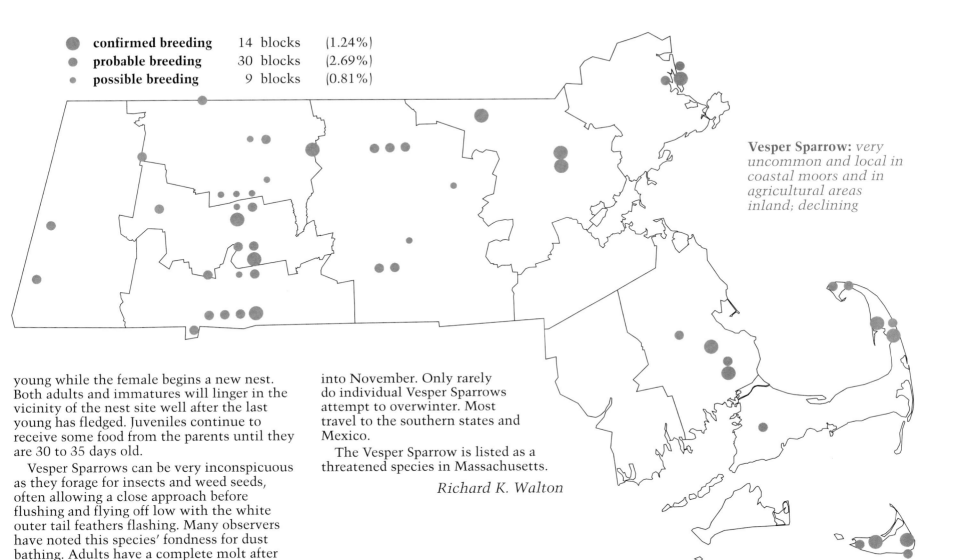

Vesper Sparrow: *very uncommon and local in coastal moors and in agricultural areas inland; declining*

young while the female begins a new nest. Both adults and immatures will linger in the vicinity of the nest site well after the last young has fledged. Juveniles continue to receive some food from the parents until they are 30 to 35 days old.

Vesper Sparrows can be very inconspicuous as they forage for insects and weed seeds, often allowing a close approach before flushing and flying off low with the white outer tail feathers flashing. Many observers have noted this species' fondness for dust bathing. Adults have a complete molt after nesting is completed, and the young assume winter plumage during this time. Twenty-five years ago, relatively large flocks of Vesper Sparrows were noted during autumn migration. Present-day sightings during September and October, away from known nesting sites, are usually limited to individual birds or groups of two to three. Migration continues into November. Only rarely do individual Vesper Sparrows attempt to overwinter. Most travel to the southern states and Mexico.

The Vesper Sparrow is listed as a threatened species in Massachusetts.

Richard K. Walton

Savannah Sparrow
Passerculus sandwichensis

Egg dates: May 21 to June 29.

Number of broods: one or two.

The deforested New England of the nineteenth century provided endless acres of suitable habitat for birds that nested in fields, meadows, and pastures, and the populations of species, such as the Savannah Sparrow, that preferred these habitats must have been legion. With the decline of agriculture in Massachusetts in the late nineteenth and throughout the twentieth century, those species that required open spaces became more and more limited to remnants of this once-expansive ecosystem as agrarian landscapes were gradually transformed into forests or suburban developments.

Today, the Savannah Sparrow is found most commonly as a breeding resident in coastal dunes and heaths, the edges of salt marshes, the farmlands of the Connecticut River valley, and the hillside pastures of the Berkshires. Elsewhere, Savannah Sparrows can be found on the grassy verges of airstrips and on the scattered, but ever fewer farms and pastures throughout the state.

Small numbers of Savannah Sparrows are found each winter in Massachusetts, especially along the coast and Islands, but the bulk of the New England population winters in the southeastern United States. Many leave for their northern breeding grounds long before the close of winter, with Massachusetts birds arriving during April. Savannah Sparrows are common migrants in Massachusetts during both spring and fall, a reflection of their widespread distribution and abundance in the northern United States and eastern Canada.

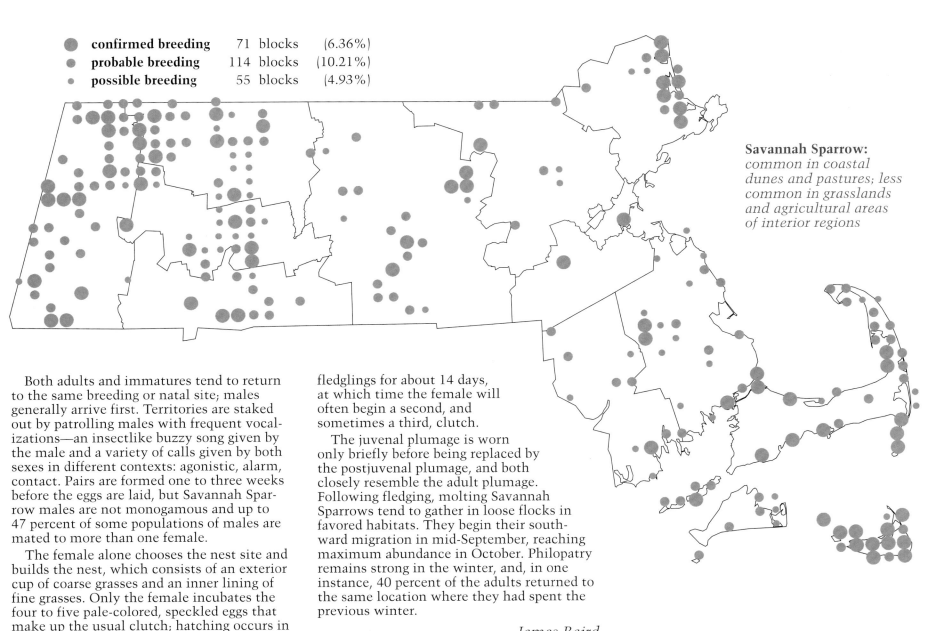

Savannah Sparrow: *common in coastal dunes and pastures; less common in grasslands and agricultural areas of interior regions*

- confirmed breeding — 71 blocks (6.36%)
- probable breeding — 114 blocks (10.21%)
- possible breeding — 55 blocks (4.93%)

Both adults and immatures tend to return to the same breeding or natal site; males generally arrive first. Territories are staked out by patrolling males with frequent vocalizations—an insectlike buzzy song given by the male and a variety of calls given by both sexes in different contexts: agonistic, alarm, contact. Pairs are formed one to three weeks before the eggs are laid, but Savannah Sparrow males are not monogamous and up to 47 percent of some populations of males are mated to more than one female.

The female alone chooses the nest site and builds the nest, which consists of an exterior cup of coarse grasses and an inner lining of fine grasses. Only the female incubates the four to five pale-colored, speckled eggs that make up the usual clutch; hatching occurs in 12 days. Brooding of the nestlings is mostly by the female, but both the male and the female feed the young, which fledge in about 11 days. Both parents feed and care for the fledglings for about 14 days, at which time the female will often begin a second, and sometimes a third, clutch.

The juvenal plumage is worn only briefly before being replaced by the postjuvenal plumage, and both closely resemble the adult plumage. Following fledging, molting Savannah Sparrows tend to gather in loose flocks in favored habitats. They begin their southward migration in mid-September, reaching maximum abundance in October. Philopatry remains strong in the winter, and, in one instance, 40 percent of the adults returned to the same location where they had spent the previous winter.

James Baird

Grasshopper Sparrow
Ammodramus savannarum

Egg dates: June 8 to June 26.

Number of broods: one.

The Grasshopper Sparrow was described in Forbush (1929) as a "queer, sombre-colored, big-headed, short-tailed, unobtrusive little bird." It is an uncommon breeding species in Massachusetts, near the northern edge of its range, and nests at scattered locations in the state. A bird of the coastal plain and river valleys, it is seldom found in Massachusetts above elevations of 1,000 feet. The state's breeding population is believed to be declining due to loss of grassland habitat.

Formerly abundant on Nantucket and Martha's Vineyard and in eastern Massachusetts, the species is now uncommon in these areas. Nashawena Island probably supports the state's largest breeding concentration. Grassland areas adjacent to airfields continue to provide habitat used by Grasshopper Sparrows at several sites in the Connecticut River valley and elsewhere.

The Grasshopper Sparrow returns to Massachusetts in mid- to late May. It is a grassland bird, preferring dry sandy fields vegetated with bunch grasses and characterized by relatively low stem densities and a limited accumulation of ground litter. It is less likely to be found in fertile fields with lush, dense growths of sod-forming grasses and is generally absent from fields with greater than 35 percent cover in shrubs. Although management to maintain grassland habitat is essential to the conservation of Grasshopper Sparrows, mowing schedules must be manipulated to minimize disturbance to nesting birds during June.

Upon arrival on their nesting grounds, male Grasshopper Sparrows establish territories. Territorial songs are delivered from the highest perch available: a tall weed, small bush, or fence post. The bird's name is derived from its grasshopperlike song—two short, low, introductory notes preceding a thin, dry buzz, *tsick, tsick, tsurrrr.* A more musical vocalization heard less frequently is a rolling trill.

Nesting takes place from late May through July. One brood per season is raised in Massachusetts. Nests are built on the ground and are well concealed at the base of clumps of grass or other vegetation. The nest is constructed of stems and blades of grass and is lined with fine grass and rootlets. The clutch size ranges from three to six eggs, with four and five as the most common numbers. A Worcester County nest contained five eggs on June 25 (DKW). Nests with eggs were found on June 8, June 11, and June 26 at Westover Air Force Base, Chicopee. Eggs are white and sparingly spotted or blotched with brown, gray, or purple. Incubation occurs at about

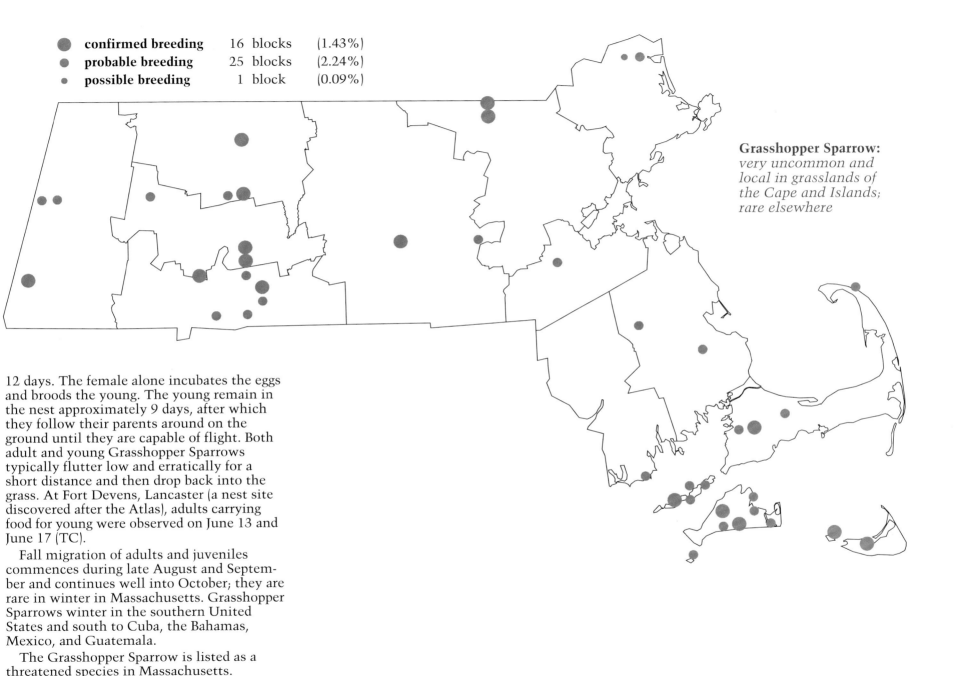

● **confirmed breeding**	16 blocks	(1.43%)
● **probable breeding**	25 blocks	(2.24%)
• **possible breeding**	1 block	(0.09%)

Grasshopper Sparrow: *very uncommon and local in grasslands of the Cape and Islands; rare elsewhere*

12 days. The female alone incubates the eggs and broods the young. The young remain in the nest approximately 9 days, after which they follow their parents around on the ground until they are capable of flight. Both adult and young Grasshopper Sparrows typically flutter low and erratically for a short distance and then drop back into the grass. At Fort Devens, Lancaster (a nest site discovered after the Atlas), adults carrying food for young were observed on June 13 and June 17 (TC).

Fall migration of adults and juveniles commences during late August and September and continues well into October; they are rare in winter in Massachusetts. Grasshopper Sparrows winter in the southern United States and south to Cuba, the Bahamas, Mexico, and Guatemala.

The Grasshopper Sparrow is listed as a threatened species in Massachusetts.

Scott M. Melvin

Henslow's Sparrow
Ammodramus henslowii

Egg dates: May 20 to August 11.

Number of broods: two or perhaps three.

The Henslow's Sparrow in Massachusetts is a rare to very rare, erratic, and local breeder at the eastern edge of its historic range. Though described by Forbush in 1929 as a "rare to common local resident," a steady decline began as early as 1935 and a major recession set in about 1950. The Henslow's Sparrow is now a protected bird in Massachusetts and is listed as endangered by the Massachusetts Natural Heritage and Endangered Species Program. It is also a candidate for listing under the federal Endangered Species Act because its numbers are reduced throughout its range.

There have been few breeding records in the state since 1970. Two pairs definitely bred at the Leicester Airport in 1973 and 1974, and at least four singing males probably bred in West Newbury from May 18 to August 10, 1974. In 1983, four males were singing in Windsor in August and early September. The once-a-decade pattern continued when in Lincoln, in 1994, a pair was discovered building a nest on July 24, with evidence of a hatch on August 11 and of nestling care until August 20. The Lincoln male sang periodically from June 28 through September 6.

Loss of habitat is probably a major cause for the decline of this secretive and easily overlooked sparrow. Many grassland habitats have been developed or have become too overgrown, and others are often mowed too early, pastured too intensively, or converted to crops, turf, Alfalfa, etc. Furthermore, since the species is said to best thrive in fields of more than 75 to 250 acres, fragmentation may impede successful nesting. There is data that suggests that the breeding density, even in large fields, has been reduced in recent years, but the relative importance of predation, competition, weather, fragmentation, and human disturbance is poorly understood.

Breeding does at times occur on smaller plots. For example, the 1994 Lincoln pair nested in an isolated hayfield of 17 acres, of which 5 acres remained uncut at the time of the observed nesting in late July.

Breeding habitats include a variety of grasslands with tall, dense grass and herbaceous vegetation. In the Northeast, the species uses hay fields, pastures, wet meadows, dry salt marsh areas, and old fields. The sparrow can also be found in damp fields heavily overgrown with tall grasses, sedges, and low shrubs. The Lincoln pair was observed in an active hay field (about five years since replanting) of tall Orchard-grass and Timothy with scattered areas of Alfalfa, vetch, and dock; by the time of nesting in late July, the hay had become dense with matted patches.

Historically, the Henslow's Sparrow generally arrived about May 6 and left by late September and October (with three winter records). Spring records since 1965 are May 16, 1976, and April 30, 1983. It is a small "mouse bird," hiding and running in the dense grass and flying infrequently and only for short distances. During some of the breeding season, however, the male will occasionally perch on a weed or a fence post, throw its head back, and sing—obsessively, compulsively, every six to ten seconds for minutes

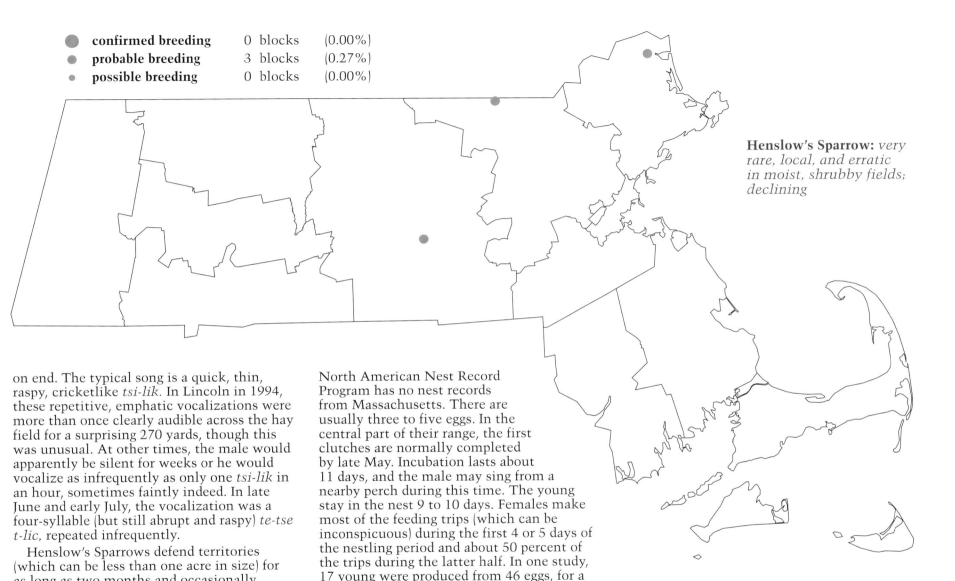

- confirmed breeding   0 blocks   (0.00%)
- probable breeding   3 blocks   (0.27%)
- possible breeding   0 blocks   (0.00%)

Henslow's Sparrow: *very rare, local, and erratic in moist, shrubby fields; declining*

on end. The typical song is a quick, thin, raspy, cricketlike *tsi-lik*. In Lincoln in 1994, these repetitive, emphatic vocalizations were more than once clearly audible across the hay field for a surprising 270 yards, though this was unusual. At other times, the male would apparently be silent for weeks or he would vocalize as infrequently as only one *tsi-lik* in an hour, sometimes faintly indeed. In late June and early July, the vocalization was a four-syllable (but still abrupt and raspy) *te-tse t-lic*, repeated infrequently.

Henslow's Sparrows defend territories (which can be less than one acre in size) for as long as two months and occasionally during the breeding season change their location. Nests are typically constructed of woven grasses on or near the ground and are sometimes domed. For example, a nest may be tucked among the stems at the base of a clump of dense Orchard-grass. The females take 5 to 6 days to construct the nest. The North American Nest Record Program has no nest records from Massachusetts. There are usually three to five eggs. In the central part of their range, the first clutches are normally completed by late May. Incubation lasts about 11 days, and the male may sing from a nearby perch during this time. The young stay in the nest 9 to 10 days. Females make most of the feeding trips (which can be inconspicuous) during the first 4 or 5 days of the nestling period and about 50 percent of the trips during the latter half. In one study, 17 young were produced from 46 eggs, for a prefledging mortality of 63 percent. Henslow's Sparrows usually raise two broods of young per breeding season, and perhaps three. Second nests are initiated in July and August with some extending into September (with young still in the nest). They may remain on their breeding grounds as late as the end of October before they migrate to the southeastern United States for the winter.

Steven Ells

Saltmarsh Sharp-tailed Sparrow
Ammodramus caudacutus

Egg dates: May 24 to July 14 (Allen 1909).

Number of broods: one; may renest if first attempt fails.

The Saltmarsh Sharp-tailed Sparrow is a locally common summer breeding resident of Massachusetts, where it is strictly confined to the immediate coastline. Because it nests exclusively in salt marshes, its distribution along the coast is discontinuous as a result of the presence of sandy beaches, rocky shores, and human intrusions, such as harbor dredging and the presence of marinas. Furthermore, wetland disturbances over three and a half centuries have eliminated many salt marshes where colonies of sharp-tailed sparrows were once documented. In some areas, nesting habitat may be only a few hundred yards in width along the shore, but in other places it may follow estuaries and inlets a short distance inland. Such a limited distribution results in the species being little known to most people and very vulnerable to habitat alteration.

Though there are a few earlier stragglers, most migrant Saltmarsh Sharp-tailed Sparrows arrive during the last week in May. Nesting starts almost immediately in the wide, flat swales of the soft green Salt-hay. Even in pure stands of Salt-hay, the birds are not ubiquitous because they tend to form loose colonies while ignoring apparently identical areas in between. At Plum Island, Barnstable, Monomoy, and Dartmouth, colonies consisted of 3 to 15 pairs, with 1 to 1.5 pairs per acre, and colonies were .5 to 1 mile apart (ACB).

There is little or no courtship behavior prior to nesting. The song consists only of a buzzing hiss preceded or followed by several sharp notes: *gshshhh swik wick*. The nests are neat, cup-shaped structures composed of fine grasses and are about 3 inches in diameter. They may be placed on the ground, tucked under tufts of grass, often with a short runway as an entrance, or may be several inches up in stems. Seven nests in Revere were located as follows: 1 on the ground in a clump of rushes, 1 on the ground in sedges (covered from above), and 5 from 1 to 4 inches up in sedge stems. In Falmouth, nests were 2 to 3 inches up in grasses, and in Barnstable nests were in Salt-hay 2 to 3 inches up and shielded from above by dead grass (ACB). Seven state nests from another sample were located as follows: Salt-grass (2 nests), sandy area in grass (2 nests), Bayberry (3 nests). Four of these nests were on the ground, 1 was at a height of 6 inches, and 2 were at a height of 1 foot (CNR).

Typical clutches consist of three to five eggs. Clutch sizes for 5 state nests were three eggs (2 nests), four eggs (3 nests) (CNR, ACB). Egg dates range from late May to early July, and later ones probably represent second nesting attempts because the species is believed generally to be single brooded in this state. Incubation lasts 11 days and is performed entirely by the female; the males are

promiscuous and roam over the ranges of several females. Nestlings have been reported in Massachusetts from June 9 to August 2 (ACB, CNR). Brood sizes for 7 state nests were one young (1 nest), three young (3 nests), four young (3 nests) (CNR, ACB, BOEM). The young fledge in 10 days and

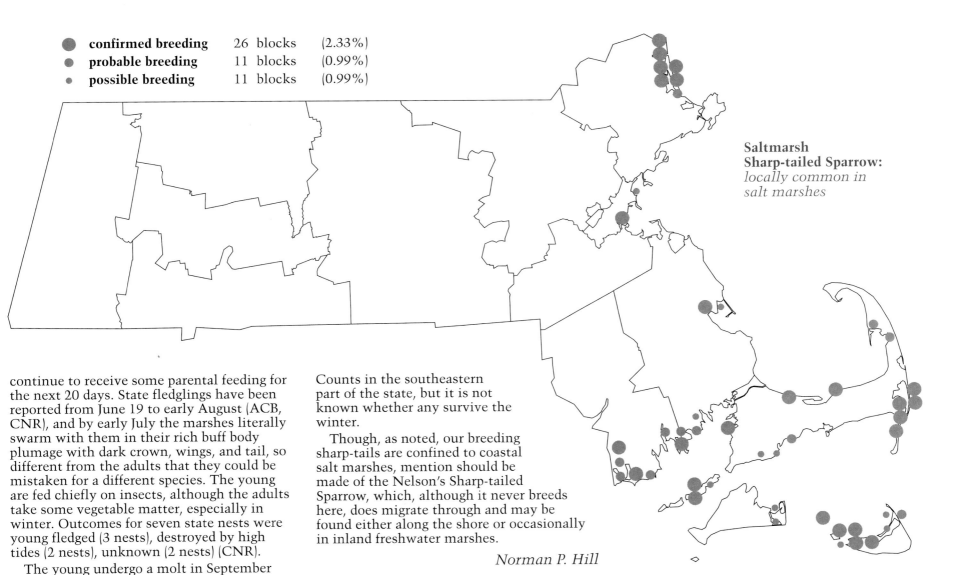

Saltmarsh Sharp-tailed Sparrow: *locally common in salt marshes*

- confirmed breeding — 26 blocks (2.33%)
- probable breeding — 11 blocks (0.99%)
- possible breeding — 11 blocks (0.99%)

continue to receive some parental feeding for the next 20 days. State fledglings have been reported from June 19 to early August (ACB, CNR), and by early July the marshes literally swarm with them in their rich buff body plumage with dark crown, wings, and tail, so different from the adults that they could be mistaken for a different species. The young are fed chiefly on insects, although the adults take some vegetable matter, especially in winter. Outcomes for seven state nests were young fledged (3 nests), destroyed by high tides (2 nests), unknown (2 nests) (CNR).

The young undergo a molt in September and October. Adults molt after nesting and also have a prenuptial molt. By mid-September, the birds gradually unobtrusively withdraw southward, and the leisurely departure lasts into November. After that a few stragglers may be recorded on Christmas Bird Counts in the southeastern part of the state, but it is not known whether any survive the winter.

Though, as noted, our breeding sharp-tails are confined to coastal salt marshes, mention should be made of the Nelson's Sharp-tailed Sparrow, which, although it never breeds here, does migrate through and may be found either along the shore or occasionally in inland freshwater marshes.

Norman P. Hill

Seaside Sparrow
Ammodramus maritimus

Egg dates: May 25 to August 11.

Number of broods: one; may renest if first attempt fails.

The Seaside Sparrow is an uncommon and extremely local summer resident in Massachusetts whose status does not seem to have changed much since the turn of the century. Small but stable populations have long been reported from the salt marshes of the Parker River in Essex County, Barnstable on the Cape, and the Westport area of southeastern Massachusetts, where the first Massachusetts nest was discovered in 1896. Only occasional nestings have been reported elsewhere.

Both the common and scientific names of this sparrow are appropriate because it is rarely seen away from a salt marsh. Although distinctively plumaged, the Seaside Sparrow bears a more than superficial resemblance in other characters to its congener, the Saltmarsh Sharp-tailed Sparrow, which resides in the same coastal habitat. The two species have similar songs, and both are very elusive, sharing behaviors of remaining hidden from view, darting from one clump of grass to the next, and, when forced to fly, making a short, rapid flight that ends by plummeting into the marsh. The Seaside Sparrow has a more southern distribution than the Saltmarsh Sharp-tailed Sparrow, with Massachusetts near its northern limit.

The first residents generally arrive during the first week in May, and by the first week in June pairs are established on territory. The song is a repetitive *kreeuk-tzzeee-eee, kreeeuk-tzzzeeeee*, with an accent on the *zeeee*. The second note sounds like it is being exhaled. Another variation is *churck-uk-lee*. The Seaside Sparrow sometimes vocalizes on the wing but more typically sings while perched on a bush or grass stalk. It is decidedly most vocal at dusk. The call notes are a scolding *chip-tip-tip-tip*, having the quality of a muted Red-winged Blackbird call. The nest is usually built on the ground in the salt marsh just above the high water mark and is therefore subject to destruction by flooding in storms. Some nests are suspended between salt marsh grass stalks, and, rarely, some are placed in low bushes. In a South Dartmouth marsh, 82 percent of 60 nests found were in Saltwater Cordgrass areas with irregular flooding and 54 percent were in medium-height Saltwater Cordgrass along creek banks. One half of the nests found were destroyed by flood tides (Marshall & Reinert 1990). Four to six grayish eggs, coarsely blotched with reddish brown, are laid in a cup nest made of grasses. The mean clutch size in the South Dartmouth study area was four eggs (range three to five) for 30 nests in 1985 and four eggs (range two to five) for 25 nests in 1986 (Marshall & Reinert 1990). The female incubates for a period of about 12 days, but both adults

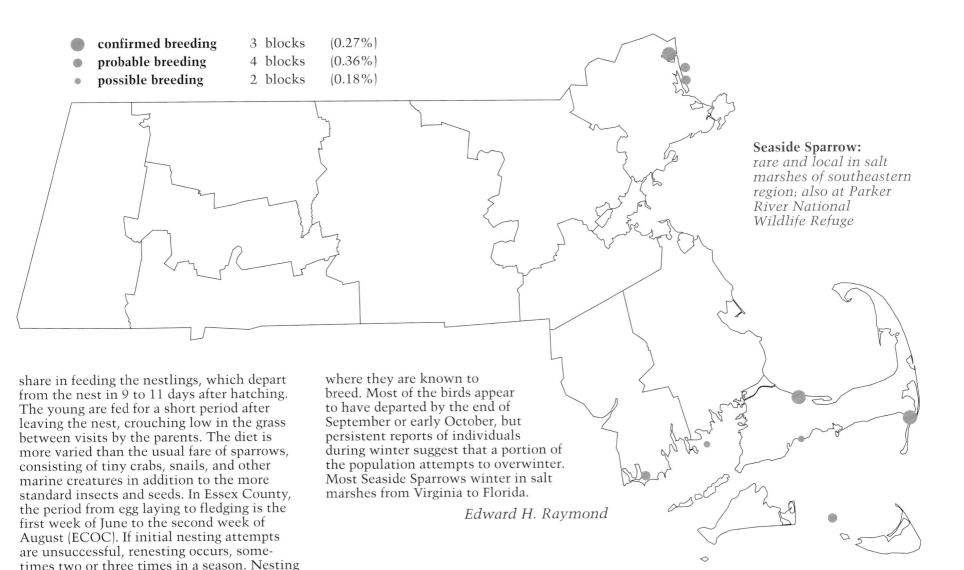

Seaside Sparrow: *rare and local in salt marshes of southeastern region; also at Parker River National Wildlife Refuge*

confirmed breeding	3 blocks	(0.27%)
probable breeding	4 blocks	(0.36%)
possible breeding	2 blocks	(0.18%)

share in feeding the nestlings, which depart from the nest in 9 to 11 days after hatching. The young are fed for a short period after leaving the nest, crouching low in the grass between visits by the parents. The diet is more varied than the usual fare of sparrows, consisting of tiny crabs, snails, and other marine creatures in addition to the more standard insects and seeds. In Essex County, the period from egg laying to fledging is the first week of June to the second week of August (ECOC). If initial nesting attempts are unsuccessful, renesting occurs, sometimes two or three times in a season. Nesting becomes highly synchronized after a flood tide wipes out the active nests, when rapid renesting allows the cycle to be completed before the next tide (Marshall & Reinert 1990). There is evidence that some dispersal occurs in late summer, from mid-August to September, because there are frequent sightings of individuals in areas away from where they are known to breed. Most of the birds appear to have departed by the end of September or early October, but persistent reports of individuals during winter suggest that a portion of the population attempts to overwinter. Most Seaside Sparrows winter in salt marshes from Virginia to Florida.

Edward H. Raymond

Song Sparrow
Melospiza melodia

Egg dates: April 30 to August 17.

Number of broods: two; possibly sometimes three.

Though the Song Sparrow is one of the most common birds in Massachusetts, it is not very familiar to the general public. It is a widespread breeding species, occupying a variety of habitats from coastal dunes to river floodplains to forest edge. The common element wherever it is found is brushy, rank growth interspersed with grassy openings. Wet meadows and sites bordering water are particularly favorable. Although Song Sparrows are nearly ubiquitous, they are not found in forests or open fields. They adapt well to the disturbed habitats created by humans, and a glance at the accompanying map will show that they are as evenly distributed throughout urbanized eastern Massachusetts as they are on Cape Cod and the Islands and the uplands of the western sections of the state.

Song Sparrows are present in the state throughout the year, but the wintering population is sparse. The first spring migrants appear in early March, and, by the end of April, most residents are on well-established territories. The vocal effort of the Song Sparrow is such a simple roundelay that one wonders why the bird was so named. Perhaps it is the quantity, not the quality, of the song, because from March through the long nesting season the males sing frequently. Vocalization is most intense shortly after the males begin to establish territories but continues well into August. Simple though it may be, the basic song has apparently endless variations that consist of short notes, often with a trill at the end or inserted in the middle of the song. Over 50 variations have been recorded or described in the literature, and each male may use several different versions of song. The common call notes of both sexes are a sharp *chimp* and a sibilant *tsst*.

Territories are generally small (usually less than an acre), and males defend them by singing from perches, by performing a stiff-winged flight, and by chasing and fighting, with much excited chipping. During courtship, the females may be chased. The nest, a cup of leaves, grass stems, and strips of bark lined with fine grasses or hairs, is constructed by the female on the ground, generally within a clump of weeds or dead grasses, or deep within small bushes or conifers a few feet high.

Nesting habitats for 19 Massachusetts sample nests were suburban (6 nests); urban (2 nests); old field (3 nests); wood edge (2 nests); cemetery, pond border, marsh, power line, beach dune, and mowed dike (1 nest each) (CNR). Bent mentions a nest found in a salt marsh in Gloucester (ACB). Other Massachusetts nests have been located on bare ground, on the ground in grass, in weed clumps and Timothy, and in bushes, evergreen shrubs, Red Cedar, Pasture-juniper, Arbor Vitae, viburnum, rose, cultivated hedges, raspberry, blackberry, and English Ivy (CNR, Meservey). A Brookline nest discovered on the ground under a brush pile was constructed of dried grasses and was lined with loose hair. Heights for 23 state nests were ground (9 nests), 1 foot (2 nests), 2 feet (3 nests), 3 feet (3 nests), 4 feet (3 nests), 5 feet (2 nests), 7 feet (1 nest) (CNR, Meservey).

The female incubates the three to six eggs for 12 to 13 days, and the August egg dates indicate that there may be three broods in a season. A clutch of five eggs in a North Adams nest was laid from May 4 to May 8 (CNR). Clutch sizes for 27 state nests were three eggs (2 nests), four eggs (7 nests), five eggs (17 nests), six eggs (1 nest) (DKW, CNR,

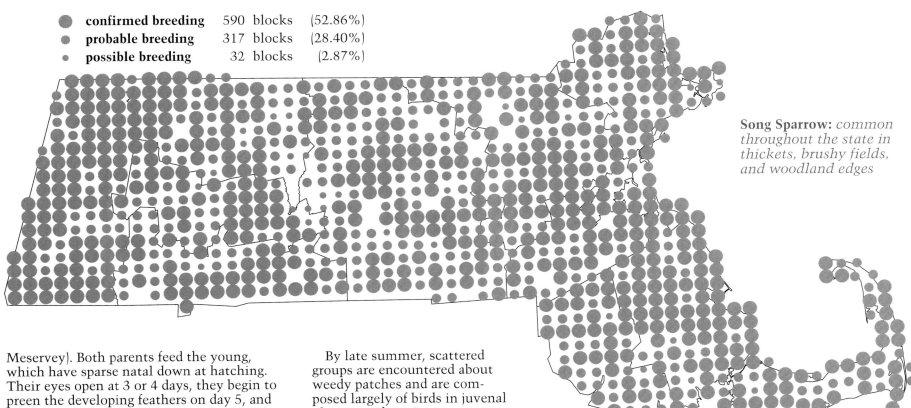

Song Sparrow: *common throughout the state in thickets, brushy fields, and woodland edges*

- confirmed breeding — 590 blocks (52.86%)
- probable breeding — 317 blocks (28.40%)
- possible breeding — 32 blocks (2.87%)

Meservey). Both parents feed the young, which have sparse natal down at hatching. Their eyes open at 3 or 4 days, they begin to preen the developing feathers on day 5, and by day 7 they perch alertly on the nest rim. By day 8 or 9, they are quite active and exercise their wings. They fledge in about 10 days, generally before they can fly, and are fed until they are 28 to 30 days of age. If a subsequent nesting begins, the care of the young falls to the male while the female lays another clutch in the same or a new nest. Brood sizes for 14 Massachusetts nests were two young (1 nest), three young (4 nests), four young (6 nests), five young (3 nests). Confirmed fledging dates were June 27 and July 4, 25, and 26 (CNR, Meservey). The egg dates indicate that the ranges for both nestlings and fledglings are greater. The outcomes for 20 sample state nests were unknown (5 nests), failed (7 nests), young fledged (8 nests) (CNR).

By late summer, scattered groups are encountered about weedy patches and are composed largely of birds in juvenal plumage. These immatures resemble the adults but are more finely streaked below and lack the characteristic central breast spot. Juvenile Swamp and White-throated sparrows, which often frequent the same habitats, share a plumage nearly identical with that of young Song Sparrows. The adults have a complete molt during August and September.

In September and October, Song Sparrows congregate at field edges and other locales where weed seeds abound. During Indian summer and also on mild winter days, muted and abbreviated songs are often heard. By the end of October, many individuals have migrated through and out of our area for wintering grounds in the southern United States. Some birds apparently remain on or near their territories all winter. Wintering birds may appear at feeders anywhere in the state but are most common in interior river valleys and milder coastal regions.

Richard A. Forster

Swamp Sparrow
Melospiza georgiana

Egg dates: May 9 to July 18.

Number of broods: one or two.

The Swamp Sparrow is a locally common resident of freshwater wetlands but is rare on the Cape and Islands. Historically, the agricultural practice of "reclaiming" wetlands in order to increase arable acreage adversely affected the Swamp Sparrow's habitat. During this century, a significant amount of suitable nesting habitat has been lost to commercial real estate ventures and suburban development. Present-day Swamp Sparrow populations have benefited from recent attempts to preserve and protect the state's wetlands.

In the springtime, usually by mid-April, Swamp Sparrows begin moving into our area from their more southerly wintering grounds. Transients continue to pass through during May. Resident males establish territories in various types of wetlands. Cattail marshes, swamps, river meadows, and pond edges are all favored residences of this species. In extensive habitat, nesting Swamp Sparrows can be numerous enough to give the impression of a loosely formed colony.

The Swamp Sparrow's song consists of a musical trill given at one pitch. The rate is usually slow enough to allow the listener to distinguish between the individual notes. Typically, the male sings this song from a conspicuous perch in his territory. The song period often extends well into August. A sharp, metallic *chink* is a frequently used call note.

Swamp Sparrows build a well-concealed, bulky nest with a fine grass lining. The nest site may be in a tussock, among cattail flags, or in a shrub. In the Quaboag River marshes in Worcester County, nests typically were placed in bushes (ACB). Heights vary from a few inches from the water's surface to 6 feet above the water level. When the nest is close to the water, late spring rains and flooding can disrupt the nesting cycle and result in renesting efforts. The clutch size is normally four to five (range three to six) eggs, which are incubated by the female for 12 to 13 days. Some pairs rear two broods per season. Little is known about Massachusetts nestings. In Worcester County, adults have been observed carrying food from June 2 to August 16 (Meservey).

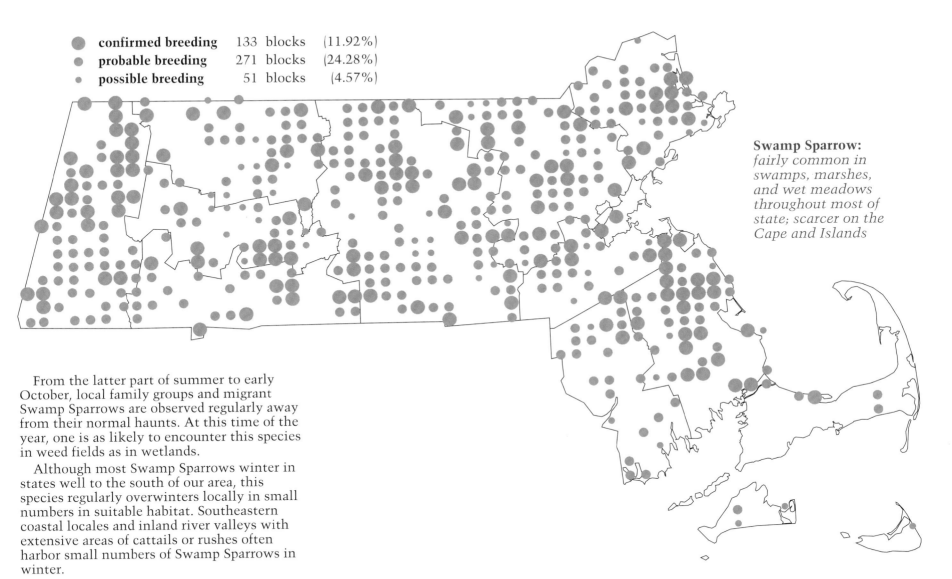

- confirmed breeding 133 blocks (11.92%)
- probable breeding 271 blocks (24.28%)
- possible breeding 51 blocks (4.57%)

Swamp Sparrow: *fairly common in swamps, marshes, and wet meadows throughout most of state; scarcer on the Cape and Islands*

From the latter part of summer to early October, local family groups and migrant Swamp Sparrows are observed regularly away from their normal haunts. At this time of the year, one is as likely to encounter this species in weed fields as in wetlands.

Although most Swamp Sparrows winter in states well to the south of our area, this species regularly overwinters locally in small numbers in suitable habitat. Southeastern coastal locales and inland river valleys with extensive areas of cattails or rushes often harbor small numbers of Swamp Sparrows in winter.

Richard K. Walton

White-throated Sparrow
Zonotrichia albicollis

Egg dates: May 25 to June 7.

Number of broods: one; may renest if first attempt fails.

The White-throated Sparrow is a fairly common nesting species in the higher elevations of the western portion of the state from Berkshire County east to Worcester County. Summer residents are markedly uncommon and local in the eastern half of the state and are absent from Cape Cod and the Islands.

The peak of the White-throated Sparrow migration occurs between mid-April and mid-May, when this species is one of the most ubiquitous and abundant of spring migrants. Residents quickly sort themselves out from the northbound migrants as they gravitate to their favored habitats. In the central and western parts of the state, White-throated Sparrows occupy a wide range of semiopen scrub and thicket habitats. They are most often found on the edges of lowland bogs and in slash areas on coniferous slopes, the scrubby growth of exposed hillsides and summits, and power-line clearings. The relatively rare residents of eastern Massachusetts occur in the cool Red Maple and Atlantic White Cedar swamps.

The *Old Sam Peabody, Peabody, Peabody* vocalization of the White-throated Sparrow is perhaps one of the better known bird songs. According to a Federation of Ontario Naturalists recording by Borror and Gunn, however, this well-known phrase is not the commonest of the 15 different song patterns attributed to this species. The most commonly given song can be represented as *Poor Peabody, Sam Peabody, Peabody, Peabody.* Most of the variations are combinations of sustained single notes and triplets. Call notes include a high *tseet* as well as a louder *chip* or *chink.*

Males claim and defend territories, and, in areas where the species is common, this may involve many chases and fights early in the season. An intriguing situation with the White-throated Sparrow is the fact that both sexes have two color morphs: one with white crown stripes and the other with tan stripes. Studies have revealed that there are behavioral differences for each morph. The more aggressive white-striped males tend to pair with the tan-striped females, leaving the white-striped females to mate with the tan-striped males.

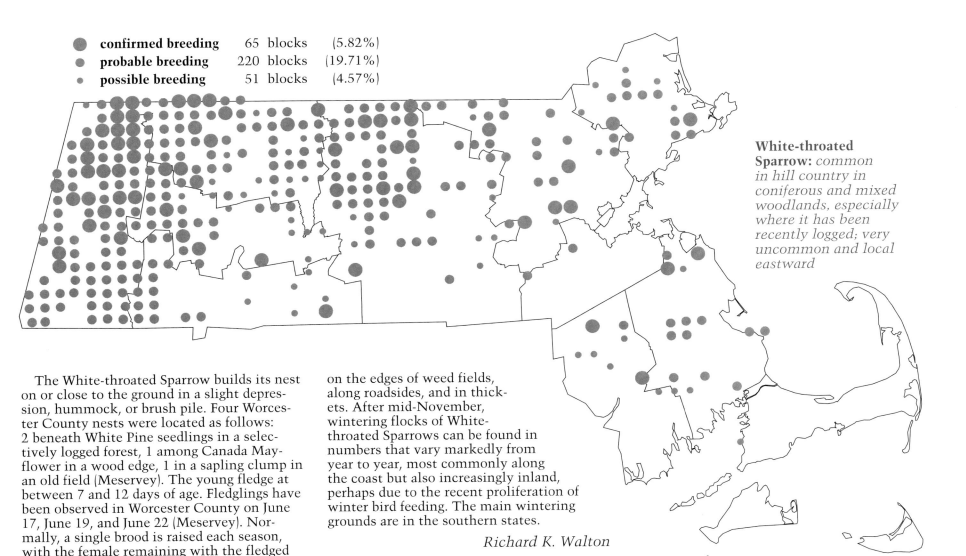

White-throated Sparrow: *common in hill country in coniferous and mixed woodlands, especially where it has been recently logged; very uncommon and local eastward*

- confirmed breeding — 65 blocks (5.82%)
- probable breeding — 220 blocks (19.71%)
- possible breeding — 51 blocks (4.57%)

The White-throated Sparrow builds its nest on or close to the ground in a slight depression, hummock, or brush pile. Four Worcester County nests were located as follows: 2 beneath White Pine seedlings in a selectively logged forest, 1 among Canada Mayflower in a wood edge, 1 in a sapling clump in an old field (Meservey). The young fledge at between 7 and 12 days of age. Fledglings have been observed in Worcester County on June 17, June 19, and June 22 (Meservey). Normally, a single brood is raised each season, with the female remaining with the fledged young even after they have left the nesting territory. Many pairs will attempt a second nesting if the first attempt fails.

Adults have a complete molt after the breeding season, and the young then attain winter plumage. Groups of migrant White-throated Sparrows are noted throughout the state from mid-September to early November. These flocks are encountered frequently on the edges of weed fields, along roadsides, and in thickets. After mid-November, wintering flocks of White-throated Sparrows can be found in numbers that vary markedly from year to year, most commonly along the coast but also increasingly inland, perhaps due to the recent proliferation of winter bird feeding. The main wintering grounds are in the southern states.

Richard K. Walton

Dark-eyed Junco
Junco hyemalis

Egg dates: late May to July 26.

Number of broods: one or two.

Even though it commonly nests in openings in coniferous and mixed woodlands and in brushy thickets of overgrown fields in the higher elevations of the western parts of the state, the junco is more familiar to Massachusetts residents as a winter visitor than as a breeding bird. It also breeds casually eastward through Worcester County into Middlesex County. In Forbush's time, the species was commoner in southern Middlesex County than it is today, and it occurred sparingly in all of the other mainland eastern counties, but these populations have now all but disappeared.

In late March, migrants that have wintered farther south begin to arrive, augmenting the local winter population, which then forsakes feeders and moves to brushy fields and forest edges. The twittering conversational notes heard in winter flocks are then accompanied by song. Northward movement peaks in mid-April, and by early May summer residents have arrived on their breeding grounds. Birds that wintered locally are often among the last to depart.

The junco's trilling song sounds like that of a melodious Chipping Sparrow but is highly variable among individuals. Males frequently sing from elevated perches, and the song period extends through late July. A twittering, warbling song is also occasionally heard. There are several call notes used in different contexts. The most familiar are a *tack* or *smack* scolding note, a *tit-tit-tit* location call and a *tchet, tchet* or *bzzz* of alarm.

Male juncos display by hopping on the ground near a female with the wings drooped and the tail fanned to show the white outer feathers. Nests are near or on the ground in some type of forest opening, often hidden under a fallen branch; sometimes they are on steep slopes. Two nests from Wendell were located as follows: 1 on bare ground mostly under a rock in a newly dug area, 1 on the ground in vegetation half under a fallen log (CNR). Occasionally, nests are placed in unusual situations. For example, Forbush reported a Townsend nest that was built in the corner of a narrow shelf inside a woodshed.

The female takes up to 12 days to make the nest, which is lined with very fine grass and hairlike roots. The three to five eggs are incubated by the female for 11 to 13 days. Both parents feed the young and the female broods them for up to 7 days, at which time their feathers begin to show. They leave the nest in about 12 days. The juveniles are partly dependent on the adults for food for three weeks. Often there is a second nesting in late June or July.

For the first few weeks out of the nest, the young bear little resemblance to the adults because in juvenal plumage they are dark brown and heavily streaked both above and below. Young male juncos molt into a plumage similar to that of the adult male while the immature female plumage is like that of the adult female but more strongly tinged with reddish brown. Adults have a complete molt in the fall.

There is little specific nesting data for the Commonwealth. In eastern Massachusetts, a male and nest were discovered on June 3, and the adults were observed carrying food during June (BOEM). One of the Wendell nests held three eggs on June 12 and four eggs on June 14, but by June 18 a predator destroyed the eggs. The second Wendell nest was being constructed on July 7. Four eggs

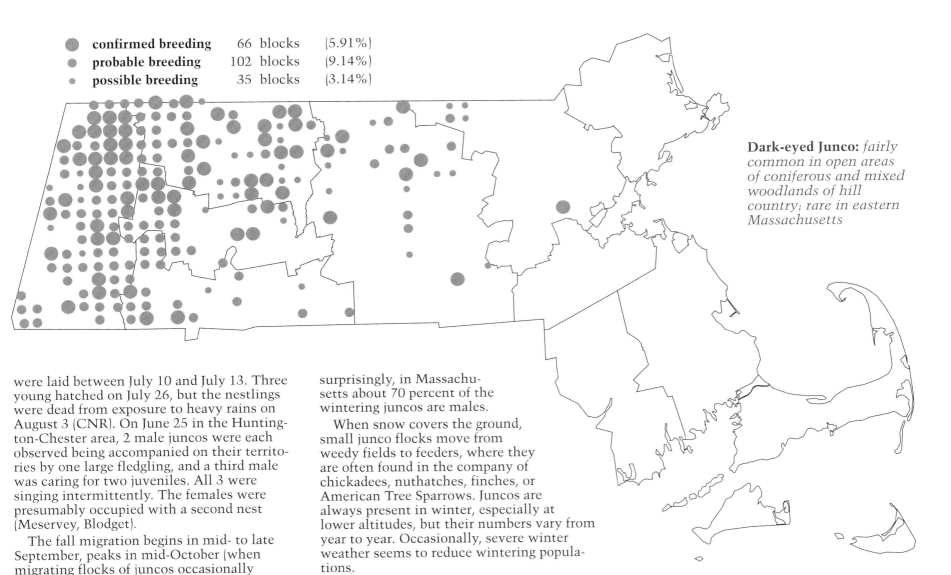

- confirmed breeding — 66 blocks (5.91%)
- probable breeding — 102 blocks (9.14%)
- possible breeding — 35 blocks (3.14%)

Dark-eyed Junco: *fairly common in open areas of coniferous and mixed woodlands of hill country; rare in eastern Massachusetts*

were laid between July 10 and July 13. Three young hatched on July 26, but the nestlings were dead from exposure to heavy rains on August 3 (CNR). On June 25 in the Huntington-Chester area, 2 male juncos were each observed being accompanied on their territories by one large fledgling, and a third male was caring for two juveniles. All 3 were singing intermittently. The females were presumably occupied with a second nest (Meservey, Blodget).

The fall migration begins in mid- to late September, peaks in mid-October (when migrating flocks of juncos occasionally number in the hundreds), and ends about Thanksgiving. Dark-eyed Juncos have an extensive winter range that includes most of the eastern United States, but the center of winter abundance is in the Southeast. It is interesting to note that males tend to winter in the northern part of the range while females are more common in the south. Not surprisingly, in Massachusetts about 70 percent of the wintering juncos are males.

When snow covers the ground, small junco flocks move from weedy fields to feeders, where they are often found in the company of chickadees, nuthatches, finches, or American Tree Sparrows. Juncos are always present in winter, especially at lower altitudes, but their numbers vary from year to year. Occasionally, severe winter weather seems to reduce wintering populations.

Robert P. Fox

Northern Cardinal
Cardinalis cardinalis

Egg dates: April 14 to August 8.

Number of broods: two; possibly sometimes three.

During recent decades, the Northern Cardinal has increased its range dramatically in New England, and today it is a common permanent resident in Massachusetts. In 1929 Forbush described this species as a "rare visitor" and mentioned one nesting record in Brookline in 1898. By 1955 Griscom reported the bird to be "pushing northward" and to be recorded annually at feeding stations. National Audubon Christmas Bird Count data shows clearly that the Northern Cardinal population has grown increasingly rapidly and that the spread into New England correlates with that of other species such as the Tufted Titmouse and Northern Mockingbird. This increase has been most pronounced since 1960, and the numbers continue to rise.

During the Atlas period, cardinals were confirmed to be nesting statewide. The species was most abundant in eastern Massachusetts and was scarce or lacking in the wooded uplands, particularly in north-central sections. The cardinal is found in a wide variety of habitats where dense cover is interspersed with relatively open areas. It is particularly common along woodland edges, brushy old fields, and wooded wetlands, and has also adapted to parks and suburban areas. Cardinals are attracted readily to feeding stations that offer sunflower seed. Indeed, winter bird feeding may be in part responsible for the northward spread of the bird.

Cardinals are basically nonmigratory, but individuals, especially the young, may disperse widely. Pair bonds are thought to be permanent, and the birds generally remain in the same home range throughout the year. During the breeding season, they are highly territorial, and both the male and female sing. The song is a loud, melodious, whistled *what-cheer, cheer, cheer,* or *sweet, sweet, sweet.* The call note is a sharp *chink*. The song can be heard at any time of year, but it is much more frequent from March through the nesting season. During courtship, the male may feed the female.

Habitats for 27 Massachusetts nests were suburban (19 nests), wooded (4 nests), swamp (2 nests), field (1 nest), rural yard (1 nest) (CNR). The nest is constructed of twigs, grass, and plant fibers. Females do most of the nest building and incubation. Nests are built in thick shrubs, dense trees, or tangles of vines. Twenty-five state nests were located as follows: unspecified bushes or shrubs (4 nests); grape and other vines (5 nests); spruce, including Norway Spruce (2 nests); Arbor Vitae (2 nests); and 1 nest each in hydrangea, Bog Rosemary, wisteria, privet, Red Cedar, rhododendron, honeysuckle,

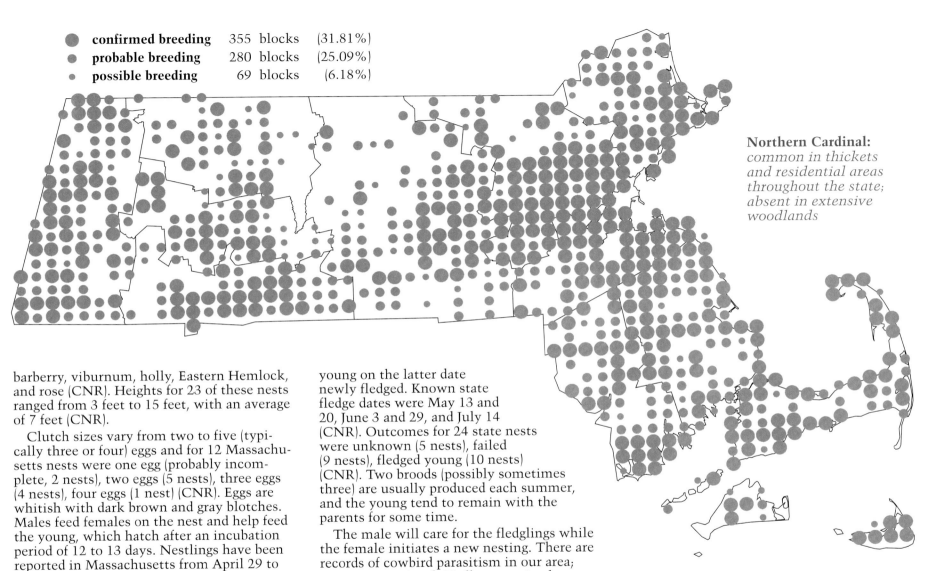

Northern Cardinal: *common in thickets and residential areas throughout the state; absent in extensive woodlands*

- confirmed breeding — 355 blocks (31.81%)
- probable breeding — 280 blocks (25.09%)
- possible breeding — 69 blocks (6.18%)

barberry, viburnum, holly, Eastern Hemlock, and rose (CNR). Heights for 23 of these nests ranged from 3 feet to 15 feet, with an average of 7 feet (CNR).

Clutch sizes vary from two to five (typically three or four) eggs and for 12 Massachusetts nests were one egg (probably incomplete, 2 nests), two eggs (5 nests), three eggs (4 nests), four eggs (1 nest) (CNR). Eggs are whitish with dark brown and gray blotches. Males feed females on the nest and help feed the young, which hatch after an incubation period of 12 to 13 days. Nestlings have been reported in Massachusetts from April 29 to July 15. Brood sizes for 14 state nests were one young (2 nests), two young (3 nests), three young (8 nests), four young (1 nest) (CNR). Known state hatch dates were May 11, June 20 and 25, and July 4 (CNR).

Young generally fledge at 9 to 11 days of age. Fledglings have been recorded in the state from May 4 to September 12, with the young on the latter date newly fledged. Known state fledge dates were May 13 and 20, June 3 and 29, and July 14 (CNR). Outcomes for 24 state nests were unknown (5 nests), failed (9 nests), fledged young (10 nests) (CNR). Two broods (possibly sometimes three) are usually produced each summer, and the young tend to remain with the parents for some time.

The male will care for the fledglings while the female initiates a new nesting. There are records of cowbird parasitism in our area; e.g., a nest in South Hadley contained a young cowbird on May 7 (CNR). Juvenile cardinals have dark bills while the adults have red bills. Young of both sexes initially resemble the adult female, but young males assume a brighter plumage before winter. Adults molt at the end of the breeding season. The Northern Cardinal is among the first bird species to be active at daybreak and among the last to roost at dusk, a fact to which feeding-station operators can attest.

John C. Kricher

Rose-breasted Grosbeak
Pheucticus ludovicianus

Egg dates: May 22 to June 11.

Number of broods: one.

Rose-breasted Grosbeaks are widespread breeders throughout Massachusetts in deciduous and mixed woodlands, in edge habitats, sometimes near water, and more recently in shade trees of parks and suburbs. The exception is the southeastern coastal plain, where they rarely nest. Forbush noted that: "Its numbers, however, fluctuate; it may appear commonly for a few years in a section and then suddenly become rare."

Soon after the first of May, we look forward to seeing the first really colorful spring migrants—Northern Orioles, Scarlet Tanagers, Indigo Buntings, and the handsome male Rose-breasted Grosbeaks, unmistakable in their black and white plumage with a rose red triangle on the upper breast. Occasionally, a few will arrive during the latter days of April, but the majority of migrants, both male and female, pass through our area to other more northerly breeding grounds during the last two weeks of May. By mid-May, the resident males have established territories and attracted mates.

The beautiful song of the male is heard commonly in late May and throughout June. It bears some resemblance to the robin's song but is much more musical and the varied phrases are longer. The call note, a sharp but penetrating metallic *eeek*, is easily recognized, and there is an alarm note that is similar but louder. Courtship can be quite spectacular, and anyone who has ever seen several males displaying before a female never forgets the sight of the males' spread, lifted, quivering wings showing the rose-colored underwing coverts to best advantage.

Both sexes share in nest building. The nest is generally 6 to 15 feet high (sometimes considerably more) and is a fairly fragile affair of small sticks, twigs, and other course material, lined with fine grasses and rootlets. A Springfield nest was being constructed on May 19 (CNR). Four state nests were located as follows: 1 at 4 feet in Elderberry; 2 at 10 feet in an unidentified shrub and hawthorn respectively; and 2 at 18 feet in an ash (EHF, CNR). Three to five pale, greenish blue and brown-speckled eggs are mostly incubated by the female. The male also regularly incubates, often singing while on the nest. It has been reported that the female also occasionally sings. When she is incubating, the

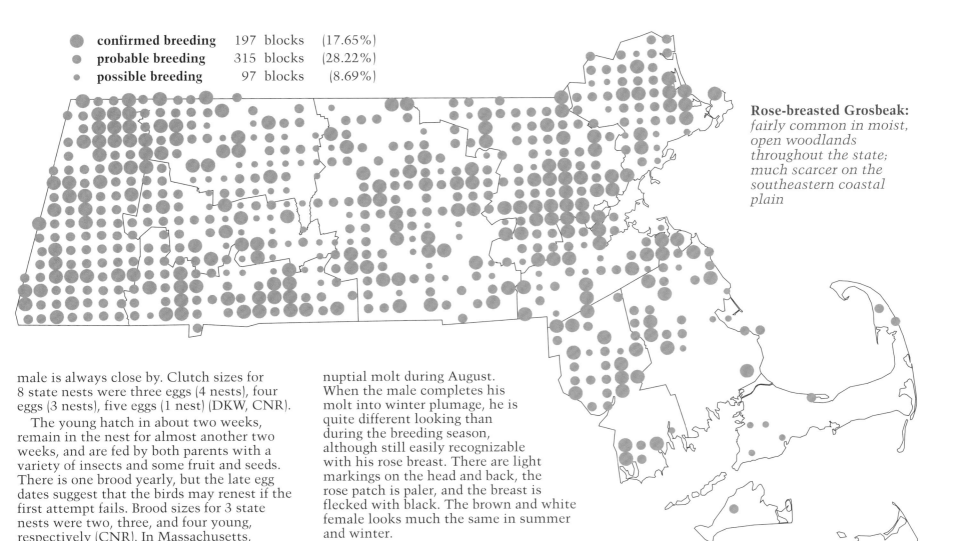

Rose-breasted Grosbeak: *fairly common in moist, open woodlands throughout the state; much scarcer on the southeastern coastal plain*

- confirmed breeding   197 blocks   (17.65%)
- probable breeding   315 blocks   (28.22%)
- possible breeding   97 blocks   (8.69%)

male is always close by. Clutch sizes for 8 state nests were three eggs (4 nests), four eggs (3 nests), five eggs (1 nest) (DKW, CNR).

The young hatch in about two weeks, remain in the nest for almost another two weeks, and are fed by both parents with a variety of insects and some fruit and seeds. There is one brood yearly, but the late egg dates suggest that the birds may renest if the first attempt fails. Brood sizes for 3 state nests were two, three, and four young, respectively (CNR). In Massachusetts, nestlings have been recorded from June 9 to June 18 and fledglings from June 15 (a known fledge date) to July 19 (CNR, Meservey). Fledglings will come to backyard feeders with their parents for sunflower seeds, which are fed to them, often by the male alone, until they can manage by themselves. At this time, the young have a soft, slurred food call.

By the end of July, the family tends to disperse, with the adults undergoing a post-nuptial molt during August. When the male completes his molt into winter plumage, he is quite different looking than during the breeding season, although still easily recognizable with his rose breast. There are light markings on the head and back, the rose patch is paler, and the breast is flecked with black. The brown and white female looks much the same in summer and winter.

The fall migration, in late August and September, is never as noticeable as that in spring because the birds move inconspicuously out of and through our area on their way to wintering grounds in Central America and northern South America. Occasionally, the metallic call note is heard from migrants and is the only indication of their presence. Most Rose-breasted Grosbeaks are gone by the end of September. There are regular reports of one or two individuals seen into December and January, but few survive a severe winter, even at feeding stations.

Charlotte E. Smith

Indigo Bunting
Passerina cyanea

Egg dates: June 1 to June 28.

Number of broods: one or two.

The Indigo Bunting was probably a more common breeding bird in Massachusetts years ago, when its favored habitat of brushy farmland was more prevalent. Now that abandoned farmland has returned to forest, the Indigo Bunting is a rather uncommon migrant and summer resident throughout the state. On Cape Cod, Nantucket, and Martha's Vineyard, it has always been rare. Its preferred habitats are woodland edges adjacent to fields and highways, overgrown pastures, power lines, and hedgerows.

Indigo Buntings do not begin arriving in Massachusetts until mid-May, but nearly every year varying numbers of premature arrivals, mostly males, turn up a month or more ahead of time, the result of southerly storms and migratory miscalculation. The males arrive first and set up territories by singing from high conspicuous perches. Singing is most vigorous in early morning and evening. First-year males may challenge older males for their territories by flying toward them. During these flights, they sing, fluff out their body feathers, and move in a slow, stalling manner with a rapid wing beat. New males may gradually adapt their song to sound like older males. For this reason, the songs of males in a given area all have similar phrases. Territories range from 2 to 6 acres.

Once the females arrive several days later, the males may chase them about persistently, singing or giving a high *zeep* call. There is still much that is not known about the courtship activity of these birds. Generally, the females are inconspicuous and stay in the cover of vegetation; in fact, in most cases it is hard to tell when they have arrived on a territory.

Nests are placed low in brambles, shrubs, or saplings, usually no more than 10 feet high. The female selects the site and does all of the building. In Massachusetts, nests have been located in raspberry, blackberry, barberry, hazelnut, ash, and rose (EHF, ACB, CNR). Heights ranged from 2 to 3 feet (EHF, CNR). The nest is a thick-walled cup of leaves, grasses, and plant stems placed in a crotch of branches. An excellent clue to the presence of a nest is that both birds utter a sharp *chip* call whenever they are approached within 40 feet.

During egg laying, the male stays closer to the female than at any other time in the breeding cycle, but once she is incubating he stays away from the nest to sing and feed. The eggs are white and unmarked, and average three to four per brood. Clutch sizes for 3 state nests were three eggs (2 nests) and four eggs (1 nest) (DKW, CNR). The female does all of the incubation for 10 to 12 days. Once the young have hatched, the female provides most of the care, and she is very secretive as she approaches and leaves the nest. The male's

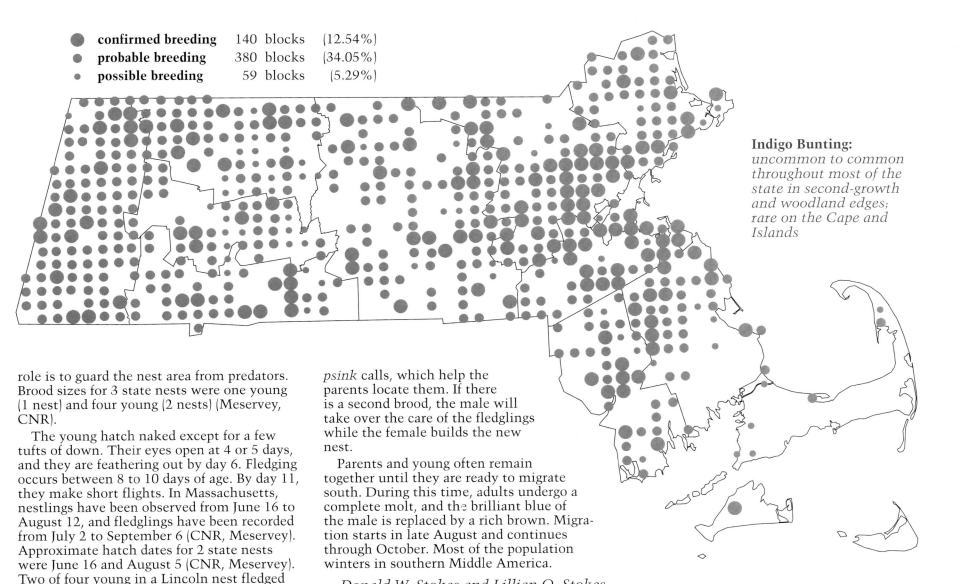

Indigo Bunting: *uncommon to common throughout most of the state in second-growth and woodland edges; rare on the Cape and Islands*

- confirmed breeding 140 blocks (12.54%)
- probable breeding 380 blocks (34.05%)
- possible breeding 59 blocks (5.29%)

role is to guard the nest area from predators. Brood sizes for 3 state nests were one young (1 nest) and four young (2 nests) (Meservey, CNR).

The young hatch naked except for a few tufts of down. Their eyes open at 4 or 5 days, and they are feathering out by day 6. Fledging occurs between 8 to 10 days of age. By day 11, they make short flights. In Massachusetts, nestlings have been observed from June 16 to August 12, and fledglings have been recorded from July 2 to September 6 (CNR, Meservey). Approximate hatch dates for 2 state nests were June 16 and August 5 (CNR, Meservey). Two of four young in a Lincoln nest fledged on July 2 (CNR), and a nest in Spencer contained a single nestling near fledging age on August 12 (Meservey).

During the fledging phase, which lasts from two to three weeks, both the male and female feed the young. The fledglings remain scattered in the shrubbery and give persistent *psink* calls, which help the parents locate them. If there is a second brood, the male will take over the care of the fledglings while the female builds the new nest.

Parents and young often remain together until they are ready to migrate south. During this time, adults undergo a complete molt, and the brilliant blue of the male is replaced by a rich brown. Migration starts in late August and continues through October. Most of the population winters in southern Middle America.

Donald W. Stokes and Lillian Q. Stokes

Bobolink
Dolichonyx oryzivorus

Egg dates: June 1 to June 25.

Number of broods: one.

The status of the Bobolink has changed from that of a common to an uncommon summer resident during the last hundred years. This is due to a number of factors, one of which was the widespread slaughter of the migrating flocks before this was prohibited by law. They were shot by the tens of thousands during both spring and fall migration because they fed on the cultivated rice crops in the southern states. These so-called "ricebirds" were then sometimes sold as gourmet restaurant items.

Furthermore, the reforestation of the cleared land of a century ago has significantly impacted nesting habitat. There has also been a reduction of the Bobolink's preferred habitat of tall hay fields and upland meadows due to suburban development and a decreasing demand for hay. Even in the few appropriate nesting areas that remain, the grass is often cut in June before the birds have completed nesting.

During the Atlas period, the Bobolink was widespread in central and western sections of the state and in Essex County. It was much scarcer to the southeast, was not confirmed nesting on Cape Cod or Martha's Vineyard, and was recorded as a possible breeder from only one block on Nantucket. Bobolinks are now most commonly seen as they migrate through in the spring or fall. They arrive in May, and a few settle in favorable areas to attempt nesting. Flocks of males appear first and sing from fields and meadows. The song is a rolling series of clear, short notes that rise higher and higher in pitch. Heard from a distance, a flock's songs sound like a pleasing chorus of musical, burbling notes.

Males sing in flight or while perched in trees or on the tops of weed stalks. Frequently, the singing is accompanied by a posturing display; the male's buff nape feathers are ruffled, his tail is spread, and his wings are partly opened. Often, the males will launch themselves high into the air while singing and then flutter downward. Males display to other males and to the females who arrive slightly later. During courtship, males chase females, sometimes pausing to do their song-flight display. The common call note is a sharp *quink*. When alarmed, females give a call that sounds like *quick*, and the males produce a somewhat lower pitched *chow*. Young call for food with a *chib* note.

Bobolinks are gregarious and flock for most of the year. They are territorial only during mating and the early part of the nesting period. After the young hatch, there is little, if any, territorial defense. Evidence indicates that these birds can be polygamous, with a male paired to more than one female.

The nest, built by the female, is well concealed in tall grass and difficult to locate. It is usually on the ground in a hollow depression or clump of weeds and is a loosely constructed cup of coarse grasses or weak stems lined with finer grasses. Two state nests were on the ground in fields of tall grasses and vetch (CNR). From four to seven eggs are laid, with an average clutch being five or six. The clutch sizes for 3 Massachusetts nests were four eggs (1 nest), five eggs (2 nests) (DKW). Eggs are gray or a pale cinnamon color with irregular splotches of

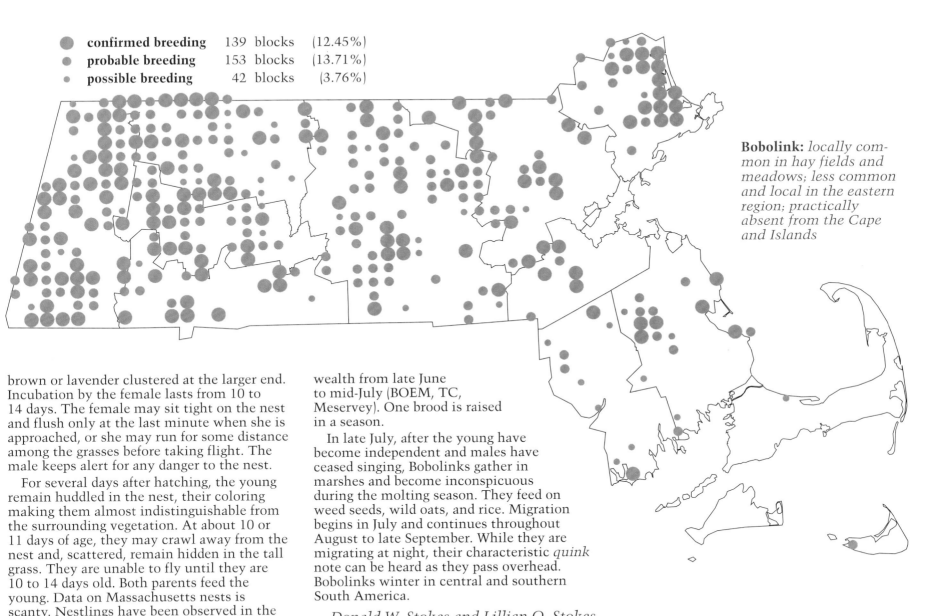

Bobolink: *locally common in hay fields and meadows; less common and local in the eastern region; practically absent from the Cape and Islands*

brown or lavender clustered at the larger end. Incubation by the female lasts from 10 to 14 days. The female may sit tight on the nest and flush only at the last minute when she is approached, or she may run for some distance among the grasses before taking flight. The male keeps alert for any danger to the nest.

For several days after hatching, the young remain huddled in the nest, their coloring making them almost indistinguishable from the surrounding vegetation. At about 10 or 11 days of age, they may crawl away from the nest and, scattered, remain hidden in the tall grass. They are unable to fly until they are 10 to 14 days old. Both parents feed the young. Data on Massachusetts nests is scanty. Nestlings have been observed in the state from June 18 to June 20 (BOEM, CNR). A nest in Hamilton contained six young a few days old on June 18, and a Rowley nest held four young on June 20 (CNR). Fledged young have been recorded in the Commonwealth from late June to mid-July (BOEM, TC, Meservey). One brood is raised in a season.

In late July, after the young have become independent and males have ceased singing, Bobolinks gather in marshes and become inconspicuous during the molting season. They feed on weed seeds, wild oats, and rice. Migration begins in July and continues throughout August to late September. While they are migrating at night, their characteristic *quink* note can be heard as they pass overhead. Bobolinks winter in central and southern South America.

Donald W. Stokes and Lillian Q. Stokes

Red-winged Blackbird
Agelaius phoeniceus

Egg dates: May 10 to June 21.

Number of broods: one; may renest if first attempt fails.

The Red-winged Blackbird is a common summer resident throughout the state, nesting in the swamps and marshes of river bottoms, the borders of streams and ponds, and sometimes in upland fields. In migration it is locally abundant, occurring in flocks in agricultural areas and other sites.

This bird is one of the earliest to arrive in the spring, with the first flocks often appearing in the last week of February. These early groups are composed almost entirely of adult males. They frequently swoop into the tops of trees and burst forth with a loud bubbling, gurgling chorus of song—a true harbinger of spring. Spring migration of both residents and migrants continues through March into early April. Resident adult males do not take up their territories until about the middle of March. The first females arrive a few weeks later.

Red-winged Blackbirds breed mainly in loose colonies in swamps and marshes. The males stake out territories, which they vigorously and noisily defend. When the females appear, there is much dashing about the marsh as the males give chase. During courtship, the males display the scarlet shoulder patches, or epaulets, by puffing themselves up with wings slightly spread and tail fanned, and they sing the familiar *ok-a-lee* or *conk-a-ree* from a prominent perch. Another display is a short vertical flight and a fluttering descent back to the perch. There are a number of other vocalizations. Males give a *zeet* call quite frequently and a *peeah* sound when they are alarmed. Nesting females characteristically produce a loud *chip-chip-chip-chip* series. The common call note is a *chuck* or *check*, often heard as the birds fly overhead.

Males are often polygamous and may have two or three mates nesting within their territories. The rather dull, brown-striped female builds the nest of cattails and other marsh vegetation, often suspended over water in the rushes, sedges, cattails, or grasses of the marsh. Sometimes nests are found in bushes or low trees and in some instances in nearby drier uplands. Sixteen Massachusetts nests were located as follows: field (8 nests), marsh (4 nests), woods (3 nests), swamp (1 nest) (CNR). The high number recorded in fields is probably a reflection of the easier accessibility of that site as compared to the wet locations. Of 17 state nests, 7 were on the ground and the others ranged from 1 to 7 feet in height, with an average of 4.6 feet (CNR).

Three to five eggs (sometimes more) may be laid, but the usual clutch consists of four. Clutch sizes for 23 Massachusetts nests were one egg, probably an incomplete clutch (1 nest); three eggs (2 nests); four eggs (17 nests), five eggs (2 nests); six eggs (1 nest) (DKW, CNR, Kroodsma). Incubation is performed entirely by the female, but once the eggs hatch in 11 or 12 days the young are cared for by both parents, with the female generally doing most of the work. The adults may travel some distance from the territory to forage. During the nesting period, the male defends the nest site from intruders, including crows and hawks. Nestlings have been observed in Massachusetts from May 24 to June 25. Known hatch dates were May 24 and 31 and June 5, 6, 7, 11, 13, and 19 (Kroodsma, CNR). Brood sizes for 14 state nests were three young (7 nests), four young (6 nests), five young (1 nest) (Kroodsma, CNR). The young may fledge at 9 days of age and clamber about the vegetation, but they generally leave the

Red-winged Blackbird: *very common in marshes, wet fields, and most other wetlands throughout the state*

- confirmed breeding — 655 blocks (58.69%)
- probable breeding — 216 blocks (19.35%)
- possible breeding — 40 blocks (3.58%)

nest in 10 or 11 days. As is usually with sexually dimorphic species, the young resemble the female. The immature males are larger but do not acquire adult plumage until the second year. State data for fledglings is scanty, but adults with dependent young have been observed at least from June 4 to June 18 (CNR, Meservey).

Interestingly, although many sources state that the Red-winged Blackbird is double brooded, the nesting records for Massachusetts fail to substantiate this, with no reports of eggs after June 21. It is known that this species, like other marsh-nesting birds, suffers high nest losses, especially from predation. Pairs will renest readily after a failed attempt, and this may account for the later records.

By late summer, after completion of the molt, Red-winged Blackbirds gather in mixed flocks with starlings, cowbirds, and grackles. During this season, they are found most commonly feeding in agricultural areas, especially cornfields. When foraging in fields, birds at the end of the flock will periodically rise to fly low over the flock and land at the front. The flock thus proceeds in this manner until the birds reach the end of the field, at which point the whole flock flies up as a unit to either perch in nearby trees or swirl around to land in another section of the field to continue foraging. Red-winged Blackbirds favor marshes as roosting sites but are not infrequently found in mixed flocks of blackbirds and starlings high in conifers.

Some Red-winged Blackbirds commence their southward flight as early as late August, and probably most residents have departed by late September. Flocks of migrants persist into late October and early November until they are ushered to the southern states by cold spells that mark the approach of winter. A few remain each year at both coastal and inland sites to challenge the vagaries of a Massachusetts winter.

Edith F. Andrews

Eastern Meadowlark
Sturnella magna

Egg dates: May 5 to July 4.

Number of broods: one.

Formerly common throughout Massachusetts and particularly abundant in hay fields and salt marshes along the coast, the Eastern Meadowlark is presently on the decline here due to reforestation and the consequent dwindling of its open-country habitat. This trend can be seen as no more than a return to the probable status that meadowlarks occupied in Massachusetts for thousands of years before European settlement, namely as local residents of coastal meadows and forest clearings that were maintained with fire by Native Americans. The mid-1800s, when 75 percent of the Commonwealth was tilled or grazed, appeared as a brief and aberrant golden age for meadowlarks and other field-inhabiting creatures. Presently, we seem to be in an equally extreme "de-pastoralization" process; someday the state's landscape may consist solely of forests, suburbs, and cities. Meadows will be few and far between.

As the map indicates, the species is presently far from rare, but over much of its range it has withdrawn into isolated local populations and retains real strongholds only in the Connecticut River valley, eastern Berkshire County, and the Middleboro-Bridgewater area in southeastern Massachusetts. On Cape Cod, where it was once abundant and widely hunted as "Marsh Quail," it is now very local.

In spring, migrant meadowlarks are seen in largest numbers from mid-March to mid-April. Males arrive two weeks before females and establish a territory averaging about 7 acres. As with other open-country birds, male meadowlarks give territorial and courtship songs from fence posts, telephone poles, or high points of vegetation within or at the margins of their territory. The song is a somewhat variable, rather twangy three- to five-note whistle rendered variously as *tsee-you tess-can* or *heetar-su-e-oo*. When the females arrive, display intensifies and includes a seldom-heard flight song that has been described as bobolinklike but less hurried. There are also a number of harsh calls including chatters, shrill whistles, and nighthawklike beeps. Male meadowlarks may mate with more than one female. Parasitism by cowbirds is apparently relatively uncommon.

Within their generalized habitat preference, meadowlarks will tolerate considerable variation from actively cultivated grain, hay, and cornfields to abandoned fields with scattered shrubs, revegetated, capped landfills, and (rarely) old orchards underlain with grasses and weeds. The nest is a rather large structure made of coarse grasses and weed

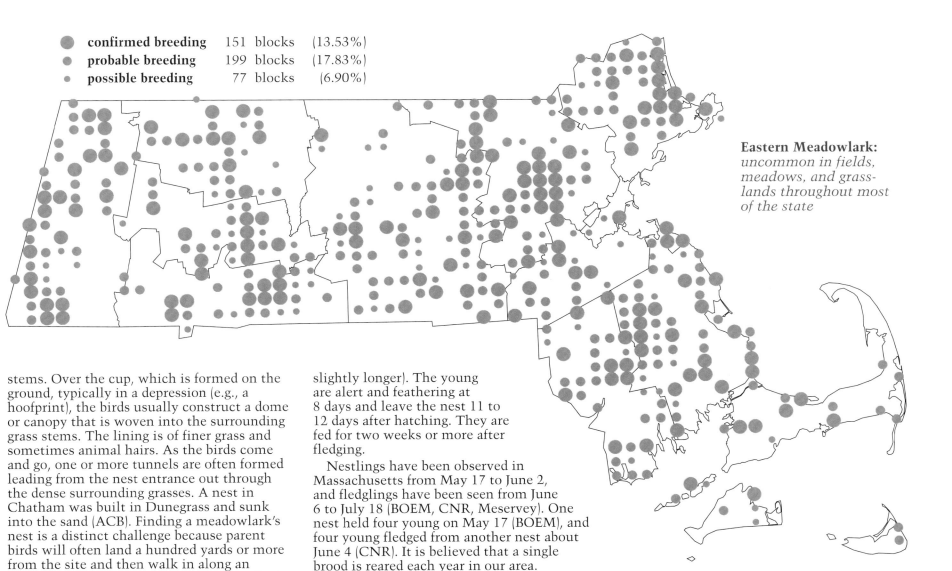

- confirmed breeding — 151 blocks (13.53%)
- probable breeding — 199 blocks (17.83%)
- possible breeding — 77 blocks (6.90%)

Eastern Meadowlark: *uncommon in fields, meadows, and grasslands throughout most of the state*

stems. Over the cup, which is formed on the ground, typically in a depression (e.g., a hoofprint), the birds usually construct a dome or canopy that is woven into the surrounding grass stems. The lining is of finer grass and sometimes animal hairs. As the birds come and go, one or more tunnels are often formed leading from the nest entrance out through the dense surrounding grasses. A nest in Chatham was built in Dunegrass and sunk into the sand (ACB). Finding a meadowlark's nest is a distinct challenge because parent birds will often land a hundred yards or more from the site and then walk in along an invisible route. The two to seven (often five) eggs are white (rarely pinkish or greenish) and finely speckled to coarsely splotched (rarely unmarked) with brownish, reddish, or purplish markings. Clutch sizes for 5 state nests were four eggs (1 nest), five eggs (4 nests) (ACB, DKW, CNR). The eggs are incubated by the female alone for 13 to 14 days (sometimes slightly longer). The young are alert and feathering at 8 days and leave the nest 11 to 12 days after hatching. They are fed for two weeks or more after fledging.

Nestlings have been observed in Massachusetts from May 17 to June 2, and fledglings have been seen from June 6 to July 18 (BOEM, CNR, Meservey). One nest held four young on May 17 (BOEM), and four young fledged from another nest about June 4 (CNR). It is believed that a single brood is reared each year in our area.

In fall, meadowlarks congregate in loose flocks in salt marshes and other open areas along the coast, with smaller numbers present in inland meadows. Migrating flocks are observed chiefly in October. Meadowlarks winter regularly in Massachusetts in variable numbers, probably surviving only the mildest winters inland but regularly occurring in small numbers locally along the coast. Most winter to the south.

Christopher W. Leahy

Rusty Blackbird
Euphagus carolinus

Egg dates: May 19.

Number of broods: one.

The Rusty Blackbird is the least frequently observed of the blackbirds that occur regularly in Massachusetts, although it is an occasional summer resident and a moderately common transient. It is known mainly as a migrant in the early spring and again in the fall, when its numbers are probably somewhat greater than its visibility would indicate. Unlike other blackbirds, Rusty Blackbirds are found infrequently in fields and pastures but instead prefer to feed quietly in wooded swamps and along borders of woodland ponds. Flocks seldom number more than 75 to 100 individuals.

The Rusty Blackbird is entirely an inhabitant of the Canadian Zone in the breeding season, nesting as far north as treeline in Ungava and Alaska and south to northern New England, wherever there are bogs, swamps, and small ponds where spruce, fir, willows, and alders are growing along the margins.

For years this blackbird has been an occasional summer resident about wetlands among the higher hills of Vermont and New Hampshire, south nearly to Massachusetts. Until recently, however, it has never been "confirmed" breeding in this state. One of the gratifying results of the Breeding Bird Atlas was the discovery of 2 nests in 1977 in the higher hills of Berkshire County. One was in a spruce sapling in a swamp near a small pond in the Savoy Mountain State Forest. The other was also in a young spruce located at the edge of an active Beaver pond in the town of Florida. Several other observations of Rusty Blackbirds were made in the breeding season in the same general area, and breeding was subsequently "confirmed" in Franklin County.

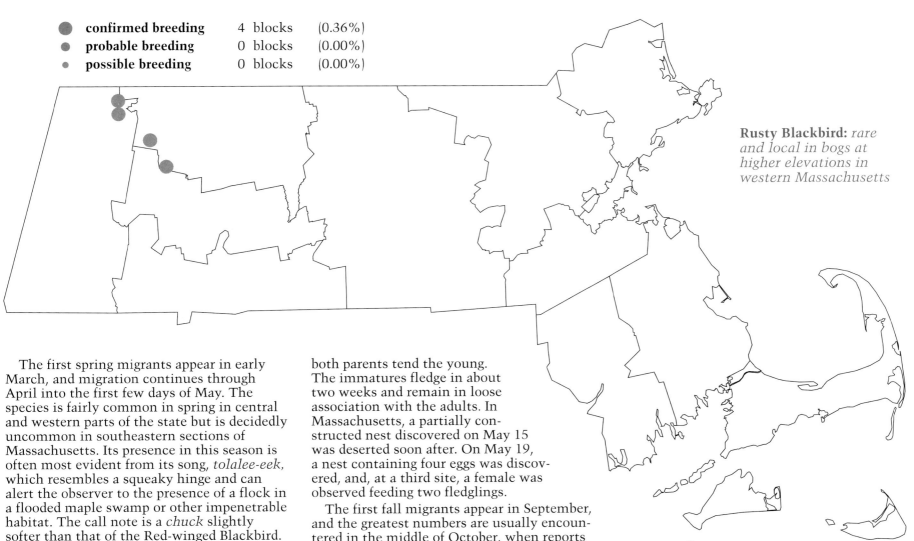

Rusty Blackbird: *rare and local in bogs at higher elevations in western Massachusetts*

- confirmed breeding — 4 blocks (0.36%)
- probable breeding — 0 blocks (0.00%)
- possible breeding — 0 blocks (0.00%)

The first spring migrants appear in early March, and migration continues through April into the first few days of May. The species is fairly common in spring in central and western parts of the state but is decidedly uncommon in southeastern sections of Massachusetts. Its presence in this season is often most evident from its song, *tolalee-eek*, which resembles a squeaky hinge and can alert the observer to the presence of a flock in a flooded maple swamp or other impenetrable habitat. The call note is a *chuck* slightly softer than that of the Red-winged Blackbird.

The nest is usually located in a spruce and is a bulky construction of twigs with the inner cup lined with fine twigs and grasses. Heights range from 2 to 20 feet. Unlike some other blackbirds, this species does not nest in colonies. The four to five eggs are incubated solely by the female for a period of about 14 days. During the incubation period, the male feeds the female, but when the eggs hatch both parents tend the young. The immatures fledge in about two weeks and remain in loose association with the adults. In Massachusetts, a partially constructed nest discovered on May 15 was deserted soon after. On May 19, a nest containing four eggs was discovered, and, at a third site, a female was observed feeding two fledglings.

The first fall migrants appear in September, and the greatest numbers are usually encountered in the middle of October, when reports are more frequent in the state's southeastern coastal sections than in spring. Most birds seen in this season are in the rust-colored plumage, which accounts for the species' name. Rusty Blackbirds occasionally frequent cornfields, weed patches, and gardens as well as the preferred wetland habitats, and they may at times flock with other blackbird species. Most winter in the swamps of the southern states, but a few may linger into winter in Massachusetts, especially in the milder eastern and southeastern sections. Some manage to survive the entire winter in years when the weather is not too harsh.

Richard L. Ferren

Common Grackle
Quiscalus quiscula

Egg dates: April 15 to June 28.

Number of broods: one; sometimes two; may renest if first attempt fails.

The Common Grackle is a survivor, able to adapt to the radical changes in environment brought about by civilization and to withstand direct human persecution. Depredation of crops during the 1800s led to a campaign of bounties and extermination in colonial days. By the first part of the twentieth century, however, the species was once again observed migrating in incredibly huge flocks. The grackle was a common breeder during the Atlas period, nesting statewide, although its numbers appear to have declined somewhat in recent decades.

Grackles often arrive during the last week of February or the first week of March. They are most commonly found in open fields and pastures but also frequent city parks, campuses, large lawns, golf courses, and salt marshes. The first grackles of spring are males, but within a week the females begin to arrive. Grackles typically nest in colonies of a few to over a hundred pairs, often in White Pine groves near water, and the same site will be used year after year. Individual territories are small and include only the nest site and its immediate vicinity. Several nests may be placed in a given tree, usually somewhat spaced but at times as close as 4 feet.

The song is a distinctive, harsh, creeky *kobuga-leek*, with a pronounced metallic quality that suggests a rusty hinge. Courting males accompany the song with a display that involves puffing up the feathers, partly opening the wings, and lowering and spreading the tail. In flight, the tail takes on a distinctive keeled shape. Several males may follow a single female, and at times they fight fiercely, alternately fluttering up at one another and locking together to grapple with bills and feet. Once a pair has formed, the male accompanies the female to and from the colony, following her closely in flight. Other common vocalizations are a loud *chuck* or *tchack* and a *tssh-shklee*; as well as a rapid series of *chuck* notes with an occasional longer *churr* interspersed.

Nest building begins in April. In Massachusetts, nests have been found in White Pine, Pitch Pine, Red Pine, Norway Spruce, Blue Spruce, Red Cedar, maple, apple, and alder, as well as in cattails and various bushes including blueberry and Buttonbush; also on the posts and cross beams of piers, on the downspout supports of buildings, and on a pipe in a wall. Nest heights are variable, ranging from 1 to 10 feet for the low vegetation and pier nests to 25 to 60 feet for the tallest tree nests (ACB, CNR, Meservey).

The nest, a bulky mass of sticks, coarse grass, and rootlets, with the interior plastered with mud and lined with fine grass, is deeply cupped. Grackles lay three to six eggs. Clutch sizes for 15 state nests ranged from three to five eggs (ACB, DKW, CNR). The female performs incubation for about 14 days. After the young hatch, the male assists the female in feeding the young, which fledge after approximately two weeks. In Massachusetts, nestlings have been reported from April 27 to July 9 (CNR, Meservey). The earliest recorded fledge date was May 10, and the peak period of fledging has been May 15 to May 28, with a second and smaller peak from June 28 to July 4 (Meservey).

Brood sizes for 16 state nests ranged from two to four young, with two representing the commonest number of young fledged. Nest failure is common, mostly due to predation by hawks and Red Squirrels, and also to high winds that knock entire nests to the ground.

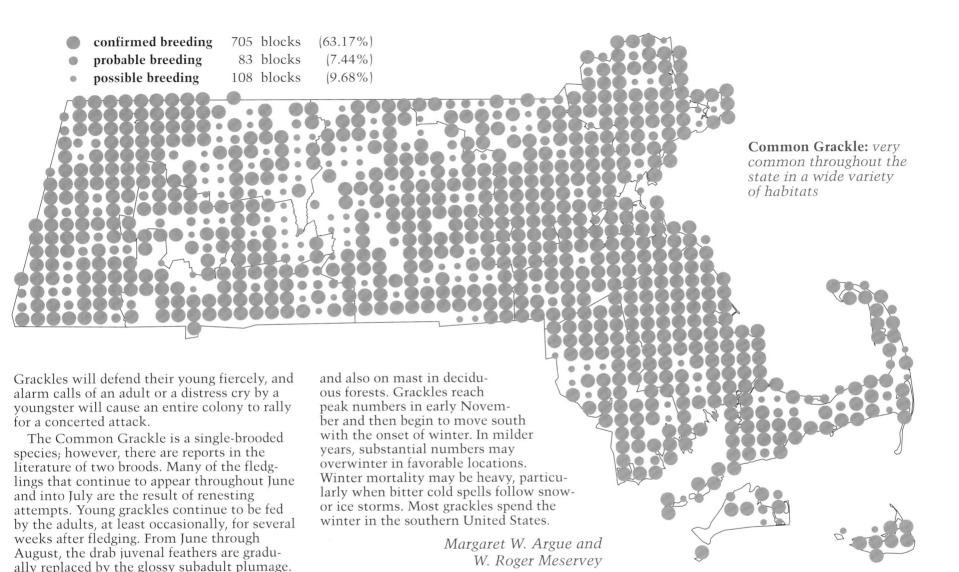

- confirmed breeding 705 blocks (63.17%)
- probable breeding 83 blocks (7.44%)
- possible breeding 108 blocks (9.68%)

Common Grackle: *very common throughout the state in a wide variety of habitats*

Grackles will defend their young fiercely, and alarm calls of an adult or a distress cry by a youngster will cause an entire colony to rally for a concerted attack.

The Common Grackle is a single-brooded species; however, there are reports in the literature of two broods. Many of the fledglings that continue to appear throughout June and into July are the result of renesting attempts. Young grackles continue to be fed by the adults, at least occasionally, for several weeks after fledging. From June through August, the drab juvenal feathers are gradually replaced by the glossy subadult plumage. Adults have a complete molt from August to October.

Beginning in mid-July, grackles begin to fly at sunset in flocks to large roosts, which they often share with other species, and by September these flocks are joined by migrant grackles. During September and October, large flocks feed in corn- and stubble fields, and also on mast in deciduous forests. Grackles reach peak numbers in early November and then begin to move south with the onset of winter. In milder years, substantial numbers may overwinter in favorable locations. Winter mortality may be heavy, particularly when bitter cold spells follow snow- or ice storms. Most grackles spend the winter in the southern United States.

*Margaret W. Argue and
W. Roger Meservey*

Brown-headed Cowbird
Molothrus ater

Egg dates: May 14 to July 1.

Number of broods: brood parasite.

The Brown-headed Cowbird, which lays its eggs in the nests of 210 North American birds (Welty & Baptista 1988), and the Bronzed Cowbird are the only native obligate brood parasites in North America. Some ornithologists speculate that the parasitic habit in cowbirds developed because they followed roaming herds of buffalo over the prairies. Unable to remain in one place long enough for nest building, the females simply deposited eggs in any available nest. Conversely, the parasitic habit may have developed first and thus permitted this species the freedom to form an association with roving herds.

The Brown-headed Cowbird is an abundant and widespread breeder and migrant in Massachusetts that expanded its range eastward as the forests were cleared for agriculture, eventually occurring over all of New England and southern Canada. During summer, the cowbird is most numerous on farmland in the vicinity of cattle, roosting and feeding in large mixed flocks with starlings and other blackbirds. The food of cowbirds is chiefly grain and weed seeds, but roughly one-fifth of the diet is insects, spiders, and a few small snails.

Spring migrants arrive in late March or early April, often in company with other blackbirds. Forbush states that "cowbirds are free lovers," that "the courtship is a happy-go-lucky affair," and that they are "neither polygamous nor polyandrous" but "just promiscuous" and "entirely unattached." However, more recently observers have found that a male cowbird will follow a given female and defend the space around her from conspecific intruders. The defended space is thus a mobile territory (Welty & Baptista 1988). Courtship behavior may occur on the ground or in a tree and begins when one or more males position themselves near a female and face her, fluffing their feathers to form a neck ruff and pointing their bills skyward. The performance culminates when the male arches his neck, spreads his wings, fans and elevates his tail, fluffs his feathers, and bows low. This performance is accompanied by a bubbly *glug, glug, glee,* which has a wide range of frequencies. Many males repeatedly return to favored display perches. Females are also vigorously pursued in flight by one or more males. These activities may continue through June.

In addition to the song of the bowing ceremony, another vocalization given by the male is a high, prolonged squeak followed by several shorter, lower-pitched sibilants: *pseeee-tsit-sit-si*. Call notes are a short *chuck* or *kuk*, a loud harsh rattle, a slightly trilled *pre-e-ah*, and a hissing *tse-e-e-e* (Saunders 1951).

Cowbirds may be observed in a diversity of habitats, wherever the nests of appropriate host species are found. What is required is a cup nest of an altricial species that produces eggs smaller than those of the cowbird. These factors will ensure that the foster parents will be capable of feeding and caring for the young cowbirds until they are ready to fly and feed on their own. The most common hosts are tyrant flycatchers, vireos, wood-warblers, finches, and sparrows. However, orioles, tanagers, mimids, thrushes, and a variety of other species may also be parasitized. Female cowbirds sometimes err in nest selection; eggs have been found in the nests of Mourning Doves, which feed their young on "pigeon milk," and in the nests of Killdeer, whose own precocial young quickly depart from the nest. Some host species (e.g., American Robin, Blue Jay, and Gray Catbird) will often puncture and eject cowbird eggs. The Yellow Warbler buries the cowbird egg under a new nest and lays another clutch. Yellow-breasted Chats desert their nests if a strange egg is present.

Species parasitized by Brown-headed Cowbirds during the Atlas period included the Least Flycatcher, Warbling Vireo, Red-eyed Vireo, Hermit Thrush, Wood Thrush, Blue-winged Warbler, Nashville Warbler, Yellow Warbler, Chestnut-sided Warbler, Black-throated Blue Warbler, Yellow-rumped Warbler, Black-throated Green Warbler, Blackburnian Warbler, Prairie Warbler, Black-and-white Warbler, American Redstart, Ovenbird, Common Yellowthroat, Eastern Towhee, Chipping Sparrow, Field Sparrow, Song Sparrow, and Northern Cardinal (Blodget, Forster, Kroodsma, Meservey, Ober, Spector).

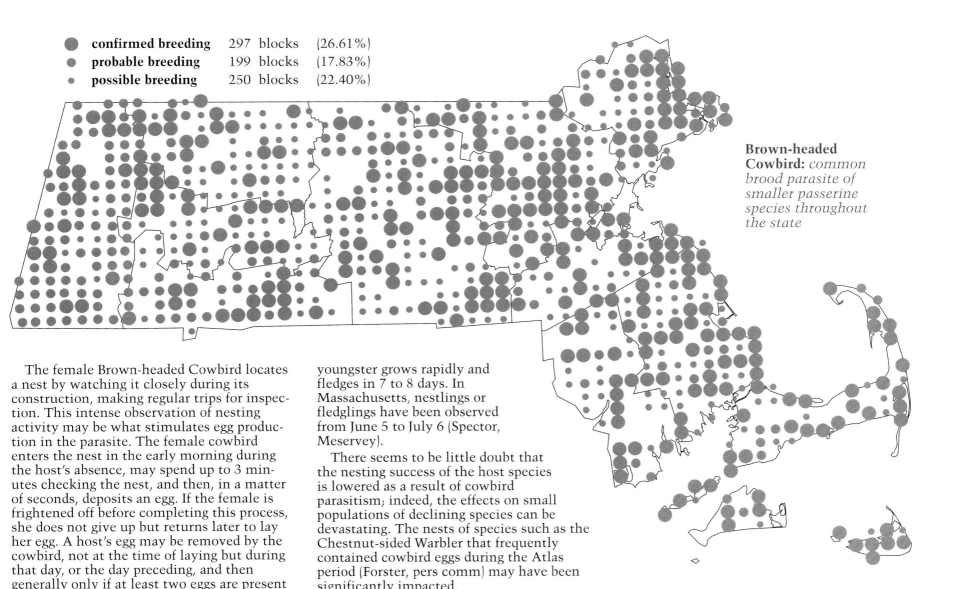

- ● **confirmed breeding** 297 blocks (26.61%)
- ● **probable breeding** 199 blocks (17.83%)
- · **possible breeding** 250 blocks (22.40%)

Brown-headed Cowbird: *common brood parasite of smaller passerine species throughout the state*

The female Brown-headed Cowbird locates a nest by watching it closely during its construction, making regular trips for inspection. This intense observation of nesting activity may be what stimulates egg production in the parasite. The female cowbird enters the nest in the early morning during the host's absence, may spend up to 3 minutes checking the nest, and then, in a matter of seconds, deposits an egg. If the female is frightened off before completing this process, she does not give up but returns later to lay her egg. A host's egg may be removed by the cowbird, not at the time of laying but during that day, or the day preceding, and then generally only if at least two eggs are present in the nest. One female cowbird may lay a total of four or five eggs (some authors say ten to twelve) in 1 or several nests at the rate of one per day. Sometimes two different cowbirds parasitize a single nest.

A cowbird egg hatches after 11 to 12 days, about 1 day ahead of the host eggs. The youngster grows rapidly and fledges in 7 to 8 days. In Massachusetts, nestlings or fledglings have been observed from June 5 to July 6 (Spector, Meservey).

There seems to be little doubt that the nesting success of the host species is lowered as a result of cowbird parasitism; indeed, the effects on small populations of declining species can be devastating. The nests of species such as the Chestnut-sided Warbler that frequently contained cowbird eggs during the Atlas period (Forster, pers comm) may have been significantly impacted.

After nesting, adult Brown-headed Cowbirds have a complete molt and gather in large flocks. Once the young have become independent, they begin to join these groups. From August to November, the immatures lose their brownish gray plumage and attain the adult plumage. Fall migration occurs from September to November. A varying number of cowbirds remain to winter in Massachusetts, but most individuals winter in the middle and southern states.

Dorothy Rodwell Arvidson

Orchard Oriole
Icterus spurius

Egg dates: May 28 to June 25.
Number of broods: one.

The Orchard Oriole is an uncommon and local summer resident in the state, where it is near the northern limit of its breeding range. Most breeding records are from coastal regions and interior river valleys because Orchard Orioles are rare or lacking in areas of higher elevations. Once this species has nested at a locality, it often returns in subsequent years, and small "colonies" consisting of several pairs may become established. Such is the case at the Heard's Pond area of Wayland, where it has nested almost annually since 1887. The distribution in Massachusetts apparently has changed little during the historical period, but the number of nesting pairs has increased since about 1960.

Migrants are seldom encountered in either spring or fall, although coastal storms of southerly origin in mid- or late April will sometimes bring individuals to southeastern coastal areas before their expected arrival. Resident males usually appear during the first week of May, at the same time or slightly before the Northern Oriole. In marked contrast to the chestnut and black older males, the first-year males resemble the yellow-olive females but have a black throat. These young males may be "floaters," wandering over a large area during May and early June in search of a mate, and, although they have been observed helping established pairs feed young, they are also capable of mating and breeding successfully. It appears that older males establish their territories fairly quickly. Orchard Orioles usually nest in rural areas but increasingly are being found in suburban residential areas; common to both is the presence of water, usually a stream or pond. Relatively open areas with trees occurring in clumps are especially favored, and dense woodlands are avoided.

The song, a quick series of varied whistled notes, is likened to those of House and Purple finches and is often the best clue to the presence of an Orchard Oriole because the birds are very retiring in nature. The call note is a low *check*. The nest, constructed solely by the female, is composed of green grasses of various lengths. It is usually placed in the

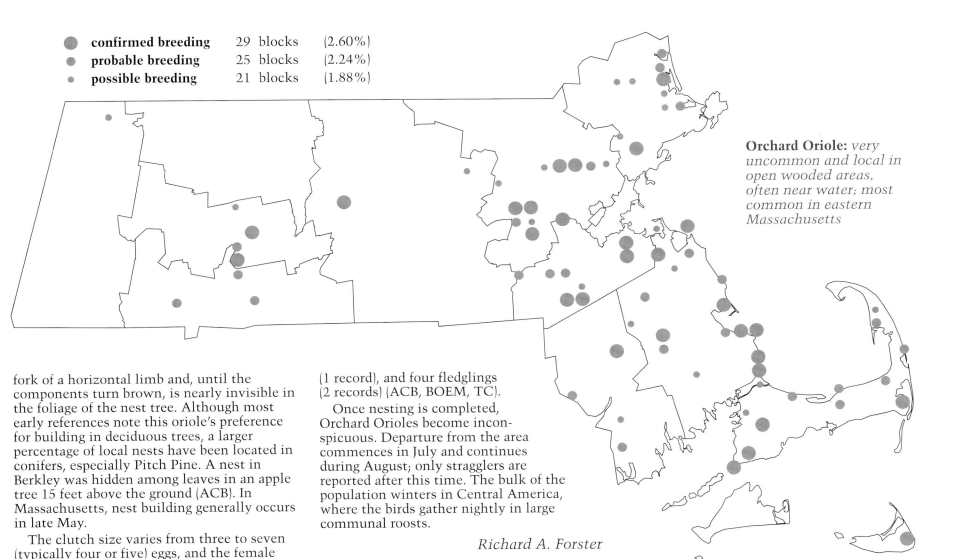

Orchard Oriole: *very uncommon and local in open wooded areas, often near water; most common in eastern Massachusetts*

● confirmed breeding	29 blocks	(2.60%)
● probable breeding	25 blocks	(2.24%)
• possible breeding	21 blocks	(1.88%)

fork of a horizontal limb and, until the components turn brown, is nearly invisible in the foliage of the nest tree. Although most early references note this oriole's preference for building in deciduous trees, a larger percentage of local nests have been located in conifers, especially Pitch Pine. A nest in Berkley was hidden among leaves in an apple tree 15 feet above the ground (ACB). In Massachusetts, nest building generally occurs in late May.

The clutch size varies from three to seven (typically four or five) eggs, and the female incubates for 12 to 14 days. When the young hatch, typically in mid-June, both parents share the feeding duties. The young remain in the nest 11 to 14 days, fledging in late June or early July in Massachusetts. Nestlings have been reported in the state from June 23 to July 15. Single adults or pairs were observed feeding one fledgling (2 records), two fledglings (1 record), three fledglings (1 record), and four fledglings (2 records) (ACB, BOEM, TC).

Once nesting is completed, Orchard Orioles become inconspicuous. Departure from the area commences in July and continues during August; only stragglers are reported after this time. The bulk of the population winters in Central America, where the birds gather nightly in large communal roosts.

Richard A. Forster

Baltimore Oriole
Icterus galbula

Egg dates: May 23 to July 4.

Number of broods: one; may renest if first attempt fails.

Historically called the Baltimore Oriole, this bird was renamed the Northern Oriole when it was found to interbreed with the Bullock's Oriole in a zone of overlap in the Great Plains, thus relegating the two distinctive forms to one species. More recently, however, the species has once again been "split." Regardless of which name it is known by, this brilliantly colored bird is common and widespread throughout the state. It is found wherever large trees are mixed with open spaces, such as farms, roadsides, parks, and suburban neighborhoods, and has a decided preference for riparian situations. It is absent only from extensive coniferous areas of the higher elevations of the state.

Except for the occasional early individual, Northern Orioles normally arrive during the first week of May. The first wave seems to occur almost simultaneously, or within a few days, across the state; later birds, mostly immatures, continue to appear throughout the remainder of the month. The musical, whistled song draws attention to the orange and black males as they establish territories in the sparse foliage of the trees. Females may arrive at the same time, or slightly later. The male courts the female with song and a bowing display with partially spread wings and tail. Males give a distinctive *hoolee* call, and both sexes produce various rattling calls and a loud chatter of alarm.

Nest construction commences in mid-May and has been observed in Massachusetts from May 14 to June 14 (CNR). The familiar pendulous nest is built hanging from the drooping tips of branches from 6 to 90 feet above the ground. Heights of 27 state nests ranged from 12 to 75 feet, with an average of 30 feet (CNR). In Massachusetts, 33 nests were located in the following tree species: maple (8 nests), apple (3 nests), Red Oak (3 nests), elm (7 nests), poplar (2 nests), White Ash (4 nests), Sycamore (1 nest), Choke Cherry (1 nest), willow (1 nest), Black Locust (1 nest), and pine (2 nests) (CNR, Meservey). Sites varied from urban parks and suburban and rural yards and orchards to various wood edges. Nests are woven in 4 to 8 days entirely by the female. A variety of plant fibers are used for the outside of the nest, and it is lined with hair, plant down, and other soft materials. Increasingly, artificially made materials

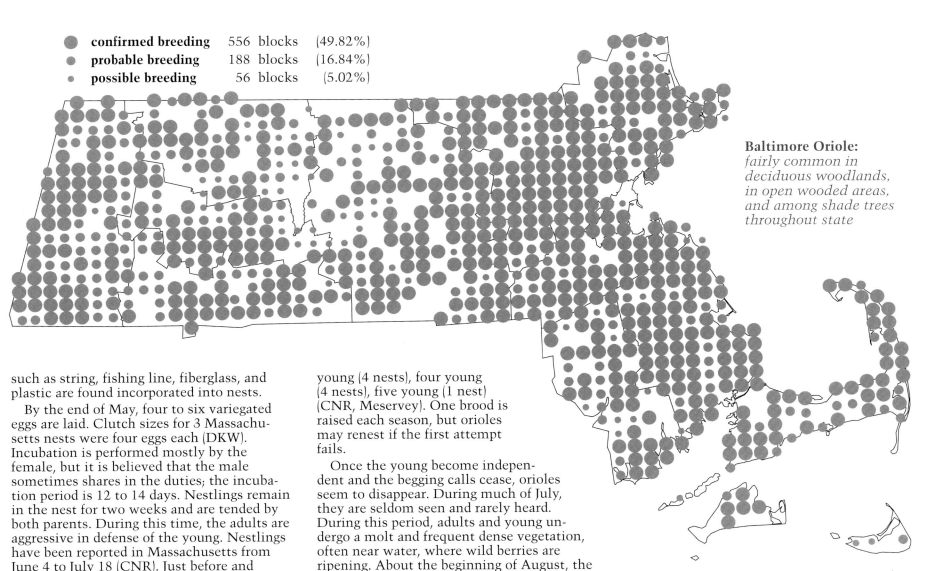

Baltimore Oriole: *fairly common in deciduous woodlands, in open wooded areas, and among shade trees throughout state*

- confirmed breeding — 556 blocks (49.82%)
- probable breeding — 188 blocks (16.84%)
- possible breeding — 56 blocks (5.02%)

such as string, fishing line, fiberglass, and plastic are found incorporated into nests.

By the end of May, four to six variegated eggs are laid. Clutch sizes for 3 Massachusetts nests were four eggs each (DKW). Incubation is performed mostly by the female, but it is believed that the male sometimes shares in the duties; the incubation period is 12 to 14 days. Nestlings remain in the nest for two weeks and are tended by both parents. During this time, the adults are aggressive in defense of the young. Nestlings have been reported in Massachusetts from June 4 to July 18 (CNR). Just before and immediately after fledging, the incessant and monotonous *tee-dee-dee, tee-dee-dee* food-begging calls of the ever-hungry youngsters are a familiar sound of the early summer. Fledglings have been observed in the state from June 19 to July 15 (CNR). Brood sizes at the times of fledging for 14 state nests were one young (1 nest), two young (4 nests), three young (4 nests), four young (4 nests), five young (1 nest) (CNR, Meservey). One brood is raised each season, but orioles may renest if the first attempt fails.

Once the young become independent and the begging calls cease, orioles seem to disappear. During much of July, they are seldom seen and rarely heard. During this period, adults and young undergo a molt and frequent dense vegetation, often near water, where wild berries are ripening. About the beginning of August, the familiar chattering call of orioles is heard once again. The shortened daylight also triggers another period of muted singing. By mid-August, the first orioles have begun to head south, and by late September all but a few stragglers are gone. Most Baltimore Orioles winter from central Mexico to northern South America, although some remain in the southern United States and the Caribbean. A few misguided lingerers may attempt to overwinter, but, even with the help afforded by bird feeders, few survive.

Mark Pokras

Purple Finch
Carpodacus purpureus

Egg dates: May 10 to July 2.

Number of broods: one.

The Purple Finch is a common nester in all the counties of the state. It seems to occupy three distinct situations: deep coniferous or mixed forest, coniferous or mixed woodland borders along swamps and pasturelands (notably in more developed areas), and parks and suburbs with Norway Spruces and other coniferous ornamentals. In the two latter areas, it will readily appear at feeding stations throughout the nesting season.

Purple Finches are erratic migrants; the first spring arrivals usually appear in numbers in mid-March. Males establish territories and maintain regular singing perches from which they sometimes launch forth into flight song. The song is a high-pitched complex warble. Notes are rapid and connected with no two in succession on the same pitch. Overall, the song has a melodious liquid quality. Courtship begins in late April. Males sing continuously and present twigs; they posture elaborately—drooping the wings, exposing the bright rump, and erecting the crown feathers—and perform short hovering flights before the female. Nest building by the female, with assistance from the male, commences shortly after copulation. Preferred nest sites are 6 to 50 feet above the ground in spruce or pine; nests are also reported less frequently in Pasture-juniper, deciduous or evergreen shrubs, and apple trees. Nests are usually straddled on a branch and consist of fine twigs and grasses lined with finer grasses and quite often hairs of various types.

Clutch sizes of three to six eggs (usually four or five) are reported. The female does virtually all of the incubating. Incubation, commencing upon completion of the clutch, lasts about 13 days, during which time the incubating female is frequently fed by the male. Young are altricial and are tended by both parents. Fledging occurs at 14 days. After leaving the nest, the young continue to beg for food from the parent birds for an undetermined period of time until independence.

After nesting, purple finches wander erratically in segregated flocks, with females and young often seeming to outnumber males. Such postnesting groups may be evident in an area one year but absent the next. In September and October, there is usually a general southward movement through and out of the Massachusetts range. Birds are usually seen moving south solitarily or in small groups. Were it not for the distinctive metallic *tick* given by birds in flight, there would be little indication of their

- confirmed breeding 211 blocks (18.91%)
- probable breeding 312 blocks (27.96%)
- possible breeding 97 blocks (8.69%)

Purple Finch: *uncommon in coniferous forests and plantings throughout the state; possibly declining*

presence. Usually from December to February, Massachusetts birds are gone. Rarely, huge nomadic flocks may suddenly materialize in winter. Notable invasions have occurred in the years 1883-1884, 1939, 1941, 1950-1951, and 1963.

Bradford G. Blodget

House Finch
Carpodacus mexicanus

Egg dates: March 31 to July 25.

Number of broods: one or two.

Originally a native of the arid western United States and Mexico, where it is found in a wide variety of habitats, the House Finch is now a common but local resident of the state. The population increase and range extension since the early 1970s are probably unrivaled by any other avian species. This is another excellent example of an introduced species that found its niche, became established, then rapidly occupied suitable areas.

The origin of House Finches in the East stems from the release of captives in the 1940s by caged-bird dealers in New York City who were illegally selling House Finches. By the spring of 1941, a remnant group of the released birds became established at a nursery on nearby Long Island. The population grew very slowly and remained very localized for the ensuing decade, during which time a small number became established along the Connecticut coast.

The avenues of invasion into Massachusetts were along the coast and in the Connecticut River valley. The expansion gained momentum in the early 1970s and exploded by the end of the decade, when House Finches became more prevalent in the higher elevations of Berkshire and Worcester counties, where their major strongholds were in highly urbanized Pittsfield and Worcester. A second population explosion occurred in the early 1980s, and they are now much more widespread in interior areas than the accompanying map would indicate. They are most common in urban and heavily residential areas, particularly newly developed regions where ornamental plantings are the vogue. Although there is some migration from our area, it is not known to what extent our breeding population is migratory. The male sings from an exposed perch such as a treetop, telephone wire, or television aerial.

The song is a musical assemblage of varied rising and falling phrases that usually end with an upward slurred *wheer* note. Both sexes have a pleasing *queet* call note, given singly or in a series.

The nest is an untidy, bulky structure located in evergreens, especially ornamental spruces; on ivy-covered walls; in potted plants; and behind porch lights. Fifteen Massachusetts nests were located as follows: Blue Spruce (3 nests), Red Cedar (2 nests), honeysuckle (2 nests), porch eaves (2 nests), light fixtures (2 nests), potted plants (2 nests), ornaments on buildings (2 nests) (CNR, Meservey). Heights ranged from 4 to 15 feet, with an average of 8 feet (CNR). Nest building in Massachusetts has been reported from

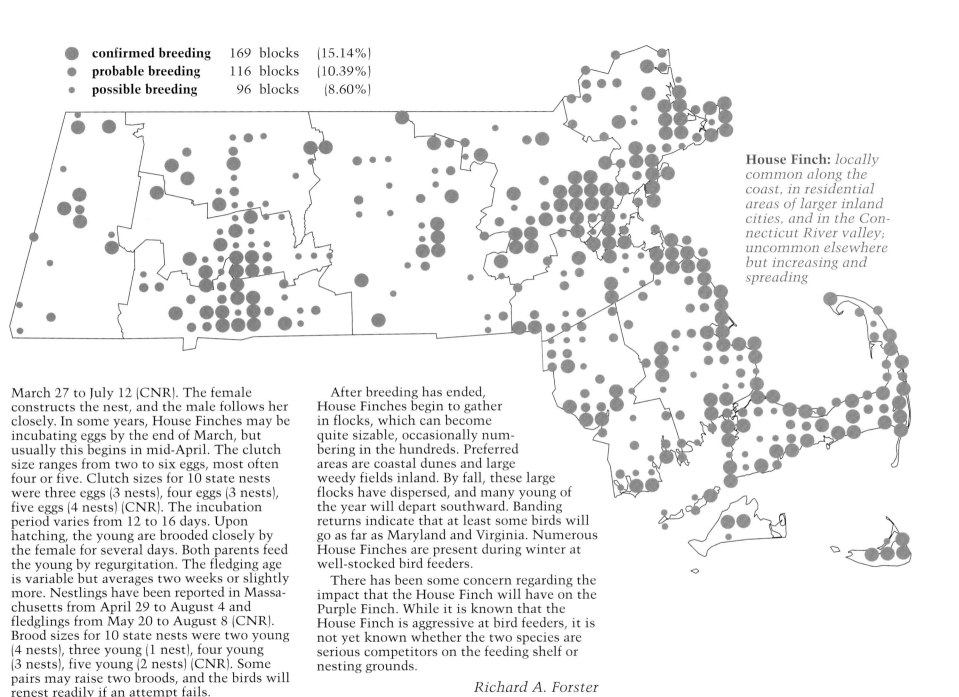

House Finch: *locally common along the coast, in residential areas of larger inland cities, and in the Connecticut River valley; uncommon elsewhere but increasing and spreading*

- confirmed breeding 169 blocks (15.14%)
- probable breeding 116 blocks (10.39%)
- possible breeding 96 blocks (8.60%)

March 27 to July 12 (CNR). The female constructs the nest, and the male follows her closely. In some years, House Finches may be incubating eggs by the end of March, but usually this begins in mid-April. The clutch size ranges from two to six eggs, most often four or five. Clutch sizes for 10 state nests were three eggs (3 nests), four eggs (3 nests), five eggs (4 nests) (CNR). The incubation period varies from 12 to 16 days. Upon hatching, the young are brooded closely by the female for several days. Both parents feed the young by regurgitation. The fledging age is variable but averages two weeks or slightly more. Nestlings have been reported in Massachusetts from April 29 to August 4 and fledglings from May 20 to August 8 (CNR). Brood sizes for 10 state nests were two young (4 nests), three young (1 nest), four young (3 nests), five young (2 nests) (CNR). Some pairs may raise two broods, and the birds will renest readily if an attempt fails.

After breeding has ended, House Finches begin to gather in flocks, which can become quite sizable, occasionally numbering in the hundreds. Preferred areas are coastal dunes and large weedy fields inland. By fall, these large flocks have dispersed, and many young of the year will depart southward. Banding returns indicate that at least some birds will go as far as Maryland and Virginia. Numerous House Finches are present during winter at well-stocked bird feeders.

There has been some concern regarding the impact that the House Finch will have on the Purple Finch. While it is known that the House Finch is aggressive at bird feeders, it is not yet known whether the two species are serious competitors on the feeding shelf or nesting grounds.

Richard A. Forster

Red Crossbill
Loxia curvirostra

Egg dates: March 18 to April 1.

Number of broods: one.

The unpredictable movements of the Red Crossbill make it difficult to determine the species' status in any given region or to define regular breeding areas throughout its nearly circumboreal range. In Massachusetts, breeding has been confirmed only a handful of times. There are several Essex County reports, and Bent mentions a nest from the Berkshires. Most records of crossbills during the Atlas project were from the northern Berkshires, a region that holds promise for regular breeding due to the larger percentage of Red and Black spruce, Balsam Fir, American Larch, and Eastern Hemlock, all preferred trees for feeding. The one confirmed Atlas nest was in the pine-oak barrens of Plymouth. Occasionally, streaked juveniles appear at feeders with adults in May or June. Although these may have been reared locally, there is always the possibility that they have wandered from other areas.

Although Red Crossbills are known to nest at virtually any time during the year, a clear majority of birds breed from February through April regardless of latitude. The song is usually a short series of single notes followed by a few two-syllable phrases, the latter notes somewhat higher pitched on the second syllable. These short songs can be given a few together or many in rapid succession, the effect being one, long, elaborate performance. Males sing from perches and also give a flight song while circling with fluttering wings. The common call note is a *kip* or *yip*.

The nest is usually placed in a conifer 10 to 40 feet above the ground. Old man's beard lichen is often a component, along with twigs and fine grass. A nest at Eastern Point in Gloucester was constructed exclusively by a female from March 2 to March 13 (Snyder 1952). Generally, three or four eggs are laid, and the female incubates for about two weeks. During this period and for the first days after the young hatch, the male feeds the female on the nest. The young fledge in about 17 days. The Gloucester nest contained eggs

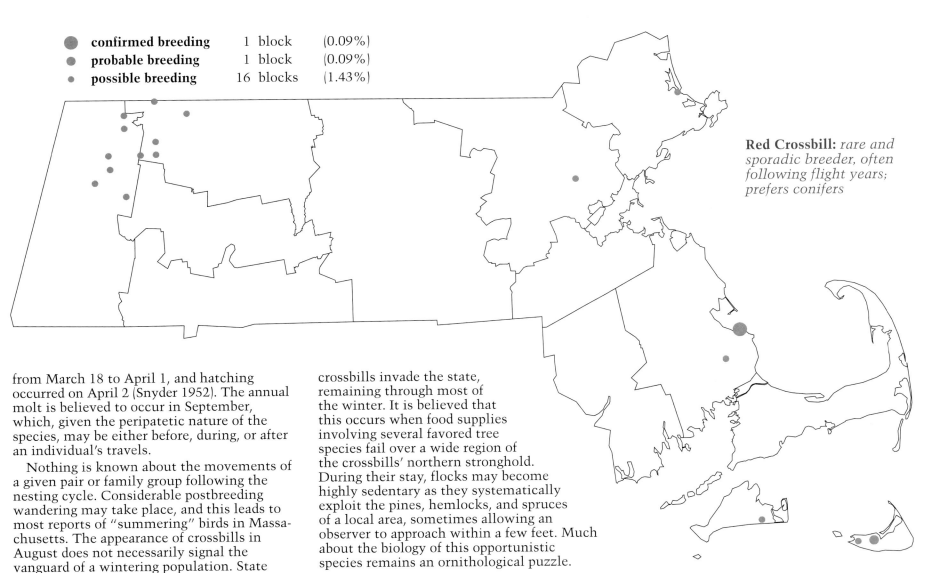

Red Crossbill: *rare and sporadic breeder, often following flight years; prefers conifers*

from March 18 to April 1, and hatching occurred on April 2 (Snyder 1952). The annual molt is believed to occur in September, which, given the peripatetic nature of the species, may be either before, during, or after an individual's travels.

Nothing is known about the movements of a given pair or family group following the nesting cycle. Considerable postbreeding wandering may take place, and this leads to most reports of "summering" birds in Massachusetts. The appearance of crossbills in August does not necessarily signal the vanguard of a wintering population. State records are most frequent in September, October, and November.

Red Crossbills feed on a variety of coniferous and deciduous tree seeds, as well as buds and insects. In most years, comparatively short movements (perhaps several in a season) are sufficient for the birds to find adequate food. Occasionally, large numbers of crossbills invade the state, remaining through most of the winter. It is believed that this occurs when food supplies involving several favored tree species fail over a wide region of the crossbills' northern stronghold. During their stay, flocks may become highly sedentary as they systematically exploit the pines, hemlocks, and spruces of a local area, sometimes allowing an observer to approach within a few feet. Much about the biology of this opportunistic species remains an ornithological puzzle.

Bruce A. Sorrie

Pine Siskin
Carduelis pinus

Egg dates: May 9 to May 29.

Number of broods: one; possibly sometimes two.

Like that of many northern finches, the Pine Siskin's status in our area is unpredictable. In some years they are common and widespread during winter, and in others they are almost totally lacking. Their presence in the state during winter is usually the result of the failure of cone and seed crops in portions of the northern forest. In some years, migrants may pass through during October and November and return in April in small numbers as they move north to breeding areas.

During their winter stay here, they feed on the seeds of a variety of trees, including alder and birch. But the increase in bird feeding has had a profound impact on winter bird populations. This has been especially true for the small finches since the advent of niger thistle seed, a tiny black variety that they particularly favor. Feeders that are well stocked with thistle seed will often play host to numerous Pine Siskins for an entire winter. These circumstances have led to most recent breeding reports. Many of the northern finches are opportunistic breeders, capitalizing on a locally abundant food supply to ensure survival of the young.

Prior to 1970, there was a mere handful of breeding records for the Pine Siskin in Massachusetts. Early confirmations include a nest found in early May 1859 in Cambridge and another on May 29, 1883, in Newton. Scattered reports continued through the 1940s. Since 1970, breeding has become widespread in years when siskins are common in winter, with numerous reports in 1971. Nestings were "confirmed" during the Atlas period in several years, including 1976, and more than half of the entries on the map occurred in only one year, 1978. The map indicates that the Pine Siskin is apt to breed almost anywhere in the state, but there is a distinct bias for the greater Boston area where feeders abound. However, nesting is not limited to the immediate location of feeders because nests have been located at Mount Tom and in the wilderness of Quabbin.

Nesting evidence is hard to detect because siskins typically feed in small flocks even during the breeding season. One clue is long bouts of singing, beginning in March. The song is goldfinchlike, a jumbled assortment of twittering interspersed with a characteristic, sharply rising *zzree* note. Common call notes are represented as *chee-ee* and *ti-er*,

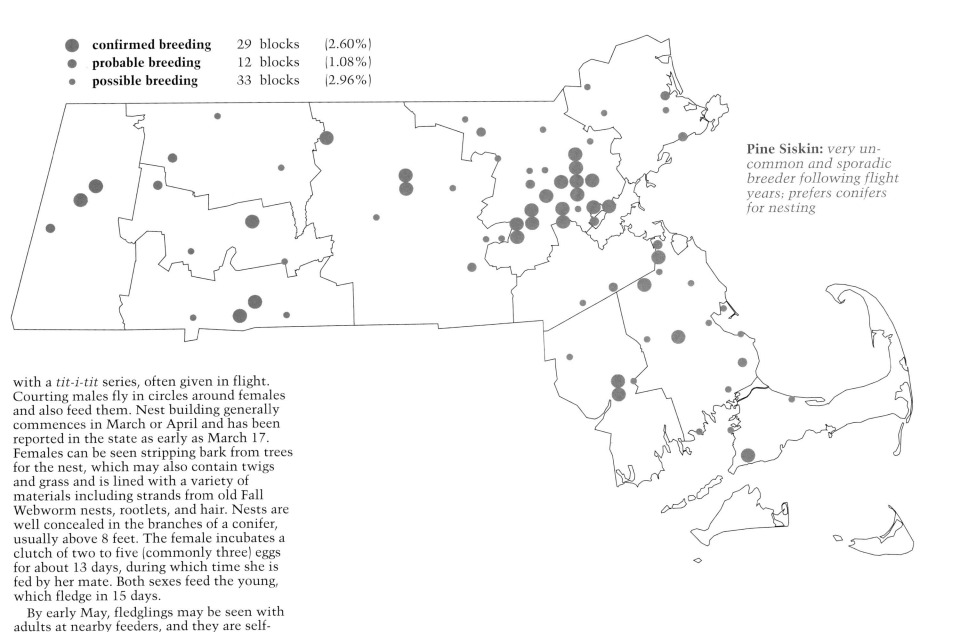

●	**confirmed breeding**	29 blocks	(2.60%)
●	**probable breeding**	12 blocks	(1.08%)
·	**possible breeding**	33 blocks	(2.96%)

Pine Siskin: *very uncommon and sporadic breeder following flight years; prefers conifers for nesting*

with a *tit-i-tit* series, often given in flight. Courting males fly in circles around females and also feed them. Nest building generally commences in March or April and has been reported in the state as early as March 17. Females can be seen stripping bark from trees for the nest, which may also contain twigs and grass and is lined with a variety of materials including strands from old Fall Webworm nests, rootlets, and hair. Nests are well concealed in the branches of a conifer, usually above 8 feet. The female incubates a clutch of two to five (commonly three) eggs for about 13 days, during which time she is fed by her mate. Both sexes feed the young, which fledge in 15 days.

By early May, fledglings may be seen with adults at nearby feeders, and they are self-sufficient by the end of the month. Adults and young then generally depart, but occasionally flocks are noted as late as mid-June.

Richard A. Forster

American Goldfinch
Carduelis tristis

Egg dates: July 10 to September 18.

Number of broods: one or two.

The American Goldfinch is a welcome year-round resident throughout Massachusetts; this bird is one of our most colorful breeders in summer and a regular visitor to feeders in winter. During the Atlas period, nesting confirmations were made throughout the state. Goldfinches breed in areas that are generally rural in character, preferring overgrown fields and wood edges where there are shrubs and sapling trees. They also seem to favor areas that are wet, near lakes or in moist meadows and roadside ditches.

The breeding schedule of the American Goldfinch is very different from that of most of our other birds. Nesting activities occur in late summer, from July into September, when an abundant supply of favorite seeds has ripened. However, several interesting features of their behavior occur in spring as well. When the birds undergo their partial molt, from March to May, the male goldfinch attains his bright yellow plumage and begins singing, often in company with other males. These periods of song are interspersed with chases between males and between males and females. The song is a long series of canarylike chirps and trills. Some believe that this is indeed courtship behavior and that the pair bond is formed at this time, even though breeding does not usually take place for a month or more.

American Goldfinches have a variety of call notes. One of the most familiar is the *per-chic-o-ree*, frequently heard as the birds fly overhead. Another call is a pleasing *chee-chee-chee-we* series. There are also short chirps, and begging young give a *chipee, chipee,…*call.

Actual nesting territories are defended starting in late June. Males sing from exposed perches and chase off other males. They also do unusual circular flights over their territories. These are unusual in that the bird flies in a flat path rather than with the usual undulating flight seen at all other times of year. The birds do not restrict their movements to the territory and may move off it to feed. Once the female initiates nest building, the territory is no longer defended and other goldfinches may even land in the nest tree without being chased.

The female does all of the nest building. This activity has been observed in Massachusetts from July 1 to August 19 (CNR). The nest is usually built from 4 to 20 feet aboveground in the crotch of a sapling, shrub, or herbaceous plant and is composed of downy fibers, such as the filaments of various composite flowers, held together with sticky threads from Fall Webworm nests. Four Massachusetts nests were located in weedy or overgrown fields, and 1 was at the edge of a salt marsh. The nests were situated as follows: 1 each in goldenrod, willow, unidentified shrub, Box Elder, and Smooth Buckthorn. Heights ranged from 3 to 9 feet, with an average of 6.4 feet (CNR).

After nest building is completed, goldfinches often leave the nest for several days to two weeks before egg laying commences. An average clutch of five pale blue eggs is laid. Clutch sizes for 7 nests in Massachusetts were four eggs (3 nests) and five eggs (4 nests) (CNR, DKW). The female does all of the incubation, during which time the male feeds her by regurgitation. The eggs hatch in 12 to 14 days. For several days, the male feeds both his mate and young, but as the nestlings get older both parents feed them seeds that are partially digested and regurgitated from the crop.

The nestlings have light gray down on the head and body when they hatch, and for the first week they are quite silent. By 10 days

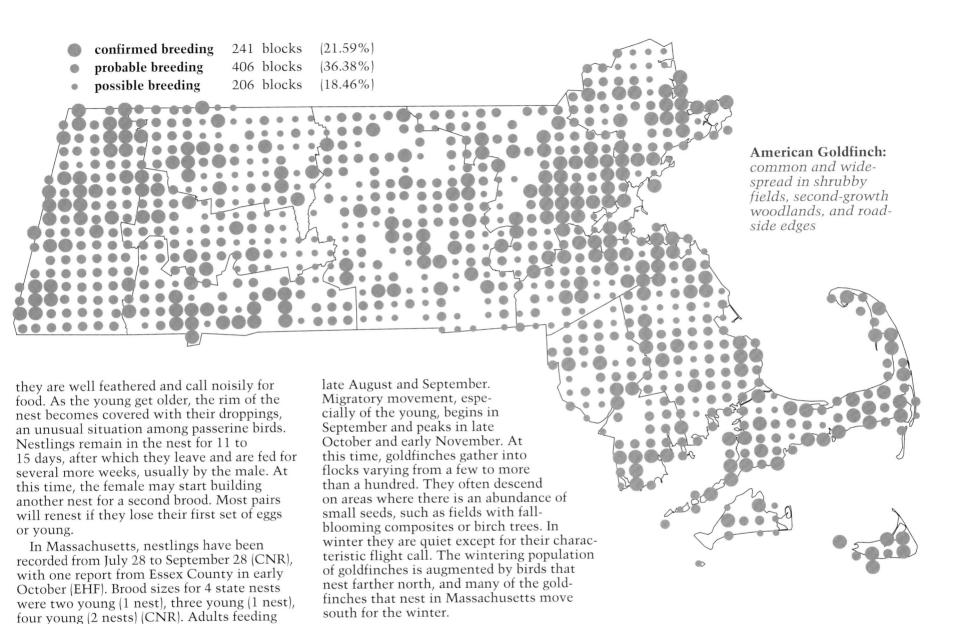

American Goldfinch: *common and widespread in shrubby fields, second-growth woodlands, and roadside edges*

- confirmed breeding — 241 blocks (21.59%)
- probable breeding — 406 blocks (36.38%)
- possible breeding — 206 blocks (18.46%)

they are well feathered and call noisily for food. As the young get older, the rim of the nest becomes covered with their droppings, an unusual situation among passerine birds. Nestlings remain in the nest for 11 to 15 days, after which they leave and are fed for several more weeks, usually by the male. At this time, the female may start building another nest for a second brood. Most pairs will renest if they lose their first set of eggs or young.

In Massachusetts, nestlings have been recorded from July 28 to September 28 (CNR), with one report from Essex County in early October (EHF). Brood sizes for 4 state nests were two young (1 nest), three young (1 nest), four young (2 nests) (CNR). Adults feeding fledglings have been observed in the Commonwealth from August 15 to September 28 (CNR).

After breeding is completed, both adult and young goldfinches undergo a molt, usually in late August and September. Migratory movement, especially of the young, begins in September and peaks in late October and early November. At this time, goldfinches gather into flocks varying from a few to more than a hundred. They often descend on areas where there is an abundance of small seeds, such as fields with fall-blooming composites or birch trees. In winter they are quiet except for their characteristic flight call. The wintering population of goldfinches is augmented by birds that nest farther north, and many of the goldfinches that nest in Massachusetts move south for the winter.

Donald W. Stokes and Lillian Q. Stokes

Evening Grosbeak
Coccothraustes vespertinus

Egg dates: not available.

Number of broods: one; possibly sometimes two.

Formerly unknown in the Northeast, the Evening Grosbeak was once a resident of the boreal forests of northwestern Canada, with other populations occurring in the western mountains south to Arizona and Mexico. About 150 years ago, the species began a slow expansion eastward. The pattern was one of winter invasions, following which some of the birds would remain behind to nest after the flocks had departed again in the spring for the Northwest. The reasons for the development of this west-to-east migration pattern are not totally clear, but it has been suggested that two factors may have played roles. First, the widespread planting of Box Elder on prairie farms provided a favorite food in the form of seeds and buds as well as a "bridge" to the eastern forests; and, second, after the 1930s, feeding stations stocked with sunflower seeds, another favored food, supplied a reliable winter food source.

The Evening Grosbeak was not known east of the Great Lakes until 1854, when it was reported in Toronto. After 1875, it began to increase in the northern midwestern states. An invasion during the winter of 1886-1887 brought the birds into Ontario and New York. The first Massachusetts records occurred during the winter of 1889-1890, when grosbeaks reached New England in numbers and ranged from southeastern Massachusetts to Orono, Maine. Small numbers subsequently were reported in winter in New England until 1910-1911, when there was another big flight. From 1920 on, the species has appeared every winter in Massachusetts in variable numbers.

By 1940 there were scattered summer records for New England and some reports of fledged young in the northern states. The eastward expansion continued, with the birds reaching the Canadian Maritimes by the mid-1960s. Nesting is now regular in the three northern New England states.

A Massachusetts nesting record at Northfield in 1937 cited by Bagg and Eliot (1937) is generally regarded as inconclusive. The first accepted state nesting followed a winter (1957-1958) when grosbeaks arrived in the Northeast in record high numbers. After the masses had departed, reports of lingerers or summerers came in during May and June, and, finally in July 1958, a female was observed feeding fledglings in Hadley.

The situation has changed little in the intervening years. Scattered summer reports, mainly from the northern and western parts of the state, follow winter records. The species remains elusive while breeding. During the Atlas period there was only one confirmation (1978), when the female of a pair was observed nest building in West Springfield. Unfortunately, the birds departed without continuing further with the nesting cycle. No one has yet found a nest with eggs or young in the state. Reports of adults feeding fledglings have come from Berkshire County (e.g., Pittsfield, 1980, 1990) and northern Worcester County (e.g., Petersham, 1990; Princeton, 1991).

Winter residents and migrants from farther south depart for more western and northern breeding grounds from March to early May, with lingerers remaining well into the latter month. Flocks move from one feeding station to another during migration, but they gradually become less interested in sunflower seeds as natural foods such as buds, seeds, and maple sap become available. Since the Evening Grosbeak is a rare nester in Massachusetts, by the end of May only scattered pairs and small groups are observed. The birds are highly gregarious outside the breeding season and apparently gather in smaller flocks to forage at times, even when nesting. In spite of their name, they are most active during the morning and are observed infrequently after early afternoon. This trait is especially noticeable in the winter, when the birds visit feeders, arriving early and bickering with one another and other species but leaving the area well before the rest.

Evening Grosbeaks have a variety of calls. While feeding, the birds may be silent, except for the biting and cracking sound of their beaks, or may keep up a *peet peet kreek peet kreek peet peet* or a *tchew-tchew-tchew*. The main contact call, a shrill *pete* or *p-teer*, is heard from birds flying overhead or from

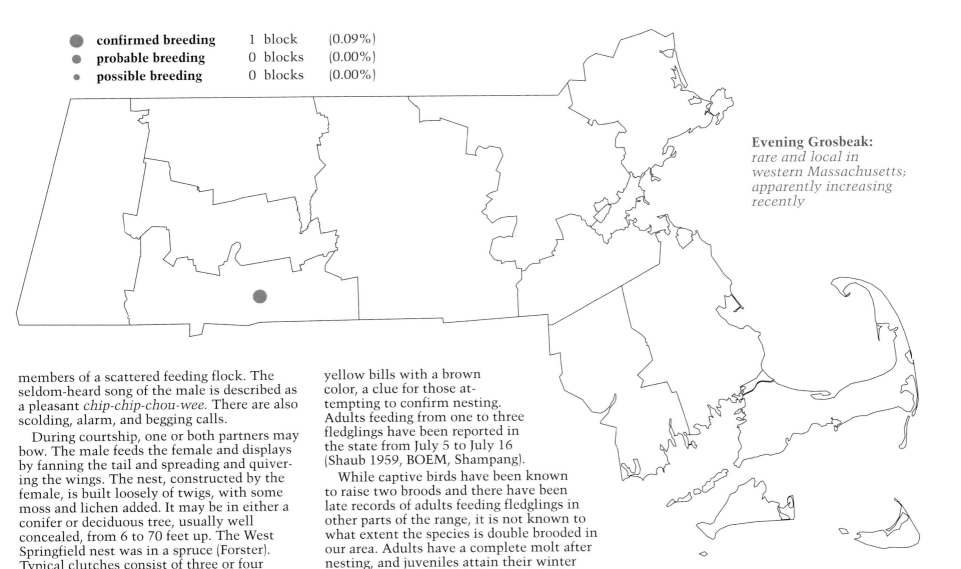

Evening Grosbeak: *rare and local in western Massachusetts; apparently increasing recently*

- confirmed breeding — 1 block (0.09%)
- probable breeding — 0 blocks (0.00%)
- possible breeding — 0 blocks (0.00%)

members of a scattered feeding flock. The seldom-heard song of the male is described as a pleasant *chip-chip-chou-wee*. There are also scolding, alarm, and begging calls.

During courtship, one or both partners may bow. The male feeds the female and displays by fanning the tail and spreading and quivering the wings. The nest, constructed by the female, is built loosely of twigs, with some moss and lichen added. It may be in either a conifer or deciduous tree, usually well concealed, from 6 to 70 feet up. The West Springfield nest was in a spruce (Forster). Typical clutches consist of three or four (range two to five) eggs. The female incubates for 12 to 14 days, during which time she is fed by the male. Both sexes feed the young, which fledge after 13 to 14 days. Parental care continues even after the young have begun feeding on their own. Adults feed insects to the nestlings and typically macerate the prey before delivering it. This stains their greenish yellow bills with a brown color, a clue for those attempting to confirm nesting. Adults feeding from one to three fledglings have been reported in the state from July 5 to July 16 (Shaub 1959, BOEM, Shampang).

While captive birds have been known to raise two broods and there have been late records of adults feeding fledglings in other parts of the range, it is not known to what extent the species is double brooded in our area. Adults have a complete molt after nesting, and juveniles attain their winter plumage by a partial molt. Nothing is known about the movements of local birds after nesting. Typically, they bring their young to a summer bird feeder for a time and then disappear from the area. Fall migrants appear in September, with heavier flights during the next two months. Some of these birds remain to winter while others move on to more southern states. In some years, the main flights do not arrive until winter is well advanced. Numbers present in Massachusetts vary greatly from year to year as well as from locale to locale.

W. Roger Meservey

House Sparrow
Passer domesticus

Egg dates: April to August.

Number of broods: two or three.

This Old World native was introduced and reintroduced at many locations in North America starting in the 1850s and by about 1910 had spread throughout nearly the entire continent. The introduced birds took advantage of urban and rural habitats, aggressively driving out native hole-nesting species such as the Purple Martin, House Wren, Tree Swallow, and Eastern Bluebird and evicting Cliff Swallows from their mud nests.

House Sparrows are in the weaver family, a large and varied group of Old World passerines characterized by heavy seed-eating bills, communal behavior, and a predilection for nesting in cavities or ovenlike nests constructed of grasses, roots, and bark.

In urban settings, House Sparrows appear to have depended greatly on horse droppings as a source of undigested grain for food. The bird was abundant in the 1920s, when horses were still relatively common in cities and towns, but subsequently declined in numbers throughout the East, probably due in part to the disappearance of horses. Large numbers of the birds still persist in poultry yards and feedlots where there is plentiful grain.

House Sparrows are not migratory, although the young wander to some extent. Though they use holes, their nesting requirements are extremely variable both as to location and season. Some individuals appear to be able to breed at nearly any time of year. Bent gives a record of a nest containing naked young in Utah on January 1, but most breeding dates range from April to September, with three broods normal as far north as Massachusetts and Maine. Early nesting is one of the major reasons that House Sparrows have been able to displace more "desirable" native birds. They are also very aggressive in laying claim to a potential nest site and will destroy the eggs and young of any competing cavity nesters.

The House Sparrow has no real song, although on occasion some individuals may produce a series of pleasing notes. The typical vocal repertoire consists of a variety of chirping notes and chatterings. One of the commonest calls is a *chissick, chissick* series, which may be repeated incessantly, especially early in the morning. The young give a similar call when they are begging for food. When agitated, the birds scold with a rapid *tut-tut-tut*.

The courtship displays of the House Sparrow are vigorous affairs, with one or more males chirping steadily, hopping about with outstretched, quivering wings before a female. If the female flies, they pursue her, vigorously battling with one another, seemingly oblivious to their surroundings. The female may be quite defensive, biting the head and wings of one or more of her suitors.

In Massachusetts, nest construction may begin in late winter, with the earliest recorded date being February 18 (CNR). Nests are placed in nest boxes, streetlight fixtures, ivy-covered areas on buildings, and a variety of other nooks, holes, and crevices on buildings, walls, etc. Eleven Massachusetts nests were located as follows: nest box (5 nests), crack in brick wall (1 nest), between air vent slats (1 nest), inside shed (1 nest), hole under eaves or roof (2 nests), house gutter (1 nest). Heights ranged from 4 to 18 feet, with an average of 8.9 feet (CNR). House Sparrows inhabit urban and suburban areas

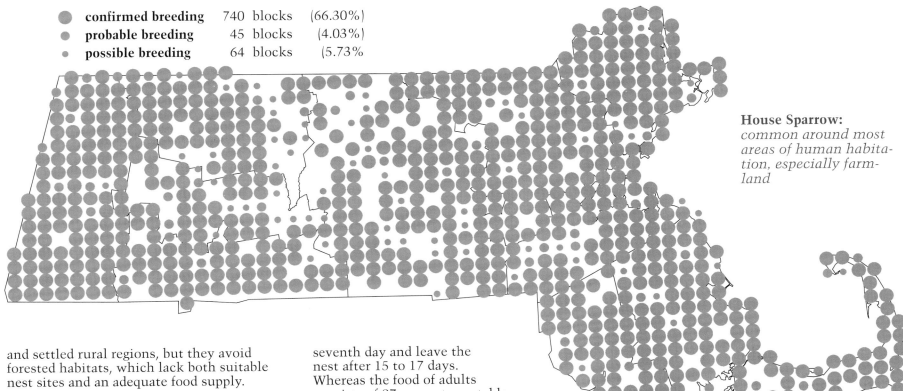

- ● **confirmed breeding** 740 blocks (66.30%)
- ● **probable breeding** 45 blocks (4.03%)
- · **possible breeding** 64 blocks (5.73%

House Sparrow:
common around most areas of human habitation, especially farmland

and settled rural regions, but they avoid forested habitats, which lack both suitable nest sites and an adequate food supply.

The nest is a bulky, ovenlike structure, constructed of coarse grasses and dried weeds then lined with finer materials such as feathers, cord, hair, string, or bits of paper. Although as many as nine eggs are possible in a clutch, the average number per nest is four or five. Clutch sizes for 5 state nests were three eggs (1 nest), four eggs (2 nests), six eggs (1 nest), eight eggs (1 nest) (CNR, DKW). Eggs are oval and variable in color, from whitish to blue or green, with overall brownish speckling. Incubation, by the female, lasts about 12 days. In Massachusetts, most records of eggs are from April to September, but there are a few earlier and later dates (EHF). Egg dates from a sample of 11 state nests ranged from May 3 to June 23 (CNR).

Hatchling House Sparrows are naked; the young begin to attain feathers by the sixth or seventh day and leave the nest after 15 to 17 days. Whereas the food of adults consists of 97 percent vegetable material, mostly seeds and grains, the young are fed almost 60 percent animal foods, mainly insects. Nestlings in the Commonwealth would be expected from early spring to early fall and were recorded from May 17 to August 7 in a sample of nests at Cornell University. Brood sizes for 7 state nests were three young (3 nests) and four young (4 nests) (CNR). Newly fledged young were observed from May 26 to July 6 (CNR), but, again, the actual range is much greater.

After nesting is completed, House Sparrows begin gathering in flocks, which persist until the next breeding season. They are particularly noisy at their communal winter roosts, where their simple, coarse *tchlay* calls combine into loud and persistent rounds.

Hundreds of birds, swarming around inside low bushes or small evergreens, suggest the presence of a predator, but close inspection reveals only an excited group of sparrows.

Soheil Zendeh

Appendices

Appendix 1: Additional Breeding Bird Atlas Species Accounts

The Massachusetts Breeding Bird Atlas project was designed to determine the composition and distribution of breeding bird species in Massachusetts during the years 1974 to 1979. In any such defined period of effort, there will inevitably be bird species that are suspected to be breeding during the Atlas period but that are simply not confirmable. In addition, there will be some species that historically nested in Massachusetts prior to the Atlas period or else have bred in the Commonwealth subsequently. The accounts that follow provide a brief summary of the status of bird species in these categories, with the exception of the now-extinct Heath Hen and Passenger Pigeon. Unlike the more detailed accounts in the main body of the text, these summaries are intended to provide only the essential details of the breeding history of each species in Massachusetts.

Manx Shearwater
Puffinus puffinus

Of all the species that have nested in Massachusetts, none is more surprising than the Manx Shearwater. Although this wide-ranging seabird breeds abundantly in Iceland, Great Britain, Brittany, Madeira, and the Azores, its first "confirmed" nesting in the western North Atlantic was not established until June 4, 1973. On that date, Augustus Ben David II fortuitously turned over a plank and discovered an adult shearwater incubating an egg (AB) at Penikese Island at the mouth of Buzzards Bay, Dukes County. On August 16, a single, well-grown chick appeared close to fledging (BOEM), an event that normally occurs approximately 70 days after hatching.

Although no further breeding confirmations for Massachusetts have been obtained, Manx Shearwaters were discovered nesting in Newfoundland in 1977, and, based on the increase of records in Gulf of Maine waters in recent years, future breeding attempts in the western Atlantic Ocean can be expected. Any Manx Shearwaters seen at dusk or heard vocalizing near offshore islands should be viewed as potential breeders. Also, birders having the opportunity to spend time at night on remote Bay State islands should keep an ear out for the nocturnal howls and screams that are characteristic of this species on its breeding grounds. Since Manx Shearwaters nest in burrows in the ground, any suspicious freshly excavated holes of appropriate size should be monitored closely. To date, the Massachusetts nesting stands as the only such record in the United States; however, the species historically nested in Bermuda.

As nonbreeders, Manx Shearwaters are regularly present over cool, surface, ocean waters north and east of Massachusetts from May to October, with a few records outside this period. They tend to be most numerous from midsummer to midfall and are most often encountered with flocks of other shearwater species.

Great Cormorant
Phalacrocorax carbo

The only nesting record for Massachusetts of this northern species was a nest containing three eggs discovered on June 4, 1984, on the Weepecket Islands off Naushon Island, Dukes County, in Buzzards Bay (Hatch).

Despite the dramatic explosion of Double-crested Cormorant breeding numbers along the Atlantic coast, no subsequent Massachusetts breeding records of the Great Cormorant have been documented, although small numbers of nonbreeding individuals are regularly noted along the coast every summer. In the future, special attention should be paid to areas where Double-crested Cormorants are breeding. At present, the Buzzards Bay nesting is the southernmost breeding record for North America and the only one south of Maine, where the species was first conclusively recorded as nesting off Isle au Haut in 1983 (AB).

Great Cormorants are most numerous in Massachusetts in winter when single-day count totals sometimes reach several hundred.

Black Vulture
Coragyps atratus

One of the most recent additions to the list of Massachusetts breeding birds is the Black Vulture. Norman Smith confirmed the first nesting of this species in the Common-

Black Vulture

The status of Black Vultures in Massachusetts prior to the first successful nesting in 1998 was that of rare to very uncommon nearly annual visitors, at least since the 1950s. The species can appear in any season of the year and has occurred in every county in the state. Although most records are in midsummer and fall, there are a few winter records, including one to two in Turkey Vulture roosts in Randolph during the mid-1990s, not far from the Blue Hills nesting location.

wealth in the Blue Hills Reservation, Milton, Norfolk County, in 1998. The continued presence of two to three suspicious-acting adults near a rocky ledge eventually led to the discovery of a nest on June 23. A single young fledged on August 2.

Black Vultures have been gradually expanding their breeding range northward from the south. The species was first confirmed as a breeder in New Jersey in 1981 (Walsh et al 1999), and by 1995 nesting was suspected in western Connecticut (AB). The regular presence of several Black Vultures in the Sheffield area of southern Berkshire County during the summers of 1998 and 1999 could presage further Bay State nesting attempts. Surprisingly, this species has not been documented as a breeding species in New York.

American Wigeon
Anas americana

The American Wigeon was not "confirmed" as a breeder during the Atlas period, although it has nested in Massachusetts both prior to the project and in subsequent years. Widely scattered and relatively uncommon to rare breeders in the Northeast, most wigeons breed in artificially impounded wildlife refuges or waterfowl management areas. The first Massachusetts breeding confirmation occurred when a female with seven young was observed at Penikese Island, Dukes County, near the entrance to Buzzards Bay, on July 1, 1972 (Nisbet).

Subsequent Massachusetts breeding confirmations were made at Monomoy National Wildlife Refuge, Barnstable County; during the summer of 1981 a female with young was observed on June 24 (French), and in 1983 breeding was again reported (Lortie). It is quite likely that this species has nested at the latter location in other years as well, and breeding has also been suspected at the Parker River National Wildlife Refuge, Essex County. Due to the difficulties of accessing suitable nesting areas at the Monomoy and Parker River national wildlife refuges, this species has probably gone undetected as a breeder more often than Massachusetts records would indicate.

As migrants and wintering birds, American Wigeons are fairly common and widespread in Massachusetts, especially in autumn when they tend to prefer coastal ponds or inland wetlands that contain an abundance of their favorite pondweeds. In many years, Great Meadows National Wildlife Refuge attracts large fall concentrations.

Bald Eagle

Bald Eagle
Haliaeetus leucocephalus

The early historical breeding status of the Bald Eagle in Massachusetts is a matter of some uncertainty. Although eagles may have nested more commonly in the Commonwealth prior to the region's extensive settlement, "confirmed" breeding during the nineteenth and twentieth centuries was, until 1989, decidedly sporadic and scattered.

Appendix 1: Additional Species 423

Reasonably solid historical evidence suggests that this species nested, at least occasionally, in Cheshire, Berkshire County; on Mount Tom, Hampden County; on Mount Toby, Franklin County; in Winchendon, Worcester County; and in Sandwich, Barnstable County. The last "confirmed" historical nesting took place at Snake Pond in Sandwich from 1900 to 1905.

The reduction and extirpation of Bald Eagle populations in some areas of the eastern United States between the 1940s and 1970s has been well documented. Various contributing factors were involved, but most critical was the increased bioaccumulation of various pesticides, especially DDT, in the reproductive tissues of adult eagles. The result was eggshell thinning, which ultimately resulted in reduced reproductive success. By the late 1960s and through the 1970s, eastern eagle populations had reached alarmingly low levels.

With the ban of DDT in 1972, a slow recovery in the health of surviving adult Bald Eagle populations began. In an effort to accelerate this gradual recovery, raptor biologists began experimenting with techniques that would assist the beleaguered birds on their road to recovery. In 1976, the first Bald Eagle hacking program in the United States began in the state of New York, where young eagles taken from nests in the Great Lakes states were returned to the wild.

In 1982, with funding provided by the Massachusetts Audubon Society, the Massachusetts Division of Fisheries and Wildlife initiated a Bald Eagle hacking program at Quabbin Reservoir in central Massachusetts. From 1982 to 1988, nineteen young Bald Eagles variously taken from nests in Michigan and Canada were hand-raised and released at a hack tower at Quabbin. In 1989, the first wild Bald Eagles to nest in Massachusetts for over 75 years were recorded at the reservoir when two pairs produced three young. Since 1989, the number of pairs has increased statewide, and, as of 1999, there were 11 active nests in the Commonwealth, most of them concentrated at Quabbin Reservoir and along the Connecticut River, with one pair in eastern Massachusetts at Assawompset Pond, Middleborough, Plymouth County.

Bald Eagles may be encountered throughout the year in Massachusetts. The greatest concentrations occur at Quabbin Reservoir in midwinter when individuals occasionally numbering up to 50 birds, mostly from points north, join local eagles to feed on deer carcasses on the frozen reservoir or to hunt fish and waterfowl in the open water. Lesser winter concentrations also occur along the Connecticut River, on the lower Merrimack River in Essex County, and at several locations in Plymouth and Barnstable counties.

Peregrine Falcon
Falco peregrinus

The history of nesting Peregrine Falcons in Massachusetts in some ways parallels that of Bald Eagles, except that the record is clearer and the documentation for the local demise more complete. In *Birds of Massachusetts and Other New England States* (1927), Forbush depicted the location of 11 historic Peregrine Falcon aeries. During the course of his duties to protect Massachusetts Peregrine Falcons from egg collectors and falconers in the 1930s, State Ornithologist Joseph A. Hagar visited most of the sites described by Forbush, in addition to locating four new aeries that had never been documented previously. By the 1940s, the threats had diminished considerably; however, beginning late in that decade, Hagar began to notice a significant decline in falcon productivity. Eventually, eggshell thinning was documented, and DDT was identified as the

Peregrine Falcon

causative agent, just as it had been for declining Bald Eagle populations during the same period. By the mid-1950s, reproduction by Massachusetts Peregrine Falcons had ceased.

In an effort to restore the beleaguered Peregrine Falcon population, in 1973 the Cornell Laboratory of Ornithology initiated a captive breeding program that would eventually make it possible to return captive-raised Peregrine Falcons to the wild.

Early attempts with this technique at Mount Tom in Massachusetts during the late 1970s were unsuccessful. Starting in 1984, a different approach was attempted when six young Peregrine Falcons were placed in a hacking tower atop Boston's McCormack Building. Although the success of some of these initial hacking attempts was mixed, in 1989, after six years of hacking efforts, "wild" Peregrine Falcons started nesting once again in Boston and Springfield. By 1999, there was a successful nesting pair in Boston, Suffolk County; Springfield, Hampden County; and Fall River, Bristol County.

Black-headed Gull
Larus ridibundus

The legacy of this species in Massachusetts is interesting in that the first specimen for North America was collected at Newburyport in 1930 and the first breeding attempt in the United States occurred at Monomoy National Wildlife Refuge in 1984 (Holt et al 1986). Following the appearance of a single territorial adult Black-headed Gull at Monomoy in 1983, a pair built a nest and produced two eggs the following season; however, the chicks eventually succumbed to exposure to heavy late-spring rains. No further evidence of nesting in Massachusetts has been obtained, although several subsequent nesting attempts have been recorded at Petit Manan, Maine.

The overall status of the Black-headed Gull in Massachusetts is that of a locally uncommon migrant and coastal winter visitor; it is very uncommon in summer and is rare inland.

Ring-billed Gull
Larus delawarensis

The first and only nesting confirmation for this species in Massachusetts was the discovery of 10 nests containing a total of sixteen eggs at Cunningham Ledge in the Wachusett Reservoir, Boylston, Worcester County, June 23, 1997 (Blodget). Due to concerns over possible water contamination in the reservoir resulting from the presence of the gulls, the eggs were oiled and several adults were removed in an effort to discourage further attempts at nesting. Considering the nearby presence of a massive breeding population of Ring-billed Gulls on Lake Champlain, Vermont, it seems likely that additional Bay State breeding records will occur.

Ring-billed Gulls are abundant migrants in Massachusetts and can currently be found in large numbers practically throughout the year, especially along the coast.

Ring-billed Gull

Forster's Tern
Sterna forsteri

A Forster's Tern nest containing two eggs was discovered in the Plum Island salt marshes at Parker River National Wildlife Refuge, Essex County, on June 10, 1991 (Rimmer & Hopping 1991). This record represents the first for Massachusetts and the northernmost documented nesting attempt on the Atlantic coast. Breeding was first suspected at Plum Island in 1990, yet, despite the fact that Forster's Terns have probably nested irregularly in the area since the initial discovery, there has been no absolute confirmation since 1991. In 1981, the species first began breeding as far north as Long Island, New York; that population is currently composed of fewer than 100 pairs (Levine 1998).

Forster's Terns are best characterized as variably uncommon to locally common fall migrants along the coast. Inland and during other seasons they are rare, except occasionally in the aftermath of hurricanes that transport them northward in large numbers.

Chuck-will's-widow
Caprimulgus carolinensis

Uncertainty and frustration best characterize the breeding status of the Chuck-will's-widow in Massachusetts. Despite the fact that this robust nightjar has gradually been extending its breeding range northward during the last half of the twentieth century, the species has never been "confirmed" as a breeder in the Bay State. Chuck-will's-widows arrived on Long Island, New York, as breeders in 1975 (Bull 1976); and, in Massachusetts, calling birds have been heard more or less regularly and continuously at Nantucket, Martha's Vineyard, and Chappaquiddick Island since the early 1970s. Despite several serious attempts to confirm nesting through the years, no definitive breeding evidence has ever been established beyond the persistent calling of individuals late into the breeding season. Even though this species was listed as a "probable" breeder during the Atlas period, final confirmation of its Massachusetts breeding will likely be the result of a serendipitous discovery of a nest or young.

Aside from those summering on the Islands, Chuck-will's-widows are rare but regular late-spring and early-summer visitors that occur most frequently in southeastern Massachusetts, although they occasionally show up elsewhere in the state.

Yellow-bellied Flycatcher
Empidonax flaviventris

The Yellow-bellied Flycatcher has a tantalizing history of summer occurrences in Massachusetts, much like the Chuck-will's-widow. As early as 1883, ornithologist William Brewster recorded a pair of these unobtrusive flycatchers in late June on Mount Greylock, Berkshire County. Since then, including the Atlas period, there have been a number of summer occurrences of singing individuals or pairs on the mountain, as well as at one or two other high-elevation bogs in Berkshire County. However, despite these summer occurrences—and the fact that the species has been found breeding in southern Vermont—there is to date no conclusive evidence of nesting in Massachusetts.

As migrants in Massachusetts, Yellow-bellied Flycatchers are inconspicuous and uncommon in late spring and late summer through early fall.

Loggerhead Shrike
Lanius ludovicianus

The Loggerhead Shrike is a species whose breeding status in Massachusetts, as well as throughout all of southern New England, has always been that of a rare, local, and sporadic nester. Documented historic breeding in western Massachusetts includes three nests located in Berkshire County between 1883 and 1886, and probably also in Savoy in 1913 and 1914. Additionally, there is a "confirmed" nesting record for Greenfield, Franklin County, in 1901 (Snyder 1956).

The next fully documented Bay State breeding records did not occur until 1956, when successful nesting took place in Danvers and Newburyport, Essex County (Snyder 1956); and in Berlin, Worcester County (Veit & Petersen 1993). Loggerhead Shrikes also nested successfully at Danvers and Newburyport in 1957. The cluster of breeding attempts during the mid-1950s is inexplicable. The last known breeding of this species in Massachusetts occurred in Newbury, Essex County, in 1971 (Veit & Petersen 1993). This nesting was observed by numerous individuals and photographically documented by Joseph Kenneally.

Since the last nesting in 1971, the Loggerhead Shrike has steadily declined as a migrant in Massachusetts, as well as throughout much of the Northeast. Forty to 50 years ago, this species was a scarce spring migrant, typically appearing in April, and an uncommon but regular fall migrant from late August to October. At the dawn of the twenty-first century, however, the species is best described as an extirpated breeder in Massachusetts and a generally rare visitor in any season. The Loggerhead Shrike is listed as an endangered species in Massachusetts.

Common Raven
Corvus corax

Common Ravens may have nested in our area when the Pilgrims arrived; however, the species must have met an early demise with the spread of European colonization. The only specific reference to what may have been a historic Massachusetts nesting

Common Raven

locality was Ragged Mountain in Adams, Berkshire County, during the late 1800s (Forbush 1927). It was not until 1982 that the first definitive confirmation of nesting in Massachusetts was obtained, when recently fledged young were observed being fed at Quabbin Reservoir, Worcester County (Gagnon). Since then, raven numbers have steadily increased in central and western Massachusetts, and in 1999 breeding was established in Ashland, Middlesex County (French). Increasingly frequent reports of ravens in eastern Massachusetts, especially Essex County, suggest that further expansion may still be taking place. In Massachusetts, Common Ravens seem to have a preference for nesting on cliff ledges, in quarry pits, and in the stone spillways of dammed rivers, although tree nests have also been found.

Common Ravens are often seen at favored hawk-watching sites such as Mount Watatic and Wachusett Mountain in Worcester County, and in winter at Quabbin Reservoir, where they can occasionally be observed on the ice feeding with Bald Eagles.

Bicknell's Thrush
Catharus bicknelli

The Bicknell's Thrush, considered a subspecies of the Gray-cheeked Thrush until 1995, went unrecorded during the MBBA, even as a "possible" breeder in the state. This is regrettable in that from 1888, when the species was first described as breeding at Mount Greylock, Berkshire County (Faxon 1889), until the early 1970s, the ethereal song of this boreal songster was a regular twilight sound near the mountain's summit. Since 1972, however, none have been seen or heard at Mount Greylock or along the nearby Saddleball Trail—another historic breeding station. What caused the demise of this relict population is open to question, but increased human disturbance at the summit of the mountain may have been a contributing factor.

Determining the migratory status of the Bicknell's Thrush in Massachusetts is clouded by difficulties associated with separating the species from the Gray-cheeked Thrush. Banding data and reliable sight reports suggest that Bicknell's Thrushes are probably scarce late-spring and early-fall transients.

Tennessee Warbler
Vermivora peregrina

The Tennessee Warbler has never been proven to nest in the Bay State; however, the presence of a singing male throughout the breeding season at a station in Berkshire County was sufficient to list the species as "possible" in the *Massachusetts Breeding Bird Atlas*. It is unlikely that the species has ever nested in Massachusetts, although there are several summer records of singing birds well south of the nearest breeding localities in northern New England.

During migration Tennessee Warblers are irregularly common spring and generally uncommon fall transients in Massachusetts.

Cerulean Warbler
Dendroica cerulea

The breeding of the Cerulean Warbler in Massachusetts is a relatively recent event, coinciding with a serious decline throughout most of its range (Dunn & Garrett 1997). Despite the scattered, irregular presence of singing males since the 1950s, mostly in central and western Massachusetts, this species was not "confirmed" as a breeder until 1989, when adults were discovered feeding young at Quabbin Reservoir, Petersham, Worcester County, from July 2 to 9 (Brownrigg & Brownrigg 1989). Since that initial confirmation, a very few pairs of

Cerulean Warbler

Cerulean Warblers have nested more or less annually in the Quabbin Reservoir area and possibly at one or two sites in the Connecticut River valley. The total Massachusetts breeding population is probably under 10 pairs.

At times other than the breeding season, the Cerulean Warbler is a regular but rare spring and late-summer to early-fall transient.

Prothonotary Warbler
Protonotaria citrea

The claim that this southern warbler is a breeding species in Massachusetts, while tenuous, is probably legitimate. The first suggestion of state breeding occurred in Concord, Middlesex County, in 1886 when William Brewster collected a male, female, and juvenile "in a swamp along the Sudbury

River" between May 9 and August 17. Brewster was apparently convinced he had missed establishing a definite breeding record, despite compelling evidence to the contrary (Griscom 1949).

More enigmatic was an apparently unmated female Prothonotary Warbler that built a nest in a garage in Hawley, Franklin County, in 1979 but was never joined by a male (Kellogg). A bona fide breeding attempt occurred in Sharon, Norfolk County, from May 27 to June 21, 1982, when a pair built a nest and was incubating eggs before the eggs were eventually destroyed by House Wrens (BOEM). Although there have been a number of reports through the years of persistently singing male Prothonotary Warblers in suitable breeding habitat into early summer, none have led to further evidence of nesting.

Prothonotary Warbler

A rare but regular visitor to Massachusetts in late April and May and late August and September, this species will likely attempt to breed in the state again, especially since it has nested as close to Bay State borders as Connecticut and Long Island, New York.

Dickcissel
Spiza americana

The historic record of the nesting status of the Dickcissel in Massachusetts suggests that the species was probably erratic and never abundant. Thomas Nuttall, referring to the early 1800s, stated that "they are not uncommon in this part of New England, dwelling here, however, almost exclusively in the high, fresh meadows near the saltmarshes" (Nuttall 1905). Brewster, however, stated that by the turn of the twentieth century, it was "unquestionably one of the rarest species known to breed within this region" (Minot 1903). Whatever the Dickcissel's precise historical status may have been, it apparently disappeared as a nesting bird from the Massachusetts landscape around 1877, and there has not been the slightest suggestion of breeding in Massachusetts during the 1900s.

Dickcissels are rare spring and uncommon and irregular fall migrants in Massachusetts, and a few occasionally attempt to overwinter at bird feeding stations.

Lincoln's Sparrow
Melospiza lincolnii

Although the Lincoln's Sparrow nests in southern Vermont and in the central and western Adirondacks in New York, the five

Lincoln's Sparrow

known breeding records of the Lincoln's Sparrow in Massachusetts place the Bay State at or very near the southernmost extremity of the breeding range in eastern North America. Even though they were not found during the MBBA, Lincoln's Sparrows were "confirmed" breeding for the first time in the state in Florida, Berkshire County, during June and July 1981 (Stemple 1981). Subsequent breeding confirmation at the Florida location was obtained in 1993 and 1994, and nesting was also proven at the Moran Wildlife Management Area, Windsor, Berkshire County, in 1996 and 1998.

Lincoln's Sparrows are uncommon spring and fall migrants throughout Massachusetts and are very rare in early winter, seemingly appearing with increasing frequency on Christmas Bird Counts.

White-winged Crossbill
Loxia leucoptera

The White-winged Crossbill is the most recent addition to the list of birds known to have bred in Massachusetts. Although there have been sporadic summer reports of White-winged Crossbills in western Massachusetts for a number years (Griscom & Snyder 1955, Veit & Petersen 1993), until 2001 there was never conclusive evidence of local nesting. Since crossbills are nomadic breeders throughout much of their range, the presence of individuals in summer, even flying young, makes confirmation of local nesting problematic. For example, in northern New England, positive proof of nesting was not obtained during breeding bird atlas efforts in Vermont in the 1970s (Laughlin & Kibbe 1985) and New Hampshire in the 1980s (Foss 1994), despite the fact that the species almost certainly breeds in these states, at least irregularly. Although the White-winged Crossbill is a more or less regular breeder in northern Maine (Adamus 1988), as well as an irregular breeder in the Adirondack region and Appalachian plateau of New York (Levine 1998), Massachusetts is south of the species' regular breeding range elsewhere in North America, other than in central New York (American Ornithologist's Union 1998).

During the summer of 2000, White-winged Crossbills were numerous in spruce areas in northern Berkshire County and were variously observed singing and engaging in courtship activity in the towns of Ashfield, Dalton, Savoy, and Windsor. By the winter of 2000-2001, before actual nesting confirmation was established, the continued presence of many singing and courting birds

White-winged Crossbill

in this region provided increased optimism for a first Bay State breeding confirmation. This confirmation was finally obtained on February 22, 2001, when Geoffrey LeBaron observed a rosy male White-winged Crossbill feeding four barely able-to-fly juveniles in Windsor, Berkshire County (BOEM).

Although it is likely that small numbers of crossbills nested in the Berkshire Hills during the winter of 2000-2001, the erratic and nomadic breeding behavior of this species may cause it not to breed again in the Bay State for many years. In general this species is best characterized as an irregular winter visitor in Massachusetts, occasionally occurring in major winter invasions yet often entirely absent.

Appendix 2: Plant and Animal Species Mentioned in the Text

Animals Other than Birds

Vertebrates
Bat Vespertilionidae
Bear *Ursus* sp
Beaver *Castor canadensis*
Bobcat *Felis rufus*
Cat, Domestic *Felis catus*
Chipmunk, Eastern *Tamias striatus*
Dog *Canis lupus familiaris*
Fox, Gray *Urocyon cinereoargenteus*
Fox, Red *Vulpes vulpes*
Frog Anura
Horse *Equus caballus*
Mouse, Deer *Peromyscus maniculatus*
Mouse, White-footed *Peromyscus leucopus*
Muskrat *Ondatra zibethicus*
Opossum *Didelphis virginiana*
Porcupine *Erethizon dorsatum*
Raccoon *Procyon lotor*
Rat Muridae
Salamander Caudata
Shrew Soricidae
Skunk, Striped *Mephitis mephitis*
Snake Squamata
Squirrel, Flying *Glaucomys* spp
Squirrel, Red *Tamiasciurus hudsonicus*
Squirrel *Sciurus/Tamiasciurus* spp
Tadpole Anura
Toad *Bufo* sp
Treefrog, Gray *Hyla versicolor*
Vole *Microtus/Clethrionomys* spp
Vole, Red-backed *Clethrionomys gapperi*
Wolf *Canis lupus*

Insects
Ant Formicidae
Ant, Carpenter *Camponotus pennsylvanicus*
Beetles Coleoptera
Blackfly *Simulium* spp
Budworm, Spruce *Choristoneura fumiferana*
Butterflies Lepidoptera
Caterpillar Lepidoptera
Caterpillar, Eastern Tent *Malacosoma americana*
Dragonfly Odonata
Grasshopper Orthoptera
Inchworm Geometridae
Mayfly Ephemeroptera
Mosquito Culicidea
Moth Lepidoptera
Moth, Gypsy *Porthetria dispar*
Moth, Saturnid Saturniidae
Webworm, Fall *Hyphantria cunea*

Other Invertebrates
Centipede Chilopoda
Crab, Fiddler *Uca* sp
Crayfish *Cambaras* sp
Earthworm *Lumbricus* sp
Eel, Sand *Ammodytes americanus*
Mollusk Mollusca
Mussel, Blue *Mytilus edulis*
Periwinkle, Common *Littorina littorea*
Sandworm *Nereis* sp
Shrimp Natantia
Slug Pulmonata
Snail Pulmonata
Spider Aranae

Fungus
Dutch Elm Disease *Ceratostomella ulmi*

Plants

Alder *Alnus* spp
Alder, Speckled *Alnus incana*
Alfalfa *Medicago sativa*
Apple *Malus/Pyrus* spp
Arbor Vitae (Northern White Cedar) *Thuja occidentalis*
Arrowwood *Viburnum dentatum*
Ash *Fraxinus* spp
Ash, White *Fraxinus americana*
Aspen *Populus* sp
Aspen, Quaking *Populus tremuloides*
Azalea, Swamp *Rhododendron viscosum*

Barberry *Berberis* spp
Bayberry *Myrica pensylvanica*
Beauty-bush *Kolkwitzia amabilis*
Beech, American *Fagus granifolia*
Birch *Betula* spp
Birch, Black *Betula lenta*
Birch, Gray *Betula populifolia*
Birch, White *Betula papyrifera*
Birch, Yellow *Betula alleghaniensis*
Blackberry *Rubus* spp
Blueberry *Vaccinium* spp
Blueberry, Highbush *Vaccinium corymbosum/fuscatum*
Blueberry, Lowbush *Vaccinium angustifolium/pallidum*
Bluejoint *Calamagrotis* sp
Bog Rosemary *Andromeda poliofolia*
Bracken *Pteridium aquilinum*
Buckthorn *Rhamnus* sp
Buckthorn, Smooth *Rhamnus frangula*
Bulrush *Scirpus* sp
Buttonbush *Cephalanthus occidentalis*

Canary-grass, Reed *Phalaris arundinacea*
Catbrier *Smilax rotundifolia*
Cattail *Typha* spp
Cattail, Broad-leaved *Typha latifolia*
Cattail, Narrow-leaved *Typha angustifolia*
Cedar, Atlantic White *Chamaecyparis thyoides*
Cedar, Red *Juniperus virginiana*
Cherry *Prunus* sp
Cherry, Black *Prunus serotina*
Cherry, Choke *Prunus virginiana*
Chestnut, American *Castanea dentata*
Cordgrass, Saltwater *Spartina alterniflora*
Cottonwood *Populus deltoides*
Crabapple *Pyrus* spp
Cranberry *Vaccinium* spp

Dock *Rumex* spp
Dogwood *Cornus* spp
Dunegrass *Ammophila breviligulata*

Eelgrass *Zostera marina*
Elder, Box *Acer negundo*
Elderberry, [Common] *Sambucus canadensis*
Elm *Ulmus* spp
Elm, American *Ulmus Americana*
Elm, Slippery *Ulmus rubra*

Fern, Cinnamon *Osmunda cinnamomea*
Fern, Sweet *Comptonia peregrina*
Fir, Balsam *Abies balsamea*
Flag, Sweet *Acorus americanus*
Forsythia *Forsythia* sp

Gale, Sweet *Myrica gale*
Goldenrod *Solidago* spp
Grape *Vitus* spp
Gum, Black *Nyssa sylvatica*

Hardhack (Steeplebush) *Spiraea tomentosa*
Hawthorn *Crataegus* sp
Hazelnut *Corylus* sp
Hemlock, Eastern *Tsuga canadensis*
Hickory *Carya* sp
Holly, American *Ilex opaca*
Honeysuckle *Lonicera* spp
Horsetail *Equisetum* spp
Huckleberry *Gaylussacia* spp
Hydrangea *Hydrangea* spp

Ivy, English *Hedera helix*
Ivy, Poison *Toxicodendron radicans*

Appendix 2: Plant and Animal Species Mentioned in the Text

Juneberry *Amelanchier* sp

Kelp *Laminariaceae* spp

Larch *Larix* sp
Larch, American *Larix laricina*
Laurel *Kalmia* sp
Laurel, Mountain *Kalmia latifolia*
Lichen, Reindeer *Cladonia* sp
Lilac *Syringa* sp
Linden *Tilia* sp
Locust, Black *Robinia pseudoacacia*
Loosestrife, Purple *Lythrum salicaria*

Magnolia *Magnolia* sp
Maple *Acer* spp
Maple, Red *Acer rubrum*
Maple, Silver *Acer saccharinum*
Maple, Sugar *Acer saccharum*
Mayflower, Canada *Maianthemum canadense*
Meadowsweet *Spiraea alba*
Milkweed *Asclepias* spp
Mock-orange *Philadelphus* sp
Moss, Hairy-cap *Polytrichum* spp
Moss *Hypnum* spp
Moss *Sphagnum* sp

Oak, Chinquapin *Quercus muehlenbergii*
Oak, Red *Quercus rubra*
Oak, Scrub *Quercus ilicifolia*
Oak, White *Quercus alba*
Oats, Wild *Avena fatua*
Old Man's Beard (Lichen) *Usnea* spp
Orchard-grass *Dactylis glomerata*

Pasture-juniper *Juniperus communis*
Pea, Beach *Lathyrus japonicus*
Pear *Pyrus communis*
Pine, Jack *Pinus banksiana*
Pine, Pitch *Pinus rigida*
Pine, Red *Pinus resinosa*
Pine, Scotch *Pinus sylvestris*

Pine, White *Pinus strobes*
Plum *Prunus* sp
Plum, Beach *Prunus maritime*
Poplar *Populus* sp
Privet *Ligustrum* sp
Pyracantha *Pyracantha* sp

Raspberry *Rubus* spp
Reed, Common *Phragmites australis*
Rhododendron *Rhododendron* spp
Rice, Wild *Zizania aquatica*
Rockweed *Fucaceae* spp
Rose *Rosa* spp
Rose, Multiflora *Rosa multiflora*
Rush *Juncus* spp
Rye *Secale cereale*

Salt-grass *Distichlis spicata*
Salt-hay *Spartina patens*
Smartweed *Polygonum* sp
Spiraea *Spiraea* sp
Spruce *Picea* spp
Spruce, Black *Picea mariana*
Spruce, Blue *Picea pungens*
Spruce, Norway *Picea abies*
Spruce, Red *Picea rubens*
Solomon's Seal, False *Maianthemum racemosum*
Strawberry *Fragaria* spp
Sycamore *Platanus occidentalis*

Timothy *Phleum pratense*
Tulip-tree *Liriodendron tulipifera*

Vetch *Vicia* sp
Viburnum *Viburnum* spp

Walnut *Juglans* sp
Willow *Salix* spp
Willow, Weeping *Salix babylonica*
Wisteria *Wisteria* sp

Yew *Taxus* sp

Bibliography

Adamus, P.R. 1988. *Atlas of Breeding Birds in Maine, 1978-1983.* Maine Dept. Inland Fisheries and Wildlife, Augusta. 366 pp.

Allen, F.H. 1909. Breeding of the mockingbird near Boston. *Auk* 26: 433-434.

Allen, G.M. 1908. Rare New England Birds. Boston Society of Natural History. *Auk* 25:233-235.

Allen, G.M. 1909. Fauna of New England. No. 11. *List of the Aves.* Boston Society of Natural History, Boston, Occ. Papers 7. 230 pp.

American Ornithologists' Union. 1983. *The A.O.U. Check-list of North American Birds*, Sixth Edition. American Ornithologists' Union, Washington, DC. 877 pp.

American Ornithologists' Union. 1998. *The A.O.U. Check-list of North American Birds*, Seventh Edition. American Ornithologists' Union, Washington, DC. 829 pp.

Anderson, J.R. 1970. Major land uses. Map scale 1:7,500,000. *The National Atlas of the United States of America.* Reston, VA: US Geological Survey. pp 158-159.

Andrews, R. (comp.). 1990. *Coastal waterbird colonies: Maine to Virginia 1984-85—An update of an atlas based on 1977 data showing colony locations, species, and nesting pairs at both time periods. Part I. Maine to Connecticut.* US Fish and Wildlife Service, Newton Corner, Massachusetts.

Bagg, A.C., and S.A. Eliot, Jr. 1937. *Birds of the Connecticut Valley in Massachusetts.* Northampton: The Hampshire Bookshop. 813 pp.

Bailey, W. 1955. *Birds in Massachusetts. When and Where to Find Them.* South Lancaster: The College Press. 234 pp.

Bent, A.C. 1919. *Life Histories of North American Diving Birds. US National Museum Bulletin* 107. Washington, DC. Reprinted 1963, Dover Publ., New York. 239 pp.

Bent, A.C. 1921. *Life Histories of North American Gulls and Terns. US National Museum Bulletin* 113. Washington, DC. Reprinted 1963, Dover Publ., New York. 337 pp.

Bent, A.C. 1923. *Life Histories of North American Wild Fowl.* Part 1. *US National Museum Bulletin* 126. Washington, DC. Reprinted 1962, Dover Publ., New York. 244 pp.

Bent, A.C. 1925. *Life Histories of North American Wild Fowl.* Part 2. *US National Museum Bulletin* 130. Washington, DC. Reprinted 1963, Dover Publ., New York. 314 pp.

Bent, A.C. 1926. *Life Histories of North American Marsh Birds. US National Museum Bulletin* 135. Washington, DC. Reprinted 1963, Dover Publ., New York. 392 pp.

Bent, A.C. 1927. *Life Histories of North American Shore Birds.* Part 1. *US National Museum Bulletin* 142. Washington, DC. Reprinted 1962, Dover Publ., New York. 359 pp.

Bent, A.C. 1929. *Life Histories of North American Shore Birds.* Part 2. *US National Museum Bulletin* 146. Washington, DC. Reprinted 1963, Dover Publ., New York. 340 pp.

Bent, A.C. 1932. *Life Histories of North American Gallinaceous Birds. US National Museum Bulletin* 162. Washington, DC. Reprinted 1963, Dover Publ., New York. 490 pp.

Bent, A.C. 1937. *Life Histories of North American Birds of Prey.* Part 1. *US National Museum Bulletin* 170. Washington, DC. Reprinted 1961, Dover Publ., New York. 409 pp.

Bent, A.C. 1938. *Life Histories of North American Birds of Prey.* Part 2. *US National Museum Bulletin* 170. Washington, DC. Reprinted 1961, Dover Publ., New York. 466 pp.

Bent, A.C. 1939. *Life Histories of North American Woodpeckers. US National Museum Bulletin* 174. Washington, DC Reprinted 1964, Dover Publ., New York. 334 pp.

Bent, A.C. 1940. *Life Histories of North American Cuckoos, Goatsuckers, Hummingbirds, and Their Allies. US National Museum Bulletin* 176. Washington, DC. Reprinted in 2 parts, 1964, Dover Publ., New York. 506 pp.

Bent, A.C. 1942. *Life Histories of North American Flycatchers, Larks, Swallows, and Their Allies. US National Museum Bulletin* 179. Washington, DC. Reprinted 1963, Dover Publ., New York. 555 pp.

Bent, A.C. 1946. *Life Histories of North American Jays, Crows, and Titmice. US National Museum Bulletin* 191. Washington, DC. Reprinted in 2 parts, 1964, Dover Publ., New York. 495 pp.

Bent, A.C. 1948. *Life Histories of North American Nuthatches, Wrens, Thrashers, and Their Allies. US National Museum Bulletin* 195. Washington, DC. Reprinted 1964, Dover Publ., New York. 475 pp.

Bent, A.C. 1949. *Life Histories of North American Thrushes, Kinglets, and Their Allies. US National Museum Bulletin* 196. Washington, DC. Reprinted 1964, Dover Publ., New York. 452 pp.

Bent, A.C. 1953. *Life Histories of North American Wood Warblers. US National Museum Bulletin* 203. Washington, DC. Reprinted in 2 parts, 1963, Dover Publ., New York. 734 pp.

Bent, A.C. 1958. *Life Histories of North American Blackbirds, Orioles, Tanagers, and Allies. US National Museum Bulletin* 211. Washington, DC. Reprinted 1963, Dover Publ., New York. 549 pp.

Bent, A.C. 1965. *Life Histories of North American Wagtails, Shrikes, Vireos, and Their Allies. US National Museum Bulletin* 197. Washington, DC. Reprinted 1965, Dover Publ., New York. 411 pp.

Bent, A.C. 1968. *Life Histories of North American Cardinals, Grosbeaks, Buntings, Towhees, Finches, Sparrows, and Allies.* Parts 1-3. O.L. Austin, Jr., (ed.). *US National Museum Bulletin* 237. Washington, DC. Reprinted 1968, Dover Publ., New York. 1,889 pp.

Blake, S.F. 1916. Breeding of the golden-crowned kinglet in Norfolk County, Massachusetts. *Auk* 33: 326-327.

Blodget, B.G. 1978. *List of the Birds of Massachusetts*. First Edition. Massachusetts Division of Fisheries and Wildlife.

Borror, D.J., and W.W.H. Gunn. *Sounds of Nature: Finches* (record). Federation of Ontario Naturalists. Boston: Houghton Mifflin.

Brewster, W. 1888. Notes on the birds of Winchendon, Worcester County, Massachusetts. *Auk* 5: 386-393.

Brewster, W. 1906. *The Birds of the Cambridge Region of Massachusetts*. Memoirs of the Nuttall Ornithological Club. No. IV. Cambridge, Nuttall Ornithological Club. 426 pp.

Brownrigg, D., and J.T. Brownrigg. 1989. First confirmed nesting of Cerulean Warbler in Massachusetts. *Bird Observer* 17: 317-318.

Bull, J. 1974. *Birds of New York State*. Garden City: Natural History Press. 655 pp.

Bull, J. 1976. *Supplement to Birds of New York State by John Bull*. Special Publication of The Federation of New York State Bird Clubs. Cortland, NY: Wilkins Printers. 52 pp.

Chandler, E.H. 1953. A breeding record for the Ring-necked Duck in Massachusetts. *Auk* 70: 86.

Davis, W.E., Jr. 1997. Breeding bird atlases—a chance for birders to make important contributions. *Birding* 29: 194-205.

Dorst, J. 1962. *The Migrations of Birds*. Boston: Houghton Mifflin Co. 476 pp.

Dunn, J., and K. Garrett. 1997. *A Field Guide to Warblers of North America*. Boston: Houghton Mifflin Co. 656 pp.

Faxon, W. 1889. On the summer birds of Berkshire County, Massachusetts. *Auk* 6: 39-46, 99-107.

Faxon, W., and R. Hoffmann. 1900. *The Birds of Berkshire County, Massachusetts*. Privately printed. 60 pp.

Finch, D.W. 1973. Northeastern maritime region. *Audubon Field Notes* 27: 1020.

Forbush, E.H. 1925. *Birds of Massachusetts and Other New England States*. Massachusetts Department of Agriculture. Vol 1. 481 pp.

Forbush, E.H. 1927. *Birds of Massachusetts and Other New England States*. Massachusetts Department of Agriculture. Vol 2. 461 pp.

Forbush, E.H. 1929. *Birds of Massachusetts and Other New England States*. Massachusetts Department of Agriculture. Vol 3. 366 pp.

Foss, C.R. (ed.). 1994. *Atlas of Breeding Birds in New Hampshire*. Audubon Society of New Hampshire by Arcadia, Dover, NH. 414 pp.

Grice, Daniel, and David Grice. 1942. Tenants of an Old Stub. *The Bulletin of the Massachusetts Audubon Society* 26: 177-184.

Griffith, G.E., C.W. Kiilsgaard, J.M. Omernik, and S.M. Pierson. 1994. *The Massachusetts Ecological Regions Project*. Prepared for the US Environmental Protection Agency Environmental Research Laboratory: Corvallis, OR. 56 pp.

Griscom, L. 1949. *The Birds of Concord*. Cambridge: Harvard University Press. 340 pp.

Griscom, L., and G. Emerson. 1959. *The Birds of Martha's Vineyard*. Sponsored by the Massachusetts Audubon Society. Portland, Maine: The Anthoenson Press. 164 pp.

Griscom, L., and E.V. Folger. 1948. *The Birds of Nantucket*. Cambridge: Harvard University Press. 156 pp.

Griscom, L., and D.E. Snyder. 1955. *The Birds of Massachusetts. An Annotated and Revised Check List.* Salem: Peabody Museum. 295 pp.

Halliwell, D.B., W.A. Kimball, and A.J. Screpetis. 1982. *Massachusetts stream classification program. Part 1. Inventory of rivers and streams.* Mass. Div. Water Pollution Control. 129 pp.

Hammond, E.H. 1970. Classes of land-surface form. Map scale 1:7,500,000. *The National Atlas of the United States of America*. Washington, DC: US Geological Survey. pp 62-63.

Hill, N.P., 1965. *The Birds of Cape Cod, Massachusetts.* New York: William Morrow and Company. 364 pp.

Holt, D.W., J.P. Lortie, B.J. Nikula, and R.C. Humphrey. 1986. First record of Common Black-headed Gulls breeding in the United States. *American Birds* 40: 204-206.

Howe, R.H., Jr., and G.M. Allen. 1901. *The Birds of Massachusetts*. Cambridge: Nuttall Ornithological Club. 154 pp.

Kingsley, N.P. 1974. The timber resources of southern New England. *USDA Forest Service Resource Bulletin* NE-36. Upper Darby, PA: Northeastern Forest Experiment Station. 50 pp.

Klimkiewicz, M.K., and J.K. Solem. 1978. The Breeding Bird Atlas of Montgomery and Howard Counties, Maryland. *Maryland Birdlife* 34: 3-39.

Kuchler, A.W. 1970. Potential natural vegetation. Scale 1:7,500,000. *The National Atlas of the United States of America*. Washington, DC: US Geological Survey. pp 89-91.

Laughlin, S.B., and D.P. Kibbe (eds.). 1985. *The Atlas of Breeding Birds of Vermont*. Hanover: University Press of New England. 456 pp.

Leahy, C., J.H. Mitchell, and T. Conuel. 1996. *The Nature of Massachusetts*. Reading, Massachusetts: Addison-Wesley Publishing Company, Inc. 226 pp.

LeBaron, G.S. 2002. First documented nesting of White-winged Crossbill (*Loxia leucoptera*) in Massachusetts: The Invasion of the Cone Slashers. *Bird Observer* 30: 28-32.

Levine, E. (ed.). 1998. *Bull's Birds of New York State*. Ithaca: Cornell University Press. 622 pp.

MacIvor, L.H. 1990. *Population dynamics, breeding ecology, and management of piping plovers on outer Cape Cod, Massachusetts*. MS thesis, University of Massachusetts, Amherst, MA.

Marshall, R.M., and S.E. Reinert. 1990. Breeding ecology of Seaside Sparrows in a Massachusetts saltmarsh. *Wilson Bulletin* 102: 501-513.

Minot, H.D. 1903. *The Land—Birds and Game—Birds of New England*. Boston: Houghton, Mifflin and Company. 492 pp.

Nice, M.M. 1926. A study of nesting of Magnolia Warblers (*Dendroica magnolia*). *Wilson Bulletin* 38: 185-199.

Nice, M.M. 1926. Behavior of Blackburnian, Myrtle, and Black-throated Blue Warblers with young. *Wilson Bulletin* 38: 82-83.

Nice, M.M. 1926. Nesting of Mourning Doves during September, 1925, in Norman, Oklahoma. *Auk* 43: 94-95.

Nice, M.M. 1927. Further notes on the singing of the Magnolia Warbler. *Wilson Bulletin* 39: 236-237.

Nice, M.M. 1928. Late Nesting of Indigo Buntings and Field Sparrows in Southeastern Ohio. *Auk* 45: 102.

Nice, M.M. 1928. Magnolia Warblers in Pelham, Massachusetts, in 1928. *Wilson Bulletin* 40: 252-253.

Nice, M.M. 1930. A Study of a Nesting of Black-throated Blue Warblers. *Auk* 47: 338-345.

Nice, M.M. 1930. Observations at a Nest of Myrtle Warblers. *Wilson Bulletin* 42: 60-61.

Nice, M.M. 1931. A study of two nests of the Ovenbird. *Auk* 48: 215-228.

Nice, M.M. 1932. Habits of the Blackburnian Warbler in Pelham, Massachusetts. *Auk* 49: 92-93.

Nice, M.M. 1933. Female Quail 'Bobwhiting.' *Auk* 50: 97.

Nice, M.M. 1933. Robins and Carolina Chickadees renesting. *Bird-Banding* 4: 157.

Nice, M.M. 1933. Summer Birds of Pelham, Massachusetts. *The Bulletin of the Massachusetts Audubon Society* 17:13-15.

Nice, M.M. 1936. Late Nesting of Myrtle and Black-Throated Green Warblers in Pelham, Massachusetts. *Auk* 53: 89

Nice, M.M., and L.B. Nice. 1932. A study of two nests of the Black-throated Green Warbler. *Bird-Banding*. 3: 95-105, 157-172.

Nuttall, T. 1905. *A Popular Handbook of the Birds of the United States and Canada*. Boston: Little, Brown and Company. 431 pp.

Omernik, J.M. 1987. *Ecoregions of the conterminous United States*. Annals of the Association of American Geographers 77: 118-125.

Palmer, R.S. (ed.). 1962. *Handbook of North American Birds*. Vol I. New Haven: Yale University Press. 567 pp.

Perring, F.H., and S.M. Walters. 1962. *Atlas of the British Flora*. London: Botanical Society for the British Isles.

Petersen, W.R. 1998. New England region. *Field Notes* 52: 435.

Poole, A.F. 1989. *Ospreys: a natural and unnatural history*. Cambridge University Press, Cambridge.

Price, J., S. Droege, and A. Price. 1995. *The Summer Atlas of North American Birds*. London: Academic Press. 364 pp.

Rimmer, D., and R. Hopping. 1991. Forster's Tern nesting in Plum Island marshes. *Bird Observer of Eastern Massachusetts* 19: 308-309.

Ripley, T.H. 1954. New England Bobwhite Quail, History and Investigations of Present Populations on Martha's Vineyard and Cape Cod, Massachusetts. MS thesis, University of Massachusetts, Amherst.

Robbins, C.S., J.W. Fitzpatrick, and P.B. Hamel. 1992. A warbler in trouble: *Dendroica cerulea*. In *Ecology and Conservation of Neotropical Migrant Landbirds*, J.M. Hagan III and D.W. Johnston (eds.). Washington, DC: Smithsonian Institution Press. 609 pp.

Root, T. 1988. *Atlas of Wintering North American Birds—An Analysis of Christmas Bird Count Data*. Chicago: University of Chicago Press. 312 pp.

Rosenberg, K.V., and J.V. Welles. 1995. *Importance of Geographic Areas to Neotropical Migrant Birds in the Northeast*. Ithaca, NY: Cornell Laboratory of Ornithology. 126 pp.

Saunders, A.A. 1951. *A Guide to Bird Songs*. Garden City: Doubleday & Co., Inc. 307 pp.

Sharrock, J.T.R. (comp.) 1976. *The Atlas of Breeding Birds in Britain and Ireland*. Berkhamsted, England: T. and A.D. Poyser. 479 pp.

Shaub, M.S. 1956. Effect of native foods on Evening Grosbeak incursions. *The Bulletin of the Massachusetts Audubon Society* 40: 481-488.

Shaub, M.S. 1959. Evening Grosbeak juveniles at Hadley, Massachusetts—July, 1958. *Bird-Banding* 30: 226-229.

Sheldon, W.G. 1967. *The Book of the American Woodcock*. Amherst, MA: University of Massachusetts Press. 227 pp.

Sibley, C.G., and B.L. Monroe, Jr. 1990. *Distribution and Taxonomy of Birds of the World*. New Haven: Yale University Press. 1,111 pp.

Smith, C.E. 1964. Nesting of Worm-eating Warbler and Slate-colored Junco in eastern Massachusetts. *Auk* 51: 96-97.

Smith, C.R. 1990. *Handbook for Atlasing American Breeding Birds*. Woodstock, VT: Vermont Institute of Natural Science. 68 pp.

Smith, S.M. 1991. *The Black-capped Chickadee—Behavioral Ecology and Natural History*. Ithaca, NY: Comstock-Cornell. 362 pp.

Snyder, D.E. 1952. Red-Crossbills. *The Bulletin of the Massachusetts Audubon Society*. 36: 383-386.

Snyder, D.E. 1956. Shrikes nest in Essex County. *The Bulletin of the Massachusetts Audubon Society* 40: 431-435.

Sorrie, B.A., and P. Somers. 1999. *The Vascular Plants of Massachusetts: A County Checklist*. Westborough, MA: Natural Heritage and Endangered Species Program, Massachusetts Division of Fisheries and Wildlife. 186 pp.

Stein, R.C. 1958. The behavioral, ecological and morphological characteristics of two populations of the Alder Flycatcher, *Empidonax traillii* (Audubon). *New York State Museum and Science Service Bulletin*. 371: 1-63.

Stemple, D. 1981. Lincoln's Sparrow nests in Massachusetts. *Bird News of Western Massachusetts* 20: 41-42.

Stokes, D.W., and Lillian Q. Stokes. 1983. *A Guide to Bird Behavior*. Vol II. Boston: Little, Brown and Co. 334 pp.

Stymeist, R.H. 1973. The Bird Observer summary for July. *Bird Observer of Eastern Massachusetts* 1: 114.

Stymeist, R.H. 1982. The Bird Observer Field Notes for August. *Bird Observer of Eastern Massachusetts* 10: 214.

Sutton, G.M. 1928. Notes on the flight of the chimney swift. *Cardinal* 2: 85-92.

Swain, P.C., and J.B. Kearsley. 2000. *Classification of the Natural Communities of Massachusetts*. Westborough, MA: Natural Heritage and Endangered Species Program, Massachusetts Division of Fisheries and Wildlife. 240 pp.

Telfair, R.C. II. 1994. Cattle Egret (*Bubulcus ibis*). *The Birds of North America*, No. 113 (A. Poole and F. Gill, eds.). Philadelphia: The Academy of Natural Sciences.

Tingley, S.I. 1982. Northeastern maritime region. *American Birds* 36: 955.

Tingley, S.I. 1983. Northeastern maritime region. *American Birds* 37: 967.

Townsend, C.W. 1905. *The Birds of Essex County, Massachusetts*. Memoirs of the Nuttall Ornithological Club. No. III. Cambridge: Nuttall Ornithological Club. 352 pp.

Vander-Haegen, W.M. 1987. Population Dynamics and Habitat Utilization of the Eastern Wild Turkey in Western Massachusetts, MS thesis, University of Massachusetts, Amherst.

Veit, R.R., and W.R. Petersen. 1993. *Birds of Massachusetts.* Massachusetts Audubon Society. 514 pp.

Walsh, J., V. Elia, R. Kane, and T. Halliwell. 1999. *Birds of New Jersey*. Bernardsville: New Jersey Audubon Society. 704 pp.

Wandell, W.N. 1942. Progress report of the ring-necked pheasant investigation in theConnecticut River Valley, Massachusetts. Mass Div. Fish Gam Res. Bulletin 5. 118 pp.

Wander, S.A., and W. Wander. 1985. Observations at a Northern Waterthrush nest. *Journal of Field Ornithology* 56: 69.

Welty, J.C., and L.F. Baptista. 1988. *The Life of Birds*. Fourth Edition. Philadelphia: Saunders.

Wetherbee, D.K. 1945. *The Birds and Mammals of Worcester County, Massachusetts*. Worcester: Century Press. 192 pp.

White, R.P., and S.M. Melvin. 1985. Rare grassland birds and management recommendations for Camp Edwards/Otis Air National Guard Base. Unpublished report prepared for Massachusetts National Guard. Geography Department, University of Wisconsin, Madison, and Massachusetts Division of Fisheries and Wildlife, Westborough. 29 pp.

Index

Accipiter
  cooperii, 96
  gentilis, 98
  striatus, 94
Actitis macularia, 136
Aegolius acadicus, 184
Agelaius phoeniceus, 392
Aix sponsa, 62
Ammodramus
  caudacutus, 372
  henslowii, 370
  maritimus, 374
  savannarum, 368
Anas
  acuta, 74
  americana, 423
  clypeata, 72
  crecca, 76
  discors, 70
  platyrhynchos, 68
  rubripes, 66
  strepera, 64
Archilochus colubris, 192
Ardea herodias, 36
Asio
  flammeus, 182
  otus, 180
Aythya collaris, 78

Bartramia longicauda, 138
Bittern
  American, 32
  Least, 34
Blackbird
  Red-winged, 392
  Rusty, 396
Bluebird, Eastern, 284
Bobolink, 390
Bobwhite, Northern, 114
Bombycilla cedrorum, 304
Bonasa umbellus, 110

Botaurus lentiginosus, 32
Branta canadensis, 58
Bubo virginianus, 176
Bubulcus ibis, 46
Bunting, Indigo, 388
Buteo
  jamaicensis, 104
  lineatus, 100
  platypterus, 102
Butorides virescens, 48

Calidris minutilla, 140
Caprimulgus
  carolinensis, 425
  vociferus, 188
Cardinal, Northern, 384
Cardinalis cardinalis, 384
Carduelis
  pinus, 412
  tristis, 414
Carpodacus
  mexicanus, 408
  purpureus, 406
Casmerodius albus, 38
Catbird, Gray, 296
Cathartes aura, 56
Catharus
  bicknelli, 427
  fuscescens, 286
  guttatus, 290
  ustulatus, 288
Catoptrophorus semipalmatus, 134
Certhia americana, 266
Ceryle alcyon, 194
Chaetura pelagica, 190
Charadrius
  melodus, 128
  vociferus, 130
Chat, Yellow-breasted, 354
Chickadee, Black-capped, 258
Chordeiles minor, 186
Chuck-will's-widow, 425
Circus cyaneus, 92
Cistothorus

  palustris, 276
  platensis, 274
Coccothraustes vespertinus, 416
Coccyzus
  americanus, 170
  erythropthalmus, 168
Colaptes auratus, 206
Colinus virginianus, 114
Columba livia, 164
Contopus
  borealis, 210
  virens, 212
Coot, American, 126
Coragyps atratus, 422
Cormorant
  Double-crested, 30
  Great, 422
Corvus
  brachyrhynchos, 240
  corax, 426
  ossifragus, 242
Cowbird, Brown-headed, 400
Creeper, Brown, 266
Crossbill
  Red, 410
  White-winged, 429
Crow
  American, 240
  Fish, 242
Cuckoo
  Black-billed, 168
  Yellow-billed, 170
Cyanocitta cristata, 238
Cygnus olor, 60

Dendroica
  caerulescens, 320
  cerulea, 427
  coronata, 322
  discolor, 330
  fusca, 326
  magnolia, 318
  pensylvanica, 316
  petechia, 314
  pinus, 328

  striata, 332
  virens, 324
Dickcissel, 428
Dolichonyx oryzivorus, 390
Dove
  Mourning, 166
  Rock, 164
Dryocopus pileatus, 208
Duck
  American Black, 66
  Ring-necked, 78
  Ruddy, 88
  Wood, 62
Dumetella carolinensis, 296

Eagle, Bald, 423
Egret
  Cattle, 46
  Great, 38
  Snowy, 40
Egretta
  caerulea, 42
  thula, 40
  tricolor, 44
Eider, Common, 80
Empidonax
  alnorum, 216
  flaviventris, 426
  minimus, 220
  traillii, 218
  virescens, 214
Eremophila alpestris, 244
Euphagus carolinus, 396

Falco
  peregrinus, 424
  sparverius, 106
Falcon, Peregrine, 424
Finch
  House, 408
  Purple, 406
Flicker, Northern, 206
Flycatcher
  Acadian, 214
  Alder, 216

Great Crested, 224
Least, 220
Olive-sided, 210
Willow, 218
Yellow-bellied, 426
Fulica americana, 126

Gadwall, 64
Gallinago delicata, 142
Gallinula chloropus, 124
Gavia immer, 24
Geothlypis trichas, 348
Gnatcatcher, Blue-gray, 282
Goldfinch, American, 414
Goose, Canada, 58
Goshawk, Northern, 98
Grackle, Common, 398
Grebe, Pied-billed, 26
Grosbeak
    Evening, 416
    Rose-breasted, 386
Grouse, Ruffed, 110
Gull
    Black-headed, 425
    Great Black-backed, 152
    Herring, 150
    Laughing, 148
    Ring-billed, 425

Haematopus palliatus, 132
Haliaeetus leucocephalus, 423
Harrier, Northern, 92
Hawk
    Broad-winged, 102
    Cooper's, 96
    Red-shouldered, 100
    Red-tailed, 104
    Sharp-shinned, 94
Helmitheros vermivorus, 338
Heron
    Great Blue, 36
    Green, 48
    Little Blue, 42
    Tricolored, 44
Hirundo

pyrrhonota, 254
rustica, 256
Hummingbird, Ruby-throated, 192
Hylocichla mustelina, 292

Ibis, Glossy, 54
Icteria virens, 354
Icterus
    galbula, 404
    spurius, 402
Ixobrychus exilis, 34

Jay, Blue, 238
Junco, Dark-eyed, 382
Junco hyemalis, 382

Kestrel, American, 106
Killdeer, 130
Kingbird, Eastern, 226
Kingfisher, Belted, 194
Kinglet
    Golden-crowned, 278
    Ruby-crowned, 280

Lanius ludovicianus, 426
Lark, Horned, 244
Larus
    argentatus, 150
    atricilla, 148
    delawarensis, 425
    marinus, 152
    ridibundus, 425
Loon, Common, 24
Lophodytes cucullatus, 82
Loxia
    curvirostra, 410
    leucoptera, 429

Mallard, 68
Martin, Purple, 246
Meadowlark, Eastern, 394
Melanerpes
    carolinus, 198
    erythrocephalus, 196
Meleagris gallopavo, 112

Melospiza
    georgiana, 378
    lincolnii, 428
    melodia, 376
Merganser
    Common, 84
    Hooded, 82
    Red-breasted, 86
Mergus
    merganser, 84
    serrator, 86
Mimus polyglottos, 298
Mniotilta varia, 334
Mockingbird, Northern, 298
Molothrus ater, 400
Moorhen, Common, 124
Myiarchus crinitus, 224

Nighthawk, Common, 186
Night-Heron
    Black-crowned, 50
    Yellow-crowned, 52
Nuthatch
    Red-breasted, 262
    White-breasted, 264
Nyctanassa violacea, 52
Nycticorax nycticorax, 50

Oceanodroma leucorhoa, 28
Oporornis philadelphia, 346
Oriole
    Baltimore, 404
    Orchard, 402
Osprey, 90
Otus asio, 174
Ovenbird, 340
Owl
    Barn, 172
    Barred, 178
    Great Horned, 176
    Long-eared, 180
    Northern Saw-whet, 184
    Short-eared, 182
Oxyura jamaicensis, 88
Oystercatcher, American, 132

Pandion haliaetus, 90
Parula americana, 312
Parula, Northern, 312
Parus
    atricapillus, 258
    bicolor, 260
Passer domesticus, 418
Passerculus sandwichensis, 366
Passerina cyanea, 388
Phalacrocorax
    auritus, 30
    carbo, 422
Phalarope, Wilson's, 146
Phalaropus tricolor, 146
Phasianus colchicus, 108
Pheasant, Ring-necked, 108
Pheucticus ludovicianus, 386
Phoebe, Eastern, 222
Picoides
    pubescens, 202
    villosus, 204
Pintail, Northern, 74
Pipilo erythrophthalmus, 358
Piranga olivacea, 356
Plegadis falcinellus, 54
Plover, Piping, 128
Podilymbus podiceps, 26
Polioptila caerulea, 282
Pooecetes gramineus, 364
Porzana carolina, 122
Progne subis, 246
Protonotaria citrea, 427
Puffinus puffinus, 422

Quiscalus quiscula, 398

Rail
    Clapper, 116
    King, 118
    Virginia, 120
Rallus
    elegans, 118
    limicola, 120
    longirostris, 116
Raven, Common, 426

Index 439

Redstart, American, 336
Regulus
   calendula, 280
   satrapa, 278
Riparia riparia, 252
Robin, American, 294
Rynchops niger, 162

Sandpiper
   Least, 140
   Spotted, 136
   Upland, 138
Sapsucker, Yellow-bellied, 200
Sayornis phoebe, 222
Scolopax minor, 144
Screech-Owl, Eastern, 174
Seiurus
   aurocapillus, 340
   motacilla, 344
   noveboracensis, 342
Setophaga ruticilla, 336
Shearwater, Manx, 422
Shoveler, Northern, 72
Shrike, Loggerhead, 426
Sialia sialis, 284
Siskin, Pine, 412
Sitta
   canadensis, 262
   carolinensis, 264
Skimmer, Black, 162
Snipe, Wilson's, 142
Somateria mollissima, 80
Sora, 122
Sparrow
   Chipping, 360
   Field, 362
   Grasshopper, 368
   Henslow's, 370
   House, 418
   Lincoln's, 428
   Savannah, 366
   Seaside, 374
   Saltmarsh Sharp-tailed, 372
   Song, 376
   Swamp, 378

Vesper, 364
   White-throated, 380
Sphyrapicus varius, 200
Spiza americana, 428
Spizella
   passerina, 360
   pusilla, 362
Starling, European, 302
Stelgidopteryx serripennis, 250
Sterna
   antillarum, 160
   dougallii, 154
   forsteri, 425
   hirundo, 156
   paradisaea, 158
Storm-Petrel, Leach's, 28
Strix varia, 178
Sturnella magna, 394
Sturnus vulgaris, 302
Swallow
   Bank, 252
   Barn, 256
   Cliff, 254
   Northern Rough-winged, 250
   Tree, 248
Swan, Mute, 60
Swift, Chimney, 190

Tachycineta bicolor, 248
Tanager, Scarlet, 356
Teal
   Blue-winged, 70
   Green-winged, 76
Tern
   Arctic, 158
   Common, 156
   Forster's, 425
   Least, 160
   Roseate, 154
Thrasher, Brown, 300
Thrush
   Bicknell's, 427
   Hermit, 290
   Swainson's, 288
   Wood, 292

Thryothorus ludovicianus, 268
Titmouse, Tufted, 260
Towhee, Eastern, 358
Toxostoma rufum, 300
Troglodytes
   aedon, 270
   troglodytes, 272
Turdus migratorius, 294
Turkey, Wild, 112
Tyrannus tyrannus, 226
Tyto alba, 172

Veery, 286
Vermivora
   chrysoptera, 308
   peregrina, 427
   pinus, 306
   ruficapilla, 310
Vireo
   Blue-headed, 232
   Red-eyed, 236
   Warbling, 234
   White-eyed, 228
   Yellow-throated, 230
Vireo
   flavifrons, 230
   gilvus, 234
   griseus, 228
   olivaceus, 236
   solitarius, 232
Vulture
   Black, 422
   Turkey, 56

Warbler
   Black-and-white, 334
   Blackburnian, 326
   Blackpoll, 332
   Black-throated Blue, 320
   Black-throated Green, 324
   Blue-winged, 306
   Canada, 352
   Cerulean, 427
   Chestnut-sided, 316
   Golden-winged, 308

   Hooded, 350
   Magnolia, 318
   Mourning, 346
   Nashville, 310
   Pine, 328
   Prairie, 330
   Prothonotary, 427
   Tennessee, 427
   Worm-eating, 338
   Yellow, 314
   Yellow-rumped, 322
Waterthrush
   Louisiana, 344
   Northern, 342
Waxwing, Cedar, 304
Whip-poor-will, 188
Wigeon, American, 423
Willet, 134
Wilsonia
   canadensis, 352
   citrina, 350
Woodcock, American, 144
Woodpecker
   Downy, 202
   Hairy, 204
   Pileated, 208
   Red-bellied, 198
   Red-headed, 196
Wood-Pewee, Eastern, 212
Wren
   Carolina, 268
   House, 270
   Marsh, 276
   Sedge, 274
   Winter, 272

Yellowthroat, Common, 348

Zenaida macroura, 166
Zonotrichia albicollis, 380